T0135290

Lecture Notes in Networks and Systems 643

The series "Lecture Notes in Networks and Systems" publishes the latest developments in Networks and Systems—quickly, informally and with high quality. Original research reported in proceedings and post-proceedings represents the core of LNNS.

Volumes published in LNNS embrace all aspects and subfields of, as well as new challenges in, Networks and Systems.

The series contains proceedings and edited volumes in systems and networks, spanning the areas of Cyber-Physical Systems, Autonomous Systems, Sensor Networks, Control Systems, Energy Systems, Automotive Systems, Biological Systems, Vehicular Networking and Connected Vehicles, Aerospace Systems, Automation, Manufacturing, Smart Grids, Nonlinear Systems, Power Systems, Robotics, Social Systems, Economic Systems and other. Of particular value to both the contributors and the readership are the short publication timeframe and the world-wide distribution and exposure which enable both a wide and rapid dissemination of research output.

The series covers the theory, applications, and perspectives on the state of the art and future developments relevant to systems and networks, decision making, control, complex processes and related areas, as embedded in the fields of interdisciplinary and applied sciences, engineering, computer science, physics, economics, social, and life sciences, as well as the paradigms and methodologies behind them.

Indexed by SCOPUS, INSPEC, WTI Frankfurt eG, zbMATH, SCImago.

All books published in the series are submitted for consideration in Web of Science.

For proposals from Asia please contact Aninda Bose (aninda.bose@springer.com).

Fausto Pedro García Márquez · Akhtar Jamil ·
Süleyman Eken · Alaa Ali Hameed
Editors

Computational Intelligence, Data Analytics and Applications

Selected papers from the International Conference on Computing, Intelligence and Data Analytics (ICCIDA)

Springer

Editors
Fausto Pedro García Márquez ⓘ
University of Castilla-La Mancha
Ciudad Real, Spain

Süleyman Eken
Information Systems Engineering
Kocaeli University
Kocaeli, Turkey

Akhtar Jamil
National University of Computer
and Emerging Sciences
Islamabad, Pakistan

Alaa Ali Hameed
Department of Computer Engineering
Istinye University
Istanbul, Turkey

ISSN 2367-3370   ISSN 2367-3389 (electronic)
Lecture Notes in Networks and Systems
ISBN 978-3-031-27101-4   ISBN 978-3-031-27099-4 (eBook)
https://doi.org/10.1007/978-3-031-27099-4

This Springer imprint is published by the registered company Springer Nature Switzerland AG
The registered company address is: Gewerbestrasse 11, 6330 Cham, Switzerland

Preface

Computational Intelligence, Data Analytics and Applications

Selected Papers from the International Conference on Computing, Intelligence and Data Analytics (ICCIDA)

This book is a compilation of the selected papers presented at the International Conference on Computing, Intelligence and Data Analytics (ICCIDA) in 2022.

The conference was hosted in Kocaeli, Turkey, and organized by Information Systems Engineering of the Kocaeli University, Turkey, on September 16–17, 2022. The conference served as an interdisciplinary forum that helped the research community to take a step forward and share the research findings. In addition, it provided an arena where researchers, scholars, professionals, students and academicians may be able to foster working relationships and gain access to the latest research results.

The book highlights some of the latest research advances and cutting-edge analysis of real-world case studies on Computational Intelligence, Data Analytics and Applications from a wide range of international contexts. It also identified business applications and the latest findings and innovations in Operations Management and the Decision Sciences, e.g.:

Data analysis and visualization

- Exploratory data analysis
- Statistical and mathematical modeling
- Business intelligence
- Big data analysis
- Data mining
- Cloud computing architecture and systems
- ETL and big data warehousing
- Business intelligence
- Data visualization
- Statistical analysis

Computer vision

- Document analysis
- Biometrics and pattern recognition
- Remote sensing & GIS
- Medical image processing
- Image and video retrieval
- Motion analysis
- Structure from motion
- Object detection and recognition
- Image restoration

- Speech and audio processing
- Signal processing

  Artificial intelligence

- Machine learning
- Pattern recognition
- Deep learning
- Human–computer interactions
- Medical image processing
- Image and video retrieval
- Audio video processing
- Text analytics
- Natural language processing
- Information retrieval
- Robotics applications

  Internet of Things

- 3D printing
- Securing IoT infrastructure
- Future of IoT and big data
- Internet of Things
- Intelligent systems for IoT
- Security, privacy and trust
- Visual analytics IoT
- Data compression for IoT devices
- IoT services and applications
- Education and learning
- Social networks analysis

  Communication systems and networks

- Antennas, propagation and RF design
- Transmission and communication theory
- Wireless/radio access technologies
- Optical networks and NGN
- 5G & 6G cellular systems and SON
- Sensor networks
- Multimedia and New Media
- High-speed communication.
- Computational intelligence in telecommunications

  Software engineering

- Requirements engineering
- Security aspects
- Agile software engineering
- Software evolution & reuse
- Reverse engineering

- Software dependability
- Data & AI monetization and products
- Data as a Service/Platform
- Biomedical experiments and simulations
- Decision support systems

Fausto Pedro García Márquez
Akhtar Jamil
Süleyman Eken
Alaa Ali Hameed

Organization

General Chairs

Fausto Pedro Garcia Marquez	University of Castilla–La Mancha, Spain
Süleyman Eken	Information Systems Engineering, Kocaeli University, Turkey

Technical Program Chairs

Zeynep Hilal Kilimci	Information Systems Engineering, Kocaeli University, Turkey
Alaa Ali Hameed	Department of Computer Engineering, Istinye University, Turkey
Akhtar Jamil	National University of Computer and Emerging Sciences, Pakistan

Publication and Publicity Chairs

Serdar Solak	Information Systems Engineering, Kocaeli University, Turkey
Mustafa Hikmet Bilgehan Uçar	Information Systems Engineering, Kocaeli University, Turkey

Organizing Committee

Mehmet Yıldırım	Information Systems Engineering, Kocaeli University, Turkey
Hikmet Hakan Gürel	Information Systems Engineering, Kocaeli University, Turkey
Halil Yiğit	Information Systems Engineering, Kocaeli University, Turkey
Adnan Sondaş	Information Systems Engineering, Kocaeli University, Turkey
Önder Yakut	Information Systems Engineering, Kocaeli University, Turkey

Yavuz Selim Fatihoğlu   Information Systems Engineering, Kocaeli University, Turkey

Alper Metin   Information Systems Engineering, Kocaeli University, Turkey

Registration Committee

Seda Balta Kaç   Information Systems Engineering, Kocaeli University, Turkey

Zeynep Sarı   Information Systems Engineering, Kocaeli University, Turkey

M. M. Enes Yurtsever   Information Systems Engineering, Kocaeli University, Turkey

Technical Program Committee

Gabriella Casalino   Department of Informatics, Università degli Studi di Bari Aldo Moro Bari, Italy

Faezeh Soleimani   Department of Mathematical Sciences, Ball State University, USA

Fatma Bozyiğit   Department of Computer Engineering, İzmir Bakırçay University, Turkey

Imran Ahmed Siddiqi   Department of Computer Science, Bahria University, Pakistan

Musa Balta   Department of Computer Engineering, Sakarya University, Turkey

Mehak Khan   Department of Computer Science, Oslo Metropolitan University, Oslo, Norway

Kiran Sood   Chitkara Business School, Chitkara University, India

Mariyam Ouaissa   Moulay Ismail University Meknes, Morocco

Enkeleda Lulaj   University Haxhi Zeka, Kosovo

Farzad Kiani   Department of Computer Engineering, Istinye University, Istanbul, Turkey

Muhammad Abdul Basit   Montana Technological University, Butte Montana, USA

Marta Cimitile   Department of Economy and Law, Unitelma Sapienza University of Rome, Italy

Muhammad Ilyas   Department of Electrical and Electronics Engineering, Altinbas University, Turkey

Subhan Ullah — National University of Computer and Emerging Sciences, Pakistan

Adem Ozyavaş — Department of Computer Engineering, Istanbul Atlas University, Turkey

Atta Ur Rehman — College of Engineering and IT, Ajman University, United Arab Emirates

Ahmet Gürhanli — Department of Computer Engineering, Istanbul Aydin University, Turkey

Chawki Djeddi — Department of Information Systems, University of Rouen, France

Cafer Avcı — Transportation Engineering, Aalto University, Finland

Adem Tuncer — Computer Engineering, Yalova University, Turkey

Costin Bădică — Computer Science, University of Craiova, Romania

Deniz Balta — Department of Software Engineering, Sakarya University, Turkey

Alexandra Vultureanu — Computer Science, University of Craiova, Romania

Aysegul Ucar — Mechatronics Engineering, Fırat University, Turkey

Mirjana Ivanovic — Faculty of Sciences, University of Novi Sad, Serbia

Enis Karaaslan — Computer Engineering, Mugla Sitki Kocman University, Turkey

Sara del Río García — RHEA Group for European Space Agency, Spain

Ahmet Şakir Dokuz — Computer Engineering, Nigde Omer Halisdemir University, Turkey

Muhammed Davud — Department of Computer Engineering, Istanbul Sabahattin Zaim University, Turkey

Sibel Senan — Department of Computer Engineering, Istanbul University, Turkey

Waleed Ead — Faculty of Computers and Artificial Intelligence, Beni-Suef University, Egypt

Zeynep Orman — Department of Computer Engineering, Istanbul University, Turkey

Pelin Angın — Department of Computer Engineering, Middle East Technical University, Turkey

Şebnem Özdemir — Management Information Systems, Istinye University, Turkey

Maytham Alabbas — Department of Computer Science, University of Basrah, Iraq

Sultan Zavrak — Computer Engineering, Duzce University, Turkey

Ekin Ekinci	Computer Engineering, Sakarya Applied Sciences University, Turkey
Meryem Uzun Per	Computer Engineering, Istanbul Health and Technology University, Turkey
İbrahim Delibaşoğlu	Sakarya University, Turkey
Ali Can Karaca	Yıldız Technical University, Turkey
Gülşen Akman	Kocaeli University, Turkey
Faruk Aktaş	Kocaeli University, Turkey
İsmail Koç	Konya Technical University, Turkey
Nuh Azgınoglu	Kayseri University, Turkey
Fidan Kaya Gülağız	Kocaeli University, Turkey
Alev Mutlu	Kocaeli University, Turkey
Muhammet Damar	Dokuz Eylül University, Turkey
Pinar Onay Durdu	Kocaeli University, Turkey
Zekeriya Anıl Güven	Ege University, Turkey
Furkan Göz	Kocaeli University, Turkey
Mustafa S. Aljumaily	University of Tennessee, USA
Claudia Catalina Caro Ruiz	Del Rosario University, Colombia
Oscar Julián Perdomo Charry	Del Rosario University, Colombia
Iyad Abu Doush	American University of Kuwait, Kuwait
Mohammad Shukri Salman	American University of the Middle East, Kuwait
Sani Salisu	Ahmadu Bello University, Nigeria
Bashir Sadiq Olaniyi	Kampala International University, Uganda
Abubakar M. Ashir	Tishk International University, Iraq
Ahmed Burhan Mohammed	University of Kirkuk, Iraq
Ghaida A. Al-Suhail	University of Basrah, Iraq
Imad Alshawi	University of Basrah, Iraq
Husham L. Swadi Roomi	University of Basrah, Iraq
Abbas H. Hassin Alasadi	University of Basrah, Iraq
Osman Altay	Manisa Celal Bayar University, Turkey
Kevser Şahinbaş	Medipol University, Turkey
Mohammed Alkrunz	Istanbul Aydin University, Turkey
Abbas UĞurenver	Istanbul Aydin University, Turkey
Wadhah Zeyad Tareq Tareq	Istinye University, Turkey
Amir Seyyedabbasi	Istinye University, Turkey

Contents

About the Editors

Fausto Pedro Garcia Marquez Fausto works at UCLM as Full Professor (Accredited as Full Professor from 2013), Spain, Honorary Senior Research Fellow at Birmingham University, UK, Lecturer at the Postgraduate European Institute, and he has been Senior Manager in Accenture (2013–2014). He obtained his European PhD with a maximum distinction. He has been distinguished with the prices: Grand Prize (2021), Runner Prize (2020) and Advancement Prize (2018) for Management Science and Engineering Management Nominated Prize, First International Business Ideas Competition 2017 Award (2017); Runner (2015), Advancement (2013) and Silver (2012) by the International Society of Management Science and Engineering Management (ICMSEM). He has published more than 170 papers (103 JCR(49Q1; 27Q2; 24Q3; 3Q4), 65 % ISI, 30% JCR and 92% internationals), some recognized as "Applied Energy" (Q1, IF 9.746, as "Best Paper 2020"), "Renewable Energy" (Q1, IF 8.001, as "Best Paper 2014"); "ICMSEM" (as "excellent"), "Int. J. of Automation and Computing" and "IMechE Part F: J. of Rail and Rapid Transit" (most downloaded), etc. He is Author and Editor of more than 40 books (Elsevier, Springer, Pearson, Mc-GrawHill, Intech, IGI, Marcombo, AlfaOmega,…), 90 international chapters and six patents. He is Editor of five international journals and Committee Member of more than 60 international conferences. He has been Principal Investigator in four European projects, eight national projects and more than 150 projects for universities, companies, etc.

He is being Expert in the European Union in AI4People (EISMD), and ESF.; Director of www.ingeniumgroup.eu.; Senior Member at IEEE, 2021-…; Honored Honorary Member of the Research Council of Indian Institute of Finance, 2021-…; Committee Chair of The International Society for Management Science and Engineering Management (ISMSEM), 2020-…. His main interests are artificial intelligence, maintenance, management, renewable energy, transport, advanced analytics and data science.

Dr. Akhtar Jamil is Associate Professor in the Department of Computer Science at the National University of Computer and Emerging Sciences, Islamabad, Pakistan. Before joining FAST, he served as Assistant Professor and Vice Head of the Computer Engineering Department at Istanbul Sabahattin Zaim University, Istanbul, Turkey. He also served as Lecturer at COMSATS University Islamabad. He has also worked in the industry as Developer for several years. He received his Ph.D. from Yildiz Technical University, Istanbul, Turkey, in machine learning and remote sensing. He has published more than 50 high-quality papers in well-known journals and top conferences. He received a fully funded Ph.D. scholarship from the Turkish Government. He is Founding Member of the ICMI, ICAETA and ICCIDA conferences. He serves as Reviewer for several journals and conferences. He focuses on applied research for solving real-world problems. His current research interests include statistical machine learning, deep learning, pattern recognition, data analytics, image classification and remote sensing.

Süleyman Eken received his MS degree and PhD degree in Computer Engineering at the Kocaeli University. He works as Associate Professor of Information Systems Engineering, Kocaeli University, Izmit, Turkey. Dr. Eken serves as Reviewer for more than 50 journals such as the IEEE Transactions on Industrial Informatics, JIS, Soft Computing, IEEE Access, International Journal of Advanced Robotic Systems, Concurrency and Computation: Practice and Experience, Journal of Ambient Intelligence and Humanized Computing, Imaging Science Journal, Journal of Network and Computer Applications, Turkish Journal of Electrical Engineering & Computer Sciences and Peer-to-Peer Networking and Applications. He has ten chapters and around 150 papers. He served as Guest Editor for three issues. Also he serves as Associate Editor for Cluster Computing (Springer) and Concurrency and Computation: Practice and Experience (Wiley).

Alaa Ali Hameed received his Master's degree in Computer Engineering from Eastern Mediterranean University, North Cyprus, in 2012, and his Ph.D. degree from the Department of Computer Engineering at Selcuk University, Turkey, in 2017. He worked as Assistant Professor in the Department of Computer Engineering, at Istanbul Aydin University, Turkey, from 2017 to 2019. He then moved to Istanbul Sabahattin Zaim University, Turkey, where he worked as Assistant Professor in the Department of Computer Engineering from 2019 to 2022. Currently, he is Assistant Professor in the Department of Computer Engineering at Istinye University, Turkey. He has published more than 60 technical articles in top international journals and conferences in a short span of time. He has served as Program Chair and Technical Program Chair member for many international conferences, also he has served as Guest Editor for many SCIE journals. His research interests include digital signal and image processing, adaptive filters, adaptive computing, data mining, machine and deep learning, big data and data analytics, neural networks and self-learning systems and artificial intelligence.

Priority-Based Pollution Management ITS

Youssef Aboulyousr$^{(\boxtimes)}$, Farida Azab , Hassan Soubra ,
and Mariam Zaky

The German University in Cairo, Cairo, Egypt
youssef.aboulyousr@yahoo.com

Abstract. Today's road networks are highly complex and dynamic.
This is largely because they must support the needs of an ever-growing
population, while also taking into account a variety of safety considerations. As such, these road networks require complex routing algorithms to
ensure that vehicles can efficiently and smoothly travel from origin to destination. Moreover, vehicle pollution is a growing problem in many cities
around the world. In order to combat this, many cities are now routing
vehicles in an effort to reduce emissions. This process involves re-routing
vehicles away from heavily populated areas and onto less traveled roads.
By doing this, cities are able to reduce the amount of vehicle pollution
that enters the atmosphere. In this paper, we will explore the challenges
of routing and re-routing road vehicles, introducing and implementing
a priority based constrained routing algorithm for road vehicles where
priorities will be assigned both statically and dynamically following certain criteria. The results of our routing algorithm will be benchmarked
against the most commonly used shortest-path routing algorithm, where
we will be able to evaluate the impact of our algorithm on pollution and
vehicle's time spent on traffic. Quantitative results that are both computed and generated by the simulation, alongside with different scenarios
will be analysed which will help us understand the necessity of having
such algorithm for routing road vehicles.

Keywords: Intelligent transportation system · Smart cities · Vehicle
routing · Priority based vehicle routing · Pollution

1 Introduction

Information and communication technologies are used in ITS (intelligent transportation systems) to create a better driving environment. Smart cities employ
ITS, commonly referred to as smart roads or smart road infrastructure, to monitor and control traffic on their roads. As the number of vehicles on the road
rises, it is increasingly crucial to make sure that they can navigate the city with
ease and don't contribute to traffic. For instance, a school bus could need to
drop off its students at a specific spot, or maybe there will be a special event in
town on a Saturday afternoon, so locals will need to know where to park their

F. P. García Márquez et al. (Eds.): ICCIDA 2022, LNNS 643, pp. 1–16, 2023.
https://doi.org/10.1007/978-3-031-27099-4_1

cars. We require a system that can optimally route these road vehicles in both scenarios without negatively impacting the rest of the traffic flow.

In order to distribute pollution across the entire network and not just in a specific area (typically inside cities) of the network, this paper aims to develop an algorithm that assigns different priorities to vehicles (statically and dynamically) while also routing the vehicles in the network accordingly. In order to mimic vehicles being routed using our method, we also updated and rebuilt an open source simulator to account for the modifications. We aim to contain the amount of pollution in certain areas by using the provided measurement information of both the greenhouse gas emissions caused by multiple identified cars and also the amount of pollution existing already in a required area, we will be able to reroute cars inside the city. By using the measurement data of both the greenhouse gas emissions caused by multiple identified cars and the amount of pollution already present in a required area, we aim to contain the amount of pollution in certain areas. A modified version of Dijkstra's algorithm is suggested to route vehicles by using the intersection points of the streets as nodes and the amount of pollution in the streets as the weight of the edges, as opposed to the edge weight in the original algorithm being determined by the distance between the two nodes.

The rest of the paper is structured as follows: Sect. 2 presents related works on vehicle routing and priority-based routing. Sections 3 and 4 present our approach and implementation respectively. Section 5 presents the results and limitations, and finally; Sect. 6 presents the conclusions and a discussion of future work.

2   Related Work

2.1   Vehicle Routing

An ITS (Intelligent Transportation System) is proposed to monitor and manage air and noise pollution caused by road vehicles in cities. The suggested system dynamically directs a vehicle based on its particle emissions and noise indicators on the one hand, and a city's pollution levels and defined thresholds on the other. They developed an app which when given the car's id and starting and ending point provides two routes. Vehicle routes are based on detected particle emissions and noise levels using real-time pollution data. 1-the pollution-optimized route, 2-the 'standard' shortest path, based on Google Maps' Routing API (Application Programming Interface) [4].

There were multiple attempts to solve the vehicle routing problem for public transportation where in [5], the topic of determining the shortest route was examined using the vehicle's oil consumption to produce the Oil Consumption Weight (OCW) with weighted computation in each stage of the journey and the ideal route is automatically designed and prepared after given starting and ending location points. As a result, driving time is saved and oil consumption is reduced.Moreover, [6] used the Sweep Algorithm, to build an application, which is then tested using current route data Both forward and backward sweeps yield two similar routes with the same links This paper also indicates that forward sweep produces better results than backward sweep. Moreover, in [7] and [12], they addressed the vrp where the first paper compared 8 evolutionary crossover operators: order crossover, partially mapped crossover, edge recombination crossover, cycle crossover, alternating edges crossover, heuristic greedy or random or probabilistic crossover, mutation by inversion, mutation by reinsertion, and swap mutation. According to their findings, all of the studied operators are suitable for the Vehicle Routing Problem, and when combined, they offer the greatest results. The second paper respectively, used a memetic algorithm alongside the crossover operators.To their knowledge, this is the first study to use a memetic algorithm. When compared to state-of-the-art techniques, the results of their tests show that their system performs well. Furthermore, their computer investigations show that the main technique included in their suggested memetic algorithm is effective. In addition, [17] used evolutionary algorithms to find a solution to the vrp. The crossover approach was chosen intuitively. They fine-tuned the algorithm to strike a compromise between discovering a solution quickly and restricting the search space. Moreover, [18] suggest a Genetic Algorithm to solve the Green vehicle Routing Problem (GVRP). This study, unlike other vrp formulations, intends to reduce CO_2 emissions per route. The suggested model optimised both the route and schedules for a number of vehicles, allowing for the prediction of transportation within a set congestion period to prevent traffic congestion and reduce CO_2 emissions. The model needed to be more realistic, thus future research should include other specialised methods. Furthermore, [19] propose iterative improvement methods based on the concept of modifying the neighbourhood structure while searching by combining various expansions of classical models that have previously been handled sepa- rately. Their algorithms may be used in inter-active dynamic planning systems because average response times were mostly less than a second.

2.2 Priority Based Routing

With the growing use of automated guided vehicles, finding optimal routing protocols has been in question for more than a decade. Hundreds of research papers has been tackling that exact same topic in one of two ways; the classical approach where routing protocols were static and deterministic and the dynamic approach which involved assigning different priorities to vehicles. As a further matter, we are interested in the dynamic approach where different vehicles in the network are assigned priorities and are limited by certain constrains; for instance, pollution threshold for different areas of the network.

Usually, all vehicles in a vehicle network have the same priority while being routed, and this gap was filled by Naveen Chauhan et al. in their research paper where they explored priority based scheduling techniques for VANETs: Unicast, Multicast and Broadcast [1]. Also, the aspect of priority based routing has been the research question for Gaurav Gupta et al. where they discussed how routing protocols would work for evacuation scnearios in emergency situation. Any natural disaster - earthquake, tsunami, flood, and fire - can bring life to a standstill in minutes [2]. The inability of large populations to quickly and easily evacuate places of risk often leads to significant loss of life. A key challenge when design-ing an evacuation plan is how best to direct people from specific locations in an organized fashion so that everyone can be safely and fastly removed before danger strikes. When a road is closed due to an emergency, natural disaster, or planned construction work, road users need to be guided to an alternative route. However, this poses a challenge when it comes to cars that are connected and automated. Trucks or other large vehicles may not be able to take an alternative route because of the size of their load or the dimensions of their cab. In these cases, instead of simply notifying the driver about the closure and offering an alternate destination, it would make more sense to re-route them automatically so they still reach their final destination without u-turning from the blocked road.

To address the challenge of evacuating vehicles in emergency situations and help people in danger efficiently escape the area, we need an effective routing protocol for evacuation scenarios, with a well designed priority assignment for different vehicles (priorities are set depending on the individual's vehicle type and situation). This article will explain the fundamentals of such a protocol. They were statically assigning priorities to the vehicles, notably, giving the emergency vehicles (ambulances, police cars etc.) the highest priority in order for them to get routed to the desired destination in the fastest way possible [2].

The authors were able to achieve their goal by introducing a prioritized rout-ing assistant for Flow of Traffic (PRAFT) that Enables the police to maximize the traffic flow while Providing the ability to account for priorities of vehicles and routes while performing the routing following the minimum cost - maximum flow path. The priority based routing protocol was also tackled by Lucas Rivoirard where he proposed a scheme to auto-organise vehicles in vehicle networks [3] that used graph algorithms like (Dijkstra's algorithm) along with the Anchor based Street (A-STAR) and Traffic Aware routing protocol to achieve his goal of routing vehicles with different priorities.

2.3   Research Gap

All identified research papers used static priority assignment for the vehicles that are to be routed in the network whenever there is an emergency but none of the related work papers re-routed vehicles in non-emergency situations nor assigned vehicle priorities dynamically. Moreover,the aspect of air and noise pollution in cities caused by multiple vehicles was never tackled by any of the related work. Which makes our research gap clearly being a priority based constrained routing for road vehicles while taking into consideration the pollution (whether it's the real-time noise pollution and emission values or thresholds set for the edges(roads) in the network).

3   Our Approach

3.1   Using a Modified Dijkstra Algorithm to Create Least Polluted Routes

Since Air pollution is primarily caused by vehicles, vehicle routing when done efficiently would be of great help to reduce the air pollution or at least contain it under the defined city's pollution thresholds. In paper [5], they proposed a modified version of Dijkstra to solve the vehicle routing problem, where they primarily used the calculated amount of oil consumption per vehicle per street length to find the least oil-consuming routes from source node to destination node, rather than using the weight of the edges between nodes as distance. In our project, we also suggest utilising a modified Dijkstra's algorithm, but the distance or oil consumption will be swapped out for the amount of pollution the users' cars emit. Road intersections are used as nodes on the map, with the weight of the pollution placed on the edges rather than the distance between them. The system searches for all connected nodes at each intersection and, rather than determining which node to take next by adding the edge weight to the distance from the previous node to the source node, it will instead add the estimated amount of greenhouse gas emissions produced by each vehicle per length of street, while also ensuring that this estimated amount is below the city's established pollution threshold.

3.2   Assigning Priorities

As per our approach to the problem of routing road vehicles, our main goal is to create a priority based constrained routing algorithm. As shown in Fig. 1, different priorities will be assigned to different road vehicles.

Assigning Static Priorities. Setting vehicle priorities will be divided into static and dynamic priorities assignment. Starting by the static priorities; Let's first talk about emergencies, when you're driving (or being a passenger), there are so many rules that need to be followed. But what happens when something goes wrong and you can't drive any further? Roads and vehicular networks requires constant attention and awareness of road conditions. Even the most experienced drivers can find themselves in dangerous situations. So it's important to have a safety plan (which requires immediate help of professional medics and law enforcement). Emergency vehicles such as ambulance, police cars and fire trucks will always have the highest priority in the system, as they are critical vehicles that either need to arrive to a certain location or evacuate a certain location as fast possible regardless of any other factors (for example: not taking into consideration other factors such as pollution).

A low-pollution city is one that has cleaner air than most other cities in its country. So what exactly does that mean? Well, it depends on where you live. Each city has different measures of pollution based on its own unique set of standards. For example, one city could have lower nitrous oxide levels than another because it has more factories. In order to qualify as a low-pollution city, there must be measurable reductions in one or more types of pollutants for that particular location, based on the local standard. Therefore, the car type (whether it's polluting or non-polluting) is going to have an impact on the vehicle's priority. Now that we have completed the static assignment of priorities for the vehicles, we will be taking a look on how the priorities will also be dynamically assigned.

Assigning Dynamic Priorities. One of the many scenarios that are most likely to happen while being on the road (whether being the driver or a passenger), is for an emergency to happen all of a sudden, you then call law enforcement asking for help and you'll have to wait for an emergency vehicle to arrive as soon as it can. In some of these situations, the driver will be perfectly safe and the (or one of the) passengers are the ones in an emergency, at this situation, you do not have to wait for emergency vehicles to arrive, and you can drive the person in danger to the nearest hospital and evacuate him yourself. Therefore, law enforcement will be able to manually change a vehicle's priority depending on the situation to allow any vehicle to evacuate a certain area or arrive to a certain location as fast as possible regardless of other constraints.

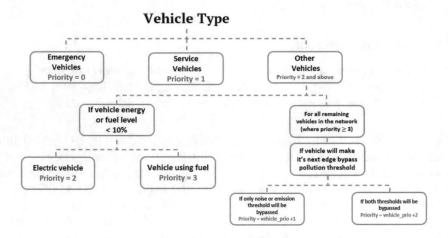

Fig. 1. Priority assignment

When concentration of a greenhouse gas such as carbon dioxide increases in the atmosphere, temperature increases not only at Earth's surface but also in the troposphere [14], so it is important to know which are the best low pollution cities to live in so that you can make an informed decision when planning your next move. Which is why a vehicle's impact on noise and air pollution thresholds will be taken into consideration. Real time values and thresholds of different zones will be monitored at all times, if a certain vehicle is to pass by a certain zone and have the zone reach or cross it's threshold, the vehicle will be re-routed for that not to happen; therefore while being routed, vehicles will have different priorities to access certain zones (constrained by real time values of noise and air pollution (taking into consideration the time of the day)). Critical fuel and energy levels of road vehicles will also be taken into consideration while assigning vehicle priorities. If a certain vehicle's energy or fuel level becomes critically low, it's priority becomes high so it reaches the nearest energy/fuel station as fast as possible.

4   Implementation

In order to visualise how our algorithm will impact vehicles in a certain network, we will need to use a simulator, in our case it's the SUMO simulator [7]. The SUMO simulator is a software that illustrates how traffic in a city can be simulated and optimized. The purpose of the simulator is to create a visualization of the effects of a pedestrian and cyclist on traffic.It also provides a visual representation of what it is like to be in a vulnerable position, such as being blind or carrying heavy items.As said, SUMO is an open source application, it will therefore enable us to implement our pre-discussed priority based routing algorithm and use it as the router which will route our vehicles.

Creating Our Network. We first need to have a map (where we will be simulating our vehicles behavior). We will be using Open Street Map (OSM) to generate our map. OSM is a collaborative project to create a free editable map of the world. The map is made using satellite imagery, GPS data, and advanced computer vision techniques. The site lets users update map information and add missing roads or buildings.

As a further matter, we need to add both air and noise pollution threshold values to the roads. We will be using a java based open street map editor (JOSM) to do the map tuning we need to. We can also edit the exported map to our liking mainly removing buildings from the map, add junctions, traffic lights and remove or add lanes to roads.

Once we have our map ready, we will need to convert our OSM map to a network by using "netconvert" which imports digital road networks from different sources and generates road networks that can be used by other tools from the package. We now need to create random trips for random vehicles in order to simulate their behaviours inside the network we just generated. We will be using a randomTrips.py python script which will generate random trips, by randomly choosing a departure vertex and an arrival vertex, choosing them randomly from the generated network file.

Choosing the Base Routing Algorithm for Priority Vehicles. SUMO simulator has a built-in router called duarouter that computes routes for vehicles that may be used by SUMO. Duarouter uses a shortest-path algorithm to compute the vehicle routes. Four routing algorithms can be used alongside with the duarouter to find the vehicle routes; Dijkstra's shortest path (Default algorithm used by Duarouter to find vehicle routes), astar search algorithm (A*), Contraction hierarchies (CH) and CHWrapper [8].

Our priority based constrained routing algorithm for road vehicles will be based on path finding computation, we therefore need to choose a suitable routing algorithm for that matter. CHWrapper and astar (A*) are variations of Contraction hierarchies and Dijkstra's shortest path respectively, therefore we will be comparing CH to Dijkstra's shortest path, the main advantage of each algorithm is shown in Table 1.

Table 1. Routing algorithms comparison

Routing Algorithms available to be used by Duarouter	
Contraction hierarchies (CH)	Dijkstra's shortest path
Advantage: Very efficient when a large number of queries is expected	Advantage: Well suited for routing in time dependent networks path
Disadvantage: Does not consider time dependent weights	Disadvantage: Slow routing algorithm

As mentioned by Guoqiang Mao et al. in their research paper: A Unified Spatio-Temporal Model for Short-Term Traffic Flow Prediction [9]; "The spatio-temporal correlation between traffic at different observation points is not stationary but varies with time of the day", which means that the travel time on a certain edge by a certain vehicle is affected by the time of the day. As seen in the comparison, Dijkstra's shortest path is well suited for routing in time dependent network, which is exactly the type of network we deal with when considering routing road vehicles which makes Dijkstra's shortest path algorithm a better fit. Now that we are settled on Dijkstra's routing algorithm, we need to choose between Dijkstra's routing algorithm and it's variant the A* search algorithm. To visualise the key differences between both algorithms, we will be using a Java graphical program called PathFinder [10] When we compare the execution of both algorithms, we can see that Dijkstra's shortest path algorithm has visited exactly 600 nodes to get to the path of cost 302 (shortest path), where the A* search has only visited 93 nodes to reach the exact same path (as shown in Fig. 2).

Normally, we would chose the A* search over the Dijkstra's shortest path algorithm as it's faster and does significantly way less computations, but during the execution of our algorithm (the priority based constrained routing for road vehicles), a vehicle's route won't necessarily be decided by the path being the shortest, but by priorities, pollution current levels and thresholds, emergencies happening and neighbor vehicles and their behavior; so visiting more nodes while finding a path is actually beneficial for our algorithm.

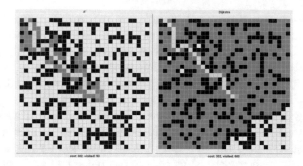

Fig. 2. Dijkstra's shortest path and A* search - PathFinder

4.1 Modifying and Re-building the SUMO Simulator

In SUMO there is a set of attributes (23 different attributes) that can be assigned to each vehicle, the ones that we are going to need in our implementation are presented in Table 2.

Table 2. Vehicle Attributes

Attribute	Value Type Date	Description
ID	String	Vehicle's name
Type	ID	Vehicle's type
Route	ID	ID of the route the vehicle shall drive along
Color	Color	Vehicle's color
Depart	Float	Simulation step at which the vehicle enters the network

Unfortunately, we are not able to add an extra attribute (being the priority) to the vehicle's class as it will force us to modify other classes, packages and libraries that we won't be needing. The work around is to combine the vehicle's priority with it's id. Each vehicle will be having an ID of the following format: $ID = X\_Y$ where X is the vehicles ID (vehicle's name) and Y is the vehicle's priority.

We need to start implementing our algorithm to build the project. Before modifying the Dijkstra algorithm itself, we need to modify how the vehicles get treated by the duarouter. Initially, a list of all vehicles in the network is created, accessed by duarouter and processed by the chosen routing algorithm. Dijkstra's shortest path is default duarouter routing algorithm and is also the one we chose for implementing our algorithm, so we won't need to add restricting parameters when calling the duarouter later on.

duarouter_main.cpp is the main file where duarouter functions are implemented, which can be found in the following path: *sumo/src/duarouter/ duarouter_main.cpp*

Whenever *duarouter* is called it calls the routing algorithm's script to be executed; in our case it's the Dijkstra's shortest path which can be located at: *sumo/src/utils/DjikstraRouter.h*. We need to sort the vehicle list passed by the duarouter to the Dijkstra's algorithm so that the vehicles list is sorted by priorities. To do that we need to override the comparer method that was predefined so that it compares vehicles according to their priorities.

We are now comparing vehicles according to their priorities and once the vehicle list gets sorted, vehicles with highest priorities will be at the head of the list. During the map edit, we have added pollution threshold to the edges (roads) right before they were converted into a network file. Now the pollution threshold (whether it's an emission or noise pollution) is consumed by higher priority network vehicles first. Once the threshold has been reached, re-route of lower priority vehicles will be needed if a higher priority vehicle needs to get into the same next edge of the lower priority; which will automatically trigger the re-route. To assign dynamic priorities, a dynamic environment needs to be implemented. Thanks to the Dynamic User Assignment implemented by SUMO. "A frequent problem with naive user assignment is that all vehicles take the fastest path under the assumption that they are alone in the network and are then jammed at bottlenecks due to the sheer amount of traffic" [11]. We therefore need to use Iterative Assignment which is also called Dynamic User Equilibrium that will determine the suitable routes considering weighted factors (example: travel time and pollution levels). To compute this dynamic user equilibrium we will be using the *duaIterate.py* tool which we will call by: $pythontools/assign/duaIterate.py - n < network - file > -t < trip - file > -l < nr - of - iterations >$.

5 Results and Limitations

5.1 Results

Quantitative Results and Analysis. In SUMO simulator, vehicles have an emission class attribute which computes how much fuel or electricity a vehicle consumes, the amount of CO_2 a vehicles emits, in PPM; and the amount of noise pollution a vehicle causes, in *db*. Emission classes in SUMO are HBEFA-based [16], in other words, SUMO has remodeled the HBEFA emission classes to be used in the simulations. The $HBEFA3/PC_GEU4$ emission class (SUMO's default emission class) will be assigned to all vehicles in our simulation to ensure that the same simulation environment is being used in all simulations that we are going to run. The least polluted route for the vehicle is computed by first taking the starting point, destination point, and the car model as inputs.

To help visualise the routes, Figs. 3 and 4 above illustrate the routes the vehicle should take in the simulator SUMO. The route is coloured blue for the shortest path using Dijkstra's algorithm and the other route is colored red presenting the least polluted route we got using the modified pollution weighted Dijkstra's algorithm proposed before. Let us take a closer look at the vehicle's path after using the modified Dijkstra algorithm and the pollutant concentrations presented

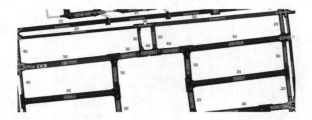

Fig. 3. Visualization of the least polluted path(using modified Dijkstra)

Fig. 4. Visualization of the shortest path without putting into consideration the pollution levels(using shortest path Dijkstra)

	A	B	C	D	E	F	G	H
1	timestep	id	eclass		CO2	fuel	electricity	noise	route
2		0 0_0	HBEFA3/PC_G_EU4	2624.72	1.13	0	55.94	!0_0

	I	J	K	L	M	N	O	P
	type	waiting	lane	pos	speed	angle	x	y
	DEFAULT_VEHTYPE	0	D2C2_0	5.1	0	270	737.7	501.6

Fig. 5. Results generated at the end of the simulation

on the graph for a more thorough view. As illustrated in Fig. 3, the first conjunction offers two options with pollution levels of 30 and 50, respectively. The system compares the lengths of both streets and calculates the pollution amount by multiplying the street length by the grams of pollution per mile caused by the vehicle model. This pollutant amount is added to both streets' present pollution levels to determine which one will produce the least amount of pollution, taking into account that neither the air pollution nor the noise pollution thresholds are exceeded. The system selects the first node in this situation. Moving forward, the same thing happens at each subsequent junction until we arrive at our final destination. After running the modified Dijkstra algorithm on our network, this route is assigned to us as the first green route to our end destination. On the other hand, when we run the system without using the modified Dijkstra we can find that yes the route chosen may be a shorter and a faster one but it is not the most optimized one according to the pollution levels causing it to increase and sometimes would go over the set pollution threshold of the streets in the city.

At each timestamp, results for each vehicle's CO_2 emissions caused, noise pollution caused, fuel or electricity consumed is calculated and stored as well as the x and y positions in the network as shown in the Fig. 5 below.

12   Y. Aboulyousr et al.

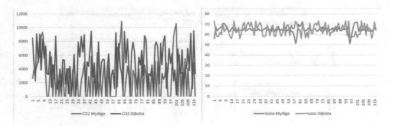

Fig. 6. Vehicle 0 CO_2 emissions and noise pollution caused

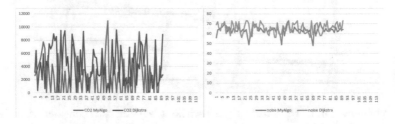

Fig. 7. Vehicle 1 CO_2 emissions and noise pollution caused

After running the simulation, emission and noise pollution caused by each vehicle as well as energy and fuel levels were measured and a results file was generated.

The Figs. 6 and 7 illustrate the simulation results of 2 arbitrarily chosen vehicles. The graphs are a representation of the total CO2 and noise pollution levels caused inside the network by each vehicle throughout the simulation run-time. No co-relation was found between total pollution values (whether it's caused by CO2 emissions or by noise pollution) caused by vehicles routed by the Priority based constrained routing and the ones caused by vehicles routed by Dijkstra's shortest path, which was expected, as the priority based constrained routing algorithm does not focus on reducing the pollution levels, but by distributing the pollution instead from being concentrated in one area, as shown in Fig. 8.

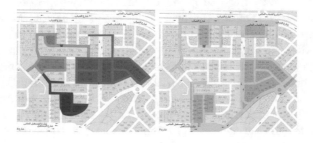

Fig. 8. Visualization of the pollution Density (Dijkstra - Left) and (Priority Based Constrained Routing Algorithm - Right)

To better benchmark both algorithms against each other, we need to generate a pollution density map from the data we got for each edge.

As we can see from the figure above, which is the representation of the same map that has been used to create the simulation's network was used to show the pollution density. Pollution values are CO_2 emissions + noise pollution caused by each one of the vehicles throughout the simulation. It's clear that the vehicles routed by Dijkstra always had the shortest path as their route, regardless of pollution thresholds, which caused the inside of the city (non high-way edges) to have the most dense red color. On the other hand, the priority based constrained routing algorithm routed the vehicles according to their priorities and according to pollution threshold levels which caused vehicles to get re-routed to longer routes (passing by high ways) and causing less pollution in the roads inside the city (the non high-way edges), but more on the high ways, which explains why the roads inside the city have less dense red color but the high way roads have a denser red color than the ones from Dijkstra.

Computational Complexity. With the development of information search, retrieval, and extraction techniques and approaches, computational complexity challenges are receiving more and more attention [15]. Now that we have assured that we have both designed and implemented an algorithm that satisfies our needs, we finally need to focus on the amount of resources in terms of time and space that our algorithm uses to compute the necessary results in comparison with the resources the default Dijkstra's shortest path algorithm would consume. In terms of the space needed, our algorithm will be using the exact same amount of space the Dijkstra's shortest path algorithm would normally use while computing the network's vehicles routes because of the way we implemented our algorithm; as we have efficiently used the exact same default Attributes for the vehicles, routes and network properties that SUMO simulator offers without requiring more attributes or data holders to be used as explained earlier in the implementation section. The time needed to compute the vehicle's routes will be harder to compare as in lot's of cases (as we have explained and seen earlier) vehicles will need to be re-routed while computing routes using our modified Dijkstra's shortest path algorithm. What we can compare is how much times it needs to compute the routes when all routes are statically computed; which is when the simulator needs the initial routes to be computed.

As Fig. 9 illustrates, in both trials the same network with same settings has been used for both Dijkstra's shortest path algorithm and our modified routing algorithm. In trial 1, using Dijkstra's shortest path, the simulation and the initial vehicle routes were computed and ready in 1162 ms in comparison with 1212 ms when using our constrained based routing. In the second trial, Dijkstra's shortest path needed 1167 ms in comparison with 1198 ms for our algorithm. In the first trial, our algorithm needed only 4.3% more time to compute initial vehicle routes and load the simulation and in the second trial it only needed 2.6% more time than the default Dijkstra's shortest path. Therefore, it's clear

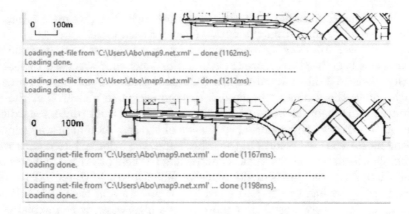

Fig. 9. Time taken by the simulator to compute initial routes (Trial 1 and Trial 2) Screenshot from SUMO simulator

that when computing the vehicle routes and loading the simulation we need the same amount of time with either algorithms and that the very slight increase in time of 2.6% to 4.3% is negligible.

5.2 Limitations

Due to the way the Priority based constrained routing algorithm was implemented with, there is a limitation that appears whenever the following scenario occurs. If a vehicle gets introduced to the network at any time step, while having a priority lower than any other vehicle's priority in the network and while all edges thresholds have been reached, the newly introduced vehicle won't be able to get routed at the time it gets introduced to the network, but will have to wait until any route to its destination is clear. Moreover, another limitation encountered was that there were no data about pollution levels of the streets in real time to connect to our database to get more accurate results from the routes constructed.

6   Conclusion and Future Work

To find the least polluted routes, a modified version of Dijkstra method is proposed in this paper, using pollution levels as weights of edges, converting all routing variables to air and noise pollution emission instead of distance between the connected nodes. The algorithm breaks down the map into nodes and edges before calculating the best green path. One or more vehicles are routed to the least polluting path, with the system using the first in, first out strategy, with each car arriving being routed first, followed by the next car arriving, and so on, until all vehicles are routed using the least polluted routes possible.

A priority based constrained routing algorithm for road vehicles was also implemented. We have discussed that it is very important to have such an algorithm to route road vehicles with, specially when pollution is a topic being discussed all the time. We have also analyzed the importance of having different priorities attributed to different vehicles as well as how will these priorities be assigned (both statically and dynamically). The open source traffic simulator SUMO was the main tool being used in this project which was modified and the whole simulator project was re-built to accommodate our implementations.

Ideally, sensors could be installed around the city to gather pollution levels every second and connect to the database, which would be updated with the most recent information every few minutes.

References

1. Dubey, B.B., Chauhan, N., Chand, N., Awasthi, L.K.: Priority based efficient data scheduling technique for VANETs. Wirel. Netw. **22**(5), 1641–1657 (2015). https://doi.org/10.1007/s11276-015-1051-8
2. Gupta, G., et al.: A prioritized routing assistant for flow of traffic. In: 21st IEEE International Conference on Intelligent Transportation Systems. Hawaii, USA (2018)
3. Rivoirard, L.: Modèle d'auto-organisation pour les protocoles de routage dans les réseaux ad hoc de véhicules : application á la perception élargie et á la localisation coopératives. Université de Lille. Lille, France (2019)
4. Zaky, M., et al.: An intelligent transportation system for air and noise pollution management in cities. In: Proceedings of the 7th International Conference on Vehicle Technology and Intelligent Transport Systems - VEHITS (2021)
5. Zhang, J., et al.: Vehicle routing in urban areas based on the oil consumption Weight-Dijkstra algorithm. IET Intell. Transp. Syst. **10**(7), 495–502 (2016)
6. Nurcahyo, G.W., Sap, M.N.M., et al.: Sweep algorithm in vehicle routing problem for public transport. Asia-Pacific J. Inf. Technol. Multimedia **2**, 51–64 2002
7. Puljic, K., et al.: Comparison of eight evolutionary crossover operators for the vehicle routing problem. Math. Commun. **18**, 11 (2013)
8. Duarouter - sumo documentation. https://sumo.dlr.de/docs/Simulation/Routing.html. Accessed 27 June 2022
9. Mao, G., et al.: A unified spatio temporal model for short-term traffic flow prediction. IEEE Trans. Intell. Transp. Syst. **20**, 3212–3223 (2019)
10. Wang, K.: Pathfinder. https://github.com/kevinwang1975/PathFinder. Accessed 8 June 2022
11. Dynamic User Assignment. https://sumo.dlr.de/docs/Demand/Dynamic_User_Assignment.html. Accessed 25 June 2022
12. Peng, B., et al.: A memetic algorithm for the green vehicle routing problem (2019)
13. Veres, M., Moussa, M.: Deep Learning for intelligent transportation systems: a survey of emerging trends. IEEE Trans. Intell. Transp. Syst. **21**, 3152–3168 (2020)
14. Manabe, S.: Role of greenhouse gas in climate change (2019)
15. Sivaram, M., et al.: The real problem through a selection making an algorithm that minimizes the computational complexity. Int. J. Eng. Adv. Technol. (IJEAT) **8**(2), 95–100 (2018)
16. Emissions. https://sumo.dlr.de/docs/Models/Emissions.html. Accessed 23 June 2022

17. Larsen : Arab Republic of Egypt - Cost of Environmental Degradation : Air and Water Pollution
18. de Oliveira da Costa, P.R., et al.: A genetic algorithm for a green vehicle routing problem (2017)
19. Goel, A., et al.: A General Vehicle Routing Problem (2006)

Autonomous Mobile Robot Mapping and Exploration of an Indoor Environment

Raghad Mando$^{(\boxtimes)}$ ⓘ, Eylül Özer, and Burak İnner ⓘ

Faculty of Engineering, Department of Computer Engineering, Kocaeli University, Kocaeli, Turkey
{195112020,225112006,binner}@kocaeli.edu.tr

Abstract. The main challenge for the robot is to interact with its environment. Generally, a robot can interact with its environment by achieving the necessary contact configuration and the subsequent motion required by the task. SLAM solves the challenge of robots exploring an unknown area. The robot's goal is to collect information while exploring the environment to develop a map it, so it wants to use its map to determine its location. The current study examines multiple 2D SLAM laser-based algorithms available in the Robotic Operating System to determine the most efficient technique for creating a map in a specific current world using an Autonomous Mobile Robot (AMR) equipped with two lidar 2D scanners in a customs warehouse environment. A good map is essential for the robot to efficiently interact with its surroundings. From this viewpoint, selecting the optimal SLAM algorithm is critical since it relies on the outcome of the final map. This paper discusses, the comparison of SLAM Toolbox, G-mapping, Hector SLAM and Karto-SLAM methodologies through real-world testing in a dedicated environment, a discussion the output maps of SLAM algorithms, their advantages and disadvantages are presented. After examining and scrutinizing the maps, the G-mapping file gives a more precise map with more defined boundaries and obstructions than the rest. The G-mapping map file shows the map without slip, tilt, or skew. However, depending on the math work, the calculation approach shows that Hector Slam provides a value error lower than the G-mapping designation.

Keywords: SLAM · SLAM-toolbox · G-mapping · Hector-SLAM · Gazebo · ROS

1 Introduction

Technological advancements and more powerful embedded systems have developed mobile robot capabilities [1]. Moving around the environment is one of the most fundamental things a robot can do for navigation. For the Mobile Robot to reach its navigation targets using its current processing capacity, it must calculate a path plan and kinematic parameters like speed and orientation to avoid obstacles. ROS is a truly open meta-operating system that is used in the development of robotic technology ROS is a shared platform that may run on several machines at the same time. As a result, it opens the

© The Author(s), under exclusive license to Springer Nature Switzerland AG 2023
F. P. García Márquez et al. (Eds.): ICCIDA 2022, LNNS 643, pp. 17–26, 2023.
https://doi.org/10.1007/978-3-031-27099-4_2

way for developing new mobile robots with greater navigation abilities. For the robot's autonomous navigation, the ROS navigation stack is employed, which is most essential functions of a robot. The robot must first comprehend where it is (its starting point) and where it should proceed (its goal). This job is often accomplished by providing the robot with a map of its surroundings as well as a beginning and ending point. The SLAM technique is used by robots to construct a map of an unfamiliar area while detecting an official's location within it. It is a technique for mapping an environment while the robot is moving. Specifically, when a robot moves about an area, it gathers data from its sensors and creates a map. Therefore, a mobile robot may create a map of an unknown region and update an existing map.

This paper's structure is divided into five parts. Section 2 contains the previously completed research that examined the performance of the ROS, as well as a brief description of the SLAM algorithms. The experimental setup and system are outlined in part 3 before the experiment and findings Sect. 4 discusses them. Section 5 consists of the conclusion and suggestion for future.

2   Related Work

The difficulty of constructing a map for an unfamiliar setting using sensing while also resolving the localization issue is referred to as simultaneous localization and mapping (SLAM). [2] SLAM is a critical challenge in robotics, and numerous solutions have been developed to address it. It is vital to provide insight into the shortcomings and strengths of these strategies as they relate to their desired ultimate use [17]. The filter-based SLAM is a common method for this challenge. This approach is founded on the idea of Bayesian filtering. It consists of two essential phases: the prognostication phase, in which previous information about the system state and input control commands are used to update robot clustering and map condition; and the quantification step, in which current sensor information is compared to the expected scheme country to make new scheme state predictions; the generated map is critical for providing navigation information to the robot, allowing it to move around the environment. Murphy [3] developed the Rao-Blackwellized particle filter technique in 2000 that address the SLAM challenge in a dynamical indoor scenario. Rao-Blackwellized particle filters usually need a large amount of dynamic measurement data and minimal measurement noise. However, this increases the computer's complexity and CPU rate. As a result, discovering ways to decrease the number of particles is a major task for academics in optimizing the particle filter algorithm. The SIR (Sampling Importance Resampling) filter in particle filters is the most utilized algorithm. The algorithm is carried out in the four steps stated below. First is the prediction stage: Initially, the particle filter generates many samples based on the state transition function prediction. These are particles, and the weighted sum of their probabilities is used to calculate the posterior probability density. The second stage is the correction: the relevant significance weights for each particle are calculated at each observation. This weight reflects the probability of noticing the predicted pose when the first particle is recorded. The higher the weight, the more frequently the particles are noticed. Third, the collection particles were resampled according to weight proportions during the oversampling process. Because of the restricted number of particles in a continuous distribution, this is critical. The collection of resampled particles is then entered into the transition probability equations in the following filter cycle, and new projected particles are created as the fourth map estimates. Based on the trajectory and observation of the sample, the relevant map estimate is computed for each particle retrieved from the sample. The SIR method must re-evaluate the particle weights when new data are discovered. The computing cost of this operation grows with the length of the trajectory over time. As a result, the vital probability density function is limited to produce a recursive formula for weighted computing significance. A 2D map must be prepared after completing a mapping technique. Many SLAM algorithms are available, depending on the purpose of the map. The most popular SLAM algorithms are G-Mapping, Hector-SLAM, Core SLAM, and Cartographer. The g-mapping technique employs a particle filter pairing strategy. Hector SLAM employs the scanning matching technique, Cartographer is a scan matching algorithm with loop detection, and RGB-D is a depth image mapping algorithm. These are some of the most well-known and widely utilized algorithms. G-mapping is a raster-mapping SLAM method that uses lasers. [3, 4]. G-mapping is the most used SLAM algorithm, and it is now the standard technique for PR2 (a popular mobile copying platform), with an implementation accessible on openSLAM.org. [4] was the first to explain this strategy. The fundamental concept is to anticipate the state transition function using Rao blackwellized particle filters (RBPFs). The technique is also known as the RBPF SLAM algorithm, which is called after using Rao-Blackwellized particle filters. [3] Optimizing the proposal distributions and integrating dynamic interpolation has enhanced the technique and made it more acceptable for practical uses. [5] Because grid maps are used, it is referred to as G-mapping (G for grid). G-mapping employs particle filters. Hector-SLAM [6] varies from previous grid-based mapping algorithms in that it uses laser data and an a priori map rather than odometer data. It works based on scan matching. The Core-SLAM technique is a SLAM-based SLAM method that necessitates a mobile robot having odometry data and a static laser rangefinder positioned horizontally. Core-SLAM employs a straightforward Monte Carlo method and necessitates loop closure. Cartographer Google's SLAM solution is a graph optimization approach. Cartographer and Cartographer ROS are two parts of Google's open-source code. A cartographer's job is to construct a map from data from lidar, s IMU, and odometer. The ROS cartographer uses the ROS communication method to collect sensor data. It changes it to Cartographer format for Cartographer processing before releasing the output for display or storage. The Cartographer algorithm raises the processor's computing workload. When the amount of data to be reviewed gets too enormous, particle-based algorithms become useless [7]. If the user has a suitable map, the Cartographer provides a pure localization mode. It also supports data serialization to save just processed partial maps. Google Maps, however, has been effectively abandoned and is no longer updated or supported by Google [8]. It cannot create adequate maps for mapping and localizing robotic systems with other localization software packages without excellent odometry. The SLAM Toolbox expands on the heritage of Open Karto, SRI International's open-source library, by providing efficient mapping approaches and a variety of additional tools and innovations. SLAM Asynchronous and synchronous mapping modes, kinematics map merger, multi-session mapping, a location mode, better graph efficiency, greatly decreased computing time, and prototypes for lifetime distributed mapping applications are all included in the toolbox. [8, 9]. This study examines a variety of SLAM algorithms, both freshly created and long-used. An experiment was carried out to evaluate and compare all approaches. While the kind of sensor had the biggest influence on the SLAM algorithm's effectiveness [18], an ideal technique for producing an acceptable map in a bonded warehouse environment was discovered by utilizing an autonomous mobile robot with several lidar scanners.

3   System and Setup

3.1   System and Robot

Figure 1 Shows the robot utilized in the study. Two LIDAR sensors were installed on the robot. Additionally, the differential drive also allowed the experimenters to change the position and direction of the robot. The experiment was conducted using an Intel® Corei7–8700 processor with RAM 8 GB and Ubuntu 20 using ROS Noetic. All four SLAM techniques discussed here were validated using real-world tests. The experiment was carried out with a robotic system in a real-world environment to investigate the performance of several SLAM tools in the absence of a perfect algorithm. Even under ideal conditions, such as noise-free odometry and distance measurement data, the algorithms of SLAM may produce imperfect results due to measurement uncertainties. The robot was remotely controlled in each experiment using twist-based ROS robotics and a normal joystick.

Fig. 1. AMR with two 2D laser scanners.

3.2   Slam Algorithm

This section briefly overviews four SLAM techniques: SLAM toolbox, G-mapping, hector SLAM, and kart SLAM.

   SLAM-toolbox: The SLAM-toolbox is a graph-based method. It uses the same concept as Google Cartographer in solving 2D SLAM: a local and a global approach [8]. ROS There are two types of nodes in the SLAM -toolbox: synchronous and asynchronous. They are two unique nodes that consist of two different blades. Even if new scans arrive while the system is still estimating scan modes, which implies system latency, the system computes all scan modes that have passed the traffic filter in the synchronous node. The status of the most recent scan is not determined until the status of the previous scans is complete. In contrast, asynchronous nodes never delay. If the system receives a new scan while working on earlier scan poses, it discards some earlier posts to catch up and compute the latest scan. Even if it still has tasks to compute earlier poses, it will always try to compute the pose of the latest scan. Also, during the motion filtering phase, this node does not check the timestamps of the scans. It only considers translational and rotational motions.

G-mapping the most extensively used SLAM program for mobile robots is G-mapping. The PF algorithm family frequently requires many particles to produce successful results, resulting in significant computational load. Additionally, the method's efficiency is influenced by the PF re-sampling process's depletion. As the uncertainty is minimized due by the scan matching technique, the number of particles required decreases. In our experiments, G-mapping was used with 80 particles.

Hector SLAM: It combines robust 2D scan-matching technology based on SLAM with 3D navigation technology based on inertial sensors. [12] Hector SLAM can produce a high-quality 2D map and perform localization. [13] SLAM is regarded as a breakthrough in particle filter-based mapping. This SLAM technique can be utilized in the absence of an odometer and on stations equipped with sensor rolling or pitching. Even though the method does not give clear loop closure, it is adequate in several real-world situations [14]. However, when the only reduced scanning is available and relatively good odometry computations are not used, it may cause problems. The 2D position is decided by improving the laser endpoint alignment with the previously obtained map. The map is projected using the endpoints, and the probability of occupancy is computed. A Gauss-Newton formula is used to solve the scan alignment, which figures out the stiff change that leads to the requirement for the laser pointers with map. To avoid being caught in local minima, a multi-resolution map format is often used.

Karto SLAM: SRI International company Karto Robotics has developed Karto SLAM, a graph SLAM. It performs well in practice and is not much influenced by noise. The solution Karto SLAM is a very efficient and Cholesky matrix reduction without iteration for sparse linear systems [10]. Sparse Pose Adjustment (SPA) manages scan matching and loop closure in its ROS version [14]. The greater number of landmarks in a SLAM, the more memory is required, particularly when keeping a map of a big region. However, because it only supports one pose graph, it is useful for Karto SLAM.

4 Experiments and Discussion of Findings

4.1 Experiment Environment

Mapping tasks were completed by developing configuration and start files for each mapping approach. ROS Noetic was used to analyze and execute all algorithms, and the experimental data were gathered under the same conditions and on the same CPU. Figure 2 depicts the natural prepared environment. It is an industrial setting with static barriers; also, the environment design has been produced; the environment design may be seen in Fig. 3

Fig. 2. Real-world environment

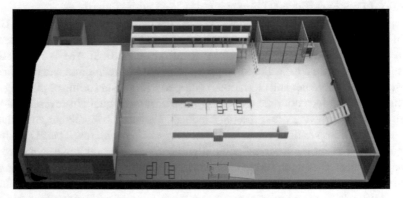

Fig. 3. Real-world environment design

4.2 Discussion of Findings

SLAM Toolbox: Slam Toolbox is a new technology with multiple functions such as integrating kinetic maps, location mode, session map, improved graph optimization, greatly decreasing computing time, and prototyping lifetime distributed map applications are all features of this softwares. Through our research on the multi-sensor robot in a custom industrial environment. We discovered that the SLAM Toolbox provides two mapping modes and many features. Despite this, we noticed that the SLAM toolbox has a problem moving between areas when it should take action to move left or right in a custom factory environment using mobile automation robots. It only acts when moving straight, not when turning. Moreover, this was not charming since we had hoped to succeed with this algorithm. Figure 3 shows the map created using the SLAM toolbox approach. Also, Fig. 4 shows the SLAM toolbox after cleaning the unnecessary parts to make the compression.

    G-mapping: In 2007, one of the most extensively used SLAM libraries was released. G-mapping is not well adapted to vast expanses and fails to conclude loops correctly on an industrial level. However, we discovered that the best map was created using the G-mapping technique. The G-mapping algorithm has been discovered to map rapidly and efficiently by executing operations of Odom data based on scan data. Figure 5 shows the map created using the G-Mapping approach. Also, Fig. 9 shows the G-mapping map after making zoom in.

Karto SLAM: In the Karto SLAM algorithm, each node in the ROS reflects the robot's location relative to its trajectory and sensor dataset. Node connections represent movements between close places together. The map is generated for each new node by finding the geographically best node arrangement. In rare circumstances, difficulties may arise during installation due to inadequate package documentation, a restricted number of adjustable options, and the necessity of an odometry application. To address the issue of nonlinear matrix direct solutions, SLAM Karto offers optimization via sparse pose configuration. Figure 6 displays the map produced by the Karto SLAM algorithm. Despite using a significant amount of memory, environment mapping is carried out swiftly and effortlessly. Because of its memory needs, it is not one of the top-performing algorithms for our environment. Also, Fig. 10 Shows the Karto SLAM map after making zoom in.

Hector SLAM: Hector SLAM does not employ odometry values, that might result in erroneous location and mapping updates when doing laser scanning at a slower pace or even in large or featureless areas. It likewise does not provide loop closure but generates a map in our environment fast and efficiently. So, while the Hector SLAM approach excels in tiny environments with obstacles, we discovered that it produces a smaller map than previous SLAM methods. Figure 7 depicts the map produced by Hector SLAM. Figure 11 also displays the Hector SLAM map after zooming in (Fig. 8).

Fig. 4. Generated map with SLAM toolbox   **Fig. 5.** Generated map with G-mapping   **Fig. 6.** Generated map with Karto SLAM   **Fig. 7.** Generated map with Hector SLAM

Math Work

A study found the Harsdorf distance as a quantitative metric between ground reality and simulation maps [10]. Furthermore, other studies, such as [15], measured the map's accuracy by dividing the noise value by the matching value. They also utilized an ADNN-based error analysis in [11] to compute the average distance to the nearest neighbor (ADNN), which was determined as the sum of all distances divided by the number of filled cells on the map. They compare the maps in [16] by matching the size information in the Gazebo simulation environment to the map information gained from the experiment. The map's degree of overlap with the natural world may be assessed. In this way, the geometric objects were measured in meters in the Gazebo simulation environment, representing the real world.

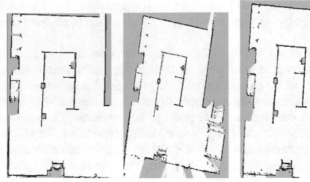

Fig. 8. Generated map with SLAM toolbox   **Fig. 9.** Generated map with G-mapping   **Fig. 10.** Generated map with Karto SLAM   **Fig. 11.** Generated map with Hector SLAM

To compare the maps, we calculated the number of black pixels in each map created by the algorithms. We also calculated the black pixels on the actual map. After calculating the pixels, we convert them to meter values and compare them to actual map measurements. Finally, as shown in Table 1, we calculated the error rate for each map deviation in m using an actual size map. The error and precision numbers for algorithms are shown in Fig. 12.

Table 1. Map comparison of experimental measurement information

Heading level	The calculated size of obstacles	Mapping error
Hector SLAM	3.2546395833 m	7.14%
Karto SLAM	1.20888125 m	65.48%
G-mapping	1.21046875 m	65.71%
SLAM toolbox	0.94244583333 m	72.85%

Based on the statistics in Table 1, the variation in 1 m is visually shown in Fig. 13. The mapping error with Karto SLAM was 0.5416 m per 1 m, the mapping error with G-mapping was 0.5428 m per 1 m, and the mapping error with the Slam toolbox was 0.7880 m per 1 m. The mapping error for Hector SLAM is 0.02188 m in 1 m.

We defined that Hector Slam's error rate as the most cost-effective. The size of the obstacles was comparable to the actual size when using Hector. The Karto SLAM ERROR was like the G-mapping Mapping ERROR, but the SLAM toolbox ERROR rate was unexpected. The SLAM toolbox had the highest error rate based on math work.

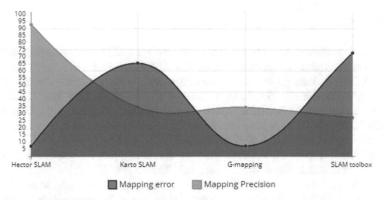

Fig. 12. Mapping Error and mapping precision in actual maps size

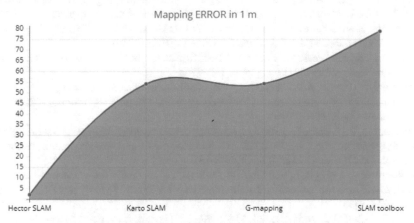

Fig. 13. Mapping error in 1 m

5 Conclusion

This paper describes four sample SLAMs: Hector SLAM, SLAM toolbox, Karto SLAM, and the G-mapping. The results of the SLAM execution were studied and compared using metrics. Each solution's faults and merits are also discussed. The major findings of our research: G-mapping provides solid map-building solutions, whereas Hector SLAM is the most computationally efficient. Comparing different SLAM methods is challenging, nevertheless, because the effectiveness of the SLAM approach is determined by its configuration, and the best configuration values may change depending on the scenario.

Nonetheless, this study provides some insight into selecting a suitable SLAM technique for a specific application. Based on the findings of this paper, more testing will be undertaken on different parametric configurations and data and maintain with noise. Future research will go toward evaluating using a robot in dynamic real-world circumstances.

References

1. Kopčík, M., Jadlovský, J.: Embedded control system for mobile robots with differential drive. Acta Electrotechnica et Informatica **17**(3), 42–47 (2017)
2. Cadena, C., et al.: Past, present, and future of simultaneous localization and mapping: Toward the robust-perception age. IEEE Transactions on robotics **32**(6), 1309–1332 (2016)
3. Murphy, K., Russell, S.: Rao-Blackwellised particle filtering for dynamic Bayesian networks. Sequential Monte Carlo methods in practice. Springer, New York, NY, pp. 499–515 (2001). https://doi.org/10.1007/978-1-4757-3437-9_24
4. Grisetti, G., Stachniss, C., Burgard, W.: Improved techniques for grid mapping with rao-blackwellized particle filters. IEEE Trans. Rob. **23**(1), 34–46 (2007)
5. Koenig, N., Howard, A.: Design and use paradigms for gazebo, an open-source multi-robot simulator. In: 2004 IEEE/RSJ International Conference on Intelligent Robots and Systems (IROS), IEEE Cat. No. 04CH37566, Vol. 3. IEEE (2004)
6. Kohlbrecher, S., et al.: A flexible and scalable SLAM system with full 3D motion estimation. In: 2011 IEEE international symposium on safety, security, and rescue robotics. IEEE (2011)
7. Hess, W., et al.: Real-time loop closure in 2D LIDAR SLAM. In: 2016 IEEE International Conference on Robotics and Automation (ICRA). IEEE (2016)
8. Macenski, S., Jambrecic, I.: SLAM Toolbox: SLAM for the dynamic world. J. Open-Source Software **6**(61), 2783 (2021)
9. Agarwal, S., Mierle, K., Others. (n.d.): Ceres Solver. http://ceres-solver.org
10. Turnage, D.M.: Simulation results for localization and mapping algorithms. In: 2016 Winter Simulation Conference (2016)
11. Yagafarova, R., Ivanou, M., Afanasyev, I.: Map comparison of lidar-based 2d slam algorithms using precise ground truth. In: 2018 15th International Conference on Control, Automation, Robotics and Vision (ICARCV). IEEE (2018)
12. Kohlbrecher, S., Meyer, J., Von Stryk, O., Klingauf, U.: A flexible and scalable SLAM system with full 3D motion estimation. In: the Int. Symp. on Safety, Security and Rescue Robotics (SSRR) (2011)
13. Saat, S., Airini, A.N.M.F., Saealal, M.S., Wan Norhisyam, A.R., Fares Ezwan, M.S.: Hector SLAM 2D mapping for simultaneous localization and mapping (SLAM). J. Eng. Appl. Sci. **14**, 5610–5615
14. Digani, V., Sabattini, L., Secchi, C., Fantuzzi, C.: Ensemble coordination approach in multi-AGV systems applied to industrial warehouses. IEEE Transactions on Automation Science and Eng. **12**(3), 922–934 (2015). A reference
15. Ouellette, R., Hirasawa, K.: A comparison of SLAM implementations for indoor mobile robots. In: 2007 IEEE/RSJ international conference on intelligent robots and systems. IEEE (2007)
16. Aydemir, H., Tekerek, M., Mehmet, G.Ö.K.: Examining the effect of geometric objects on SLAM performance using ROS and Gazebo. El-Cezeri **8**(3), 1441–1454 (2021)
17. Fan, X., Wang, Y., Zhang, Z.: An evaluation of Lidar-based 2D SLAM techniques with an exploration mode. J. Physics: Conference Series **1905**, 1–7 (2021)
18. Qu, P., et al.: Mapping performance comparison of 2D SLAM algorithms based on different sensor combinations. Journal of Physics: Conference Series. Vol. 2024. No. 1. IOP Publishing (2021)

Medical Image Transmission Using a Secure Cryptographic Approach for IoMT Applications

Renjith V. Ravi[1](✉)(iD), S. B. Goyal[2](iD), Chawki Djeddi[3,4](iD), and Vladimir Kustov[5](iD)

[1] Department of Electronics and Communication Engineering,
M.E.A Engineering College, Kerala, India
`renjithravi@meaec.edu.in`
[2] Faculty of Information Technology, City University, Petaling Jaya 46100, Malaysia
`sb.goyal@city.edu.my`
[3] Department of Mathematics and Computer Science,
Larbi Tebessi University, Tebessa, Algeria
`c.djeddi@univ-tebessa.dz`
[4] LITIS Lab, Rouen University, Rouen, France
[5] Saint Petersburg Railway Transport University of Emperor Alexander I,
Saint Petersburg, Russia
`kvnvika@mail.ru`

Abstract. Image security is very important in the field of medical imaging. To protect medical healthcare images, many studies have been undertaken. Because it eliminates data loss, encryption is the greatest solution for image secrecy. Traditional encryption methods, on the other hand, are difficult to apply to electronic health data due to data size limitations, redundancy, and scalability, especially when the data of a patient is delivered via different networks. As a result, since images are distinct from the text in terms of data loss and confidentiality, patients may loose the confidentiality and privacy of their data. Researchers discovered these security flaws and proposed different image encryption ways to remedy the problem. In order to build a safe image encryption solution for the healthcare business, this research proposes an efficient, lightweight encryption algorithm. The suggested lightweight encryption algorithm first separates the image into multiple clusters and then employs cluster-based permutations. Other procedures like diffusion and modulation could be used after that. The suggested technique is studied, analysed, and compared to generally encrypted ones in terms of information security and time complexity. The performance of the proposed method was assessed using a range of test images. Several experiments show that the proposed image cryptosystems methodology is more efficient than existing approaches.

Keywords: Image security · Hyperchaos system · Image clustering · Logistic adjusted sine map · Modulation encryption

© The Author(s), under exclusive license to Springer Nature Switzerland AG 2023
F. P. García Márquez et al. (Eds.): ICCIDA 2022, LNNS 643, pp. 27–38, 2023.
https://doi.org/10.1007/978-3-031-27099-4_3

1   Introduction

The Internet of Medical Things (IoMT) [7] is a multidisciplinary field in which Internet of Things (IoT) technology [10] is applied to medicine. Many forms of medical imaging equipment, such as MRI for brain tumour detection and CT for identification of lung nodule, have been more extensively linked and employed due to the development of IoMT.

Three factors must be considered while managing IoMT-enabled cybersecurity. Recognise the gadgets as networked devices first. Second, figure out and enforce how other systems, apps, and tools connect with one another [16]. Finally, if anything goes wrong, make sure that these units of IoMT have not interacted with the other terminal devices within the network or organisations. IoMT applications global accessibility enables the diagnostic, copying, data, and retrieval of many digital photos. As a consequence, it often leads to the manufacture of illegal copies or usage in concerns. As a result, several researchers have concentrated on creating image security approaches for IoMT applications [7]. As a result, cryptography provides one of the best ways to protect biomedical images.

Traditional cryptographic strategies have used DES, AES, and RSA algorithms to address the issues of low capacity while taking into account small data sizes and high redundancy [18]. Numerous reversing approaches have resulted in the publication of various algorithms or strategies for converting, replacing, and inverting the images. The simple image was employed in a transposing mechanism that used the random reordering method. Permutation may be done for bits, pixel density, or blocks. Each element in the plaintext image is mapped to another property in the cypher image via replacement, also known as significance methods. Positioning permutation and pixel value diffusion are two main particles used in transcription techniques.

The major contribution of this research is a novel lightweight encryption technology that is offered to preserve the privacy of patients' medical images.

2   Materials and Methods

2.1   Hyperchaos System

A hyperchaotic structure is a technologically sophisticated chaotic system. However, compared to other chaotic systems, its dynamical structure is more intricate [13]. Hyperchaos also promotes a greater amount of randomness and unpredictability. By combining two or more positive Lyapunov exponents, hyperchaotic systems may be separated from chaos [21]. In four-dimensional nonlinear systems, hyperchaos exists. Because of its simple formula and increased efficacy, the chaotic system does have a small key space and less complexity [4].

As a consequence, the chaotic system offers less assurance in terms of security. The hyperchaotic framework, on the other hand, contains many state variables, a larger keyspace, and more sophisticated nonlinear behaviour, which together help to boost security measures. As a result, the hyperchaotic system may be identified as [16] [3] in Eq. 1:

$$\dot{a}_1 = \alpha\,(a_2 - a_1) + \lambda_1 a_4,$$
$$\dot{a}_2 = \xi a_1 - a_1 a_3 + \lambda_2 a_4,$$
$$\dot{a}_3 = -\beta a_3 + a_1 a_2 + \lambda_3 a_4, \tag{1}$$
$$a_4 = -\tau a_1,$$

where $\alpha, \beta, \varepsilon, \tau, \lambda_1, \lambda_2,$ and λ_3 represents the control parameters of the four dimensional(4D) hyperchaotic system. When the control variables are $\alpha = 35, \beta = 3, \varepsilon = 35, \tau = 5, \lambda_1 = 1, \lambda_2 = 0.2,$ and $\lambda_3 = 0.3$, the system exhibits hyperchaotic behaviour.

2.2   Pre-processing of Chaotic Sequences for Image Clustering

We may employ the hyperchaotic system described in [3,16] to produce the pseudorandom sequences. This is important for cryptography because chaotic maps may create pseudorandom sequences due to their nonlinearity and random behavior. The sequence generation is expressed in the following phases [16]:

1. Pre-iterate the hyperchaotic model N_0 times to eliminate cross effects and increase security.
2. After N0 iterations, the framework is iterated $M \times N$ times again. Four state variables $a_1^j, a_2^j, a_3^j, a_4^j$ are sorted in each iteration j, where j indicates the index of iteration;
3. Each of the four-state variables $a_1^j, a_2^j, a_3^j, a_4^j$ is utilized to produce two new key values $((y_i^1)^j \in [0, 255], i = 1, 2, 3, 4$ and $(y_i^2)^j \in [0, 255])$ throughout each cycle 2.

$$(y_i^1)^j = \text{mod}\left\{floor\left(\left[\left(\left|a_i^j\right| - floor\left(\left|a_i^j\right|\right)\right) \times 10^{15}\right]/10^8\right), 256\right\}, \tag{2}$$

$$(y_i^2)^j = \text{mod}\left(floor\left(\text{mod}\left\{\left[\left(\left|a_i^j\right| - \text{floor}\left(\left|a_i^j\right|\right)\right) \times 10^{15}\right]/10^8\right\}\right), 256\right) \tag{3}$$

where in Eq. 2 and Eq. 3 modulo operation is shown as $mod(.)$ and $i = 1, \ldots, 4$;
4. Concatenating the equations, Eq. 2 with Eq. 3 to get y^j in Eq. 4:

$$y^j = \left[\left(y_1^1\right)^j, \left(y_2^1\right)^j, \left(y_3^1\right)^j, \left(y_4^1\right)^j, \left(y_1^2\right)^j, \left(y_2^2\right)^j, \left(y_3^2\right)^j, \left(y_4^2\right)^j\right] \tag{4}$$

5. After the whole repetition, these sequences are combined using Eq. 5 to yield K:

$$K = [y^1, y^2, \ldots\ldots, y^{M \times N}] \tag{5}$$

K has one component that is denoted by $K_i, i \in [1, 8MN]$.

2.3 Two-dimensional Logistic-adjusted-Sine Map

Two-dimensional Logistic-adjusted-Sine map (2DLASM) is proposed in [9, 22] is shown in Eq. 6.

$$a_{i+1} = \sin\left(\pi\mu\left(b_i + 3\right)a_i\left(1 - a_i\right)\right)$$
$$b_{i+1} = \sin\left(\pi\mu\left(a_{i+1} + 3\right)b_i(1 - b_i)\right) \tag{6}$$

where $\mu \in [0, 1.0]$. If $\mu \in [0.370, 0.380] \cup [0.40, 0.420] \cup [0.440, 0.930] \cup 1.0$ is set, then the 2D-LASM will present the chaotic dynamic behavior. The starting values with this chaotic map are as follows: $\mu = 0.81160, a_0 = 0.13070$ and $b_0 = 0.41260$.

2.4 Permutation Operation

Two middle variables, c and d, are calculated (Eq. 7) as [22], using the old keys a_0, b_0, a_0, and b_0 (before they were modified).

$$\begin{cases} c = \lceil (a_0 + b_0 + 1) \times 10^7 \rceil \mod e + 1 \\ d = \lceil (a_0' + b_0' + 2) \times 10^7 \rceil \mod f + 1 \end{cases} \tag{7}$$

where $[a]$ rounds a's elements to the closest integers in the direction of negative infinity. The cth row of P is then picked as a vector x. In contrast, the dth column is y. The variable x with length f and y with length e are changed as Eq. 8 to save time using the sort algorithm.

$$\begin{cases} x' = \lceil x \times 10^{14} \rceil \mod e \\ y' = \lceil y \times 10^{14} \rceil \mod f, \end{cases} \tag{8}$$

where x represents circular permutation [20, 22] in the column by column and y represents it in the direction of the row. A permuted image C is generated by doing permutation to the plaintext image A. Incidentally, the value of information entropy after and before the permutation is the same: $I(A) = I(B)$. As a result, the keystreams vary for various plaintext images, and no further transmission is required.

2.5 Modulation and Inverse Modulation Operation

The approach will be subject to numerous attacks, if the distribution of gray level stays intact prior to the diffusion procedure. As a result, in this phase, a modulation operation [22] is used to modify the statistical feature of the permuted image C.

Firstly, a matrix M of dimension $e \times f$ is created based on the plaintext image size as in Eq. 9:

$$M(p, j) = (ef + p + q) \mod 256$$
$$p = 1, 2, \ldots, e; q = 1, 2, \ldots, f. \tag{9}$$

The modulation procedure for the permuted image B by M is then performed as in Eq. 10:

$$R = B + M \bmod 256 \qquad (10)$$

The Inverse modulation is expressed mathematically as in Eq. 11

$$B = R - M \bmod 256 \qquad (11)$$

2.6 Diffusion

The diffusion [22] is to acquire the avalanche effect, it is applied to image R. Using the key before updation, a_0, b_0, a_0', and b_0', an update function is designed by

The avalanche effect is implemented using diffusion [22], and diffusion encryption is performed to image R. An updating function is constructed using the previous keys (before updating) a_0, b_0, a_0', and b_0' as in Eq. 12.

$$\begin{cases} a_0' \doteq a_0' + \frac{1}{a_0+b_0+1} & \bmod 1 \\ \bar{Y}_0' \pm b_0' + \frac{2}{a_0+b_0+2} & \bmod 1 \end{cases} \qquad (12)$$

Correspondingly, the keys a_0 and b_0 forms a union with the keys a_0' and b_0'. After that, the chaotic map will be iterated to produce a random matrix K which is having similar size of the plaintext image. To fulfill the gray level, it is then refined using the Eq. 13.

$$K = \left[K \times 10^{14} \right] \bmod 256 \qquad (13)$$

2.7 Proposed Algorithm for Secure Image Communication

The proposed algorithm can be used to encrypt RGB color images. Firstly, the color image will be divided into its three colour planes, which were processed separately. Here is the hyperchaotic system in Sect. 2.1. This hyperchaotic system will generate chaotic sequences and plays a vital role in almost all the operations during the encryption. The next operation is to cluster the three components according to the chaotic sequences from the hyperchaotic system. In this algorithm, we are using three shuffling (Permutation) operations. The first one is intra cluster shuffling, and the second one is inter-cluster shuffling. The pixels in the images will be shuffled inside the cluster and between the different clusters. This gives a highly efficient permutation mechanism. Further, these clusters will be integrated to form the three color planes. This permuted image is in an unreadable format for human eyes. However, its histograms will remain the same, which may cause the algorithm to weaken for statistical attacks. For this, we perform diffusion by XORing these color planes with the chaotic sequence from the hyperchaotic system. During this XoR process, the values of all the pixels will be changed. To enhance the security further, we will use pixel permutation as mentioned in Sect. 2.4. The next step is modulation, as mentioned in Sect. 2.5. For this, a random image of the same size has been generated from the plaintext image to do the modulation operation. Finally, the three color planes will be joined to form the ciphertext image for the modulation. All of these processes are depicted in Fig. 1 and written step by step in the preseeding subsection.

Encryption Process

1. Let $M \times N \times 3$ denote the entire dimension of the contributing colour image P.
2. From the input image, extract the colour planes R, G, and B;
3. Partition each channel into $p \times q$ number of clusters in both the row and column direction, where $p = 02, 04, 08, 016$ and $q = 02, 04, 08, 016$ arbitrarily based on the chaos sequence formed. When $p = 02$ and $q = 04$, for example, the resultant image is divided into eight different clusters.
4. Making use of the chaotic sequence K, shuffle all the clusters.
5. Rearrange the pixels in each cluster by applying a chaotic sequence to them.
6. Create a permuted image that is the same size as the new coloured image by combining these shuffled clusters.
7. Perform Diffusion Operation
8. Perform Permutation operation as mentioned in
9. Perform Modulation Operation.
10. Concatenate R, G, and B channels

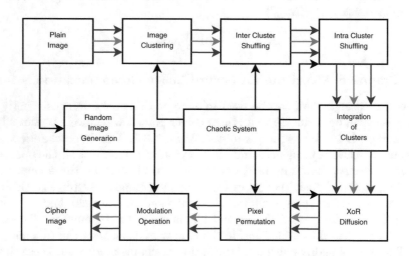

Fig. 1. Proposed Encryption Process

Decryption Process: As this algorithm performs symmetric encryption, all the operations carried out during the encryption process will be performed in reverse order in the decryption process to recover the deciphered image.

However throughout decryption process, the ciphertext image in RGB format will be separated into its three color components. Then, the inverse modulation will be applied at first using the mask image generated from the plaintext image. Except for this modulation operation, all other inverse operations will be carried out in accordance with the chaotic signals produced by the hyperchaotic system. The inverse of pixel permutation and inverse diffusion is the next operations. Further, the image will be divided into clusters for performing the intra-cluster and inter-cluster permutations. After these operations, the clusters will be joined to form the plaintext image. All the operations in the decryption process are depicted in Fig. 2.

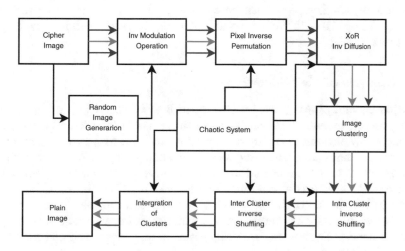

Fig. 2. Decryption of Proposed Encryption Process

3   Results and Discussion

In this work, we have used different types of medical images as the test data. All of these test images are listed below in Fig. 3.

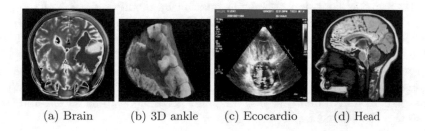

(a) Brain   (b) 3D ankle   (c) Ecocardio   (d) Head

Fig. 3. Test images

3.1   Key Space

The starting conditions of hyper-chaotic systems, a_1, a_2, a_3, a_4, s_1, and s_2, and logistic sine map, a_0, b_0, a'_0, b'_0 . The range of values were, $a_1 \in (-40, 40)$, $a_2 \in (-40, 40)$, and $a_3 \in (1, 80)$, each one with a step size of, $10^{(-13)}$, and , $a_4 \in (-250.0, 250.0)$, with a step size of, 10^{-12}. s_1 and s_2 are two 8-bit random integers with a single step size and a value range of $[0, 255]$. The encryption and decryption processes employed these parameters as undisclosed keys. As a result, the suggested algorithm's keyspace is 1.6777×10^{64}. To prevent brute-force assaults, the keyspace must be exceedingly big; hence, cracking the system would take 2.034510×10^{52} days. As an outcome, the recommended cryptosystem is immune to brute-force attacks.

3.2   Analysis of Key Sensitivity

By adjusting and altering one bit [16] in the created secret key, a_1, a_2, a_3, a_4, s_1, and s_2, we can witness the sensitivities of the created secret key in the suggested approach. Any element of the secret key a_1, a_2, a_3, or a_4 may be changed by 10^{-13} or 10^{-12}, and any value of s_1 and s_2 can be changed by 1. A different ciphered image results from this minor adjustment. As a result, the ciphered image would be completely unlike the original input image. To show the sensitivity of difference in key, we used the proposed algorithm to crypt the plaintext images twice. In the first step of encryption, the true secret key was used to encrypt the plaintext images. During the next encryption step, though, the plaintext images are enciphered with the incorrect key. The incorrect key was created from the original keys with making a slight difference on it. The Fig. 4 illustrate the findings of the algorithm on the test image *Brain*. The Fig. 4a shows, the enciphered image with right key and Fig. 4b and Fig. 4c shows the enciphered images with two different wrong keys. From Fig. 4, we can identify that these images are not looks like the same. This shows that the encryption method is very sensitive to the value of secret keys.

3.3   Differential Attack

An attacker might get important information by modifying a few of the pixels in the original plaintext image. Against test the encrypted plaintext image's durability to differential assaults, the unified averaged changed intensity (UACI) and the number of changing pixel rates (NPCR) are often utilised. We employed the specified encryption approach with an unifying secret key to cipher I_1 and I_2 to generate their corresponding ciphertext images, E_1 and E_2, where I_2 indicates a one-pixel modification in the original plaintext image I_1. The NPCR and UACI are shown Eq. 14 and Eq. 15 respectively.

$$\text{NPCR} = \frac{1}{W} \sum_{i=1}^{M} \sum_{j=1}^{N} |\text{Sign}\left(\text{E}_1(i, j) - \text{E}_2(i, j)\right)| \times 100\% \tag{14}$$

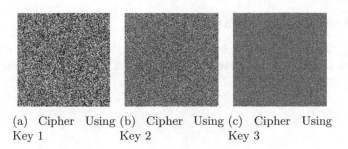

(a) Cipher Using Key 1   (b) Cipher Using Key 2   (c) Cipher Using Key 3

Fig. 4. Key Sensitivity Analysis

$$\text{UACl} = \frac{1}{W} \sum_{i=1}^{M} \sum_{j=1}^{N} \frac{|\text{Sign}\,(\text{E}_1(i,j) - \text{E}_2(i,j))|}{256} \times 100\% \qquad (15)$$

Table 1. Values of UACI and NPCR Obtained for test Images

Image	NPCR (%)			b	UACI (%)	
	R	G	B	R	G	B
Brain	99.768	99.897	99.986	34.895	34.981	34.973
3D Ankle	99.891	99.869	99.889	34.976	34.958	34.889
Ecocardiagram	99.868	99.745	99.881	34.456	34.825	34.453
Head	99.745	99.986	99.896	34.542	34.654	34.678

Table 2. The *UACI* and *NPCR* values for the Fig. *3D Ankle* with the literature

Image	NPCR (%)			UACI (%)		
	R	G	B	R	G	B
Proposed Work	99.89100	99.86900	99.88900	34.97600	34.95800	34.88900
Gafsi et al. [5]	99.89427	99.83960	99.89441	34.07125	34.08311	34.19787
Hfsa et al. [6]	99.69900	99.68700	99.89800	33.89600	33.89500	33.6780
Ravi et al. [18]	99.89800	99.85900	99.89700	34.96400	34.94700	34.87700

In each of the original input images, the value of one pixel was changed ten times randomly. After that, the NPCR and UACI of the test images were calculated. The findings are summarised in Table 1 and Table 2, revealing that our new encryption algorithm is very responsive to single-pixel value fluctuations in the source image. In [5,6,18] NPCR and UACI get close to theoretical findings, although their noise attack efficacy is lower. When compared to the proposed crypto algorithm, it is clear that the algorithm performs admirably by producing average results that are extremely close to their theoretical results.

3.4   Entropy of Information

The Shannon's entropy or information entropy, typically indicates the uncertainty of digital image, was represented by $E(t)$ of matrix t. It was assesed using the following equation Eq. 16.

$$E(t) = -\sum_{i=0}^{255} P_b(t_i) \log_2(P_b(t_i)) \tag{16}$$

The probability of t_i is denoted by $P_b(t_i)$. A precise random image will produce 256-pixel values with the same probability. As a result, the hypothetical information entropy number is nearer to 8. Table 3 shows that the entropy levels of cypher images are near to the theoretical results. As a result, the results in Table 4 shows that the suggested cryptosystem is capable of effectively counteracting information entropy.

Table 3. Information entropy Analysis in R, G and B planes

Image	Plaintext image			Ciphertext image		
	R	G	B	R	G	B
Brain	6.14452	6.47563	6.09944	7.99652	7.99743	7.99712
3D Ankle	4.86450	4.96910	5.20520	7.99420	7.99280	7.99310
Ecocardiagram	5.2436	5.2489	5.6473	7.9965	7.9958	7.9986
Head	6.69615	6.16902	6.81389	7.99687	7.99773	7.99711

Table 4. In formation Entropy- Comparison with literature for the Fig. *3D Ankle*

Reference	R	G	B
Proposed Method	07.9942	07.9928	07.9931
Gafsi et al. [5]	07.9563	07.9559	07.9559
Hfsa et al. [6]	07.9563	07.9559	07.9559
Ravi et al. [18]	07.9942	07.9928	07.9931

4   Conclusion

This paper provides an encryption technique for transmitting medical images securely through IoMT application scenarios. Image clustering, inter cluster scrambling, intra cluster permutation, diffusion, and modulation functions are the key methods employed in this approach. In this approach, two chaotic systems are employed to generate the keys. One is a four-dimensional hyperchaotic system, whereas the other is a two-dimensional logistic sine dynamic system. The proposed encryption method was utilized to the three color channels of a plaintext image in order to increase the plain image's uncertainty. An investigation of the numerical data shows that the proposed method is impermeable to a range of attacks, including known plaintext, differential cypher image, brute force and entropy attacks. It also has a large keyspace, making it vulnerable to even modest modifications in the key chosen. The encryption approach was also put to the test against previously released encryption algorithms, and it was shown to be superior.

References

1. Abd El-Latif, A.A., Niu, X.: A hybrid chaotic system and cyclic elliptic curve for image encryption. AEU-Int. J. Electron. Commun. **67**(2), 136–143 (2013). https://doi.org/10.1016/j.aeue.2012.07.004
2. Çavuşoğlu, Ü., Kaçar, S., Pehlivan, I., Zengin, A.: Secure image encryption algorithm design using a novel chaos based s-box. Chaos, Solitons Fractals **95**, 92–101 (2017). https://doi.org/10.1016/j.chaos.2016.12.018
3. Chun-Lai, L., Si-Min, Y.: A new hyperchaotic system and its adaptive tracking control. Acta Physica Sinica **61**(4) (2012). https://doi.org/10.7498/aps.61.040504
4. Fang, D., Sun, S.: A new secure image encryption algorithm based on a 5d hyperchaotic map. PLoS ONE **15**(11), e0242110 (2020). https://doi.org/10.1371/journal.pone.0242110
5. Gafsi, M., Abbassi, N., Hajjaji, M.A., Malek, J., Mtibaa, A.: Improved chaos-based cryptosystem for medical image encryption and decryption. Scientific Programming **2020** (2020). https://doi.org/10.1155/2020/6612390
6. Hafsa, A., Gafsi, M., Malek, J., Machhout, M.: Fpga implementation of improved security approach for medical image encryption and decryption. Scientific Programming **2021** (2021). https://doi.org/10.1155/2021/6610655
7. Hasan, M.K., Islam, S., Sulaiman, R., Khan, S., Hashim, A.H.A., Habib, S., Islam, M., Alyahya, S., Ahmed, M.M., Kamil, S., et al.: Lightweight encryption technique to enhance medical image security on internet of medical things applications. IEEE Access **9**, 47731–47742 (2021). https://doi.org/10.1109/access.2021.3061710
8. Hu, G., Xiao, D., Wang, Y., Li, X.: Cryptanalysis of a chaotic image cipher using latin square-based confusion and diffusion. Nonlinear Dyn. **88**(2), 1305–1316 (2017). https://doi.org/10.1007/s11071-016-3311-2
9. Hua, Z., Zhou, Y.: Image encryption using 2d logistic-adjusted-sine map. Inf. Sci. **339**, 237–253 (2016). https://doi.org/10.1016/j.ins.2016.01.017
10. Jang, W., Lee, S.Y.: Partial image encryption using format-preserving encryption in image processing systems for internet of things environment. Int. J. Distrib. Sens. Netw. **16**(3), 1550147720914779 (2020). https://doi.org/10.1177/1550147720914779
11. Laiphrakpam, D.S., Khumanthem, M.S.: Medical image encryption based on improved elgamal encryption technique. Optik **147**, 88–102 (2017). https://doi.org/10.1016/j.ijleo.2017.08.028
12. Liu, H., Liu, Y.: Cryptanalyzing an image encryption scheme based on hybrid chaotic system and cyclic elliptic curve. Opt. Laser Technol. **56**, 15–19 (2014). https://doi.org/10.1016/j.optlastec.2013.07.009
13. Liu, X., Xiao, D., Liu, C.: Quantum image encryption algorithm based on bit-plane permutation and sine logistic map. Quant. Inf. Proc. **19**(8), 1–23 (2020). https://doi.org/10.1007/s11128-020-02739-w

14. Mattioli, M., Shackelford, S.J., Myers, S., Brady, A., Wang, Y., Wong, S.: Securing the internet of healthcare. Minn. J Law Sci. Technol. **19**, 405 (2018)
15. Mitra, A., Rao, Y.S., Prasanna, S., et al.: A new image encryption approach using combinational permutation techniques. Int. J. Comput. Sci. **1**(2), 127–131 (2006)
16. Mohamed, H.G., ElKamchouchi, D.H., Moussa, K.H.: A novel color image encryption algorithm based on hyperchaotic maps and mitochondrial DNA sequences. Entropy **22**(2), 158 (2020). https://doi.org/10.3390/e22020158
17. Priya, S., Santhi, B.: A novel visual medical image encryption for secure transmission of authenticated watermarked medical images. Mob. Netw. Appl, pp. 1–8 (2019) https://doi.org/10.1007/s11036-019-01213-x
18. Ravi, R.V., Goyal, S.B., Djeddi, C.: A New Medical Image Encryption Algorithm for IoMT Applications. In: Djeddi, C., Siddiqi, I., Jamil, A., Ali Hameed, A., Kucuk, İ (eds.) MedPRAI 2021. CCIS, vol. 1543, pp. 145–157. Springer, Cham (2022). https://doi.org/10.1007/978-3-031-04112-9_11
19. Toughi, S., Fathi, M.H., Sekhavat, Y.A.: An image encryption scheme based on elliptic curve pseudo random and advanced encryption system. Signal Proc. **141**, 217–227 (2017). https://doi.org/10.1016/j.sigpro.2017.06.010
20. Wang, X.Y., Gu, S.X., Zhang, Y.Q.: Novel image encryption algorithm based on cycle shift and chaotic system. Opt. Lasers Eng. **68**, 126–134 (2015). https://doi.org/10.1016/j.optlaseng.2014.12.025
21. Yang, Q., Bai, M.: A new 5D hyperchaotic system based on modified generalized lorenz system. Nonlinear Dyn. **88**(1), 189–221 (2017). https://doi.org/10.1007/s11071-016-3238-7
22. Ye, G., Pan, C., Huang, X., Zhao, Z., He, J.: A chaotic image encryption algorithm based on information entropy. Int. J. Bifurcat. Chaos **28**(01), 1850010 (2018). https://doi.org/10.1142/s0218127418500104

Applied Decision Support System Using TOPSIS – AHP, and ICT Newhouse Indicators for Evaluation of Courses at University of Economics Ho Chi Minh City (UEH), Vietnam

Viet Phuong Truong[ID] and Quoc Hung Nguyen[✉][ID]

University of Economics Ho Chi Minh City (UEH), Ho Chi Minh City, Vietnam
hungngq@ueh.edu.vn

Abstract. Evaluating courses' quality in education is an important part of the entire training process, to ensure that the program is successfully practiced. In the process of researching and improving teaching materials to enhance courses' quality, understand students' expectations and objectives, the problem they are facing and give the appropriate advice, courses evaluation based on traditional methods such as valuing assessment data by averaging results will not apply to criteria of different importance levels. Proposal to build a decision support system using TOPSIS (Technique for Order of Preference by Similarity to Ideal Solution) - AHP (Analytic Hierarchy Process) evaluates course quality based on the Newhouse ICT criteria set and set of criteria from the University of Economics Ho Chi Minh City (UEH). Experimental results showed that the proposed model using TOPSIS - AHP integration method improved over traditional methods, contributing to helping managers make the right decision with multiple purpose criteria.

Keywords: Decision support system · TOPSIS- AHP · Quality assessment · Smart University

1 Introduction

In a new time of development and integration already going on, with science and technology breakthroughs, the 4th industrial revolution was formed on the foundation of the knowledge economy [1], and the trend of active integration has strongly impacted all sectors. Education and training are considered the key to unlocking the development of human potential and are the strongest lever in providing human resources and talents for the development of science and technology. On the other hand, the development of science and technology impacts the entire structure and quality of the education system [2]. Therefore, improving the quality of education and training is a requirement and a matter of top concern in this field.

Evaluating the quality of courses in universities is an indispensable stage in the development of improving the quality of education and training. The implementation of quality assessment before, during, and after courses implemented in recent years has had a positive impact on the quality of the general education system [3] The evaluation of the quality of courses is the first step in the overall quality inspection activities to help see the advantages and limitations of the training activities undertaken by the training institutions, thereby promoting and adjusting the process to suit the students, creating a prerequisite for improving the quality of the next course program.

Developing a decision support model provides leaders with counseling results to assess course quality and improve counseling outcomes. The proposed model combines TOPSIS and AHP to evaluate course quality based on the Newhouse set of ICT criteria [4] and the set of criteria from UEH. Specific objectives and results of the study are completed as follows:

1. Propose a system model to help make decisions to evaluate the quality of the course using the TOPSIS algorithm - AHP integrated
2. Install a help system that determines course quality assessment on the Windows platform
3. Experimental application of 3 courses at the UEH and evaluate the quality of the course after the end of the program to obtain experimental results and validate the proposed model

The paper is organized as follows: Sect. 2 overviews the theoretical basis of the decision support system, the set of criteria, and algorithms TOPSIS, AHP. In Sect. 3, we proposed a decision support system using the TOPSIS - AHP model to evaluate the quality of the course. Some experimental and simulated results in Sect. 4 are presented. Finally, we conclude and give some ideas for future works.

2 Background Research

2.1 Decision Support System

In the early 1970s, Gorry and Scott-Morton (1971) defined Decision Support Systems (DSS) as interactive computer systems that helped decision-makers use data and models to solve non-structured problems [5]. In which:

- Data governance: includes databases that contain data related to a situation and are managed by software that is database management (management and exploitation).
- Model management: enables the exploitation and management of various dosing (processing) models, providing analytical capabilities for the system.
- Dialogue administration: provides the interface for the user to communicate with and dictionary orders the Decision Support System.
- Knowledge management: acts as an independent component, or can assist any of the three systems mentioned above.
- Helper data sources decide: inside data, external data, private data.

2.2 Topsis

TOPSIS is multi-criterion decision-making (MCDM) method, that selects the alternative that is the closest to the positive ideal solution and farthest from a negative ideal alternative. The algorithm was first introduced in 1980 by Kwangsun Yoon and Hwang Ching-Lai. The technique was later further developed by other scientists and completed in 1993 [6]. Some authors are interested in the TOPSIS technique, as it is perhaps the most widely used multi-criteria technique [7]. Its use is diverse [8], e.g. in tourism [9], transport) [10], insurance [11], agriculture [12], in assessing the competitiveness of countries [13], or when evaluating cloud service providers [14].

The main idea of the TOPSIS method is to add a good ideal solution (PIS) and a bad ideal solution (NIS). For an alternative to being called best, its alternatives should be as close to the PIS as possible and as far away from the NIS as possible. The distance between these alternatives to these good and bad ideal alternatives is calculated in terms of the geometrical sense. TOPSIS uses a decision matrix which is a graph in which m rows represent alternatives and n columns represent selection criteria. The intersection of rows and columns represents the performance of a selection solution according to a certain criterion, so when applying TOPSIS is the selection of a set of criteria.

2.3 AHP Technique

AHP is one of the multi-criteria decision-making methods. It is a method of evaluating an overview of the sort order of choices and from which we find the most reasonable final option [15]. Based on mathematics and psychology, AHP was developed by Thomas L. Saaty in the 1970s [16] and has been expanded and supplemented to date. AHP provides a structured method for solving complex decision-making issues. Generally, AHP consists of four steps: 1. Modeling; 2. Review; 3. Prioritization, and 4. Aggregation. Step 3 is the most important of multiplicative AHP [17]. AHP combines closely with the decision-making standard. AHP provides a mechanism to prioritize decision makers for alternatives or criteria using a pairwise comparison matrix based on the usefulness of the best alternatives in a set of options [18]. In nature, AHP is often used in two main fields of research:

1. First, assess the nature of objects, things, and things through the criteria of the nature of the subjects themselves.
2. Second, compare "entities" with each other through a variety of relevant criteria.

2.4 ICT Newhouse Criteria Set

Many factors affect and influence modern education these days. Information technology is increasingly being applied strongly in teaching and learning.

In a recent study, Paul Newhouse [4] – Western Australia's Department of Education has shown the relationship and impact of information technology (ICT) on teaching in Australian schools. The author evaluates and classifies the positive impacts of ICT on learning and improvement in student learning results. There are nine positive impacts that ICT has brought to significant improvements in the learning methodology of students, namely:

1. Ability to conduct practical surveys and obtain knowledge
2. Promoting positive learning and authentic evaluation
3. Attract students by motivations and challenges
4. Provide tools to increase academic productivity
5. Provide tools to support high thinking
6. Increase the independence of the student
7. Strengthen collaboration and partnership
8. Curriculum design for students
9. Overcoming physical defects

Based on this set of criteria, we build a research model that integrates the techniques of decision support to evaluate courses and improve the quality of courses in universities, professional secondary schools, and high schools.

3 The Proposed Model

3.1 Detail Diagram of Model Architecture

The TOPSIS-AHP model uses the ICT Newhouse [4] set of criteria to assess the course quality described as shown below:

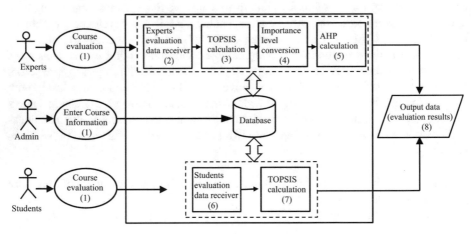

Fig. 1. Proposed overall architecture

The proposed model as shown in Fig. 1 consists of 8 main modules of the following:

1. **Input data**: Includes course information, expert or student evaluation
2. **Experts/Students data receiver**: Receive data as input values – corresponding to each criterion in the database.
3. **TOPSIS calculation**: Algorithm calculated by TOPSIS to give the order of sorting and evaluating inputs of experts and students

4. **Importance level conversion**: Calculating and converting data from "expert data receiving mechanisms" to corresponding importance levels
5. **AHP calculation**: Use the AHP algorithm to evaluate the course
6. **Students evaluation data receiver**: Users will participate in the survey.
7. **TOPSIS calculation**: Build the decisive matrix and Calculate the distance from the expert assessment plan.
8. **Output data**: Is the most suitable course or assessment results most suitable for the course

3.2 Describe the Steps in the TOPSIS-AHP Integration Proposal Model

- **Step 1: Course assessment survey**
  Users will participate in the survey according to the ICT criteria and the criteria set by the UEH, the results of the survey will be used as input for the TOPIS algorithm.
- **Step 2: Implement the TOPSIS algorithm**

  - *Step 2.1: Build the decisive matrix, suppose we call*
    G_i: Expert assesses
    v_j: Criterion j-th
    x_{ij}: Expert assessment level for course A based on criteria v_j.
    W_j: Weighting criteria j, with $i = 1,2, \ldots, m$ and $j = 1,2, \ldots, n$.

  Representative values are represented in the decision matrix as follows:

  $$D = \begin{bmatrix} x_{11} & x_{12} & \cdots & x_{1n} \\ x_{21} & x_{22} & \cdots & x_{2n} \\ \vdots & \vdots & & \vdots \\ x_{m1} & x_{m2} & \cdots & x_{mn} \end{bmatrix} \tag{1}$$

  $W = [\, W_1, W_2, \ldots, W_n \,]$ với $. \sum_{j=1}^{n} W_j = 1$

  - *Step 2.2: Standardize the matrix*
    From the built-in D matrix, standardize into an R matrix in the form of:

  $$R = \begin{bmatrix} r_{11} & r_{12} & \cdots & r_{1n} \\ r_{21} & r_{22} & \cdots & r_{2n} \\ \vdots & \vdots & \cdots & \vdots \\ r_{m1} & r_{m2} & \cdots & r_{mn} \end{bmatrix} \tag{2}$$

  To standardize from matrix D to matrix R using formulas: $R_{ij} = \dfrac{x_{ij}}{\sqrt{\sum_{i=1}^{m} x_{ij}^2}}$
  - *Step 2.3: Build a weighted decision matrix*

By weighing the W_j of each criterion with the R standardized matrix, building a weighted V decision matrix

$$V = \begin{bmatrix} v_{11} & v_{12} & \cdots & v_{1n} \\ v_{21} & v_{22} & \cdots & v_{2n} \\ \vdots & \vdots & \cdots & \vdots \\ v_{m1} & v_{m2} & \cdots & v_{mn} \end{bmatrix} = \begin{bmatrix} w_1 r_{11} & w_2 r_{12} & \cdots & w_n r_{1n} \\ w_1 r_{21} & w_2 r_{22} & \cdots & w_n r_{2n} \\ \vdots & \vdots & \cdots & \vdots \\ w_1 r_{m1} & w_2 r_{m2} & \cdots & w_n r_{mn} \end{bmatrix} \tag{3}$$

- *Step 2.4.* Establish the FPIS (A^*) a positive ideal solution and the FNIS (A^-) negative ideal solution

$$A^* = \left(v_1^*, v_2^*, \ldots v_n^*\right) \tag{4}$$

$$A^- = \left(v_1^-, \left(v_2^-, \ldots v_n^-\right)\right) \tag{5}$$

which:
$$v_j^* = \{(\max v_{ij} v_{ij} \mid i = 1,2,\ldots m \mid j = 1,2, \ldots n)\}$$
$$v_j^- = \{(\min v_{ij} v_{ij} \mid i = 1,2,\ldots m \mid j = 1,2, \ldots n)\}$$
- *Step 2.5: Calculate the distance from the expert assessment plan to FPIS (A^*) and FNIS (A^-)*
  The formula for calculating the distance from assessment options to A^* and A^-

$$S^*(v_j^*, v_{ij}) = \sqrt{\sum_{j=1}^{n}\left(v_j^* - v_{ij}\right)^2} \tag{6}$$

$$S^-(v_j^-, v_{ij}) = \sqrt{\sum_{j=1}^{n}\left(v_j^- - v_{ij}\right)^2} \tag{7}$$

- *Step 2.6*: Calculate the closeness and rating of assessment options
  After calculating the distance of the options to the best solution and worst solution. Next, determine the number of proximity to determine the direction according to the following formula:

$$CC_i = \frac{S_i^-}{S_i^* + S_j^-} \tag{8}$$

- Step 3: Implement the AHP algorithm
  Use expert evaluation options G_i for A course based on criteria v_j.

  - *Step 3.1: Determine priority for criteria*
    With n criteria, the building matrix compares with values a_{ij} calculated according to formulas: $a_{ij} = \frac{S_i}{S_j}$ with S_i, S_j is the rating of the criteria x_i, x_j.

- *Step 3.2: Determine the weight of the evaluation criteria*
  Determine weights by proportioning components by rows and columns

$$W_{ij} = \frac{a_{ij}}{\sum_1^n a_{ij}} \tag{9}$$

- *Step 3.3: Determine the highest characteristic value by formula*

$$\lambda_{max} = \frac{1}{n} \sum_{i=1}^n (\sum a_{ij} \times w_i) \times \frac{1}{w_i} \tag{10}$$

- *Step 3.4: Calculate consistent indicators*

$$CR = \frac{CI}{RI} \tag{11}$$

which is:
$CI = \frac{(\lambda_{max}-1)}{n-1}$ the smaller the test process, the more accurate it is
RI is a random me index that corresponds to each criteria (Table 1)

Table 1. RI value

n	1	2	3	4	5	6	7	8	9	10	11	12
RI	0	0	0.58	0.90	1.12	1.24	1.32	1.41	1.45	1.49	1.51	1.54

If CR < 0.1, the results are consistent, high-reliability rating

4 Experimental Results

4.1 Experimental Configuration

To conduct the experiment we have established the environment as follows:

- **Installed environment**: The proposed method is installed on C# build language Visual Studio 2019
- **Experimental dataset**: Running 3 courses at the UEH including:

  - ST (Softwave Technology)
  - EC (E-commerce)
  - MIS (Management Information System)

- Criteria for evaluating algorithms

  - The construction system ensures the functions and has been installed TOPSIS - AHP algorithm as analyzed in the problem.
  - The result of the input for an assessment is by the input values.
  - The program is quite flexible, the interface is easy to use from data entry, expert options, rating options

4.2   Experimental Illustration of the Proposed Model

Application of TOPSIS - AHP algorithm to course counseling suitable for professionals

- **Step 1: Identify inputs and course data to be evaluated**
  Input: List of courses to be evaluated (Table 2)

Table 2. Set of 9 criteria for ICT Newhouse

Symbol	Criteria	Description
B1	*Discover the truth and develop knowledge*	Explore more of the real world, using tools and updates to build a deep and broad knowledge platform
B2	*Support positive learning and real assessment*	Learning by action is more than just listening and reading
B3	*Actively engage students with encouragement and challenges*	Create motivation and challenges to encourage students to participate more actively in the lesson
B4	*Provide tools to improve the quality of students' products*	Providing writing, drawing, and calculating tools for slow-moving NH, the basis for developing higher thinking skills
B5	*Provide ladders for higher thinking skills*	Development of higher thinking skills such as application, analysis and synthesizing, and creativity
B6	*Enhancing the independence of students*	Providing learning experiences where and when necessary allows students to learn at their own pace
B7	*Strengthening cooperation and coordination*	Collaborate with many students inside and outside the school and with teachers, developing a thriving learning community
B8	Attach learning to the student	Personalize the NH through the use of intelligent tutoring systems or computerized learning management systems
B9	*Overcoming physical difficulties*	Providing, an ICT environment, and opportunities for students with disabilities to participate in learning activities like other normal students

- **Step 2: Collect comments from experts**

In this step there are 2 cases:

- *Case 1: there is an expert participating in supplier evaluation, the expert's evaluation data set will be also the standard data set*
- *Case 2: with 2 or more experts, we will take an additional step of pre-processing data to get the best set of evaluation data*

Let's say we have 5 experts here: G1, G2, G3, G4, G5, and 9 ICT impact criteria are A1, A2, A3… A9. The rating is given on a scale of 9. That is: For each ICT impact criterion, the higher the level (inferior) in impact, the closer the score to that criterion.

For example, G1 experts evaluate the K1 course with the following scores: GK1 (5/4/7/6/8/6/3/2/5) i.e.

Criterion impact level 1: is 5
Criterion impact level 2: is 4
Criterion impact level 3: is 7
…
Criterion impact level 9: is 5

Collecting opinions from experts on us receives a table of figures as follows (Table 3):

Table 3. Table of assessments from experts

	A1	A2	A3	A4	A5	A6	A7	A8	A9
G1	5	4	4	5	6	3	5	3	5
G2	4	4	3	4	5	2	5	4	6
G3	4	5	4	3	4	2	4	5	5
G4	3	4	3	4	6	3	6	3	4
G5	3	5	3	3	5	3	4	3	5
Weight	0.05	0.1	0.05	0.05	0.2	0.05	0.2	0.1	0.2

The weight set above, given by analysts, will decide which of the 9 impacts of ICT which have the strongest impact on the quality of the course. Total weighting value for 9 ICT impact criteria equals 1 (Table 4, 5).

- **Step 3: Process the data obtained from the above table using the topsis method**

  - *Step 3.1: Standardize values*
  - *Step 3.2: Calculate values by weight*

- *Step 3.3: Calculate the ideal solution*
  $A^* = (0.029, 0.051, 0.026, 0.029, 0.102, 0.0255, 0.11, 0.061, 0.106)$
  $A^- = (0.0175, 0.04, 0.0195, 0.0175, 0.068, 0.017, 0.074, 0.036, 0.07)$
- *Step 3.4: Calculate the distance to the ideal solution*
  $S_i^* = (0.036, 0.032, 0.055, 0.046, 0.053)$
  $S_i^- = (0.046, 0.045, 0.032, 0.051, 0.027)$

Table 4. Standardized values table

	A1	A2	A3	A4	A5	A6	A7	A8	A9
G1	0.58	0.4	0.52	0.58	0.51	0.51	0.46	0.36	0.44
G2	0.46	0.4	0.39	0.46	0.43	0.34	0.46	0.49	0.53
G3	0.46	0.51	0.52	0.35	0.34	0.34	0.37	0.61	0.44
G4	0.35	0.4	0.39	0.46	0.51	0.51	0.55	0.36	0.35
G5	0.35	0.51	0.39	0.35	0.43	0.51	0.37	0.36	0.44
Weight	0.05	0.1	0.05	0.05	0.2	0.05	0.2	0.1	0.2

Table 5. Values calculated by weight

	A1	A2	A3	A4	A5	A6	A7	A8	A9
G1	0.029	0.04	0.026	0.029	0.102	0.0255	0.092	0.036	0.088
G2	0.023	0.04	0.0195	0.023	0.086	0.017	0.092	0.049	0.106
G3	0.023	0.051	0.026	0.0175	0.068	0.017	0.074	0.061	0.088
G4	0.0175	0.04	0.0195	0.023	0.102	0.0255	0.11	0.036	0.07
G5	0.0175	0.051	0.0195	0.0175	0.086	0.0255	0.074	0.036	0.088
Weight	0.05	0.1	0.05	0.05	0.2	0.05	0.2	0.1	0.2

- *Step 3.5: Similar measurements to ideal solutions*
  $C_i^* = (0.439, 0.415, 0.627, 0.475, 0.657)$

  - **Conclusion: According to the above results, the positive rating (from best to worst) of experts will be G5 > G3 > G4 > G1 > G2 (0.675 > 0.627 > 0.475 > 0.439 > 0.415)**
  - *Cons consyn cons*: taking the results of the G5 expert's assessment will give us the most positive assessment. But not always the most positive is the best. In order for the problem to be reasonable and suitable for the majority of assessments from experts, we will choose the evaluation of G4 experts (assess the impact of 9 impact criteria in moderated).

- **Step 4: Set up the Impact Matrix**

  - *Step 4.1: Set up an impact matrix between elements*

  From the data obtained by A4 experts we have the following (Table 6):
  Converted to a matrix that compares the criteria (ICT impact) as follows (Table 7):
  Formulas that convert importance from the evaluation:

$$a_{ij} = \frac{S_i}{S_i}$$

Table 6. Expert assessment figures A4

	A1	A2	A3	A4	A5	A6	A7	A8	A9
G4	3	4	3	4	6	3	6	3	4

Table 7. Example of how to change importance from a set of evaluation figures

	A1	A2	A3	A4	A5	A6	A7	A8	A9
A1	1	1	1	1	1/2	1	1/2	1	1
A2	1	1	1	1	1	1	1	1	1
A3	1	1	1	1	1/2	1	1/2	1	1
A4	1	1	1	1	2	1	1	1	1
A5	2	2	3	1	1	2	1	2	2
A6	1	1	1	1	1/2	1	1/2	1	1
A7	2	3	2	2	1	2	1	2	2
A8	1	1	1	1	1/2	1	1/2	1	1
A9	3	1	1	1	2	1	2	1	1

which: S_i, S_i is the rating of the criteria of x_i, x_j, with the value of the Matrix calculated as follows (Table 8):

Table 8. Example of how to change importance from a set of evaluation figures

	A1	A2	A3	A4	A5
A1	1	1	1	1	1/2
A2	1	1	1	1	1
A3	1	1	1	1	1/2
A4	1	1	1	1	2
A5	2	2	3	1	1

The value of row A3 column A4 will be the A3/A4 = 3/4 = 0.75 ~ 1.

(Note: According to the proposed AHP versus Satty algorithm, the degree of comparison of the importance of each criterion is an inverse – or inverse of an inverse, so the number of traded values will be rounded).

- *Step 4.2: Set up the Impact matrix between assessment level selections and each criterion.* The rating is as follows (Table 9):

Table 9. Ratings

Ratings	Notation	Definition
Excellent	M1	Highest quality evaluation
Good	M2	High-quality evaluation
Quite good	M3	Moderate quality evaluation
Normal	M4	Acceptable quality evaluation
Bad	M5	Unacceptable quality evaluation

Now we go to the matrix to compare the evaluation results for each criterion (In total we have to set 9 Matrix corresponding to 9 criteria).

Example: With the first criterion (A1). The assessor will give the following importance score:

- Very good: 6
- Good: 4
- Pretty good: 5
- Normal: 2
- Bad: 3

Applying the importance conversion formula, we have the following matrix (Table 10):

Table 10. Matrix comparison between the level of assessment for criterion A1

Criterion X1	M1	M2	M3	M4	M5
M1	1	1	1	3	2
M2	1	1	1	2	1
M3	1	1	1	3	2
M4	1/3	1/2	2/5	1	1
M5	1/2	1	1	2	1

To ensure each with different criteria, the matrix compares to different results, so we go to set the key with 8 criteria rest. We get the matrix results presented in the table as follows (Table 11):

- **Step 5: Use the algorithm of AHP to solve this problem**
- Apply the formula for calculating the eigenvectors (9), we find the eigenvectors corresponding to the 9 matrix medium are set. Eigenvectors of the matrix comparison between the criteria:

w = (0.1798, 0.1345, 0.1232, 0.1056, 0.1341, 0.1342, 0.1259, 0.1214, 0.1488)

Table 11. Matrix comparison between the level of assessment for the criteria Who

Criteria impact ICT	Matrix					
A1	A1	M1	M2	M3	M4	M5
	M1	1	1	1	3	2
	M2	1	1	1	2	1
	M3	1	1	1	3	2
	M4	1/3	1/2	2/5	1	1
	M5	1/2	1	1	2	1
A2	A2	M1	M2	M3	M4	M5
	M1	1	1	2	1/2	1/2
	M2	1	1	2	1/3	1/2
	M3	3	5	1	1/5	1/3
	M4	2	3	1/2	1	2
	M5	2	3	1/3	1/2	1
A3	A3	M1	M2	M3	M4	M5
	M1	1	1	1	3	5
	M2	1	1	1	1	1
	M3	1	1	1	1/3	1/3
	M4	1/3	1/3	1/3	1	1
	M5	1/5	1/3	1/5	1	1
A4	A4	M1	M2	M3	M4	M5
	M1	1	1	2	1	2
	M2	2	1	1	1	1
	M3	2	1	1	2	2
	M4	1	1	1/2	1	1
	M5	1/2	1/2	1/2	1	1
A5	A5	M1	M2	M3	M4	M5
	M1	1	1	1/2	1	2
	M2	1	1	1	1	3
	M3	2	1	1	3	3

(continued)

Eigenvectors of the matrix comparison between the choices (assessment level) corresponding to the 9 impacts (Table 12).

Vector priority W = [w][w1 w2 w3... w9] = (0.252, 0.258, 0.239, 0.160, 0.091). That is, the likelihood of the course being evaluated at a very good level is 29%, the

Table 11. (*continued*)

Criteria impact ICT	Matrix					
	M4	1	2	1	1	2
	M5	1	1/3	1	1/2	1
A6	**A6**	**M1**	**M2**	**M3**	**M4**	**M5**
	M1	1	1	1/2	3	2
	M2	3	1	1	1	2
	M3	1	1	1	2	2
	M4	1	2	1/2	1	1
	M5	1/2	1/2	1/2	1	1
A7	**A7**	**M1**	**M2**	**M3**	**M4**	**M5**
	M1	1	2	1/2	1/2	1
	M2	1	1	1	1	2
	M3	1/2	1	1	3	2
	M4	2	3	1	1	3
	M5	1	1/2	1/2	1/3	1
A8	**A8**	**M1**	**M2**	**M3**	**M4**	**M5**
	M1	1	1	3	1	2
	M2	2	1	1	1/3	3
	M3	1/2	1	1	1	3
	M4	1	1	1	1	3
	M5	1/2	1/3	1/3	1/3	1
A9	**A9**	**M1**	**M2**	**M3**	**M4**	**M5**
	M1	1	1	1/2	2	3
	M2	1	1	2	1/3	3
	M3	1/2	1/2	1	2	1
	M4	1/2	1/3	1/2	1	1
	M5	1/3	1/3	1/3	1	1

good level is 24%, the level is quite good is 28%, the normal level is 12%, and the level is poor is 7%. So we can evaluate this course at a good level.

5   Conclusions

This study provides a TOPSIS-AHP model based on the Newhouse ICT criteria and the UEH set of criteria that combines knowledge from experts/lecturers/students to come up with an assessment process. The experimental results showed that the proposed model for

Table 12. Individual vector values corresponding to 9 matrix comparing rating

Eigenvectors	Values
w1	0.1439, 0.2182, 0.2506, 0.2718, 0.1155
w_2	0.3399, 0.2830, 0.1867, 0.1203, 0.0700
w_3	0.2899, 0.2618, 0.2899, 0.0873, 0.2830
w_4	0.2916, 0.3163, 0.1924, 0.1079, 0.0918
w_5	0.1882, 0.2344, 0.2693, 0.2162, 0.3163
w_6	0.2232, 0.2232, 0.2563, 0.1691, 0.1282
w_7	0.3041, 0.3266, 0.2182, 0.1066, 0.0568
w_8	0.2152, 0.2334, 0.0711, 0.2334, 0.0844
w_9	0.3399, 0.2830, 0.1867, 0.1203, 0.0700

assessing the quality of the course showed the multi-dimensional impacts on the positive performance of the expert/student in the assessment model. This assessment method combined with traditional assessment will give the right assessment for the courses and suggest to the manager in improving the quality of the course at the university in particular and the training unit in general. Expanding this research model, We continue to expand the TOPSIS-AHP-Kansei model with more intuitive evaluation results. The knowledge base will be built to store expert knowledge to reduce service costs and improve the quality of consultation for the quality courses.

Acknowledgment. This research is supported by University of Economics Ho Chi Minh City (UEH), Ho Chi Minh City, Vietnam under project CTD-2022–11.

References

1. Popkova, E.G., Ragulina, Y.V., Bogoviz, A.V.: Industry 4.0: Industrial Revolution of the 21st Century. Springer (2019). https://doi.org/10.1007/978-3-319-94310-7
2. Li, S.-T.T., Klein, M.D., Balmer, D.F., Gusic, M.E.: Scholarly evaluation of curricula and educational programs: using a systematic approach to produce publishable scholarship. Acad. Pediatr. **20**(8), 1083–1093 (2020)
3. Somasundaram, M., Junaid, K.M., Mangadu, S.: Artificial intelligence (AI) enabled intelligent quality management system (IQMS) for personalized learning path. Procedia Computer Science **172**, 438–442 (2020)
4. Newhouse, C., Clarkson, B., Trinidad, S.: A framework for leading school change in using ICT. Journal of Computer Assisted Learning (2005)
5. Akoka, J.: A framework for decision support systems evaluation. Inf. Manage. **4**(3), 133–141 (1981)
6. Hwang, C.-L., Yoon, K.: Methods for multiple attribute decision making. In: Multiple attribute decision making, pp. 58–191 (1981). Springer https://doi.org/10.1007/978-3-642-48318-9_3
7. Kumar, A., Baldea, M., Edgar, T.F., Ezekoye, O.A.: Smart manufacturing approach for efficient operation of industrial steam-methane reformers. Industrial Eng. Chemistry Res. **54**(16), 4360–4370 (2015)

8. Vavrek, R., Bečica, J.: Population size and transport company efficiency–Evidence from Czech Republic. Transportation Res. Interdisciplinary Perspectives **6**, 100145 (2020)
9. Yin, J., Yang, X., Zheng, X., Jiao, N.: Analysis of the investment security of the accommodation industry for countries along the B&R: an empirical study based on panel data. Tour. Econ. **23**(7), 1437–1450 (2017)
10. Marković, L., Milić Marković, L., Mitrović, S., Stanarević, S.: Vrednovanje varijantnih rešenja trase autoputa E-763 Beograd-Južni Jadran: studija slučaja u Srbiji. Tehnički vjesnik **24**(6), 1951–1958 (2017)
11. Mandić, K., Delibašić, B., Knežević, S., Benković, S.: Analysis of the efficiency of insurance companies in Serbia using the fuzzy AHP and TOPSIS methods. Economic Research-Ekonomska istraživanja **30**(1), 550–565 (2017)
12. Seyedmohammadi, J., Sarmadian, F., Jafarzadeh, A.A., Ghorbani, M.A., Shahbazi, F.: Application of SAW, TOPSIS and fuzzy TOPSIS models in cultivation priority planning for maize, rapeseed and soybean crops. Geoderma **310**, 178–190 (2018)
13. Kaynak, S., Altuntas, S., Dereli, T.: Comparing the innovation performance of EU candidate countries: an entropy-based TOPSIS approach. Economic research-Ekonomska istraživanja **30**(1), 31–54 (2017)
14. Rădulescu, C.Z., Rădulescu, I.C.: An extended TOPSIS approach for ranking cloud service providers. Studies in Informatics Control **26**(2), 183–192 (2017)
15. Chang, D.-Y.: Applications of the extent analysis method on fuzzy AHP. Eur. J. Oper. Res. **95**(3), 649–655 (1996)
16. Saaty, T.L.: The analytic hierarchy process: decision making in complex environments. In: Avenhaus, R., Huber, R.K. (eds.) Quantitative Assessment in Arms Control, pp. 285–308. Springer, Boston (1984). https://doi.org/10.1007/978-1-4613-2805-6_12
17. Lin, C., Kou, G.: A heuristic method to rank the alternatives in the AHP synthesis. Appl. Soft Comput. **100**, 106916 (2021)
18. Oliva, G., Setola, R., Scala, A.: Sparse and distributed analytic hierarchy process. Automatica **85**, 211–220 (2017)

Assessment of Grey Wolf Optimizer and Its Variants on Benchmark Functions

Elif Varol Altay$^{(\boxtimes)}$ (iD) and Osman Altay (iD)

Manisa Celal Bayar University, Manisa 45400, Turkey
{elif.altay,osman.altay}@cbu.edu.tr

Abstract. One of the most current metaheuristic swarm intelligence algorithms is the Grey Wolf Optimizer (GWO). Since the number of parameters is small and there is no need for information during the first search, GWO has been adapted to different optimization problems, giving it superiority over other metaheuristic methods. At the same time, it is easy to use, simple, scalable, and adaptable, with its unique ability to provide the ideal balance throughout the search, leading to positive convergence. As a result, the GWO has lately attracted a large research audience from a variety of areas in a very short amount of time. However, it has some disadvantages, such as a slow convergence rate, low sensitivity, and so on, which are seen in the vast majority of metaheuristic methods. Various versions of the current GWO have been proposed to eliminate them. In this article, GWO, improved GWO (IGWO), and augmented GWO (AGWO) methods are examined, and the performances of these methods are discussed in CEC'20 functions and analyzed statistically. The results of the studies demonstrated that IGWO outperformed standard GWO and AGWO.

Keywords: Swarm intelligence · Metaheuristic · Grey wolf optimizer · Global optimization · CEC'20 functions

1 Introduction

The traditional, deterministic methods of optimization have been surpassed by stochastic optimization strategies throughout the course of the last decade for a number of different reasons. Deterministic optimization and search approaches are notoriously ineffective when it comes to tackling NP-hard issues, which is one of the primary reasons why. In point of fact, many NP-hard problems do not yet have any deterministic solutions available at this time. One other reason is that many deterministic algorithms are dependent on the calculation of the problem's derivation in order to function properly. Such approaches are unable to be used for the challenges due to the fact that the derivation is ill-defined or difficult to get [1].

In contrast, stochastic optimization algorithms may find near-optimal solutions to NP-hard problems in a reasonable period of time. Additionally, the majority of them regard optimization issues to be a "black box" that can be solved without any derivation being necessary. This indicates that the core computer program or mathematical model

F. P. García Márquez et al. (Eds.): ICCIDA 2022, LNNS 643, pp. 55–66, 2023.
https://doi.org/10.1007/978-3-031-27099-4_5

of an optimization issue is not necessary in order to apply the same approach to several problems.

Techniques that take their cues from the natural world are included in a collection of prominent stochastic optimization algorithms that have gained a lot of traction recently. These approaches simulate natural intelligence and give problem-solving strategies that are inspired by the natural world. The Genetic Algorithm (GA), which is held in high esteem, is considered to be one of the pioneering algorithms in this field. The concept of biological evolution served as the basis for the development of this program, which recreates the process of evolution using a computer. In point of fact, in order to accomplish this goal, this method has been outfitted with recombination, selection, and mutation operators. The GA is a member of the evolutionary algorithm family. Evolution Strategy, which is sometimes called EA, and Differential Evolution are two other well-known approaches to evolution.

Several more kinds of nature-inspired algorithms emerged after the GA's proposal. The communal and collective intelligence of living things in nature served as the major inspiration for one of the courses, called Swam Intelligence. Ant colony optimization, for instance, imitates the way ants use intelligence to determine the shortest route from their colony to a food source. Artificial bee colony algorithm and particle swarm optimization are two other well-known algorithms in this family. Numerous applications have made use of swarm intelligence techniques. These approaches are the oldest and best known approaches. However, many methods have been proposed recently. Some of them are Harris hawks optimization [2], Archimedes optimization algorithm [3], COOT optimization algorithm [4], Aquila optimizer [5], Equilibrium optimizer [6], Arithmetic optimization algorithm [7], Chimp optimization algorithm [8], Gradient-based optimizer [9], Henry gas solubility optimization [10], Red fox optimization algorithm [11], and Horse herd optimization algorithm [12]. There are also studies comparing the performance of metaheuristic methods [13, 14].

In recent years, Grey Wolf Optimizer (GWO) is one of the commonly used methods to obtain optimal results in optimization problems [15]. Since the number of parameters is small and there is no need for information during the first search, GWO has been adapted to different optimization problems. It also has the unique ability to maintain an optimal balance of exploration and exploitation throughout the search, resulting in positive convergence. It is straightforward, adaptive, simple to use, and scalable. As a consequence, in a very short amount of time, the GWO has recently drawn a sizable research audience from several fields. However, it has significant drawbacks that are present in the great majority of metaheuristic approaches, like a sluggish convergence rate, limited sensitivity, etc. To get rid of them, many variants of the present GWO have been suggested. The performance of proposed GWO and its variants was not tested at CEC'20 functions. The performance of the GWO, improved GWO (IGWO), and augmented GWO (AGWO) approaches is addressed in CEC'20 functions and statistically studied in this work.

There are many studies done with GWO. Some of those; Koc et al. proposed two different mathematical models to estimate Turkey's energy demand using GWO [16]. Altay et al. predicted the reservoir temperature by using the standard GWO in the training phase of the artificial neural network [17]. Altay used GWO and different metaheuristic methods to solve five different engineering design problems [18]. Zhang and Hong forecast the electric charge using a hybrid of GWO and SVR methods [19].

In the second section of this study, the working principles of the current metaheuristic method GWO and its variants IGWO and AGWO are explained. In the third section, 10 different functions selected from the CEC'20 test suites are explained and the performances of the GWO and its variants, IGWO and AGWO, are compared using the CEC'20 test suites. The fourth section includes the results section.

2   Grey Wolf Optimizer and Its Variants

In this section, GWO and its variants, improved GWO and augmented GWO, are examined.

2.1   Grey Wolf Optimizer

The GWO method is a recent metaheuristic algorithm that was described by Mirjalili et al. [15]. It is designed to simulate the social behavior of grey wolves, who are known to dwell in packs that range in size from 5 to 12 individuals. The pack of grey wolves has been taught to follow a rigid dominating hierarchy, in which the pack has a leader known as alpha (α), followed by subordinate wolves known as beta (β), who aid alpha in decision making, followed by δ and ω as illustrated in Fig. 1.

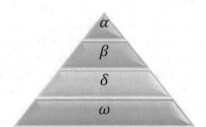

Fig. 1. Hierarchy in the grey wolf swarm

Grey wolves hunt prey in the following ways: looking for the prey, encircle the prey, hunting the prey, and finally attack the prey. The following is the mathematical expression of the model for encircling the prey:

$$D = \left| X_p(t) \times C - X(t) \right| \tag{1}$$

$$X(t+1) = X_p(t) - D \times A \tag{2}$$

where X and X_p are the position vectors of grey wolf and prey, respectively. C and A are vectors generated as follows:

$$a(t) = 2 - (2 \times t)/MaxIter \tag{3}$$

$$C = 2 \times r_2 \tag{4}$$

$$A = 2 \times a \times r_1 - a(t) \tag{5}$$

r_1 and r_2 are randomly generated vectors with a uniform distribution between 0 and 1, and a is a constant that decreases linearly from 2 to 0 as the iteration number (t) increases up to the maximum possible iteration count (*MaxIter*). The ability to achieve divergence of search agents, which can be achieved when $|A|$ is more than 1, allows for the exploration of the prey's position. This can be accomplished when $|A|$ is greater than 1. The exploitation of the prey, also known as an attack, may be accomplished by the convergence of search agents, which is researched when $|A|$ is less than 1. The search is directed by agents and aided by agents as well as help from and agents as in (6)–(8).

$$D_a = |X_a \times C_1 - X(t)|, D_\beta = |X_\beta \times C_2 - X(t)|, D_\delta = |X_\delta \times C_3 - X(t)| \tag{6}$$

$$X_{i1}(t) = X_a(t) - A_{i1} \times D_\alpha(t), X_{i2}(t) = X_\beta(t) - A_{i2} \times D_\beta(t), X_{i3}(t)$$
$$= X_\delta(t) - A_{i3} \times D_\delta(t) \tag{7}$$

$$X(t+1) = \frac{X_{i1}(t) + X_{i2}(t) + X_{i3}(t)}{3} \tag{8}$$

2.2 Improved Grey Wolf Optimizer

In GWO, α, β, and δ lead ω and guide wolves to regions of the search space that hold promise for finding the best answer. This conduct could trap one in a locally optimum solution. Another negative impact is a decrease in population variety, which makes GWO fall towards the local optimum. An improved GWO (IGWO) has been suggested to address these problems [20]. A new search method linked to a choosing and updating stage is one of the changes. The IGWO then includes three distinct stages, which are called initializing, moving, and selecting and updating, respectively. Initializing phase: Using Eq. (9), N wolves are scattered at random in the search space between $[l_i, u_j]$ during this phase.

$$X_{ij} = l_j + rand_j[0, 1] \times (u_j - l_j), i \in [1, N], j \in [1, D] \tag{9}$$

A vector of real numbers of the form represents as follows:

$$X_i(t) = x_{i1}, x_{i2}, \dots, x_{iD}$$

where D is the problem's dimension number. A Pop matrix with N rows and D columns is used to hold the whole wolf population. The fitness function, $f(X_i(t))$, determines the fitness value of $X_i(t)$.

Moving phase: One of the most fascinating aspects of the social lives of grey wolves is the individual hunting they do in addition to the group hunting they do. The IGWO includes a new mobility technique known as the dimension learning based hunting (DLH) search approach. Each wolf in DLH is acknowledged by its peers as a prospective candidate for the new position of $X_i(t)$. The methods below illustrate how the standard DLH search and GWO algorithms yield two distinct candidates.

DLH search strategy: In the standard GWO, three Pop leader wolves create a new post for each wolf. This slows GWO convergence, reduces population diversity, and traps wolves in local optima. The DLH search approach takes neighbor-learned individual wolf hunting into account to fix these shortcomings.

In DLH, each dimension of wolf $X_i(t)$'s location is determined using Eq. (12), which uses this wolf's neighbors and a randomly chosen wolf from *Pop*. The DLH search method also produces $X_{i-DLH}(t + 1)$ in addition to $X_{i-GWO}(t + 1)$. First, a radius $R_i(t)$ is determined using the Euclidean distance between $X_i(t)$ and $X_{i-GWO}(t+1)$ by Eq. (10).

$$R_i(t) = \|X_i(t) - X_{i-GWO}(t + 1)\| \tag{10}$$

Then, using Eq. (11) with regard to the radius $R_i(t)$, the neighbors of $X_i(t)$ represented by $N_i(t)$ are created, where D_i is the Euclidean distance between $X_i(t)$ and $X_j(t)$.

$$N_i(t) = \{X_j(t)|D_i(X_i(t), X_j(t) \leq R_i(t), X_j(t) \in Pop\} \tag{11}$$

Once the neighborhood of $X_i(t)$ has been established, a random neighbor $X_{n,d}(t)$ selected from $N_i(t)$ and a random wolf $X_{r,d}(t)$ selected from Pop are used to calculate the d-th dimension of $X_{i-DLH,d}(t + 1)$.

$$X_{i-DLH,d}(t + 1) = rand \times (X_{n,d}(t) + X_{i,d}(t) - X_{r,d}(t)) \tag{12}$$

The superior candidate is chosen in this phase by comparing $X_{i-DLH}(t + 1)$ and $X_{i-GWO}(t + 1)$, using Eq. (13).

$$X_i(t + 1) = \begin{cases} X_{i-GWO}(t + 1), \text{ iff } (X_{i-GWO}) < f(X_{i-DLH}) \\ X_{i-DLH}(t + 1), \text{ otherwise} \end{cases} \tag{13}$$

2.3 Augmented Grey Wolf Optimizer

It has been suggested that the GWO method may benefit from a new tweak that would enhance exploration without compromising its ease of use, versatility, or capacity for global optimization. As explained in the original GWO algorithm, parameter A is the most important one for exploration and exploitation. Equation (4) shows that parameter A depends mostly on parameter a. The exploration and exploitation of the GWO method are both controlled by the behavior of a parameter called a, which in the first version of the GWO method changes linearly from a value of 2 to a value of 0. The parameter a shifts from 2 to 1 in a nonlinear and unpredictable fashion in the method for the proposed augmentation (AGWO) [21], as can be seen in Eq. (14).

$$a(t) = 2 - \cos(\text{rand}) \times t/MaxIter \tag{14}$$

The GWO algorithm's searching and decision-making are dependent on the frequency at which alphas (α), betas (β), and deltas (δ) are updated, as shown in (6)– (8).

On the other hand, in the AGWO algorithm that has been presented, the hunting will simply rely on α and β as in (15)–(17).

$$D_a = |C_1 \times X_a - X(t)|, D_\beta = |C_2 \times X_\beta - X(t)| \tag{15}$$

$$X_{i1}(t) = X_a(t) - A_{i1} \times D_\alpha(t), X_{i2}(t) = X_\beta(t) - A_{i2} \times D_\beta(t) \tag{16}$$

$$X(t+1) = \frac{X_{i1}(t) + X_{i2}(t)}{2} \tag{17}$$

3 Experimental Results

The IEEE Congress on Evolutionary Computation (CEC) test suites were selected to analyze the performance of GWO, IGWO, and AGWO methods. CEC'20 test suites include ten test functions, including multimodal, unimodal, composition, and hybrid functions. Unimodal functions are used to assess the algorithm's convergence performance; multimodal functions are used to determine if the method has early convergence and local fixation issues. Hybrid and composition functions are used to measure the ability to avoid local optima, which can have more than one, and the balance between exploration and exploitation. Table 1 lists the properties of the CEC'20 test suites. f_i^* denotes the function's optimal global value. More detailed information can be obtained from the relevant study for the CEC'20 test suites [13].

Table 1. CEC'20 test suites

No	f_i^*	Features
f01	100	Unimodal Function
f02	1100	Basic Functions
f03	700	
f04	1900	
f05	1700	Hybrid Functions
f06	1600	
f07	2100	
f08	2200	Composition Functions
f09	2400	
f10	2500	

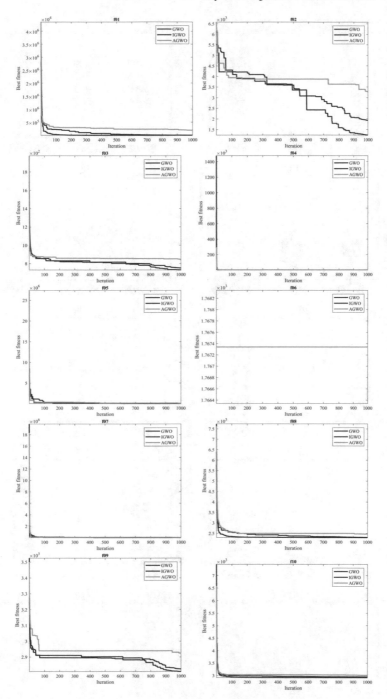

Fig. 2. Converge curve of the compared methods

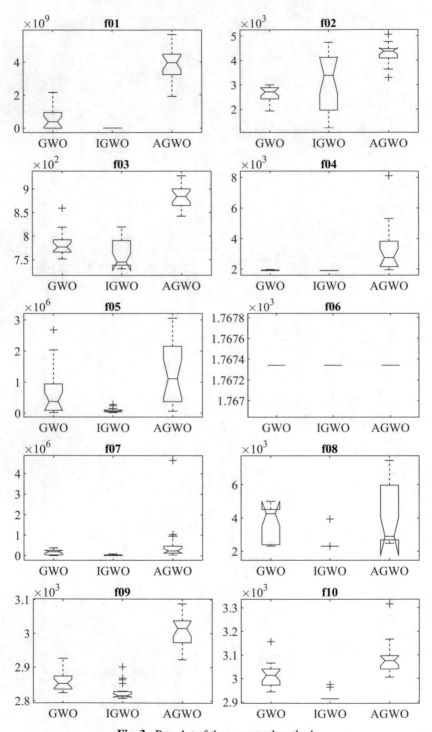

Fig. 3. Boxplot of the compared methods

All experiments were conducted on the MATLAB 2021a platform with the Windows 10, 32 GB of RAM and a CPU of an Intel (R) core i9–10900 k (3.7 GHz) processor. The control parameters of the methods were chosen to be the same as the values used in the original articles. In order to make a fair evaluation under equal conditions, the number of maximum iterations was chosen as 1000, dimension as 20, and the number of population as 30. Algorithms were run 20 times in all experiments, and the results of mean, maximum (Max), standard deviation (SD), minimum (Min), and median values are presented in Table 2 in a comparative way.

Table 2. Experimental results of GWO and its variants

Function	Metric	GWO	IGWO	AGWO
F1	Mean	6.07E + 08	**4.99E + 04**	3.86E + 09
	SD	6.91E + 08	**3.38E + 04**	1.01E + 09
	Max	2.16E + 09	**1.24E + 05**	5.66E + 09
	Min	4.07E + 04	**9.71E + 03**	1.91E + 09
	Median	3.87E + 08	**3.71E + 04**	3.95E + 09
F2	Mean	**2.64E + 03**	3.13E + 03	4.29E + 03
	SD	**3.08E + 02**	1.17E + 03	3.93E + 02
	Max	**3.00E + 03**	4.74E + 03	5.06E + 03
	Min	1.93E + 03	**1.24E + 03**	3.29E + 03
	Median	**2.71E + 03**	3.39E + 03	4.37E + 03
F3	Mean	7.83E + 02	**7.60E + 02**	8.85E + 02
	SD	2.63E + 01	3.14E + 01	**2.49E + 01**
	Max	8.59E + 02	**8.19E + 02**	9.28E + 02
	Min	7.52E + 02	**7.30E + 02**	8.42E + 02
	Median	7.77E + 02	**7.45E + 02**	8.84E + 02
F4	Mean	1.93E + 03	**1.91E + 03**	3.32E + 03
	SD	2.07E + 01	**2.88E + 00**	1.53E + 03
	Max	1.97E + 03	**1.91E + 03**	8.11E + 03
	Min	**1.90E + 03**	1.90E + 03	1.94E + 03
	Median	1.92E + 03	**1.91E + 03**	2.75E + 03
F5	Mean	6.58E + 05	**8.67E + 04**	1.23E + 06
	SD	7.34E + 05	**7.06E + 04**	9.63E + 05
	Max	2.68E + 06	**2.80E + 05**	3.05E + 06
	Min	2.55E + 04	**2.31E + 04**	6.79E + 04

(*continued*)

Table 2. (*continued*)

Function	Metric	GWO	IGWO	AGWO
	Median	3.78E + 05	**6.22E + 04**	1.11E + 06
F6	Mean	**1.77E + 03**	**1.77E + 03**	**1.77E + 03**
	SD	**0.00E + 00**	**0.00E + 00**	**0.00E + 00**
	Max	**1.77E + 03**	**1.77E + 03**	**1.77E + 03**
	Min	**1.77E + 03**	**1.77E + 03**	**1.77E + 03**
	Median	**1.77E + 03**	**1.77E + 03**	**1.77E + 03**
F7	Mean	1.65E + 05	**2.73E + 04**	5.31E + 05
	SD	1.25E + 05	**2.55E + 04**	1.01E + 06
	Max	3.84E + 05	**8.65E + 04**	4.65E + 06
	Min	7.98E + 03	**5.79E + 03**	2.93E + 04
	Median	1.95E + 05	**1.48E + 04**	2.33E + 05
F8	Mean	3.72E + 03	**2.38E + 03**	4.28E + 03
	SD	1.04E + 03	**3.65E + 02**	1.86E + 03
	Max	5.00E + 03	**3.93E + 03**	7.45E + 03
	Min	2.31E + 03	**2.30E + 03**	2.47E + 03
	Median	4.26E + 03	**2.30E + 03**	2.89E + 03
F9	Mean	2.86E + 03	**2.83E + 03**	3.01E + 03
	SD	3.09E + 01	**2.49E + 01**	4.55E + 01
	Max	2.93E + 03	**2.90E + 03**	3.09E + 03
	Min	2.82E + 03	**2.81E + 03**	2.92E + 03
	Median	2.85E + 03	**2.82E + 03**	3.02E + 03
F10	Mean	3.01E + 03	**2.92E + 03**	3.09E + 03
	SD	5.11E + 01	**1.70E + 01**	7.17E + 01
	Max	3.16E + 03	**2.97E + 03**	3.32E + 03
	Min	2.94E + 03	**2.91E + 03**	3.01E + 03
	Median	3.01E + 03	**2.91E + 03**	3.08E + 03

Figure 2 also illustrates the convergence performance of the methods on the CEC'20 test suites based on the best values of the methods. When Fig. 2 is examined, it is seen that IGWO gives the best values and convergence in f01, f03, f07, f08, f09, and f10. Figure 3 demonstrates the boxplot of the compared methods. When Fig. 3 is examined, it is seen that IGWO is more consistent and superior to other methods in f01, f04, f05, f07, f08, f09, and f10 functions.

When Table 1 is examined, if we consider all test suites, IGWO ranks first in nine out of ten test suites, followed by standard GWO with two best average results. The AGWO did not achieve the best results in any of the ten test suites at the mean value. When we looked at its minimum value, it showed its superiority by reaching the best result in all of the IGWO test suites.

When investigating the exploitation capabilities of algorithms, unimodal functions have been shown to be useful. IGWO earned the best results possible across all assessment criteria when it came to unimodal functions. On the basis of this, we are able to conclude that IGWO performs better than other algorithms when it comes to unimodal functions. Similarly, we can conclude that IGWO outperforms other methods in multimodal hybrid and composition functions.

4 Conclusions

Since there is no algorithm that performs best for every problem among metaheuristic methods, it remains up-to-date for researchers. Therefore, new methods are being proposed and it seems that they will continue to be suggested. The increase in the application areas of the methods in parallel with the proposed methods also positively affects the popularity of this field. In this study, the GWO method and its variants, which have been proposed in recent years and used in the solution of many problems, have been selected and applied to the CEC functions in the literature in analyzing the performance of optimization problems. Among the CEC functions, different types of problems were selected to show the different capabilities of the metaheuristic optimization algorithms in the CEC'20 test suites. The results of the experiments proved the superiority of IGWO over standard GWO and AGWO. In future studies, the performances of different GWO versions newly introduced to the literature and of new methods can be analyzed in CEC'20.

References

1. Mirjalili, S., Aljarah, I., Mafarja, M., Heidari, A.A., Faris, H.: Grey wolf optimizer: theory, literature review, and application in computational fluid dynamics problems. In: Mirjalili, S., Song Dong, J., Lewis, A. (eds.) Nature-Inspired Optimizers. SCI, vol. 811, pp. 87–105. Springer, Cham (2020). https://doi.org/10.1007/978-3-030-12127-3_6
2. Heidari, A.A., Mirjalili, S., Faris, H., Aljarah, I., Mafarja, M., Chen, H.: Harris Hawks optimization: algorithm and applications. Futur. Gener. Comput. Syst. **97**, 849–872 (2019). https://doi.org/10.1016/j.future.2019.02.028
3. Hashim, F.A., Hussain, K., Houssein, E.H., Mabrouk, M.S., Al-Atabany, W.: Archimedes optimization algorithm: a new metaheuristic algorithm for solving optimization problems. Appl. Intell. **51**(3), 1531–1551 (2020). https://doi.org/10.1007/s10489-020-01893-z
4. Naruei, I., Keynia, F.: A new optimization method based on COOT bird natural life model. Expert Syst. Appl. **183**, 115352 (2021). https://doi.org/10.1016/j.eswa.2021.115352
5. Abualigah, L., Yousri, D., Abd Elaziz, M., Ewees, A.A., Al-Qaness, M.A.A., Gandomi, A.H.: Aquila optimizer: a novel meta-heuristic optimization algorithm. Comput. Ind. Eng. **157**, 107250 (2021). https://doi.org/10.1016/j.cie.2021.107250
6. Faramarzi, A., Heidarinejad, M., Stephens, B., Mirjalili, S.: Equilibrium optimizer: a novel optimization algorithm. Knowledge-Based Syst. **191**, 105190 (2020). https://doi.org/10.1016/j.knosys.2019.105190

7. Abualigah, L., Diabat, A., Mirjalili, S., Abd Elaziz, M., Gandomi, A.H.: The arithmetic optimization algorithm. Comput. Methods Appl. Mech. Eng. **376**, 113609 (2021). https://doi.org/10.1016/j.cma.2020.113609

8. Khishe, M., Mosavi, M.R.: Chimp optimization algorithm. Expert Syst. Appl. **149**, 113338 (2020). https://doi.org/10.1016/j.eswa.2020.113338

9. Ahmadianfar, I., Bozorg-Haddad, O., Chu, X.: Gradient-based optimizer: a new metaheuristic optimization algorithm. Inf. Sci. (Ny) **540**, 131–159 (2020). https://doi.org/10.1016/j.ins.2020.06.037

10. Hashim, F.A., Houssein, E.H., Mabrouk, M.S., Al-Atabany, W., Mirjalili, S.: Henry gas solubility optimization: a novel physics-based algorithm. Futur. Gener. Comput. Syst. **101**, 646–667 (2019). https://doi.org/10.1016/j.future.2019.07.015

11. Połap, D., Woźniak, M.: Red fox optimization algorithm. Expert Syst. Appl. **166**, 114107 (2021). https://doi.org/10.1016/j.eswa.2020.114107

12. Miarnaeimi, F., Azizyan, G., Rashki, M.: Horse herd optimization algorithm: a nature-inspired algorithm for high-dimensional optimization problems. Knowledge-Based Syst. **213**, 106711 (2021). https://doi.org/10.1016/j.knosys.2020.106711

13. Varol Altay, E., Altay, O.: Güncel Metasezgisel Optimizasyon Algoritmalarının CEC2020 Test Fonksiyonları Ile Karşılaştırılması. DÜMF Mühendislik Derg. **5**, 729–741 (2021). https://doi.org/10.24012/dumf.1051338

14. Altay, E.V., Alatas, B.: Randomness as source for inspiring solution search methods: music based approaches. Phys. A Stat. Mech. Its Appl. **537**, 122650 (2020). https://doi.org/10.1016/j.physa.2019.122650

15. Mirjalili, S., Mirjalili, S.M., Lewis, A.: Grey wolf optimizer. Adv. Eng. Softw. **69**, 46–61 (2014). https://doi.org/10.1016/j.advengsoft.2013.12.007

16. Koc, I., Kivrak, H., Babaoglu, I.: The estimation of the energy demand in turkey using grey wolf optimizer algorithm. Ann. Fac. Eng. Hunedoara - Int. J. Eng. **17**, 113–117 (2019)

17. Varol Altay, E., Gurgenc, E., Altay, O., Dikici, A.: Hybrid artificial neural network based on a metaheuristic optimization algorithm for the prediction of reservoir temperature using hydrogeochemical data of different geothermal areas in Anatolia (Turkey). Geothermics **104**, 102476 (2022). https://doi.org/10.1016/j.geothermics.2022.102476

18. Altay, E.V.: Investigation of the performance of metaheuristic optimization algorithms used in solving real-world engineering design problems. Int. J. Appl. Innov. Eng. Manag. **6**(1), 65–74 (2022)

19. Zhang, Z., Hong, W.C.: Application of variational mode decomposition and chaotic grey wolf optimizer with support vector regression for forecasting electric loads. Knowledge-Based Syst. **228**, 107297 (2021). https://doi.org/10.1016/j.knosys.2021.107297

20. Nadimi-Shahraki, M.H., Taghian, S., Mirjalili, S.: An improved grey wolf optimizer for solving engineering problems. Expert Syst. Appl. **166**, 113917 (2021)

21. Qais, M.H., Hasanien, H.M., Alghuwainem, S.: Augmented grey wolf optimizer for grid-connected PMSG-based wind energy conversion systems. Appl. Soft Comput. J. **69**, 504–515 (2018). https://doi.org/10.1016/j.asoc.2018.05.006

Pre-production Design of a Robotic Arm Mounted on an Unmanned Aerial Vehicle (UAV)

Murat Bakirci[1](✉) and Abdullah Demiray[2]

[1] Faculty of Aeronautics and Astronautics, Tarsus University, Mersin 33400, Turkey
muratbakirci@tarsus.edu.tr
[2] Aeronautical Engineering Department, Istanbul Technical University, Istanbul 34469, Turkey
abdullah.demiray@itu.edu.tr

Abstract. It has been demonstrated through many applications that unmanned aerial vehicles, which have been enhanced with various equipment and whose performance has been increased, can perform more effective tasks. Considering the results of these successful applications, the further development of unmanned aerial vehicles with new equipment has recently gained momentum. Among these development studies, the integration of robotic systems in particular is of critical importance. In this study, a robotic arm design is presented to be mounted on unmanned aerial vehicles and used in a wide variety of critical missions. The geometric design of the robotic arm, which is adjusted to be integrated under the fuselage of a rotary wing aircraft, was carried out considering the geometrical characteristics of the unmanned system. The weight-to-load ratio of the robotic arm was arranged as 1 and the motors were selected accordingly. The limits of the geometric design were determined based on the system requirements. Within the scope of the kinematic model of the robotic arm, the general transformation matrix is also obtained based on all its joints and links.

Keywords: Robotic arm · Unmanned aerial vehicle · Conceptual design · Kinematic model · Strength analysis

1 Introduction

Unmanned aerial vehicles (UAV), whose performance and capabilities are increased by being equipped with a wide variety of software and hardware, are also pioneers in the realization of new scientific research [1]. This enables many applications that are not possible or difficult with traditional methods and increases the usage areas of these vehicles. Especially in missions where human access is not possible, the superior features and benefits of UAVs become more evident. Along with technological developments, access to these vehicles, which can be produced faster and at a lower cost, has become easier, and most scientific studies have been taken to the next level. Easy and fast system updates are another important advantage of the UAVs. Unmanned systems with superior features can be developed with software and hardware updates, which are needed in critical missions due to the demands that push the limits. These systems, which are

F. P. García Márquez et al. (Eds.): ICCIDA 2022, LNNS 643, pp. 67–77, 2023.
https://doi.org/10.1007/978-3-031-27099-4_6

frequently used in applications such as remote sensing [2], surveillance [3], atmospheric measurement and mapping [4], are also widely preferred in intelligent transportation systems [5].

Although fixed-wing UAVs have faster and more stable flight characteristics, rotary-wing systems stand out with their features such as high maneuverability, flying in tight spaces and hovering. This causes rotary wing systems to be preferred in many applications compared to fixed wing systems [6]. Quadrotors are the most popular of rotary-wing systems, but hexacopters, another important variant, continue to grow in popularity. These unmanned systems, controlled by six rotors equidistant from each other around the main body, are more advantageous than quadcopters with their ability to fly even in thin air. Moreover, they can be considered as an alternative platform to conventional surveillance platforms because of hovering feature.

Robotic arms are one of the robotic machines used in a wide variety of fields, especially in industry. Their strong structure and ability to perform their duties extremely quickly are the main reasons why they are often preferred in a wide variety of applications. Although there are many varieties, they generally have similar applications. However, they also have features that make each types specific and differentiate it from other robotic arm systems. They can even be designed just for a specific task. Articulated arms, one of the most commonly used types, are multi-purpose robotic arms with high degrees of freedom. Six-axis systems, which are a sub-type, are the most preferred robotic arms in the industry. In addition to these general examples, collaborative robot arms specially designed for hybrid applications are also quite remarkable examples [7]. Cartesian robot arms, which are generally controlled by three linear actuators, are simpler structures that operate by making translational movements in all three axes [8].

Integration of robotic arms into other robotic devices is one of the applications that can be considered new, apart from their traditional use. The most common example of this application is the robotic arms integrated into the underside of the UAV. Most of the studies on this subject in the recent past are such integrated robotic structures [9]. The dynamics of the unmanned system integrated with a multi-purpose roboarm was developed in [10] with the Euler-Lagrange formalism. In addition, the authors examined the external inputs acting on the whole system and investigated their effects on the dynamics of the UAV. Robotic arms are mounted on small-sized rotary-wing UAVs for the purpose of grasping and relocating relatively small objects. A similar study has been applied to single rotor UAVs [11]. Robotic arms with various degrees of freedom have been produced and integrated into various UAV systems for the rescue of objects or small animals stuck in tight spaces [12]. In addition to these, unmanned aerial systems containing more than one arm have also been developed [13]. Later, the use of these robotic arms as a three- or four-legged landing gear was also discussed [14]. Combining these two approaches, unmanned systems with both arms and legs have also been proposed [15]. In this way, even a simple drone has been made multi-purpose and highly useful. Especially, drones with three robotic arms or legs are very useful vehicles due to their ability to land on any type of ground.

In this study, a robotic arm design that can be integrated into hexacopters is presented to be used for various purposes. In line with this goal, first of all, the system requirements were determined and the concept design of the robotic arm that would meet these requirements was made. A geometric and kinematic model is presented in accordance with the UAV to which it will be mounted. Appropriate material options for the production of the robotic arm, of which the concept was designed, were also investigated and the effects such as stress, deflection and bending that may occur on the arm were examined, considering the amount of load expected to carry.

The remainder of the article is organized as follows: Next chapter discusses the system requirements in detail. The kinematic model required for the proper utilization of the designed arm is presented in the third chapter. Material selection for the production of the robotic arm, whose model is presented, is given in the fourth chapter. In the fifth chapter, detailed robustness analysis of the prototype to be produced is made. In the sixth and last chapter, the results of the study are discussed.

2   System Requirements

The unmanned aerial vehicle in which the robotic arm, whose design is presented, will be integrated, is a hexacopter with the most suitable features for this application. A critical advantage that makes these drones extraordinary is that they can fly at high altitudes and even in thin air. Another obvious advantage of their powerful rotors is that they can be kept under control even when exposed to intense and harsh weather conditions. Having six propellers not only increases their maneuverability, but also allows them to continue flying even if one of the rotors fails. Considering its integration with the robotic arm, hexacopters that can stay in the air are the most ideal aircraft for this and similar applications. The hexacopter selected for this study, which can fly up to a maximum altitude of 2 km, is a UAV that can perform both fully autonomous and manually controlled flight for approximately 20 min, considering the current drawn by its motors and battery capacity. The artificial intelligence supported development board on it has the capabilities of detecting the mission targets with a high resolution, wide angle and high data rate camera and extracting the location vector of the target. Flight telemetry data is transmitted to the ground station in an end-to-end encrypted manner with low latency at a maximum range of 50 km.

Considering the geometric features and engine capacities of the hexacopter, it will have a payload capacity of approximately 900 g. Based on the operations required by the robotic arm integrated into the UAV, it would be appropriate to have a maximum weight of 400 g and a load carrying capacity of 400 g. In other words, the weight-to-load ratio for the robotic arm in question is set to 1, and in this respect it differs from similar designs. As seen in Fig. 1, gaps are opened in the appropriate areas of the robotic arm which results overall weight reduction, and in this way, a more robust material can be used for production.

It should be noted that the maximum amount of payload that the UAV can carry primarily limits the mechanical and electronic design of the system. Due to the geometric design, the maximum extension of the robotic arm is approximately 47 cm. Considering the tasks planned to be performed by the robotic arm, it gives the result that the most suitable point to be mounted on the UAV is under the body. It is designed in such a way that the robotic arm is folded over itself in such cases, in order to provide convenience during take-off, landing, and stable flight, except when performing a task. There are 6

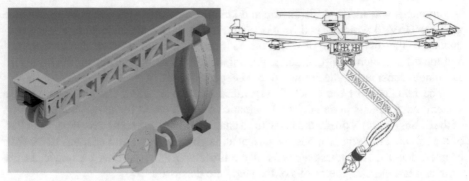

Fig. 1. CAD drawings of the robotic arm. Solid mode (left) and wireframe mode (right) integrated into a rotary-wing UAV.

motors in total in the robotic arm. A DC motor is used in the joint where the arm is connected to the UAV, and servo motors are used in the other joints. The number of motors in the joint points, the working principle of the robotic arm and the result of dynamic analysis are determined.

3 Kinematic Model

The kinematic model of the presented robotic arm, including all mechanical connections, was created using the Denavit-Hartenberg (DH) method [16]. With a similar approach, the entire dynamical model can also be constructed with the Euler-Lagrange energy equations. The kinematic parameters of the robotic arm are detailed in Fig. 2.

By design, there are no joints showing translational mechanical behavior and all mechanical joints are revolute. In this case, the position vector, $\mathbf{q}$, in the joint space is one-to-one with the velocity vector, $\mathbf{v}$, i.e. $\mathbf{q} = \begin{bmatrix} q_1 \dots q_n \end{bmatrix} = \begin{bmatrix} v_1 \dots v_n \end{bmatrix} = \mathbf{v}$. Kinematic coupling, on the other hand, can be briefly expressed as $\widehat{\mathbf{q}} = \boldsymbol{H}\boldsymbol{q}$, where $\widehat{\mathbf{q}}$ is the motor position and $\mathbf{H}$ is the matrix that transforms the joint space into motor space. The matrix $\mathbf{H}$ for the designed robotic arm is a 6×5 matrix and the determinant of this matrix is greater than zero. This shows that kinematic coupling is a reversible equation, such that $\boldsymbol{q} = \boldsymbol{H}^{-1}\widehat{\mathbf{q}}$. The amount of torque, τ_m produced by the motors under ideal conditions gives the general operating scheme of the system. However, it is the joint torques, τ_j that are important during the physical operation of the robotic arm. The direct relationship between these two parameters is by taking into account the H matrix. The amounts of torque produced by the motors, and the joint torques, are linked by the expression $\tau_m = \mathbf{H}^{-T}\tau_j$. On the other hand, for a holistic definition, after the robotic arm is produced, tests in accordance with the dynamics of the system must be carried out. With the system identification tests to be performed, the entire kinematic and dynamic scheme of the robotic system can be drawn.

The direct kinematic expression of the robotic arm is given by the transformation matrix, **T**, below [16].

$$T(q) = \begin{bmatrix} R(q) & p(q) \\ 0 & 1 \end{bmatrix} \tag{1}$$

Here, the term $R(q)$ indicates the end effector's rotation matrix, and $p(q)$ indicates the position vector on the origin reference frame. Thus, the direct kinematic equation is expressed, which gives the relationship between the combined variables and the end effector pose. If this expression is to be written clearly, the movement of the robotic arm can be expressed with a set of three matrix. That is, the transformation matrix of the endpoint of the arm with three degrees of freedom should be obtained. The robotic arm, whose model parameters are shown in Fig. 2, has a three-joint configuration. While the first joint can rotate around the y-axis, the other two joints are parallel to each other and perpendicular to the first joint. Starting from the local reference system of each joint, the transforming matrix of the robotic arm is created. The transformation matrix for the first link is depicted as:

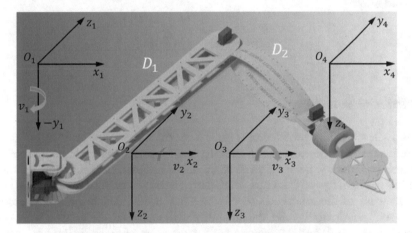

Fig. 2. Kinematic model parameters of the robotic arm.

$$^1T_2(\delta_1) = \begin{bmatrix} \cos\delta_1 & 0 & \sin\delta_1 & 0 \\ \sin\delta_1 & 0 & -\cos\delta_1 & 0 \\ 0 & 1 & 0 & 0 \\ 0 & 0 & 0 & 1 \end{bmatrix} \tag{2}$$

where δ represents the angle between links. On the other hand, the transformation matrices for the 2nd and 3rd links are as follows:

$$^2T_3(\delta_2) = \begin{bmatrix} \cos\delta_2 & -\sin\delta_2 & 0 & D_1\cos\delta_2 \\ \sin\delta_2 & \cos\delta_2 & 0 & D_1\sin\delta_2 \\ 0 & 0 & 1 & 0 \\ 0 & 0 & 0 & 1 \end{bmatrix} \tag{3}$$

$$
{}^3T_4'(\delta_3) =
\begin{bmatrix}
\cos\delta_3 & -\sin\delta_3 & 0 & D_2\cos\delta_3 \\
\sin\delta_3 & \cos\delta_3 & 0 & D_2\sin\delta_3 \\
0 & 0 & 1 & 0 \\
0 & 0 & 0 & 1
\end{bmatrix}
\tag{4}
$$

where D represents length of the links. As a result, the overall transformation matrix for the endpoint of the arm is obtained as follows.

$$
{}^1T_4' = {}^1T_2' \cdot {}^2T_3' \cdot {}^3T_4' =
\begin{bmatrix}
\cos\delta_1\cos\delta_{23} & -\cos\delta_1\sin\delta_{23} & \sin\delta_1 & \cos\delta_1(D_2\cos\delta_{23}+D_1\cos\delta_2) \\
\sin\delta_1\cos\delta_{23} & -\sin\delta_1\sin\delta_{23} & -\cos\delta_1 & \sin\delta_1(D_2\cos\delta_{23}+D_1\cos\delta_2) \\
\sin\delta_{23} & \cos\delta_{23} & 0 & D_2\sin\delta_{23}+D_1\sin\delta_2 \\
0 & 0 & 0 & 1
\end{bmatrix}
\tag{5}
$$

where $\cos\delta_{23}$ and $\sin\delta_{23}$ correspond to $\cos(\delta_2+\delta_3)$ and $\sin(\delta_2+\delta_3)$ respectively. It should be noted that the joint offsets are assumed to be zero for all three joints. The differential kinematics giving the relationship between the end effector linear, $\dot{p}_{ee}$, and angular velocity, ω_{ee}, corresponding the joint velocities is given by the geometric Jacobian matrix expressed below [16].

$$
v_{ee} = \begin{bmatrix} \dot{p}_{ee}(\mathbf{q}) \\ \omega_{ee}(\mathbf{q}) \end{bmatrix} = J(\mathbf{q})\dot{\mathbf{q}}
\tag{6}
$$

4   Material Selection

Material selection at the production stage is a critical and detailed process. For this, the requirements should be determined carefully and the selection should be made accordingly. It is important that the object to be produced has the desired strength conditions, as well as having a low cost. At this point, the material and performance indexes related to the structure of the material to be used are very useful, which give important clues such as strength, flexibility and machinability of the material. Although a wide variety of materials are used for the production of robotic arms, aluminum alloys are the most preferred ones. For advanced technology applications such as space probe, expensive and rare but high quality materials such as titanium metal are preferred. The type of task and the task to be performed by the robotic arm are the determining factors in the material selection. In this context, there is a wide variety of materials on the market, but Methacrylate monomers that can be used in 3D printers will be sufficient for the production of this robotic arm. Considering the system requirements, a robotic arm made of this material with a 3D printer will have sufficient strength. The important point here is that the first arm piece has the option to have more weight than the others. In other words, the part where the robotic arm is mounted on the body may be heavier than other parts. However, this amount of weight should gradually decrease in the next parts. In fact, this must be considered, since the entire weight of the robotic arm will affect the most severely at the point where it is connected to the UAV.

5 Strength Analysis

One of the measures of the designed robotic arm to fulfill its task effectively is to carry the desired amount of load safely. Since the weight and strength of the material to be used for the arm are generally directly proportional, the proper choice and analysis should be made so that the material to be preferred can be used for further applications. Considering the robustness and weight, an optimum choice should be made in accordance with the mission objectives. In this context, the importance of stress analysis emerges. It is quite critical to determine the changes such as deformation and flexibility in the material exposed to the forces caused by the loads to be carried by the robotic arm. In fact, the aim of this analysis is to obtain a solid structure that can carry the desired load by using the least amount of material possible. In terms of the output from here, this analysis is actually a tool to move to the next stage rather than a determined goal.

For a proper stress analysis, the geometric properties of the structure to be produced should be well defined. This seemingly simple step is actually of critical importance. After that, it should be ensured that the correct material selection is made. Another important point is how the parts are put together. The correct assembly of robotic arm parts both increases the amount of force it will withstand and enables it to work more effectively. Finally, the average and maximum forces that the structure can be subjected to within the scope of the task must be accurately determined. As a result of a stress analysis, taking into account all these stages, a map of the force distribution on the examined structure, the specific stress and strain properties of each component of the structure, and the elasticity properties of the whole structure can be inferred. In the case of moving structures such as the robotic arm, vibrations from the motors that move the system can also be taken into account. The robotic arm designed in this study is relatively light and the amount of load expected to be carried is also quite low. Considering the material used for its production and the task it will perform, only analysis of deflection characteristics will be sufficient for the robotic arm. This investigation can be done with the Euler-Bernoulli beam theory, which is a simplified form of the linear elasticity theory. For any part of the arm, the variation of the amount of deflection according to the applied load can be computed through Euler-Bernoulli expression below [17].

$$\frac{d^2}{dx^2}\left[EI\frac{d^2\alpha}{dx^2}\right] = f \tag{7}$$

where E indicates elastic modulus. The term I corresponds to the 2nd moment of area and defined as $\iint z^2 dydz$, $\alpha(x)$ is the deflection amount in vertical aspect at any position (x), and f is the load. Since the product of E and I is generally constant, rearranging the equation for i^{th} arm yields

$$EI\frac{d^4\alpha}{dx^4} = \frac{1}{D_i}fx \tag{8}$$

Here, the right hand side provides the load at a specific point on the arm where x ranges from 0 to D_i. Integrating the above equation then yields

$$EI\alpha(x) = \frac{f}{D_i}\frac{x^5}{120} + c_1\frac{x^3}{6} + c_2\frac{x^2}{2} + c_3x + c_4 \tag{9}$$

By applying the boundary conditions $\alpha(0) = 0$, $\alpha(D) = 0$, $\ddot{\alpha}(0) = 0$, and $\ddot{\alpha}(D) = 0$, the constants in the equation can be determined and the solution is obtained. Moreover, the bending moment, M, and the shear force, σ, in the robotic arm are also given as follows.

$$M = -EI\frac{d^2\alpha}{dx^2} \tag{10}$$

$$\sigma = -\frac{d}{dx}\left(EI\frac{d^2\alpha}{dx^2}\right) \tag{11}$$

For the first part of the robotic arm, the change in deflection amount according to the arm position is shown in Fig. 3. Deflection amount is calculated for different load amounts that the arm is required to carry, and normalized values are expressed. As expected, it is clearly seen that there is more deflection as the amount of load to be carried by the robotic arm increases.

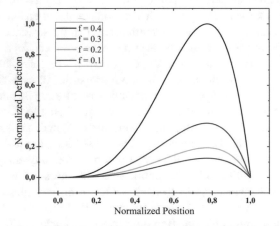

Fig. 3. Variation of deflection with position on the arm for various load values.

The f value corresponding to 0.4 corresponds to the maximum load. It is seen that the amount of deflection increases very rapidly for this load value. The amount of deflection corresponding to the load value of 0.1 is approximately 1/10 of the amount of deflection corresponding to the maximum load value, and this shows the big difference. Bending moment and shear force graphs corresponding to the same load amounts are also shown in Fig. 4.

The deflection amounts corresponding to different I values for the same arm are shown in Fig. 5. The amount of load carried by the arm was chosen as 0.25. Normalized values are given as in the previous chart. It can be seen from the figure that the amount of deflection is inversely proportional to the value of I. At the point where the deflection amount is maximum, the deflection amount corresponding to the value of $I = 1$ is approximately 49% more than the deflection amount when $I = 1.5$.

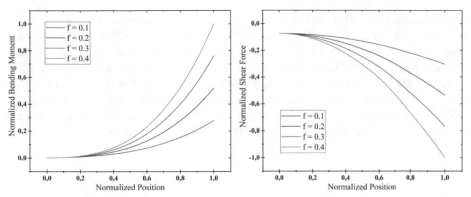

Fig. 4. Variation of normalized bending moment (left) and normalized shear force (right) with normalized position for different load values.

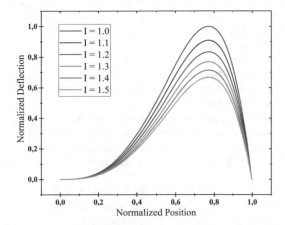

Fig. 5. Variation of deflection with position on the arm for various second moment of area of the arm's cross-section.

6   Conclusions

The pre-production design of a multi-purpose robotic arm mounted on an unmanned aerial vehicle, which can be used in variety of applications such as bomb disposal and rescue, is presented. In line with the mission objectives, the system requirements are determined and the robotic arm is designed to meet these requirements. Unlike the studies in the literature, the model is developed in a way that the ratio of the weight of the robotic arm to the maximum load carrying capacity is 1. The entire kinematic model of the arm, which has features that can be used even in model UAVs, has also been created. It has been determined that the motors at the joints increase the weight significantly and it is concluded that it is the most decisive element of the design. In addition, kinematics and dynamics complexity increases in proportion to the number of motors in the joints. This raises some problems for optimum control. The need for effective use of the robotic arm integrated into the UAV also determines the lower limit of the UAV's avionic system requirements. Material types that can be used for the production of the designed robotic arm were also investigated and methacrylate monomers, which can be used in 3D printers, were preferred. Based on the defined geometrical parameters, robustness analysis was also carried out by considering the weights that the robotic arm will carry according to the mission objectives. In this context, deflection, bending moment and shear force amounts were calculated along the robotic arm and outputs compatible with the examples in the literature were obtained. Considering the maximum load it can carry, it has been observed that the amount of flexibility on the robotic arm is tolerable.

References

1. Varbla, S., Puust, R., Ellmann, A.: Accuracy assessment of RTK-GNSS equipped UAV conducted as-built surveys for construction site modeling. Surv. Rev. **53**(381), 477–492 (2020)
2. Alvarez-Vanhard, E., Corpetti, T., Houet, T.: UAV & satellite synergies for optical remote sensing applications: a literature review. Science of Remote Sensing **3**(100019), 1–14 (2021)
3. Alladi, T., Chamola, V., Kumar, N.: PARTH: a two stage lightweight mutual authentication protocol for UAV surveillance networks. Comput. Commun. **160**, 81–90 (2020)
4. Kerle, N., Nex, F., Gerke, M., Duarte, D., Vertivel, A.: UAV-based structural damage mapping: a review. Int. J. Geo-Information **9**(14), 1–23 (2020)
5. Li, X., Tan, J., Liu, A., Vijayakumar, P., Kumar, N., Alazab, M.: A novel UAV-enabled data collection scheme for intelligent transportation system through UAV speed control. IEEE Trans. Intell. Transp. Syst. **22**(4), 2100–2110 (2021)
6. Shahsavani, H.: An aeromagnetic survey carried out using a rotary-wing UAV equipped with a low-cost magneto-inductive sensor. Int. J. Remote Sens. **42**(23), 8805–8818 (2021)
7. Singh, J., Srinivasan, A.R., Neumann, G., Kucukyilmaz, A.: Haptic-guided teleoperation of a 7-DoF collaborative robot arm with an identical twin master. IEEE Trans. Haptics **13**(1), 246–252 (2020)
8. Barnett, J., Duke, M., Au, C.K., Lim, S.H.: Work distribution of multiple cartesian robot arms for kiwifruit harvesting. Comput. Electron. Agric. **169**(105202), 1–9 (2020)
9. Wuthier, D., Kominiak, D., Kanellakis, C., Andrikopoulos, G., Fumagalli, M., Schipper, G., Nikolakopoulos, G.: On the design, modeling and control of a novel compact aerial manipulator. Mediterranean Conference on Control and Automation, pp. 665–670. Malta (2016)
10. Lippiello, V., Ruggiero, F.: Cartesian impedance control of a UAV with a robotic arm. In: 10th IFAC Symposium on Robot Control, pp. 1–6. IFAC, Croatia (2012)
11. Pounds, P.E.I., Bersak, D.R., Dollar, A.M.: Grasping from the air: hovering capture and load stability. In: IEEE International Conference on Robotics and Automation, pp. 2491–2498. IEEE, China (2011)
12. Tognon, M., Chavez, H.A.T., Gasparin, E., Sable, Q., et al.: A truly-redundant aerial manipulator system with application to push-and-slide inspection in industrial plants. IEEE Robotics and Automation Letters **4**(2), 1846–1851 (2019)
13. Suarez, A., Soria, P.R., Heredia, G., Arrue, B.C., Ollero, A.: Anthropomorphic, compliant and lightweight dual arm system for aerial manipulation. IEEE/RSJ International Conference on Intelligent Robots and Systems, pp. 992–997. IEEE, Canada (2017)
14. Sarkisov, Y.S., Yashin, G.A., Tsykunov, E.V., Tsetserukou, D.: DroneGear: a novel robotic landing gear with embedded optical torque sensors for safe multicopter landing on an uneven surface. IEEE Robotics and Automation Letters **3**(3), 1912–1917 (2018)
15. Paul, H., Miyazaki, R., Ladig, R., Shimonomura, K.: TAMS: development of a multipurpose three-arm aerial manipulator system. Adv. Robot. **35**(1), 31–47 (2021)

16. Bellicoso, C.D., Buonocore, L.R., Lippiello, V., Siciliano, B.: Design, Modeling and control of a 5-DoD light-weight robot arm for aerial manipulation. 23rd Mediterranean Conference on Control and Automation, pp. 1–6. IEEE, Spain (2015)
17. Sideris, S.A., Tsakmakis, C.: Consistent Euler-Bernoulli beam theories in statics for classical and axplicit gradient elasticities. Composite Strctures **282**(115026), 1–13 (2022)

Reliability of MEMS Accelerometers Embedded in Smart Mobile Devices for Robotics Applications

Murat Bakirci[1]([⊠]) [iD] and Mecit Cetin[2] [iD]

[1] Faculty of Aeronautics and Astronautics, Tarsus University, 33400 Mersin, Turkey
`muratbakirci@tarsus.edu.tr`
[2] Transportation Research Institute, Old Dominion University, Norfolk, VA 23529, USA
`mcetin@odu.edu`

Abstract. This study focuses on assessing the reliability of data from the accelerometer sensors embedded in smart mobile devices that may potentially be used for robotics and intelligent transportation systems (ITS) applications. It is shown how bias and noise elimination can be executed more consistently from acceleration profiles obtained from an accelerometer with a high amount of error. In cases where accurate acceleration information could not be detected through noise filtering, averaged acceleration values in specific time windows were computed and introduced as measurement values to the filtering algorithm. Thus, more consistent acceleration profiles were obtained through making better state estimations. As an alternative to one dimensional bias elimination process, bias errors were detected in six different orientations in three dimensions and subtracted from raw readings. Furthermore, ratiometricity analysis, which is important in applications that require long-term data collection, but is generally overlooked, was also performed through collecting continuous acceleration data for twelve hours. Ratiometric error was numerically quantified and completely subtracted from the raw data through computation of slopes between specific error regions with linear variation assumption between these regions.

Keywords: Smart mobile device · Accelerometer · Unmanned ground vehicle · Intelligent transportation systems · Sensor fault

1 Introduction

Micro-electromechanical-systems (MEMS) technology is an advanced procedure to fabricate micro scale integrated devices or systems. Primarily, the MEMS technology produces integrated devices which consist of micro scale mechanical and electrical components. The primary technique to create such devices is the integrated circuit batch processing [1, 2]. These devices are capable to sense, identify and react at the microscale, and generate at the macroscale [3]. Unlike the current device electronics which fabricated through different integrated circuit technology, the MEMS device components are microfabricated via micromachinery which fabricates mechanical objects in the same manner

F. P. García Márquez et al. (Eds.): ICCIDA 2022, LNNS 643, pp. 78–90, 2023.
https://doi.org/10.1007/978-3-031-27099-4_7

as integrated circuits. Research fields that previously seemed to be unrelated, such as biology and microelectronics, has emerged by the advancements in MEMS technology. Systems that integrate MEMS with other applications can be found in aerospace, biotechnology, medicine, communication, inertial sensing, and defense applications [4–7]. More recently, intelligent transportation systems (ITS) community has begun utilization of accelerometers for a number of various applications [8–10].

Current research on mobile robotics, vehicular dynamics and ITS has helped develop platforms to analyze driving behavior. These platforms have been increasingly centered on smart mobile devices (SMD) since the current technology provides a wide range of built-in MEMS sensors [11]. Through these embedded sensors, particularly accelerometers, SMDs are now capable of providing key information, such as vehicle dynamics [12], travel mode detection, etc., for many applications [13]. These advancements in mobile device technology have made numerous research applications possible.

A SMD accelerometer can provide continuous information with high data rates which does not depend on any external sources. The accelerometer data can be processed and analyzed to infer vehicle dynamics and key vehicle activities [14]. Key information about the instant state of a moving vehicle can be obtained via examining variations in acceleration readings [15]. In [16], driver phone use is detected by smartphone accelerometers which could assist ITS applications for safety regulations. The proposed approach takes advantage of accelerometer and gyroscope sensors of smartphones to determine variations in rotational acceleration of the vehicle. Another study [17] estimates vehicle speed instantly utilizing the accelerometer embedded in a mobile device in conditions where the GPS signal is weak due to various external influences.

It is important to understand to what extent MEMS sensors embedded in SMDs, which have been used in a wide variety of applications in the past few years, are reliable and successful. To gain insights, an in-depth analysis of the accelerometer data is performed. Error sources of the accelerometer were determined by laboratory tests and error removal procedure was performed from raw sensor readings. By eliminating errors such as noise and sensitivity, the quality and reliability of the accelerometer output, which is of great importance for many ITS applications, is ensured. A significant increase in accelerometer performance has been achieved. In addition, the benefit of a SMD to ITS applications with the help of an autonomous mobile robot was investigated with some performance tests.

The remainder of the article is organized as follows: An overview of the operating principle and theoretical model of accelerometers are given in the next section. In the third section, general errors in accelerometer readings and elimination techniques are discussed. An evaluation of the proposed error elimination methods through various performance tests is examined in the fourth section. In the fifth, which is the last section, the implications of the study are discussed.

2 Theoretical Model

The principle of an inertial MEMS accelerometer can be represented by a translational electro-mechanical system with a proof mass m subjected to an external force F as shown in Fig. 1. According to Newton's second law of motion:

$$d/dt(mv) = \sum F \tag{1}$$

where m and v represents mass and speed of the object, and F is the total force acting on the object. In case of a constant proof mass, the above equation is also expressed as

$$mdv/dt = \sum F \tag{2}$$

The displacement, x, is measured in proportion to a known reference point, which is generally the equilibrium position of the proof mass. To satisfy (1) and (2), the momentum and acceleration quantities must be determined through setting an inertial frame of reference. Although the momentum, acceleration, and force are vector quantities, it is assumed that the proof mass is obliged to one dimensional motion, thus, scalar equations can be written [18].

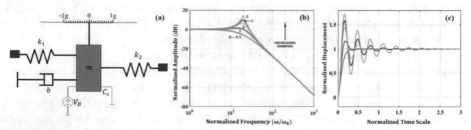

Fig. 1. Canonical 2nd order electromechanical mass/spring/dashpot system: (a) Free-body diagram; (b) Plot of the corresponding transfer function; (c) Step response of the system.

Equation (2) can be used since the proof mass is assumed to be a constant mass that does not change over time and relativistic effects are neglected. Therefore, it is possible to represent the proof mass with a simple algebraic equation which relates the acceleration and the applied force. To satisfy (2), the signs of acceleration and applied force need to be equivalent because the velocity boost will be in the same direction with the applied force. Friction elements, such as the dashpot with a damping coefficient b shown in Fig. 1, can be used to model forces that are parameters of the relative velocity between objects. Figure 2a depicts the element law for damping that varies linearly with velocity [19]. The slope of the line corresponds to damping coefficient. There are several sources for damping; however, air damping is the dominant factor for inertial sensors. Considering the proof mass moving under an applied force, the frictional force arises in the opposite direction of the motion. In case of a dashpot that has no mass, a force is generated that decelerates the external force applied on the opposite side of the mass. Therefore, the external force is transferred through the proof mass and is applied directly on the dashpot. The power dissipated by friction can then be modeled through relating the applied force with velocity of the mass.

Stiffness, which is the measure of flexibility of an object, is another important element for mechanical systems. Objects that experience an applied force, tend to resist deformation and the rate of resistance is defined by stiffness. A spring that stores mechanical energy is considered as a typical stiffness element. If the length of a spring is l_0 at the point when the spring is at rest as depicted in Fig. 1, the total length of the spring $l(t)$, is given by $l(t) = l_0 + x$, where x is the elongation of the spring resulted by the applied force. Unlike the dashpot, the algebraic relationship between x and F is nonlinear due to stiffness property as shown in Fig. 2b [18, 20]. Therefore, calculating the stiffness coefficient is rather tedious unlike calculating the damping coefficient.

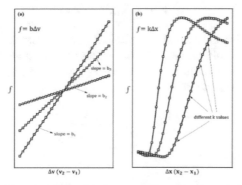

Fig. 2. Damping (a) and stiffness (b) characteristics of various dashpots and springs.

In case of a spring without mass, similar to a dashpot, a force is generated that decelerates the external force exerted on the opposite side of the mass. Therefore, the external force is transferred over the proof mass and applied directly to the spring. The potential energy stored in a stretched or compressed spring can also be rebounded to the system. Therefore, in order to analysis the complete reaction of a system, initial elongation of the spring needs to be known. A capacitor, $C(x)$, is attached to the proof mass in order to detect the displacement of the proof mass by turning mechanical alteration into electrical signal. Thus, the change in charge, $\Delta Q(x)$, stored in the capacitor due to the displacement of the proof mass is given as

$$\Delta Q = \frac{\partial C(x)}{\partial x} V_b \Delta x \tag{3}$$

where V_b is the fixed bias voltage. Considering the displacement Δx as the system's input, the expression above indicates the transformation of mechanical input into potential energy stored in the capacitor. Since this physical process is reversible, the capacitor $C(x)$ can also be used to impose an electrostatic force on the proof mass. This electrostatic force will be proportional to the stored potential energy on the capacitor. Thus, similar to the relationship between ΔQ and Δx, the expression that associates the change in force ΔF to the change in voltage Δv on the capacitor is given as

$$\Delta F = \frac{\partial C(x)}{\partial x} V_b \Delta v \tag{4}$$

Furthermore, it is possible to mention an implicit relationship between the capacitor and the electrostatic force depending on the relative displacement of the proof mass. The influence of this electrostatic force on the system can simply be modeled for small displacements by an equivalent spring constant, k_e, as shown in Fig. 1. By obtaining the element laws for the mass, spring, dashpot and capacitor and applying Newton's second law, it can be shown that the electromechanical system shown in Fig. 1 is a second order translational mechanical system (5).

$$m\ddot{x} + b\dot{x} + k_1 x + k_2 x = F \tag{5}$$

where m is the mass, b is the damping coefficient, k_1 and k_2 are the stiffness coefficients, and x is the relative displacement. The linear form of the state-variable equation can be obtained and corresponding transfer function (6) which relates system output to the input can be plotted as shown in Fig. 1b.

$$\frac{X(s)}{F(s)} = \frac{1}{s^2 + \frac{\omega_0}{Q}s + \omega_0^2} \tag{6}$$

where

$$\omega_0 = \sqrt{\frac{k_1 + k_2}{m}} \text{ and } Q = \frac{\sqrt{(k_1 + k_2)m}}{b} \tag{7}$$

The input to the electromechanical system in Fig. 1 can also be defined as a step function. In this case, the force F in the state-variable equation of the system is changed and the corresponding step response graph of the electromechanical system can be plotted for the same coefficients as shown in Fig. 1c.

3   Sensor Fault Detection and Elimination

It is a well-known fact that accelerometers are noisy and readings fluctuate over time. Therefore, raw sensor readings must be filtered to remove this noise and obtain smoother output. In the process of filtering, estimation of the current state through measured data is the primary goal to be accomplished. The filtering density $p(\alpha|\beta)$ needs to be computed for each of the measured data and this can be done through recursion which follows from Bayes' rule [21]

$$p(\alpha_\tau|\beta_{1:\tau}) = \frac{p(\beta_\tau|\alpha_\tau)p(\alpha_\tau|\beta_{1:\tau-1})}{p(\beta_\tau|\beta_{1:\tau-1})} \tag{8}$$

$$p(\beta_\tau|\beta_{1:\tau-1}) = \int p(\beta_\tau|\alpha_\tau)p(\alpha_\tau|\beta_{1:\tau-1})d\alpha_\tau \tag{9}$$

$$p(\alpha_\tau|\beta_{1:\tau-1}) = \int p(\alpha_\tau|\alpha_{\tau-1})p(\alpha_{\tau-1}|\beta_{1:\tau-1})d\alpha_{\tau-1} \tag{10}$$

where α_τ is the state at time τ and $\beta_{1:\tau} = \{\beta_1, \beta_2, \ldots, \beta_\tau\}$ contains the entire set of the measured data. The computation is theoretically straightforward, however an analytical solution is not available due to the multidimensionality of the integrals in (9) and (10). On the other hand, an approach to a solution can be made with filtering techniques such as extended Kalman filter and particle filter. The acceleration data corresponding to the constant velocity motion obtained from the accelerometer of a relatively high quality SMD are shown in Fig. 3. The figure depicts an example of the accelerometer readings before and after filtration. It can clearly be seen that the sensor readings are noticeably smoother allowing for easier analysis.

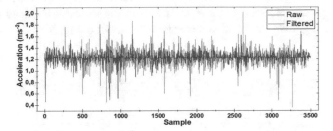

Fig. 3. Accelerometer signal before and after filtration.

Figure 4 shows the acceleration data collected by SMD of the unmanned ground vehicle (UGV), which started with a low initial velocity while at rest. It is clearly seen that the new acceleration profile obtained as a result of filtering in the first 15 s of motion does not consistently represent the real acceleration values of the UGV. Due to the high amount of noise and bias, the movement of the UGV cannot be detected correctly in the first seconds. In this and similar cases, before filtering the collected acceleration data, moving average values can be viewed by averaging the acceleration values in certain time windows. These values provide a more consistent result as they give better predictions in short intervals. For example, when the Kalman filter is used for filtering, it will make more consistent predictions in each different time interval, as the average values computed over different time windows are considered measurements. The filtration through this method is shown in blue in Fig. 4. When the time window value is selected as 8, the averages in these intervals are updated as the measurement value and fed into the Kalman filter, it is seen that a better state estimation result is obtained. It can be seen from the figure again that the new acceleration profile obtained in the first two time windows predicts the movement of the UGV much better.

Accelerometer readings consist of a small additional offset in the average signal for each component, called the bias. If there were no bias error, the sensor reading would indicate an offset of $0g$ on the x-axis when the sensor is placed on a perfectly horizontal surface. From the same theoretical aspect, the y and z components of the sensor would read $0g$ and $+1g$, respectively. However, the physical properties of MEMS sensors change over time, resulting in sensor readings that differ from these ideal outputs. The

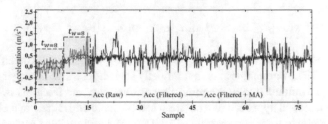

Fig. 4. Raw acceleration data and two different filtering results.

bias removal methodology starts with placing the accelerometer module on a flat surface. Considering a two dimensional accelerometer, the x and y components of the acceleration are obtained while the sensor is standing on a flat surface. Since the device is at rest, the sensor readings are expected to be zero. However, the readings obtained are different from zero, representing the accelerometer bias as shown in Fig. 5.

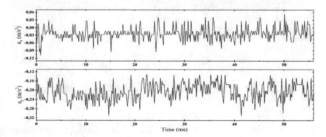

Fig. 5. x and y components of the bias values read by the accelerometer.

Accelerometer sensitivity is typically a measure of the sensor's output for a given input. Ideally, this ratio would be expected to be 1, but the sensitivity value often differs from 1 due to internal errors. To eliminate bias and sensitivity errors simultaneously, an experimental method like the one below should be executed [22]. The accelerometer module, in this case a smartphone, is placed on an approximately horizontal surface. The module is then rotated $90°$ to each of the six positions as shown in Fig. 6.

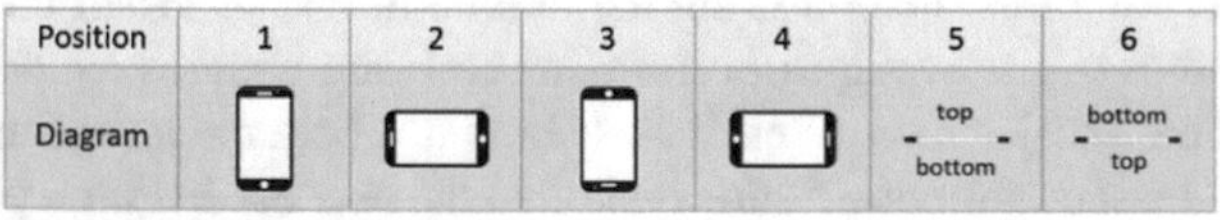

Position	1	2	3	4	5	6
Diagram	▯	▭	▯	▭	top / bottom	bottom / top

Fig. 6. Rotating a sensor module (smartphone) to six different positions on a flat surface.

Sensor readings are recorded through these positions, and bias and sensitivity errors can then be computed as follows:

$$\sigma_{xx} = (a_{x2} - a_{x4})/2 \quad \sigma_{yy} = (a_{y1} - a_{y3})/2 \quad \sigma_{zz} = (a_{z5} - a_{z6})/2 \tag{11}$$

$$[b_x]_{0g} = (a_{x1} + a_{x3} + a_{x5} + a_{x6})/4 \tag{12a}$$

$$[b_y]_{0g} = (a_{y1} + a_{y4} + a_{y5} + a_{y6})/4 \tag{12b}$$

$$[b_z]_{0g} = (a_{z1} + a_{z2} + a_{z3} + a_{z4})/4 \tag{12c}$$

where the first subscript indicates the sensor component while the second subscript indicates the position of the sensor. After recording these bias and sensitivity values, they are used in all subsequent calculations to obtain corrected acceleration values as:

$$\hat{a}_i = (a_{ij} - [b_i]_{0g})/\sigma_{ii} \tag{13}$$

where a_{ij} indicates raw acceleration values while $\hat{a}_i$ indicates corrected acceleration values. Since the smartphone is placed on a flat surface with no movement in any direction, the expected readings for the x, y, and z components of the acceleration are $0g$, $0g$, and $+1g$, , respectively. To compare the outcome of the corrected acceleration values with these actual values, the root-mean-square error (RMSE) for the three components was computed. The RMSE of 0.017, 0.031 and 0.021 was computed as an indicator for the bias error rate between actual and corrected bias profiles. These low RMSE values prove that the experimental method applied gives satisfactory results.

The output signal of MEMS accelerometers is also sensitive to alterations in power supplies. This phenomenon is known as ratiometricity, and most MEMS accelerometers are naturally ratiometric. Therefore, any fluctuation in the supply voltage will directly cause a variation in the sensor output signal. In the case of a smartphone, a variation in accelerometer output is observed as the smartphone's battery drains over time. A laboratory test can be conducted to identify and eliminate this ratiometric offset error. For most smartphones, the error can be computed as follows when operating at a nominal supply voltage of 3.8 V

$$e_r = \left(\frac{O_{V_s}}{O_{3.8V}} - \frac{V_s}{3.8V} \right) \cdot 100 \tag{14}$$

where O_{V_s} indicates the amount of offset under the supply voltage. A laboratory test was conducted for a smartphone's accelerometer to examine the variation of the output signal with respect to the supply voltage. The battery of the smartphone is replaced by a constant power supply. Voltage sweeps are performed gradually as acceleration data is collected. As depicted in Fig. 7, it turns out that the $0g$ offset values change with gradually decreasing supply voltage as expected. The linear variation of the $0g$ offset values provides an easier process for error rectification. The slope of the fitted line gives an offset of approximately $0g$, which must be compensated.

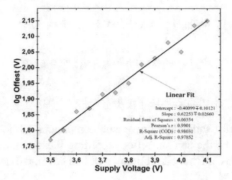

Fig. 7. Ratiometric voltage plot showing variation of 0g offset by supply voltage.

4 Performance Tests

Several performance tests were conducted to further investigate suitability of a SMD's accelerometer for robotics and ITS applications. In the first test, the SMD's accelerometer was compared to an advanced MEMS accelerometer embedded in the microcontroller board of an unmanned ground vehicle (UGV). The test UGV platform is configured as a differential drive robot, and the control board is equipped with the necessary electronics to drive the UGV. It is greatly enhanced with multiple sensors, including an accelerometer, and a total of three embedded microprocessors that sample sensors and can control various motors. The interface of the UGV is the Multi-Platform Arduino-compatible Integrated Development Environment (MPIDE).

In this test, the SMD is attached to the test UGV and it is driven in a straight line for about 2.5 min. The acceleration data were collected from both the accelerometer of the SMD and the accelerometer of the UGV during the test. Since the accelerometer of the UGV is of much higher quality, it was accepted as a reference by assuming that it would output very close to the actual acceleration values. Figure 8 shows the y-component (forward direction) of the acceleration readings from both devices. The raw acceleration output from the SMD is shown in green and has quite high boise as expected. This fact can be deduced from the random and high variation of the acceleration values along the vertical axis. Moreover, the vertical shift of the acceleration profile indicates the bias error. That is, the acceleration values differ significantly from zero in the first three seconds, when the UGV has not started its movement yet. The acceleration profile, expressed in blue, is the output obtained from the UGV's accelerometer. As the figure indicates, the acceleration profile is highly consistent and contains relatively low noise and bias. The acceleration profile obtained after bias and noise elimination through proper filtering and bias removal method presented in this trial is shown in red. As can be seen from the figure, the bias and noise errors were eliminated to a great extent and the resulting profile was highly similar to the UGV's output. This acceleration profile, which is largely free from errors, will be more useful in robotics and ITS applications and will allow more accurate analysis and evaluation.

In the second performance test, it was investigated whether the UGV's travel mode could be determined using the SMD's accelerometer. As in the previous test, the SMD

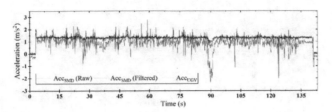

Fig. 8. Y-component (forward direction) of the acceleration output from UGV and SMD.

was fixed to the UGV with their forward directions overlap. The UGV, which was programmed to stop for a certain period of time to perform bias analysis, was given a rapid forward motion command and allowed to move at a constant speed. Raw (blue) and processed (red) acceleration profiles obtained from the SMD accelerometer are shown in Fig. 9. The raw output clearly shows that in the first 265 ms, the full stop mode of the UGV is quite difficult to detect due to the high bias error. Conversely, it can also be misconcluded that the UGV travels in the opposite direction due to negative acceleration values. However, when the errors in the acceleration profile are properly corrected (red profile), full stop, acceleration, and constant velocity motion modes can be easily detected.

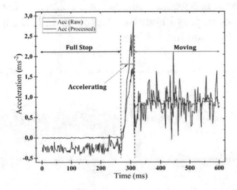

Fig. 9. Travel mode detection via acceleration data.

In previous performance tests, only bias and noise error corrections were made, as the driving tests were not long enough for ratiometric error to occur. In order to observe how the ratiometric error affects the accelerometer output, the SMD was fixed on a flat surface and acceleration data were collected continuously for approximately 12 h. During this time, other applications were left open in the background to drain the SMD's battery faster. The test result is shown in Fig. 10.

The results show a slight change in the raw acceleration data after 12 h. When the average of the first 500 data points is compared to the average of the last 500 data points, there is a decrease of approximately 11.5% from the initial acceleration values. While no change is observed in the first 4.5 h, the decrease in acceleration values begins to be evident after this period. In particular, a rapid decrease is observed in the acceleration

88   M. Bakirci and M. Cetin

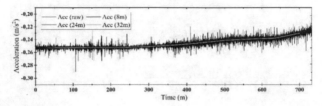

Fig. 10. Ratiometric error observed in acceleration data collected over 12 h.

values collected in the last 2 h, when the battery of the SMD drops to 20 percent and continues to decrease.

In order to better visualize and quantify this change in acceleration values, moving averages at different time windows (t_{ma} = 8m, 24m, 32m) were also computed and shown in Fig. 10. For the time window values of t_{ma} = 8m and t_{ma} = 32m, the moving averages represent the average acceleration values quite well. One of these values can be used to eliminate this observed ratiometric error. It would be more logical to choose t_{ma} = 32m in order to make the computation cheap. By finding the slopes of the lines connecting the computed consecutive moving average values, the ratiometric error increasing over time is corrected as shown in Fig. 11. An acceleration error of 0.038m/s^2 may seem trivial, given that the average acceleration value corresponding to the average vehicle speed is between 0.8m/s^2 and 1.2m/s^2. However, this small error of ratiometric origin can be very important, especially in ITS applications that require high precision such as autonomous vehicles.

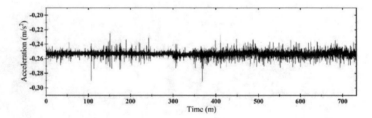

Fig. 11. Acceleration data free of ratiometric error.

5 Conclusions

New approaches and laboratory tests are presented to improve reliability of acceleration data obtained from an accelerometer embedded in a smart mobile device (SMD). It was observed that the data collected with SMD had higher bias and noise error as expected when compared to the data collected from an advanced accelerometer of an unmanned ground vehicle. It is observed that although the noise is significantly reduced with the current filtering methods, it does not produce consistent results in every application. The filtering algorithm is enhanced by the local mean values obtained in specific time windows, resulting in more accurate state estimation, and thus, consistent filtering.

Additionally, it is also seen that these values, obtained by defining proper time windows, represent the entire data to a great extent, and thus, it is concluded that an accurate noise analysis can be made with a cheaper computation through these values. Moreover, it is also observed that a more effective bias elimination can be achieved than one dimensional bias removal methods by performing a three-dimensional bias analysis of the SMD's accelerometer. Furthermore, ratiometric error is detected in applications that require long-term data collection. Although the inferred error is somewhat small, it may need to be considered depending on the weight of robotic or autonomous vehicle to be used. It is shown that this error can be eliminated numerically through clustering data points where the ratiometric error becomes evident. Finally, it should be pointed out that the driving modes of mobile robots or autonomous vehicles can be determined with high accuracy by proper processing of the acceleration data obtained from the SMD.

References

1. Tsuchiya, T., Hemmi, T., Suzuki, J., Hirai, Y., Tabata, O.: Tensile strength of silicon nanowires batch-fabricated into electrostatic MEMS testing device. Appl. Sci. **8**(6), 880–891 (2018)
2. Hafner, J., Teuschel, M., Schrattenholzer, J., Schneider, M., Schmid, U.: Optimized batch process for organic MEMS devices. Proceedings **2**(904), 1–4 (2018)
3. Zhou, G., Lim, Z.H., Qi, Y., Chau, F.S., Zhou, G.: MEMS gratings and their applications. Int. J. Optomechatronics **15**(1), 61–86 (2021)
4. Gad-el-Hak, M.: The MEMS Handbook, 2nd edn. CRC Press, Florida (2002)
5. Ejeian, F., et al.: Design and applications of MEMS flow sensors: a review. Sens. Actuators, A **295**, 483–502 (2019)
6. Ren, Z., Chang, Y., Ma, Y., Shih, K., Dong, B., Lee, C.: Leveraging of MEMS technologies for optical metamaterials applications. Advanced Optical Materials **8**(1900653), 1–20 (2020)
7. Blachowicz, T., Ehrmann, A.: 3D printed MEMS technology-recent developments and applications. Micromachines **11**(4), 1–14 (2020)
8. Liang, X., Zhang, Y., Wang, G., Xu, S.: A deep learning model for transportation mode detection based on smartphone sensing data. IEEE Trans. Intell. Transp. Syst. **21**(12), 5223–5235 (2020)
9. Alkinani, M.H., Almazroi, A.A., Adhikari, M., Menon, V.G.: Design and analysis of logistic agent-based swarm-neural network for intelligent transportation system. Alex. Eng. J. **61**(10), 8325–8334 (2022)
10. Sysoev, A., Khabıbullina, E., Kadasev, D., Voronin, N.: Heteregeneous data aggression schemes to determine traffic flow parameters in regional intelligent transportation systems. Transportation Research Procedia **45**, 507–513 (2020)
11. Guevara, L., Cheein, F.A.: The role of 5G technologies-challenges in smart cities and intelligent transportation systems. Sustainability **12**(16), 1–15 (2020)
12. Ustun, I., Cetin, M.: Speed estimation using smartphone accelerometer data. Transp. Res. Rec. **2673**(3), 65–73 (2019)
13. Carlos, M.R., Aragon, M.E., Gonzales, L.C., Escalante, H.J., Martinez, F.: Evaluation of detection approaches for road anomalies based on accelerometer readings-addressing who's who. IEEE Trans. Intell. Transp. Syst. **19**(10), 3334–3343 (2018)
14. Ahmed, U., Sahin, O., Cetin, M.: Minimizing GPS dependency for a vehicle's trajectory identification by utilizing data from smartphone inertial sensors and OBD device. Transp. Res. Rec. **2644**(1), 55–63 (2017)
15. Villanueva, J.C., Zapata, D.C., Saenz, F.T., Vela, M.V., Skarmeta, A.F.: Vehicle maneuver detection with accelerometer based classification. Sensors **16**(1618), 1–23 (2016)

16. Sanchez, S.H., Pozo, R.F., Gomez, L.A.H.: Driver identification and verification from smart-phone accelerometers using deep neural networks. IEEE Trans. Intell. Transp. Syst. **23**(1), 97–109 (2020)
17. Bo, C., Jung, T., Mao, X., Li, Y., Wang, Y.: Smartloc-sensing landmarks silently for smartphone-based metropolitan localization. EURASIP J. Wirel. Commun. Netw. **1**(111), 13–22 (2016)
18. Won, D.-J., Lee, S., Kim, J.: Analysis of liquid-type proof mass under oscillating conditions. Micro and Nano Systems Letters **8**(1), 1–7 (2020)
19. Close, C.M., Frederick, D.K., Newell, J.C.: Modeling and Analysis of Dynamic Systems, 3rd edn. John Wiley & Sons Inc., New Jersey (2002)
20. Zeng, X., Zhang, L., Yu, Y., Shi, M., Zhou, J.: The stiffness and damping characteristics of a dual-chamber air spring device applied to motion suppression of marine structures. Appl. Sci. **6**(74), 1–20 (2016)
21. Kim, D.Y., Jeong, Y.S., Kim, S.: Data filtering to avoid total data distortion in IoT networking. Symmetry **9**(16), 1–13 (2017)
22. You, J.C., Yang, P.X., Qin, Y.Y., Yan, G.M.: Modeling and calibration of the accelerometer size effect error of the SINS. Journal of Astronautics **33**(3), 311–317 (2012)

HST-Detector: A Multimodal Deep Learning System for Twitter Spam Detection

Insaf Kraidia[✉] ⓘ, Afifa Ghenai ⓘ, and Nadia Zeghib

LIRE Laboratory, Constantine2 –, Abdelhamid Mehri University, Constantine, Algeria
{insaf.kraidia,afifa.ghenai,nadia.zeghib}@univ-constantine2.dz

Abstract. Twitter is one of the most famous microblogging and Online Social Networks (OSNs). Nowadays, the unimodal spam filtering systems in Twitter achieved a good detection rate. To bypass the existing systems and reduce their recognition rate, attackers inject spam information into various portions of tweets and fuse them. Therefore, a powerful Heterogeneous Spam Tweets (HST) filtering system is necessary to fully protect users. To address the above challenge, we propose a Deep Learning (DL) system combining a Convolutional Neural Network model (CNN) and a Long Short-Term Memory model (LSTM) to classify heterogeneous malicious tweets. To ensure the efficient filtering of spam hidden in text or images, we apply a multimodal combination approach in which LSTM and CNN models deal respectively the text and image portion of heterogeneous tweets to obtain two predictions. Afterward, a fusion model is used to decide if the tweet is malicious or legitimate. This system is more efficient than the existing spam filtering systems trained on Image Spam Hunter (ISH), Dredze, and Honeypot datasets. The proposed system can achieve accuracies of 98% on heterogeneous tweet datasets.

Keywords: Deep learning · CNN · LSTM · Twitter spam detection · OSNs · Heterogeneous tweet

1 Introduction

Twitter is an Online Social Network (OSN), and a very important part of many people's daily lives. Users share many types of content in this network such as personal problems, events, and news. The amount of information shared in this network has continued to increase over the past few years. As a result, Twitter is one of the most famous microblogging, wherein people post text pieces known as "tweets" with 280 characters. According to some reports [1], Twitter has over 396.5 million users per month and posts over 560 million tweets per day.

Unfortunately, the high utilization of Twitter makes it an attractive platform for threat actors to execute many attacks. Nowadays, Twitter phishing is a favorite method used by social engineering criminals to deceive their victims, as it makes their work very simple. Many alternatives can be found for this kind of attack, such as sending spam Twitter messages or posting spam tweets. As a result, several Deep Learning (DL) and

© The Author(s), under exclusive license to Springer Nature Switzerland AG 2023
F. P. García Márquez et al. (Eds.): ICCIDA 2022, LNNS 643, pp. 91–103, 2023.
https://doi.org/10.1007/978-3-031-27099-4_8

Machine learning (ML) phishing detectors are developed to classify users and/or tweets as spam. These detectors use account-based or content-based features, such as the age of an account, number of followers, and number of tweets. Nevertheless, phishing via tweet attacks is increasingly becoming multimodal and includes multimedia with embedded text to evade text-based systems. This heterogenous spam tweet deceives people to open some malicious websites, causes malware infection, or includes some bad information. In Fig. 1, we display three attack scenarios examples via tweets. (a) contains a simple malicious tweet and (b) contains a spam tweet with legitimate images and malicious text. Whereas, (c) contains a spam tweet with malicious images and legitimate text.

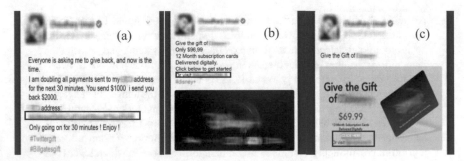

Fig. 1. Attack scenarios via Twitter

The effectiveness of unimodal filtering systems on simple tweets does not necessarily implicate its success on heterogeneous tweets. Due to the presence of multimedia content in the tweet, the existing approaches should be improved and their effectiveness increased by using the new data modality. Since there are no heterogeneous tweet filtering systems [2] in the literature, it is very important to throw light and understand how to adequately filter malicious heterogenous tweets by analyzing the behavior of the multimedia (specifical images) present in the tweet.

For image classification tasks, the Convolutional Neural Network (CNN) has been widely utilized and has higher accuracy compared to the other methods. For text classification tasks, Long Short-Term Memory (LSTM) is increasingly utilized than other approaches because of its memory capacities and time properties in Natural Language Processing (NLP) [3]. This is why we choose the LSTM and CNN approaches to deal with the multi-variety portions of heterogeneous tweets. Keeping the research line, this paper proposes an architecture to detect heterogeneous spam tweets using a multimodal Deep Learning approach. The image and text filtering models are proposed to create feature vectors from tweet images and text. These feature vectors are inputted into the heterogeneous filtering model to classify tweets. The performance of our system is verified based on recall, precision, accuracy, and f1-score. The rest of the paper is organized as follows: Sect. 2 shows an overview of related work and literature. Section 3 gives a detailed description of our contribution. Section 4 explains the experimental evaluation of our system. Section 5 displays our conclusion and future work.

2   Related Work

For a decade the phishing problem has been studied and the literature shows various studies using content-based features such as images or text.

In the image-based studies, for the identification of spam images, the authors in [4] applied the Support Vector Machine (SVM) and the Principle Component Analysis (PCA) to extract the Eigenspace of the image as a feature and used it to classify the image as spam or not. The proposed method has been evaluated using an image-based dataset called the "Image Spam Hunter" (ISH) dataset [5]. In [6] an image spam classifier using CNN, SVM, and a Multilayer Perceptron (MLP) is proposed. The CNN classifier includes 3 convolutional layers, 3 max-pooling layers, and a dropout unit. For MLP and SVM, a Canny Edge Detector was used to extract efficient edge information. The experiment showed promising results on the improved ISH dataset and challenged dataset. In [3], the authors proposed a DL approach that combines CNN and LSTM algorithms. The evaluation of this system shows a rate of 92.1% on Personal Image, Enron, and ISH datasets. [7] offered an approach based on Transfer Learning and CNN. For semantic feature extraction and classification, the authors used the ResNet and Feedforward Neural Network. Experimental results have shown an accuracy of 95%.

For the text-based studies, many Deep Learning techniques have been proposed and shown a good performance. [8] proposed a system based on DL approaches to filtering malicious accounts. The authors offered two models (text and combined classifiers) based on tweets text and user metadata. Their experiment showed promising results for two datasets. [9] proposed a Deep Learning system based on the Bidirectional Long Short-Term Memory (Bi-LSTM) method and Deep Neural Network (DNN). To represent tweets, the authors extract features from the system's hidden layers. In [10], the authors proposed a system based on DL that uses an attention-based approach to classify the differences in user tweets. Experimental results have shown an accuracy of 98.60%.

We observe that the existing studies in the filtering domain on Twitter only exploit text or images. In view of this situation, it is necessary to propose a system that overcomes the multimodality issue in heterogenous tweets classification.

3   The Proposed System

As mentioned in Sect. 2, the tweet may include spam in all data or only in some of them. So, it is indispensable to exploit a technique based on the multimodalities of data to improve the effectiveness of the tweet filtering systems. Therefore, we offer a system named Heterogeneous Spam Tweets Detector (HST-Detector) that utilizes multi-modal Deep Learning approaches which are CNN and LSTM to detect heterogeneous spam tweets as shown in Fig. 2.

The practical steps of the HST-Detector are described as follows:

- Preprocessing (see Sect. 3.1).
- Text and image datasets are utilized to train LSTM and CNN models.
- To get the predicted value of tweet image as malicious, we have re-entered the image dataset in the trained CNN Model.

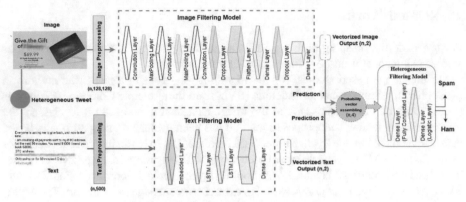

Fig. 2. HST-Detector architecture

- To get the predicted value of tweet text as malicious, we have re-entered the text dataset in the trained LSTM Model.
- The Heterogeneous Filtering Model can fuse the prediction results of all secondary models to obtain the final prediction.

After the presentation of the overall architecture of the HST-Detector, we will describe the data preprocessing and the structure of the Image Filtering Model, Text Filtering Model, and Heterogeneous Filtering Model.

3.1 Data Preprocessing

In this subsection, the preprocessing techniques used in text and image data as an initial step in the HST-Detector are presented.

- **Image preprocessing:** This consists of changing the image dimensions to a size of *(n,128,128)* pixels due to the image's variant size.
- **Text preprocessing:** To obtain the LSTM inputs *(n,500)*, we apply the following steps:
- *Case Folding*: This is the upper to lowercase converting letter process.
- *Tokenization*: This is the process of dividing sentences into words and deleting delimiters such as spaces, dots (.), numeric characters, and commas (,).
- *Stemming*: Consists of deleting suffixes from words to get a word stem. In our case, we use Porter Stemmer from the NLTK library in Python.
- *Lemmatization*: This is the algorithmic process of grouping the various inflected forms of a word depending on its meaning and context so they can be studied as a single item.
- Stop word Removal: This is the process of deleting the words that occur commonly or useless features in text documents.

3.2   Image Filtering Model

In recent, after Alex Net's distinction in ImageNet, the effectiveness of CNNs has been approved in computer vision tasks and become the first choice for spam image classification [11].

The CNN model has three convolution layers, which contain 32, 64, and 128 filters respectively. The first 2 convolutional layers are followed by the max-pooling layer. After that, dropout regularization is utilized and the flatten technology is adopted to get a one-dimensional vector result. The result is then inputted into a dense layer with a ReLU activation function and a dropout regularization. In the end, a dense layer is utilized with a sigmoid function. Figure 3 shows the CNN model output shapes and architecture.

Fig. 3. CNN architecture

In [12] the authors explained carefully the details of the CNN approach. Figure 4 gives the overall description of the image classification algorithm which inputs the image and returns the prediction r, $r \in R^{1 \times 2}$ of the image.

Input: Image *img*, size 128 × 128 RGB
Output: Image spam classification probability value r
1   Enter *img* to the convolutional layers, obtaining the results g, $g = (g_1, ..., g_{128})$;
2   Enter g to the FC layers, which contain 64 and 32 neurons, obtaining the results c, $c = (c_1, c_2, ..., c_{128})$;
3   Enter c to FC layer, and use a sigmoid activation function to obtain classification probability value r;
4   return r;

Fig. 4. Image spam classification algorithm

3.3   Text Filtering Model

To solve Recurrent Neural Network (RNN) issues in remembering long-term memories, LSTM is capable of handling the information in memory for a good period as compared to RNN. To remember or forget data, LSTM adds cell states that include cell gates with 4 parts: forget gate, memory cell, output gate, and input gate, as illustrated in Fig. 5.

Our LSTM model is formed of a word embedding layer followed by one dense layer and two LSTM layers. The steps of processing the text part of a tweet are as follows:

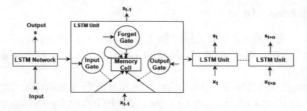

Fig. 5. LSTM architecture

Firstly, to attain the word vector representation, we utilize the word embedding method. Next, we capture features from the text part by LSTM layers. Finally, to obtain the last prediction values, we create a dense layer with a softmax activation function.

The algorithm details of the LSTM model are given in Fig. 6. The LSTM unit concatenates the output of the last unit with the provided text parts. With each inserted sentence, this procedure is frequent where the essential and significant features are kept by the LSTM units. The proposed model returns the prediction values r, $r \in R^{1 \times 2}$ of the text from the dense and LSTM layers. The LSTM algorithm details are described in [13].

	Input: Text *txt*
	Output: Text spam classification probability value r
1	Enter *txt* to a toolkit named "word2vec" to obtain the word vector k, $k = (k_1,.., k_n)$;
2	Enter k to the LSTM layer, to get the text feature vector $h = (h_1,..,h_{64})$;
3	Input h to the second LSTM layer and get better abstract text feature vector s, $s = (s_1, s_2,.., s_{32})$;
4	Input s to the last FC layer to obtain classification probability value r and use a SoftMax activation function;
5	return r.

Fig. 6. Text spam classification algorithm

3.4 Heterogeneous Filtering Model

In this system, we aim to gain better prediction results of the heterogeneous tweet. Accordingly, the heterogeneous filtering model is utilized to combine the prediction values of the tweet text part and the prediction values of the tweet image part. This model can be detailed in the following steps:

- To obtain the first format of the feature vector g, $g \in R^{1 \times 4}$, we fuse the predictions of the CNN model (z) and LSTM model (x), $z, x \in R^{1 \times 2}$.
- After concatenating both various predictions, a linear fully connected layer containing 64 filters is used to give a second comprehensive format of the feature vector.

- We input the last feature vector into a logistic layer with a logistic regression activation function to get the best accurate prediction value for the heterogenous tweet.
- For training the model, we choose as hyperparameters Adam optimizer with a batch size equal to 20, ReLU activation function, and early-stopping conditions to avoid overfitting.

The dataset classification probability inputted to the Heterogeneous Filtering Model is: $D = \{(q_1, y_1), (q_2, y_2), .., (q_v, y_v)\}$, $q_i \in R^{1 \times 4}$, $y_i \in \{0, 1\}$. In logistic regression function, the conditional probability distribution is given by the equations:

$$P(Y = 1|q) = \pi(q) = \frac{e^{-w^T \cdot q}}{1 + e^{-w^T \cdot \dot{q}}} \tag{1}$$

$$P(Y = 0|q) = 1 - \pi(q) = \frac{1}{1 + e^{-w^T \cdot \dot{q}}} \tag{2}$$

4   Implementation and Evaluation

This section shows the results of the different experiments, details the datasets used to train and assess the system, and displays the measures used for the evaluation and validation.

4.1   Experimental Setup

To develop and verify the proposed system, a free environment called Google Colaboratory is used. As software, we use Python 3 to build the system and TensorFlow and Keras libraries for the preprocessing steps.

4.2   Datasets

In our study, three public datasets are utilized and two datasets are built.

- **The Honeypot dataset [14]:** This dataset contains real data collected from Twitter. This data includes *two* types of classes: 2,353,473 malicious tweets and 3,259,693 legitimate tweets, which can precisely specify the effectiveness of our system. The honeypot dataset has six text files, just the two following files are used:

  Content_polluters_tweets: Contains malicious tweet information.
  Legitimate_tweets: Contains legitimate tweet information.

- **Image Spam Hunter (ISH dataset) [5]:** ISH is a real and public dataset that contains two types of classes which are spam and ham images. This dataset includes 926 spam images from real malicious content and 810 legitimate images that are randomly collected.

- **Dredze dataset:** The authors of [4] apply an image processing approach to spam images to make them appear more challenging. A weighted overlay method was used to blend these malicious images with the legitimate images from the ISH dataset.

Since there are no public heterogeneous tweet datasets that contain both text and image, we choose to combine the previous datasets (the fusion is not random) to get heterogeneous datasets. Table 1 shows the construction probabilities that we use based on [5] and [4] datasets as follows:

Table 1. Heterogeneous dataset probabilities

Image	Text	Final prediction
Malicious	Malicious	Malicious
Malicious	Legitimate	Malicious
Legitimate	Malicious	Malicious
Legitimate	Legitimate	legitimate

- **Heterogeneous Dataset 1 (Honeypot + ISH dataset):** This dataset contains 5,613,166 heterogeneous tweets (5,613,166 tweet text and 1,731 tweet images are formed into 5,613,166 heterogeneous tweets) in which 2,613,166 are spam tweets and 3,000,000 are ham tweets. Table 2 shows some samples taken from this dataset.
- **Heterogeneous Dataset 2 (Honeypot + Dredze dataset):** This dataset contains 5,613,166 heterogeneous tweets (5,613,166 tweet text and 5,165 tweet images are formed into 5,613,166 heterogeneous tweets) in which 2,000,000 are spam tweets and 3,613,166 are ham tweets.

Table 2. Samples of heterogeneous dataset 1

Image post	Text post	Label
	"If I tell you about my past, the biggest slap to the face is you repeating it."	Ham
	"GET 3 BEATS FOR THE PRICE OF 1 BUY 1 BEAT GET 2 FREE CLICK HERE https://t.co/px7hGKbRoP AND DOWNLOAD https://t.co/iSGfpRilxd"	Spam

4.3 Evaluation Criteria and Method

Concerning quality measures, the most utilized in the previous studies were precision, F1 score, accuracy, and recall. These criteria are determined with the following Eqs. (3)–(6):

$$Accuracy = \frac{TP + TN}{TP + TN + FP + FN} \tag{3}$$

$$Precision = \frac{TP}{TP + FP} \tag{4}$$

$$Recall = \frac{TP}{TP + FN} \tag{5}$$

$$F1 - score = \frac{2 \times Precision \times Recall}{Precision + Recall} \tag{6}$$

The specific descriptions of TN, TP, FP, and FN are given as follows:

- True Positive (TP): This is the number of heterogeneous spam tweets that are correctly classified.
- True Negative (TN): This is the number of heterogeneous legitimate tweets that are correctly classified.
- False Negative (FN): Represent the number of misclassified heterogeneous spam tweets.
- False Positive (FP): This is the number of heterogeneous legitimate tweets that are misclassified.

4.4 Experimental Results

In the present section, we show the evaluation results on five datasets and provide some explanations and analyses.

The heterogeneous dataset 1 is split into two portions. The first portion consists of 20% for testing and the last portion consists of 80% for learning. Based on the confusion matrix of the HST-Detector described (see Table 3), the accurate classification is while the normal categories are predicted as ham. Also, the proposed system has identified spam tweets with low false-positive and false-negative rates.

Table 3. The prediction results of the confusion matrix utilizing heterogeneous dataset 1

	Prediction		Number of samples
	Legitimate	Malicious	
Legitimate	217 (TN)	3 (FP)	220
Malicious	3 (FN)	177 (TP)	180
Testing data			400

The models were trained using the range of hyperparameters cited in Table 4. The grid search optimization method was utilized to obtain the best hyperparameters. For the LSTM, the Adam optimization algorithm, batch sizes equal to 20, 100 epochs, and a learning rate of 0.1 give the optimal values. For CNN, the SGD optimization algorithm, batch sizes equal to 15, 100 epochs, and a learning rate of 0.01 give the optimal values.

Table 4. HST-Detector hyperparameters range

Model	Hyperparameter			
	Learning rate	Batch size	Epochs	Optimization algorithm
Text Filtering Model	[0.001, 0.01, **0.1**]	[5, 10, **20**]	[8, **100**]	[SGD, RMSprop, **Adam**]
Image Filtering Model	[0.001, **0.01**, 0.1]	[5, **15**, 20]	[8, **100**]	[**SGD**, RMSprop, Adam]

To evaluate the effectiveness of the HST-Detector, a set of performed models are utilized for comparison by different datasets. For the ISH and Dredze datasets, HST-Detector, [3, 4, 6], and [7] are utilized for comparisons. For the Honeypot dataset, HST-Detector, [8, 10], and [9] are utilized for comparisons. For the heterogeneous datasets 1 and 2, the results of the HST-Detector were displayed to evaluate its performance.

Table 5. Comparison of different model results with image and text datasets

	ISH dataset			Dredze dataset				Honeypot dataset			
	[4]	[3]	HST-Detector	[4]	[6]	[7]	HST-Detector	[8]	[10]	[9]	HST-Detector
F1-score	–	–	**98.24**	–	–	–	**90.98**	79.56	89.7	93.75	**93.1**
Recall	–	–	**98.31**	–	–	–	**89.85**	73.82	90.8	93.44	**93.19**
Precision	–	–	**98.17**	–	–	–	**92.19**	86.5	88.6	95	**93.21**
Accuracy	97	92.1	**98.27**	70	83.13	95	**96.58**	76.68	–	91.75	**93.2**

Table 6. Comparison of different model results with heterogeneous datasets

Model	Dataset	Precision	Accuracy	Recall	F1-score
HST-Detector	Heterogeneous Dataset 1	**98.35%**	**98.35%**	**98.34%**	**98.34%**
	Heterogeneous Dataset 2	98.29%	98.3%	98.3%	98.29%

As shown in Tables 5 and 6, In unimodal image-based experiments, the results expose that HST-Detector outperforms the other models in Dredze and ISH datasets. For unimodal text-based experiments, HST-Detector gives the best precision, recall, accuracy, and f1-score compared with [8] and [10]. For the multimodal experiments, the results in two heterogenous datasets show that our system is efficient for filtering any type of tweet with an accuracy of 98%. The HTS-Detector efficiency is resulting from the use of LSTM sequential processing and the word2vec technique which use the semantic representation of the text sentence. LSTM architecture performs well on datasets than the related work approaches with the advantage of features auto extraction and designation. Also, by the sequential process of the LSTM, the input words correlate with the previous words and understand slang and new words.

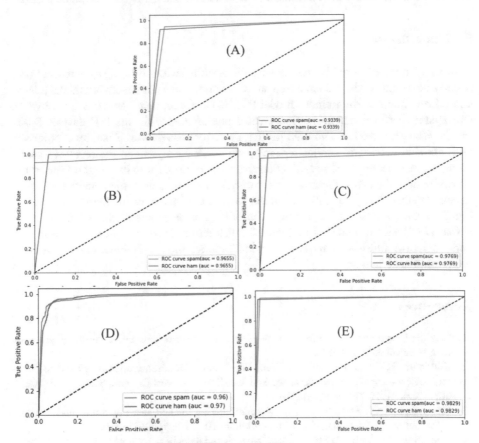

Fig. 7. ROC charts of HST-Detector

A ROC (Receiver Operating Characteristic) chart is the average value of the sensitivity for a test over all possible values of specificity or vice versa. So, we utilize a ROC chart to better clarify the effectiveness of the HST-Detector. From Fig. 7, (A) displays the honeypot ROC chart, (B) displays the Dredze ROC chart, (C) displays the ISH ROC chart, (D) displays the ROC chart for the Heterogeneous Dataset 1, and (E) displays the ROC chart for the Heterogeneous Dataset 2. The system's AUC indicator on the text-based dataset is 0.93. The system's AUC indicators on the image-based datasets and heterogeneous datasets are all greater than 0.96, which indicates that the performance of the HST-Detector is outstanding for tweet classification.

We can conclude that HST-Detector is able to filter spam tweets, even if the spam is hidden in the text, in the image, or hidden in the text and image. The performance could be better if the system is trained on more datasets. Also, non-English words can be injected sometimes into English tweets. It is a challenging case when the system is trained only for the English datasets which are not sufficient. So, the spam classification performance is affected as the model does not know how to label the non-English untrained words.

5   Conclusion

In this paper, we explored the issues with the current Twitter filtering systems and proposed a new multimodal system based on a Convolutional Neural Network model (CNN) and a Long Short-Term Memory model (LSTM) to detect heterogeneous spam tweets. Our system performance evaluation gives a rate of 0.9829 for the ISH dataset, 0.932 for the Honeypot dataset, and 0.98 for the heterogeneous dataset. Also, by comparison with some relevant and similar research, we noticed progress in the recall, f1-score, precision, and accuracy of classification from 1% to 25%. From the overall experimental evaluation, we can summarize that our system is efficient in the detection of spam tweets whether the spam is hidden in the text, in the image, or hidden in the text and image of the same tweet. The primary limitation for the generalization of these results is that HST-Detector can deal just with English terms. In future work, we aim to trait non-English or multi-language tweets and incorporate more techniques to mitigate from more intelligent attacks.

References

1. "Twitter Usage Statistics - Internet Live Stats." https://www.internetlivestats.com/twitter-sta tistics/. Accessed 4 Feb 2022
2. Kaddoura, S., Chandrasekaran, G., Elena Popescu, D., Duraisamy, J.H.: A systematic literature review on spam content detection and classification. PeerJ Comput. Sci. **8**, e830 (2022). https://doi.org/10.7717/peerj-cs.830
3. Yang, H., Liu, Q., Zhou, S., Luo, Y.: A spam filtering method based on multi-modal fusion. Appl. Sci. **9**(6), 1152 (Mar.2019). https://doi.org/10.3390/app9061152
4. Annadatha, A., Stamp, M.: Image spam analysis and detection. J. Comput. Virol. Hack. Tech. **14**(1), 39–52 (2016). https://doi.org/10.1007/s11416-016-0287-x
5. Gao, Y., et al.: Image spam hunter. In: 2008 IEEE International Conference on Acoustics, Speech and Signal Processing, Las Vegas, NV, USA, pp. 1765–1768, March 2008. https://doi.org/10.1109/ICASSP.2008.4517972
6. Sharmin, T., Di Troia, F., Potika, K., Stamp, M.: Convolutional neural networks for image spam detection. Inf. Secur. J. Glob. Perspect. **29**(3), 103–117 (May2020). https://doi.org/10.1080/19393555.2020.1722867
7. Listík, V., Šedivý, J., Hlaváč, V.: Email image spam classification based on ResNet convolutional neural network. In: Proceedings of the 6th International Conference on Information Systems Security and Privacy, Valletta, Malta, pp. 457–464 (2020). https://doi.org/10.5220/0008956704570464
8. Alom, Z., Carminati, B., Ferrari, E.: A deep learning model for Twitter spam detection. Online Soc. Netw. Media **18**, 100079 (Jul.2020). https://doi.org/10.1016/j.osnem.2020.100079

9. Ban, X., Chen, C., Liu, S., Wang, Y., Zhang, J.: Deep-learnt features for Twitter spam detection. In: 2018 International Symposium on Security and Privacy in Social Networks and Big Data (SocialSec), Santa Clara, CA, USA, pp. 208–212, December 2018. https://doi.org/10.1109/SocialSec.2018.8760377

10. Shen, H., Liu, X., Zhang, X.: boosting social spam detection via attention mechanisms on Twitter. Electronics 11(7), 1129 (Apr.2022). https://doi.org/10.3390/electronics11071129

11. Yin, W., Kann, K., Yu, M., Schütze, H.: Comparative study of CNN and RNN for natural language processing. arXiv170201923 Cs (2017). http://arxiv.org/abs/1702.01923. Accessed 31 Mar 2022

12. Bouvrie, J.: Notes on Convolutional Neural Networks, p. 8 (2006)

13. Jain, G., Sharma, M., Agarwal, B.: Optimizing semantic LSTM for spam detection. Int. J. Inf. Technol. 11(2), 239–250 (2018). https://doi.org/10.1007/s41870-018-0157-5

14. Lee, K., Caverlee, J., Webb, S.: Uncovering social spammers: social honeypots+ machine learning. In: Proceedings of the 33rd International ACM SIGIR Conference on Research and Development in Information Retrieval, pp. 435–442 (2010)

Analysis of Deep Learning Sequence Models for Short Term Load Forecasting

Oben Dağ[(⊠)] [iD] and Oğuzhan Nefesoğlu[iD]

Istanbul Arel University, Istanbul, Turkey
obendag@arel.edu.tr

Abstract. Short Term Load Forecasting (STLF) is an essential part of generator scheduling in power plants. Better scheduling is crucial for both economic and environmental aspects. In this study, two different deep learning (DL) model architectures based on Long Short Term Memory (LSTM) and Convolutional Neural Networks (CNN) were tested and compared on load datasets from Spain and Turkey. In addition, a novel Long Short Term Memory (LSTM) model with embedding layer was proposed and compared with these two models. Optimum model design choices and practices were discussed for achieving reliable and robust results. It was showed that datasets with different characteristics require different model design choices. The simulations of the proposed DL method were carried out with Python and the performance parameters were also presented.

Keywords: Short Term Load Forecasting · Deep Learning · LSTM

1 Introduction

In the near future, "electricity load forecasting" via Machine Learning (ML) will have more importance for the studies to reduce climate change effects [1]. Electricity generation is one of the main causes of air pollution and it impacts climate change. More than half of the electricity generation is achieved by fossil fuel consumption which involves combustion processes. Combustion process results emissions of Carbon Dioxide (CO_2), Sulfur Dioxide (SO_2), and Nitrous Oxides (NO_2) which causes depletion of ozone layer, acid rains, etc. [2]. After surveying 28 US and Canadian electric utilities, it was revealed that, STLF prediction by Artificial Neural Networks (ANN) would save an average of 600 000 dollars per utility [3]. Better generator scheduling and dispatching, more labor savings, more optimized power purchases and saves, and some other optimization effects would be achieved with this method [3]. Considering these economic and environmental factors; electricity load forecasting optimization and generator scheduling become more crucial.

Electricity load forecasting is predicting the electricity demand in a certain power plant, electric utility, region, or country, based on historical demand and other environmental data. There are three main types of load forecasting:

© The Author(s), under exclusive license to Springer Nature Switzerland AG 2023
F. P. García Márquez et al. (Eds.): ICCIDA 2022, LNNS 643, pp. 104–116, 2023.
https://doi.org/10.1007/978-3-031-27099-4_9

- Short Term Load Forecasting (STLF) is used to balance electricity systems with flexible demands through generator scheduling and dispatching, and to manage variable sources in power plants. It also helps power system operators to develop better price strategy. Forecasting period spans hours to several weeks.
- Medium Term Load Forecasting (MTLF) is used to obtain correct predictions for helping maintenance scheduling, and planning for outages and major works [4]. Forecasting period is up to one year.
- Long Term Load Forecasting (LTLF) is used for planning of building new power plants, and planning the capacities of already established power plants. Forecasting period is up to several years.

STLF is a complex task where the factors like socio-economic characteristic and climate/weather conditions of the particular region affect the load characteristic. Also, different patterns and seasonality are present in the load characteristic. This complex nature of the task makes it necessary to use ANN rather than classical ML algorithms.

Advancements in the hardware technology, revealed the potential of already existing Deep Learning (DL) algorithms by allowing to train deeper neural networks with bigger datasets. ANN showed superior performance to other classical ML techniques in complex sequence tasks like Natural Language Processing (NLP), Speech Recognition and Stock Market Prediction. Just like other sequence tasks, many DL techniques have been applied to STLF in recent years.

Beside these complex architectures which can encode non-linear relationships in the data, some linear models were also used in the past. Autoregressive moving average model (ARMA) was one of them. Because of its nature, it cannot extract seasonal patterns and the model expects a stationary data, which is usually not the case. Autoregressive integrated moving average model (ARIMA) is another linear model which can transform a non-stationary data into a stationary data by differencing. Yet, the seasonal patterns could not encode with these models. Therefore, Seasonal Autoregressive Integrated Moving Average model (SARIMA) become one of the most popular linear models and used in many STLF studies successfully [5, 6].

Support Vector Regression (SVR) is another well-known algorithm. Its robustness to overfitting due to the penalty term of model weights in its loss function, made it one of the most popular ML algorithms. It is also adapted to STLF task by some researchers. SVR with Particle Swarm Optimization (PSO) was used in a study to make day-ahead predictions of the load [7]. Different models were trained for different hours of the day and categorical exogenous features that helps model to learn seasonality were fed to the model by one-hot encoded vectors.

In recent years, success of sequence architectures like CNN, LSTM and Attention-Mechanism in many sequence tasks like speech recognition, NLP, stock market predictions, etc. caused many researchers to apply these sequence architectures for STLF. Gasparin et al. [8] compared different DL approaches and showed that the performance of auto-regressive approaches were worse than multiple input multiple output (MIMO) approach with architectures like LSTM and Gated Recurrent Units (GRU). This performance difference was due to the accumulated errors present in auto-regressive approach. Although most of the models benefit from exogenous variables that holds weather and calendar information, it was indicated that MIMO based models did not get a boost from introduced exogenous variables. LSTM with different model/algorithms like SARIMA, SVR were compared in a study and it was shown that LSTM is superior to the other models [9]. Non-seasonality and non-stationarity characteristics of the electric load caused SARIMA to not perform well. Univariate analysis was performed and only the past 96 h' load values were used to predict next 24 h' load values. On the other hand, long sequence and one predictor were not optimal for STLF. In another study, three blocks of convolutional and pooling layers to make 3-day ahead load predictions were used [10]. It was shown that LSTM based architecture was performing better than CNN [11]. And a two-stage architecture (where the first stage is LSTM combined with Attention Mechanism) was suggested. In one another study, it was shown that CNN, combined with LSTM architecture improves the performance of the pure LSTM model [12]. It was stated that CNN captures local trends while LSTM encode the long-term dependencies in the input. When the input variables become too much, they could degrade the model's performance through overfitting. In that case, selection of most relevant features and elimination of others was crucial; however manually selecting features before training might not be optimal since different features could be beneficial for the model in different conditions. In another recent study, researchers used attention mechanism along with LSTM architecture for feature and time-step selection [13]. This enables to select most appropriate inputs for the model and model performed its feature selection task itself dynamically. Also, attention mechanism was used before LSTM to weight different time steps. In that study, attention mechanism was constructed with two-layer perceptron.

The main contribution of our paper was to compare Deep Learning Sequence Models for Short Term Load Forecasting. Two conventional methods based on Long Short Term Memory (LSTM) and Convolutional Neural Networks (CNN) were tested and compared on load datasets from Spain and Turkey. A modified LSTM model with embedding layer was introduced. It was then compared with these two models. It was showed that datasets with different characteristics require different model design choices. The simulations on Python platform revealed that the proposed method might perform better in some cases in comparison with the other methods.

The paper is organized as follows: The topic and literature survey were given in the introduction. In the coming section, theoretical background which forms the algorithms in this paper were given. For this idea; CNN, RNN, LSTM models, attention mechanism and embedding layer structures were mentioned in respect to literature. In the third section, the proposed model was introduced. A diagram for the proposed model architecture was also illustrated. Hyperparameters and metrics for performance testing were defined next. Finally, the data-set used for the analysis were shown as a time-series plot. In the fourth section, the simulation and the test results were given. A discussion about all results and some future work were given in the conclusion section.

2   Sequence Models

Sequence models were used to encode the relationship of different data points in time (or space) or to extract patterns. They were powerful building blocks for the sequence DL models. CNN was used in reference [10], and LSTM was used in [11] and in the proposed model. Attention Mechanism was also vital to extract patterns. In this section, the sequence models will be briefly reviewed.

2.1   CNN

CNN was originally introduced to DL for computer vision applications (as a two-dimensional CNN). It was mainly inspired from animal's visual system [14]. After the success of two-dimensional CNN in image recognition benchmarks, its one-dimensional variant was also used in different sequence tasks like speech recognition and stock market prediction [14, 15].

$$y[n] = x[n] * h[n] = \sum_{i=-\infty}^{\infty} \sum_{j=0}^{c-1} x[n, j] \cdot h[n - i, j] \qquad (1)$$

Equation (1) shows a one-dimensional convolution operation. h is a fixed size learnable filter which is updated with gradient descent (GD) algorithm. Filters slide over the input sequence x with c feature channels and extract local features. Although convolutional layer is good at extracting local features, it cannot extract global features directly. It is because of the restricted receptive field of a convolutional layer. Receptive field of CNN is the size of the input that feature map's (output of convolutional layer) each element gets information from. To extract more complex and global features, convolutional layers are stacked sequentially. So, feature map of one convolutional layer becomes the input of the other one and hierarchical features are constructed through deeper layers. Although small sized filter selection and sliding the same filter on the input data restricts the receptive field of the CNN in early layers, it has advantages over Fully Connected Networks (FCN). Since the weights (filters) were shared by the input, it has lower computational complexity and had a regularization effect. Thus, overfitting is avoided to some extent. In addition, selection of this filter over the input (sliding it) makes CNN translate equivariance. Therefore, patterns that are in different location on the input, could be detected even if these patterns were not seen around that location before.

2.2   Recurrent Neural Networks (RNN) and LSTM

RNN is as ANN that can handle sequence inputs with the feedback connections. It carries the information through sequence by internal representations. These representations were usually a vector. This vector was updated as the new information arrives (processing through sequence). RNN's ability of detecting complex patterns and seasonality made them very popular in the sequence tasks like natural language processing, audio processing, and stock market prediction.

$$h_n = \sigma_h(W_{hh} \cdot h_{n-1} + W_{hx} \cdot x_n + b_n) \qquad (2)$$

$$y_n = \sigma_y(W_{yh} \cdot h_n + b_n) \qquad (3)$$

Equation (2) and Eq. (3) show the internal computations of a particular timestamp, also called RNN cell. The parameter ht is the internal representation of the information that is carried up to t-th cell, where yt is the output of the t-th cell. All the timestamps share the same weight; thus the computational complexity increases linearly with increasing the sequence size. This is an advantage of RNN over FCN since computational complexity of FCN increases exponentially with increasing the sequence size.

Although RNN performs impressive in sequence tasks, it might suffer from vanishing/exploding gradient problem when the sequence become longer [15]. Gradient vanishing or exploding in the back-propagation stage, later layers' error gets extremely smaller (or bigger) as the error propagates to the early layers. This causes extremely slow training and learning can even stop. Some non-linear functions' low slope (like hyperbolic tangent and sigmoid) and non-careful weight initializations were the main reasons of that phenomenon. To overcome this issue, LSTM was developed by Hochreiter et al. [16]. It was showed that LSTM could remember longer dependencies in longer sequences and handles the noise better than RNN. Difference between the RNN and LSTM is the computations inside the cells. Inside of the LSTM cell, there were different gate mechanisms to allow model to forget irrelevant information from past or update relevant information from current timestamp.

$$f_n = \sigma_g(W_{fx} \cdot x_n + W_{fh} \cdot h_{n-1} + b_f) \tag{4}$$

$$u_n = \sigma_g(W_{ux} \cdot x_n + W_{uh} \cdot h_{n-1} + b_u) \tag{5}$$

$$o_n = \sigma_g(W_{ox} \cdot x_n + W_{oh} \cdot h_{n-1} + b_o) \tag{6}$$

$$\hat{c}_n = \sigma_c(W_{cx} \cdot x_n + W_{ch} \cdot h_{n-1} + b_c) \tag{7}$$

$$c_n = u_n * \hat{c}_n + f_n * c_{n-1} \tag{8}$$

$$h_n = o_n * \sigma_c(c_n) \tag{9}$$

The Eqs. (4) to (9) represent the computation mechanism of a LSTM cell. The parameters σ_g and σ_c represent sigmoid and hyperbolic tangent functions respectively. Soft gates were constructed with sigmoid function. Which information to decay and which information to remember/update were determined with these gates.

Superiority of the LSTM over RNN was shown in many studies and LSTM achieved state-of-the-art results in many different sequence tasks [17, 18]. Although better performances than RNN was achieved, this architecture suffered from lost information when the sequence becomes very long. To speed up the training step and prevent vanishing/exploding gradient problem, different kind of normalization layers were proposed [19, 20]. The main idea of the normalization methods was to speed up the training phase by normalizing the activations around zero or allowing model to adjust the mean and variance of the activation before non-linear function. This shifts activation values close to the area where the non-linear activation function has higher gradient value.

2.3 Attention Mechanism

RNN models still suffer from lost information when the sequence gets bigger and it was shown that the effective receptive field of the CNN was smaller than the theoretical receptive field [21]. To improve the neural machine translation task in the long sequences, Bahdanau et al. proposed the *attention mechanism*. Unlike CNN and RNN based architectures, this model had direct access to each element of the sequence without loss of information. In addition, Vaswani et al. [22] proposed *transformer architecture with scaled dot product attention.*

$$y_j = \sum_{i=0}^{T} a_{ji} * X_i^v \qquad (10)$$

$$a_{ji} = \frac{e^{X_i^q \cdot X_j^k}}{\sum_{l=0}^{T} e^{X_l^q \cdot X_j^k}} \qquad (11)$$

In Eq. (10) and Eq. (11), self-attention mechanism was shown. X^q, X^k and X^v were query, key, and value matrices respectively. They were projected from input matrix X with W_q, W_k and W_v matrices respectively. Then the dot product between query and key vectors were taken and *softmax* function was applied to get attention weights. Finally, new set of sequence vectors were computed as weighted sum of value vectors. This process was analogous to information retrieval systems.

Attention mechanism also makes it possible to obtain the effect of the input on the output for the models with attention weights. A disadvantage of the attention is exponential computational complexity and this led researchers to develop different kind of attention mechanisms [23]. Attention mechanism was considered as a global learning algorithm and used in many different areas like computer vision, protein structure prediction and exceeded the performances of former architectures.

2.4 Embedding Layer

Embedding process converts a categorical variable into a dense vector. It was first used for NLP tasks to represents words or characters in a text. In STLF tasks, many researchers used one-hot encoded vectors or directly normalized values for categorical features. But these representations of the categorical features were not intuitive and did not represents real relationship in the feature space. For instance, one-hot encoded vectors were orthogonal to each other in the vector space and there was no relationship between them. When embedding process was used in NLP, it was showed that similar words' vectors were close to each other with similar angle and magnitudes where vectors of opposite meaning words had opposite angles with similar magnitudes. Therefore, linear relationship was present in the vector space [24]. This was more intuitive and help models to learn complex relationship of the categorical input features. Another advantage of the embedding layer was the reduced feature size compared to one-hot encoded vectors. Although there was not complex relationship involved in the categorical features of the STLF as much as in NLP tasks, more complex datasets which had high load variance could get significant benefit from embedding layer. Process of embedding was not complicated. Randomly initialized weight matrix was hold in the memory which represents different values of specific categorical feature. For instance, 16x24 matrix would be initialized for hour-of-day feature. So, every hour's embedding vector had a length of 16. Then, in the forward propagation of the training, this matrix would be dot product with the one-hot encoded vector of the categorical variable, which was essentially a column selection operation. Equation (12) shows the forward propagation process of the embedding layer. These vectors were learned end-to-end with actual training.

$$e_j = W \cdot o_j = W_j^T \tag{12}$$

3 The Proposed Model

3.1 Parameter Set-Up

The benchmark models were obtained from the studies in [10] and [11]. Input and output sequence length were both chosen as *24*. Day-ahead load predictions were performed by using the samples of last *24*-h. Since the temperature data was absent, all the input variables from reference [11], except temperature were used as inputs.

The input variables were as follows: *24*-h lagged load, a-week lagged load, hour, day of week, month of year, holiday indicator (whether it is holiday or workday), holiday to workday indicator (is today workday and was yesterday holiday) and *24*-h lagged *3*-day mean load. Since univariate forecasting yielded poor results, it was decided to feed exogenous inputs of [11] to the [10]. In [10] only load values as inputs were used, which gave poor results. For the proposed model in this paper, *24*-h lagged load, hour, day of week, day of month, and holiday indicator were used as inputs.

All the models were trained via "Adam optimizer" with learning rate of *0.01*. Learning rate was decreased when loss did not reduce for *3* epochs and training was stopped after loss did not deteriorate for *7* epochs.

3.2 Model Infrastructure

The first stage of the proposed model was the vectorization of the input variables. Categorical features were vectorized through embedding layer. In this layer, continuous variables scale their corresponding feature vector. Although load value was not a categorical feature, other higher dimensional feature vectors dominate the load input and the performance drops. So, a random vector was initialized for load input. The load value was the main feature of the proposed model and it yielded the optimum results. The length of the load vector was chosen as four times the categorical embedding vectors' length. Then in the second stage, all the feature vectors were concatenated and two-layer Multi-Layer Perceptron (MLP) obtains the features before sequence architecture. In the third stage, LSTM processes the sequence for detecting patterns/seasonality and passes the outputs to the 2-layer MLP for regression. Categorical and load input's embedding dimensions were chosen as *8* and *32*. The MLP had two layers with *32* and *16* neurons. The activation function was Rectifier Linear Unit (ReLU). Hidden dimension of the one-layer LSTM was *64*. Regressor MLP had one hidden layer with *16* neurons and ReLU activation function. It had one output layer with 1-neuron and sigmoid activation function. As a result, the model was able to predict the normalized load.

ML model has hyperparameters that must be set up to adapt the model to the dataset. This is actually a challenging task which is usually accomplished based on practice rather than theory. For the hyperparameter search, one method is grid search where every possible combinations of the hyperparameters were tested. This technique is time consuming; and it was not used in this study. The method used in this study was as follows: First some reasonable values for were chosen. This was called base-hyperparameters. Then, for every hyperparameter, a set of candidate hyperparameters were constructed. For the hyperparameter that was searched, a new model was trained until error did not reduce for 4 epochs (not full training). This was performed for each value in that hyperparameter's candidate set. This process was repeated 3 times. In that way, hyperparameter value which trained the models with lowest mean loss was chosen. Candidate sets of hyperparameter search space and chosen hyperparameters were shown in Table 1. Another option could be the grid search with a very time consuming process where every possible combination of the hyperparameters were tested.

Table 1. Hyperparameter search space and chosen hyperparameters for the proposed model.

Hyperparameters	Base hyperparameters	Candidate set	Chosen hyperparameters
Load embedding dimension	16	{4, 8, 16, 32, 64}	32
Categorical feature embedding dimension	16	{4, 8, 16, 32, 64}	8
MLP layer 1 hidden dimension size	32	{8, 16, 32, 64, 128}	32
MLP layer 2 hidden dimension size	32	{8, 16, 32, 64, 128}	16
LSTM hidden dimension size	32	{8, 16, 32, 64, 128}	16
Regressor MLP hidden size	16	{4, 8, 16, 32, 64}	16
Batch size	64	{16, 32, 64, 128, 256}	32

The proposed model was illustrated in Fig. 1. x_{-n} is the n-hour lagged feature vector. It consists of normalized continuous (float) load value and other exogenous and categorical (integer) variables. y_{-n+24} was the load value 24 h after x_{-n} was observed. $V(x)$ was the load vectorization operation and shown in Eq. (13). v_l was the randomly initialized, trainable vector which was scaled with load value in the forward pass of the model ($v_l \subset \mathbb{R}^{32}$). $E(x)$ was the embedding function, operated with different weight/embedding matrices for each variable. Embedding operation for one variable was shown in Eq. (12). Embedding matrices of size ($8 \times m$) were used in this study. The parameter m was the number of possible inputs of a variable ($W \subset \mathbb{R}^{8 \times m}$).

$$V(x) = x * v_l \tag{13}$$

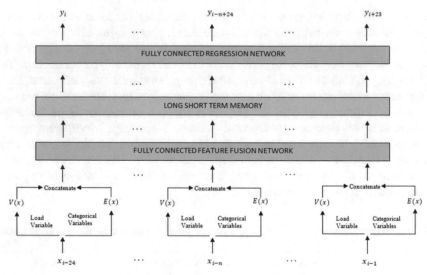

Fig. 1. Diagram of the proposed model

3.3 Metric and Loss Function

Mean Absolute Percentage Error (MAPE) was used as a scale invariant metric. Mean Absolute Error (MAE) was chosen as the loss function; because it gave more stable results than Mean Square Error (MSE). Equation (14) and (15) shows the MAPE and MAE respectively.

$$MAPE = \frac{1}{N} \sum_{i=0}^{N} \frac{|F_\theta(x_i) - y_i|}{y_i} \times 100\% \tag{14}$$

$$MAE = \frac{1}{N} \sum_{i=0}^{N} |F_\theta(x_i) - y_i| \tag{15}$$

N was the sample size of the dataset and F_θ was the forward pass function of the model with parameters θ.

3.4 Datasets

In order to test the robustness of the different models, two different datasets from Spain and Turkey were used. The datasets were illustrated in Fig. 2.

The first dataset was from Spain and it was available in European Network of Transmission System Operators (ENTSOE) website. Dataset spans the time between 2015 to 2018. Load value range was from 18MWh to 41 MWh.

The second dataset was from Turkey, which was obtained from Energy Markets Management Inc. (EPIAS). This dataset spans the time between 2016 to 2020. Load value range was between 15MWh and 55MWh.

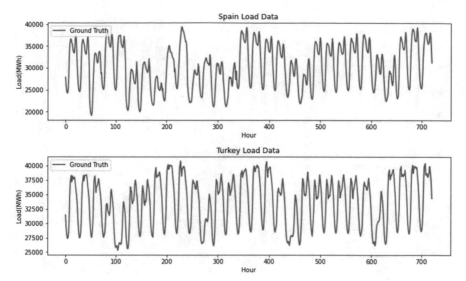

Fig. 2. A 30-day section of ground truth values for Spain and Turkey datasets

3.5   Test Results

The tests were performed for the Spain and Turkey Datasets with three different ANN models. The proposed model performance was compared with the LSTM and CNN models mentioned earlier. Coefficient of Variation (CV) and Approximate Entropy (ApEn) of load data from Spain were 0.26 and 0.785 respectively, where CV and ApEn of load data from Turkey were 0.192 and 0.772 respectively. CV and ApEn were calculated with normalized load values; thus, different scale between datasets were not critical. In addition, CV was calculated on daily mean of load data, otherwise it could give false cues about the volatility of the data where daily periodicity was present in the load data. According to accomplished simulations, if the previous day's load values were used as prediction for next day, MAPE values for Spain and Turkey data were 8.65% and 5.57% respectively. According to these observations, the data from Spain was less predictable with higher variations and higher MAPE values with same architectures could be expected. Figure 3 clearly represents the test results for each dataset. The predicted load values and the real data were compared in each plot with different markers.

The simulation results were also summarized in Table 2. In Table 2, "Only MLP" and "Only Seq" means that only the MLP or sequence part of the [11] is used. No attention means, attention layer is removed, so only the LSTM part remained. The parameter h is the hidden dimension size of the LSTM.

According to Table 2, the proposed model showed better performance than LSTM and CNN models for the "Spain load dataset". MAPE metric values were 3.99% (row 1), 5.12% (row 2) and 4.68% (row 7) for the three compared models. In comparison, the proposed model performance was 14.7% and 22% more than each model. On the other hand, the "Spain load dataset" was complex and it was more difficult to predict because of high volatility issue caused by exegenous variables mentioned earlier. The input representations with embedding layer were the main superiority of the proposed

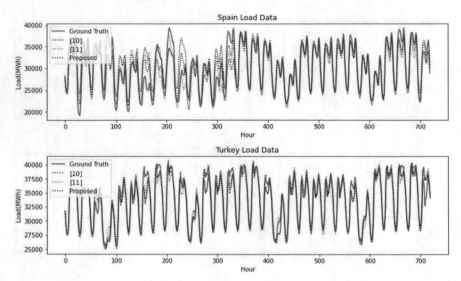

Fig. 3. A 30-day section of ground truth and predicted load values from three different models

model which makes it to capture the relation of the exegenous variables with the load demand. Therefore, the performance increased conspicuously. In order to test the effect of this fact, the embedding layer was also included to the two other models. The metric values were eventually improved by 9.2% and 15.6%. This proves that, poor input representations were the main drawbacks of the other two models.

Table 2. Comparison of the different models on two different datasets.

Row		Spain load dataset	Turkey load dataset
1	Proposed model	3.99	2.45
2	Reference [11]	5.12	2.29
3	Reference [11] (Only MLP)	5.48	2.15
4	Reference [11] (Only Seq, h = 100)	4.79	2.43
5	Reference [11] (Only Seq, h = 64, no attention)	–	2.09
6	Reference [11] (Only Seq, h = 100) + Proposed inputs and embedding	4.35	–
7	Reference [10]	4.68	2.3
8	Reference [10] + Proposed inputs and embedding	3.95	–

In "Turkey load dataset" the performance for the two models [11] and [10] were similar for MAPE values of 2.29% and 2.3% respectively (row 2 and row 3). On the other hand, the proposed model yielded 2.45% performance (row 1). In addition, "only MLP" model (row 3) yielded better results than these other networks with 2.15% of MAPE. This was because the dataset was easier to predict and this model was simpler and less prone to overfitting, to make the model simpler to remedy the overfitting, attention layer was removed and hidden dimension size was reduced from 100 to 64. Then, this model gave MAPE of 2.09% (row 5), which is 13% better than original sequence model and better than sequence sub-model of [11]. To conclude, it might be possible to interpret that simpler models work best for datasets like "Turkey load dataset".

4 Conclusion

A modified LSTM model with embedding layer was proposed in this paper or the STLF study. The proposed model was compared with two different models based on LSTM and CNN. Various architectural modifications were performed on the models according to the characteristics of the datasets. The main causes of diverse performances of the models were discussed. It was seen that, the proposed model performs better than the compared two models, particularly for the "Spain load dataset" which was difficult to interpret. In addition, the proposed model performed similar performance in the "Turkey load dataset". It is usually necessary to develop models/algorithms specific to different datasets with different characteristics. The experiments showed that, the proposed model could perform better results for particularly complicated datasets. It was also appeared that simpler models with less neurons and layers perform better with the dataset that have higher forecastability and lower volatility. In contrast, components for learning complex relationships (like embedding layer) could help models to get better results with more complex datasets.

Since, obtaining comprehensive dataset for load models is a cumbersome task, we would test the proposed model with two datasets only. In addition, there were some overtraining issue in some cases in terms of ANN training and validation performance. As a future work, it might be beneficial to develop some regularization and data augmentation methods and test the performance with more detailed datasets, to improve the results.

References

1. Rolnick, D., et al.: Tackling climate change with machine learning. ACM Comput. Surv. (CSUR) **55**(2), 1–96 (2022)
2. Rahman, S., de Castro, A.: Environmental impacts of electricity generation: a global perspective. IEEE Trans. Energy Conv. **10**(2), 307–314 (1995)
3. Hobbs, B.F., Helman, U., Jitprapaikulsarn, S., Konda, S., Maratukulam, D.: Artificial neural networks for short-term energy forecasting: accuracy and economic value. Neurocomputing **23**(1–3), 71–84 (1998)
4. Abu-Shikhah, N., Elkarmi, F.: Medium-term electric load forecasting using singular value decomposition. Energy **36**(7), 4259–4271 (2011)
5. Bozkurt, Ö.Ö., Biricik, G., Tayşi, Z.C.: Artificial neural network and SARIMA based models for power load forecasting in Turkish electricity market. PLoS ONE **12**(4), e0175915 (2017). https://doi.org/10.1371/journal.pone.0175915

6. Musbah, H., El-Hawary,M.: SARIMA model forecasting of short-term electrical load data augmented by Fast Fourier transform seasonality detection. In: 2019 IEEE Canadian Conference of Electrical and Computer Engineering (CCECE), pp. 1-4 (2019). https://doi.org/10.1109/CCECE.2019.8861542

7. Ceperic, E., Ceperic, V., Baric, A.: A strategy for short-term load forecasting by support vector regression machines. IEEE Trans. Power Syst. **28**(4), 4356–4364 (2013). https://doi.org/10.1109/TPWRS.2013.2269803

8. Gasparin, A., Lukovic, S., Alippi, C.: Deep learning for time series forecasting: the electric load case. CAAI Trans. Intell. Technol. **7**(1), 1–25 (2022)

9. Zheng, J., Xu, C., Zhang, Z., Li, X.: Electric load forecasting in smart grids using long-short-term-memory based recurrent neural network. In: 2017 51st Annual Conference on Information Sciences and Systems (CISS), pp. 1–6. IEEE (2017)

10. Kuo, P.H., Huang, C.J.: A high precision artificial neural networks model for short-term energy load forecasting. Energies **11**(1), 213 (2018)

11. Xie, Y., Ueda, Y., Sugiyama, M.: A two-stage short-term load forecasting method using long short-term memory and multilayer perceptron. Energies **14**(18), 5873 (2021)

12. Rafi, S.H., Deeba, S.R., Hossain, E.: A short-term load forecasting method using integrated CNN and LSTM network. IEEE Access **9**, 32436–32448 (2021)

13. Lin, J., Ma, J., Zhu, J., Cui, Y.: Short-term load forecasting based on LSTM networks considering attention mechanism. Int. J. Electr. Power Energy Syst. **137**, 107818 (2022)

14. Fukushima, K., Miyake, S.: Neocognitron: a new algorithm for pattern recognition tolerant of deformations and shifts in position. Pattern Recogn. **15**(6), 455–469 (1982)

15. Hochreiter, S.: The vanishing gradient problem during learning recurrent neural nets and problem solutions. Int. J. Uncertain. Fuzziness Knowl.-Based Syst. **6**(02), 107–116 (1998)

16. Hochreiter, S., Schmidhuber, J.: Long short-term memory. Neural Comput. **9**(8), 1735–1780 (1997)

17. Kai, C., Zhou, Y., Dai, F.: A LSTM-based method for stock returns prediction: a case study of China stock market. 2015 IEEE International Conference on Big Data (Big Data). IEEE (2015)

18. Huang, Z., Wei, X., Kai, Y.: Bidirectional LSTM-CRF models for sequence tagging. arXiv preprint arXiv:1508.01991 (2015)

19. Ioffe, S., Szegedy, C.: Batch normalization: accelerating deep network training by reducing internal covariate shift. In: International Conference on Machine Learning. PMLR (2015)

20. Ba, J.L., Kiros, J.R., Hinton, G.E.:Layer normalization. arXiv preprint arXiv:1607.06450 (2016)

21. Luo, W., et al.: Understanding the effective receptive field in deep convolutional neural networks. Advances in Neural Information Processing Systems, vol. 29 (2016)

22. Vaswani, A., et al.:Attention is all you need. In: Advances in Neural Information Processing Systems, vol. 30 (2017)

23. Luong, M.-T., Hieu, P., Manning, C.D.: Effective approaches to attention-based neural machine translation. arXiv preprint arXiv:1508.04025 (2015)

24. Mikolov, T., Chen, K., Corrado, G., Dean, J.: Efficient estimation of word representations in vector space. In: Proceedings of Workshop at ICLR (2013)

Security Issues for Banking Systems

Mohammed Khodayer Hassan[1]([✉]), Aymen Mohammed Khodayer[2], Ali Hassan[3], Omer Mohammed Khodayer[4], and Maryem Mahmood[5]

[1] Al-Rafidan University, Baghdad, Iraq
Aldulaimiomer187@gmail.com
[2] Al-Farhidai University, Baghdad, Iraq
[3] Iraqi Commission, Institutes for post graduate studies, Baghdad, Iraq
[4] Polteckina University of Bucharest, Bucharest, Romania
[5] Institutes for Post Graduate Studies, Baghdad, Iraq

Abstract. It's no secret that businesses of all sizes and types rely on IT to carry out their day-tō-day operations and provide the best possible service to their clientele and constituents. Since the early 2000s, banks and other financial institutions have grown increasingly reliant on computers and the internet to manage day-to-day operations, client interactions, and market activity. Damage to a bank's reputation and bottom line can result from security breaches and other dangers. Businesses throughout the world have lost billions of dollars each year due to a variety of insider and outsider assaults. When it comes to a financial institution's competitive edge, money flow, legal compliance, and commercial rules and regulations compliance, the three pillars of information security are very critical. (confidentiality, integrity and availability).This has made it necessary for financial institutes and banks to put adequate security controls and information security governance frameworks (ISGF) to ensure data accessibility to all the authorized users, and prevent intruders to gain access to the banking information system. Security measures strengthen the safeguards against all types of threats across the bank and guarantee information systems safety. The potential threats of information security banking system are discussed in this research. It proposes a framework for information security governance (ISGF). ISO 27002 is one example of a widely-accepted framework for governing information security. ISSA; FFIEC; PCI-DSS; COBIT and others are explained in this paper. The comprehensive information security governance framework (ISG) categorized into three levels which are strategic, tactical operational level and technical level. All components of the suggested ISG are listed in table. The banking system will actually use this framework. It explains the importance of each part of the ISG and how they work together to make the ISGF as strong as possible. Recommendations would be included to prevent all intrusions. In this study has been made and suggestions have included gaining robust information security governance frameworks to achieve better real banking system environment.

Keywords: ISG · ISGF · ISSA · FFIEC · PCI-DSS · COBIT · OCED

© The Author(s), under exclusive license to Springer Nature Switzerland AG 2023
F. P. García Márquez et al. (Eds.): ICCIDA 2022, LNNS 643, pp. 117–131, 2023.
https://doi.org/10.1007/978-3-031-27099-4_10

1 Introduction

Public confidence is a cornerstone in the stability and reputability of a bank and financial institutes. The banks should be proactive to identify and specify the minimum security baselines to ensure confidentiality and security of the information and client's data. If the bank loses complete control over information system security, confidential, sensitive information, personally identifiable information and critical customer data, becomes jeopardized. Poor services of the service provider will be harmful for the reputation of the bank and will harm its relation with the customers and might have direct effect on its business by giving the bank a bad reputation [1]. Computers are now widely used in virtually every part of our lives, from business and government to education and healthcare, as well as in the aerospace and defense industries. We now live in a world where information systems are at the core of contemporary banking, and information is now the most precious asset to safeguard from both internal and external threats. The ramifications of computer crime can be quite serious because of society's increasing reliance on IT. Businesses around the world lose $1.6 trillion a year due to security breaches and computer infections there is an elevated risk to private information as a result of security system weaknesses [2]. Privacy and security in the financial sector are major concerns for customers. Providing reciprocal network and information access to business partners is viewed as a key priority in terms of security, as mention in Fig. 1.

Open, accessible, available, and secure network services are typically critical to a bank's capacity to capitalize on new prospects and opportunities for expansion. With a good reputation for protecting customer information, the bank's profits and its impact on the market will be positively impacted. Because of this, banks must be held accountable for online fraud performed by their customers. In most cases, banks are required to compensate most clients for any damages they may suffer as a result of security system flaws [3]. Banking security and financial institutions are the most regular targets of phishing attacks. In order to deceive consumers and staff into handing over their passwords, cyber hackers utilize this method. Fraudsters can simply steal money from the account of a client by using this login information. Spyware, Trojan horses, and key loggers are just some of the malicious programs that a user could unwittingly acquire. Identity theft, which relies on the stolen data to perpetrate a crime, is a much more serious risk than account skimming or takeover.

Case studies and anecdotal evidence have grown into empirical studies on Information Security Governance. However, relying on examples like this to manage information security is unrealistic. Research and hypothesis must be founded on solid evidence. In industrialized nations and Europe, there are now frameworks for information security governance that are extensively used. These frameworks must be tailored to the specifics of the organization's structure and environment, since each has strengths and weaknesses [4]. There are five main parts to this paper. First section provides an overview of the study's scope and focus. Banking information security governance is examined in Sect. 2 of this paper. Information security governance structures are examined in section three. The suggested ISG framework for the financial sector is discussed in section four. Information Security Governance in Banking is briefly discussed as the article draws to a close.

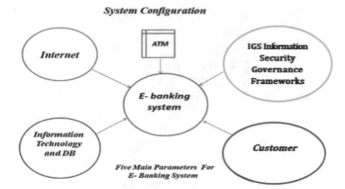

Fig. 1. Five Main parameters for e-banking system

2   Information Data Base Security Governance in Banking

In order to implement ISG, organizations must have clearly defined tasks and supervision systems, as well as a clear organizational structure with clearly defined roles and duties. Different authors use different terminology to describe what constitutes "information security governance." Information security governance (ISG) is defined by Moulton and Cole as "the process of establishing and maintaining a control environment for the purpose of detecting, investigating, and responding to threats to the confidentiality, integrity, and availability of information" [5]. The technologies, staff, and business procedures that assure security to fit a company's particular demands are together referred to as "information security governance," according to Harris [6]. Rastogi and Von Solms describe information security governance as the frameworks for decision-making and performance measurement that the board and executive management implement to preserve stakeholder value. For the sake of the paper, this definition of ISG will be used as a guide [7]. As illustrate in Fig. 2. The implementation of an ISG is driven by a desire to protect a company's most prized possessions. When it comes to implementing information security, the value of a company's data assets cannot be overstated. These four types of data are considered to be valuable resources. That need to be secured in the banking business are:

2.1   The Classification of the Information Assets to Be Protected in the Banking Industry

As explained the assets to be protected process and classification of assets of information in the Fig. 3.

- Balance information: Data that represents client or partner commercial claims against the bank (for example: deposit balances, account balances).
- Client information: all the information of the clients that could be copied. It includes information related to the name, address, data of birth including bank contact information and deposit number of the client.

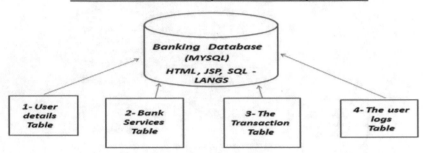

Fig. 2. Internet banking knowledge of data base

Assets to be Protected

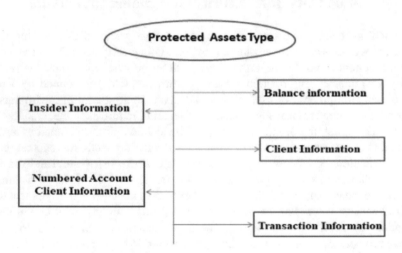

Fig. 3. Classification of assets of information

- Insider information: For the purposes of insider trading, the following are examples of the types of information that a holder might possess: internal financial data, board meeting minutes, and capital market data.
- Numbered account client information: It is information related to Client or economic beneficiary.
- Transaction information: Modifications to the bank's commercial claims against its customers or business partners (such as business events in trade, account and deposit movements) [8].

2.2 The Types of Threads Performed by Unauthorized Users

1) Malware Program Attacks.

   It's a program that can be used to make unauthorized alterations to a computer's system software over a network. Spreading malware from one computer to another is possible. Few examples are Threats such as viruses, worms, and scripted attacks should not be discounted. A malware assault might compromise the robustness, resilience, and safety of the financial system. An assault on the secrecy of a server's screen or keystrokes is referred to as a "malware attack" in this context. Unauthorized file writers, overwriting data, and alterations to the banking system's setup can all jeopardize the system's integrity and lead to undesirable outcomes [9]. When a denial of service attack happens, legal customers may be unable to use the banking system and the security system may be disabled. It involves the deletion and renaming of files and folders. There has been a lot of harm done to the financial institutions by malware attacks. While investigating ATM breaches in nations like Russia and Ukraine, Chicago-based information security and card industry vendor Trust wave uncovered malware attacks that gave criminals access to sensitive information, personal identification numbers (PINs), and even financial transactions. Our months-long investigation into the security of ATMs in Russia and Ukraine revealed that over 20 of them were infected with malware.. An inside job was the only possibility since the attacker would require physical access to an ATM [10]. Anyone with access to the ATM key might launch an assault. External attacks on the bank's information system, such as those carried out by hackers and viruses, result in the breakdown of business operations, the loss of data, and the compromise of data confidentiality [11].

2) Attack of Denial of Service

   A DOS assault is one sort of cyber-attack. As a result of this attack, customers and clients are unable to access financial services. The damage to the bank's finances will be substantial, and the cost to fix it will be high. The FBI ranks this type of attack as the third greatest threat, after terrorism and espionage. When it comes to the financial sector, distributed denial of service (DDoS) attacks is by far the most common threat. An assault on the targeted system is launched via a distributed denial-of-service attack, which utilizes hundreds of 'zombie' machines. Zombie PC gets fresh software installed. The malware has the capability of self-replicating and launching a massive attack on the network. These "zombie" send a flood of packets to the system at once, overwhelming the network and causing the legitimately requested packet to be discarded due to a timeout. The financial system's availability and consistency will be compromised as a result of such an infiltration. Customers, business partners, and vendors have all been let down by financial institutions. Fraud, errors, or a lack of services all contribute to operational risk, which can damage a company's reputation. The financial institutions face these risks. Natural disasters, such as hurricanes, tornadoes, and earthquakes, might cause a loss of service. The bank's demise is assured if there is no emergency.

3) Data Intrusion

   When there are security holes in the monetary system, data breaches can arise. Weaknesses in the bank's security system make it possible for criminals to get access and commit fraud.. As a result, banks must be aware of all threats that might compromise the security of their computer systems. There is a good chance that the integrity and confidentiality of data will be compromised when there is a data breach. Personal information of clients or financial system information can be accessed or stolen by this attacker. This might result in a breach of confidentiality for the bank and its customers. Unauthorized individuals can potentially compromise system integrity by making unauthorized changes to the data stored in the system. In order to get access to a financial institution's confidential information, thieves typically use stolen identities. Data breaches can occur if a credit card is lost or if the user's permission is inadequate. Is not authenticated or authorized in any way Intruders can gain access to any data they choose by breaking into the system unlawfully. As a result, authentication and authorization are critical components in maintaining financial institutions' data security, accuracy, and privacy [12].

4) Attack of TCP/IP Spoofing

TCP/IP spoofing is a type of online camouflage that allows an attacker to gain access to a machine or network without the owner's knowledge or consent. Spoofing is the act of forging an IP address in order to send a malicious message from a trusted computer to a victim server. The target of the hoax believed it was genuine [13]. Through the use of spoofing techniques, an attacker can send packets through a network without having them detected and blocked by the firewall. Nevertheless, a firewall's primary function is to block connections from outside IP addresses to internal network resources. If the attacker uses an IP address from inside the company's network, the firewall will not be able to detect their presence. The primary goal of this attack is to get root access to the victim server so that a backdoor may be created through the banking system. If the security system is flawed or has a hidden backdoor, the attacker can return to the target system at any time. As illustrate in Fig. 4. TCP/IP makes it simple to conceal one's IP address in a packet's header, making it a prime target for malicious actors looking to steal sensitive information, such as passwords, bank account details, credit card numbers, and personal identification numbers. The use of IP address spoofing is a popular method for online con artists to steal money from banks. Financial institutions around the world have suffered significant losses as a result of these kinds of attacks. Lloyd's Bank, First Union Bank, Bank of America, and Barclays Bank have all been targeted by phishing emails [14]. Spoofing also poses a risk of personal data being leaked. If critical information is sent to the wrong Gmail account, it could severely damage the bank's relationship with its customers. The biggest cost is lost confidence in the bank's ability to provide necessary services. This compromises the security of the customer's financial data and the bank's ability to keep it private. There is a severe issue with any breach of customer confidentiality that occurs between a bank and its clients. Given that it involves sensitive client and financial institution data, protecting their privacy is a top priority. Customer confidentiality must be safeguarded against IP spoofing [15].

5) Human Error

There are several elements that contribute to human error, including overwork, lack of training and the appropriateness of equipment, as well as a company's culture and work ethic.

6) The illegal ways to gain a lot of money by employees or clients

This is done by making use of the expense of the financial institutes through covert channels.

7) The Confidentiality Of Client Information

This can be released by foreign intelligence services may be able to gain access to sensitive client information via analyzing data and telecommunications activity. Threat operations such as this one led in a systematic accumulation of illicit information.

8) Changing Password

In order to get access to the system and make changes to passwords, the Social Engineering strategy involves either pursuing the victim and asking for their identifying information or calling the bank's help desk and pretending to be an authorized user [16].

There needs to be a heightened level of security at financial institutions and banking systems to ensure that no tampering with the systems' integrity occurs. These measures, particularly the use of authentication mechanism, which is regarded as one of the essential parameters to assure confidence and good operation of the banking system, will boost the efficiency of the financial institution or banking system Authentication.

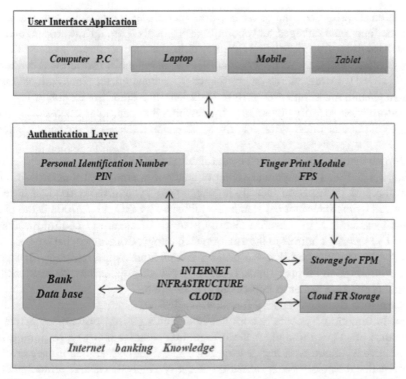

Fig. 4. The configuration of banking system with end users application interface.

3 Information Security Awareness

One way to measure someone's knowledge of and compliance with relevant information security policies, guidelines and norms is through the concept of "information security awareness" (also known as "security literacy" or "security sensitivity"). User security awareness is critical to the long-term viability of a company. When it comes to securing data, passwords and social engineering, network usage, malware, the use of personal equipment and clean workstations are some of the main issues that are common in these efforts. In information security, people are the weakest link in the chain. At work, it's common for employees to be eager to assist others. People are motivated to perform a good job for their coworkers, customers, and suppliers. Social engineers are drawn to these human traits and are looking for ways to take advantage of them. Security awareness programs are the only known protection against social engineering assaults. Organizational data can be compromised if people aren't aware of social engineering strategies and procedures [17]. Even if it's only for their own benefit, it might damage his data and assets.

4 Information Security Governance Framework

Information security governance frameworks include such elements as organizational structures, functional linkages, and operational protocols. Through the process of assigning responsibilities for Information Security Governance to different levels of the company's structure. PCI Data Security Standard, Federal Financial Institutions Examination Council, COBIT, Information Systems Security Association, and International Organization for Standardization (ISO) 27002 Data Security Standard are just a few of the many information security governance frameworks that banks can choose from. A few examples of popular information security governance frameworks are shown below.

1) International Organization for Standardization (ISO 27002)
   ISO, the world's leading creator and publisher of international standards, has taken into account information security management as one of the issues that have been covered by the international standards published by ISO. ISO 27002 is the international standard for information security practice. There are 11 control mechanisms and 130 security controls in information technology. Companies that have standardized their IT may find that the principles and fundamental concepts that govern their information security management's day-to-day operations, maintenance, and enhancement are easier to implement and adhere to.
2) Information Systems Security Association (ISSA)
   The Information Systems Security Association's (ISSA) primary objective is to advance practices throughout the information security industry, beginning at the executive level and working its way down to the individual practitioner level. The International Society for Information Security Analysts (ISSA) promotes collaboration and training to improve the security of information systems worldwide. The Information Security Professionals Association advocates for the protection, confidentiality, and accessibility of a company's valuable data (ISSA).

3) Control Objectives for Banking Information and related Technology (COBIT))
   In order to assist management and business process owners in identifying and controlling IT risks, the Information Systems Audit and Control Association & Foundation (ISACAF) developed an IT governance model. Plan and organize; acquire
   and execute; deliver, and support; finally, monitor and evaluate are the primary
   components of COBIT, (IT governing institute).
4) The Federal Financial Institutions Examination Council (FFIEC)
   Examiners from the Federal Financial Institutions Examination Council (FFIEC) use
   the FFIEC's "Information Security & IT Examination Handbook" while conducting
   compliance audits of financial institutions. The FFIEC has the authority to establish
   uniform principles, rules, and reporting formats for federal examination of financial
   institutions.
5) Data Security Standard (PCI DSS),
   American Express, Discover Financial Services, JCB International, MasterCard
   Worldwide, and Visa Inc. Are all members of the PCI Security Standards Council, an
   organization tasked with enforcing guidelines for the secure handling of credit card
   information. International designed it for operational level enhancement of payment
   account data security in a very particular and detailed manner. Requirements for
   network architecture, software design and processes, policies, and other important
   protection measures are included in the PCI DSS.
6) The Corporate Information Security Working Group (CISWG)
   The Corporate Information Security Working Group (CISWG) has developed guidelines and compiled a comprehensive list of security management resources with the
   goal of improving corporate data security. Risk assessments, technological controls,
   policy, and information security awareness are all components of information security that may be defined in this way. Security governance frameworks like this one
   help firms develop a strategy for securing their digital assets.

As shown in Table 1, many information security governance models have been widely
implemented (ISO 27002; ISSA; FFIEC; PCI-DSS; COBIT). An information security
governance structure for the banking system may be defined and constructed using this
method. Each approach's components were evaluated for their importance in terms of
essential principles or controls relating to information security. Components from several
techniques were integrated into a single category. Table 1 also demonstrates that ethical
behavior; auditor security program, trust, and corporate governance are not covered in
other frameworks, despite the fact that all four components are regarded significant by
researchers when managing information security applied to a firm.

It's worth noting that no single framework has all of the necessary elements. PCI, for
example, is a framework designed specifically for use at the operational level. Technical
security standards, such as ISO and COBIT, also exist [18]. In reality, no standards exist
that can fully regulate and meet the needs of individual businesses in the same way.
The reason for this is because of the wide range of economic operators, even within a
single industry, that exist. It is important to note that each company has a unique set of
characteristics, including its size and financial resources, as well as its culture, vision,
business strategy, and business model. Because of this, firms have varied views on the
value and significance of information security for achieving certain business objectives,

and they are ready to pay for it in different ways. Security policies and practices should be aligned with the company's long-term goals. The corporate ISG needs its own space within the larger corporate governance and IT governance framework. Each corporate governance framework has its own set of pros and cons; therefore, it is best to modify the framework so that it fits the demands of the individual company [19]. We consider X = Not Included and Y = Included.

Table 1. Information Security Governance Approach Comparison

ISG components	FFIEC	ISO 27002	COBIT	IISA	PCI
Information security strategy	X	X	Y	X	X
Leadership and sponsorships	Y	Y	Y	Y	X
Security return on investment	X	X	Y	Y	Y
Security metric and measurement	X	X	Y	Y	Y
Corporate governance	Y	X	X	X	X
Internal and external auditor information security program	X	X	X	X	X
Security program organization	Y	Y	Y	Y	X
Security policies, procedure, best practice, standards, and guidelines	Y	Y	Y	Y	Y
Compliance	Y	Y	Y	Y	X
Monitoring and auditing	Y	Y	Y	Y	X
Legal and regulatory	Y	Y	Y	Y	X
User awareness education and training	X	Y	Y	Y	Y
Ethical values and conduct	X	X	Y	Y	X
Privacy	X	X	X	X	X
Trust	X	X	X	X	
Certification against a standard	X	Y	Y	Y	X
Risk management and assessment process	Y	Y	Y	Y	Y
Best practice and baseline consideration	X	Y	Y	Y	X
Asset management	X	Y	Y	Y	Y
Physical and environmental controls	X	Y	Y	Y	Y
Technical operations	X	Y	Y	Y	Y
System acquisition, development, and maintenance	X	Y	Y	Y	Y
Incident management	X	Y	X	X	X
Business continuity planning	X	Y	Y	X	X
Disaster recovery planning	X	X	Y	X	X
User management	Y	Y	Y	X	Y

5 The Types of Threats Due to the Lack of ISG Can Be Classified as

It is possible to secure the banking sector's information assets from the dangers outlined in the literature study by adopting the first architecture of the proposed ISG framework to manage information security. Currently in a generic form, the proposed architecture for an information security governance program requires expert testing before it can be used to assure its efficacy in a real banking context. Additional components may be needed, while others may not be important as long as each organization operates under separate national and international laws and regulations. The first architecture of an information security governance framework, shown in Fig. 5 [20], is based on a mapping of information security components into the business hierarchy at the strategic, tactical, and operational levels.

1) Strategic Level
    The Strategic level incorporates the board of directors and the top executives.
    At this stage, a comprehensive literature review of the framework and all proposed texts is conducted. Information security strategies are produced by the leadership and governance subsystem to underpin effective information security programs. To ensure that the organization's objectives are met over the medium and long term, it is necessary to have an information security strategy and an IT strategy that are complementary to one another. To protect information assets at this level, an information security program needs the backing of upper management.. In reality, the ISG is an important element of the company's governance structure. All protected operations are managed and organized under the direction and supervision of the board and its leaders. According to IT governance, policies and processes are in place to manage the use of technology, while also ensuring the security of its data. Information security governance programs are evaluated at this level using the framework's measurement and notions of metrics. When it comes to measuring the performance of their information security programs in achieving the organization's goal, many businesses use metrics.
2) Tactical and Operational Level
    This level addresses user awareness; education and training as key component. This level is much related to the senior managers and operation managers. Organization for Economic Cooperation and Development (OCED) argues that one of the cornerstones is ethical behavior; trust and privacy are therefore important components that researchers have included. Invasion of privacy, sale of client data, and illegal data alteration are only a few of the risks that must be minimized by the company. As part of the company's security awareness program, personnel must adhere to certain ethical standards. This level also emphasizes the importance of trust. On one hand, it's important for managers to have faith in their employees' compliance with information security standards, but on the other hand they must have faith in their managers' ability to maintain their promises. To maintain an organization's reputation, trading partners and clients must have a mutual trust and respect for each other. This level also emphasizes the importance of privacy as a foundational element of building trust with consumers, suppliers, and other business. At this level, it's critical to think about things like program structure and compliance with applicable

regulations. Organization relates to the program's composition, reporting structure, and organizational architecture for data security. Various nations have different laws and regulations, thus legal and regulatory considerations are an important part of any program for information security governance. An organization's processes and compliance monitoring should be used to implement the recommended security policies. In order to offer management and staff with guidance and assistance, security policies comprise processes, standards, and recommendations. [21].

An example of an information security policy is an access control policy that includes regulations for e-mail, the Internet, as well as physical and environmental restrictions. In order to achieve the security policy's criteria, processes are backed by standards and guidelines, such as a standard for creating strong passwords. For effective management of sensitive data, this tier of the architecture must incorporate monitoring, compliance, and auditing. It is crucial to keep an eye on staff actions to ensure continued adherence to information security standards and prompt resolution of incidents.. Keeping tabs on employee activity might entail checking for things like the installation of unapproved software, the use of strong passwords, or even the websites visited. There must be an information security audit to make sure that the organization's policies, procedures, and controls are aligned with its goals and vision, the next year [22].

3) Technical Level

Employees of various levels of expertise are included in this category. An IT environment's security is a combination of both technological and physical measures According to several of the frameworks reviewed, information security governance programs should focus on technology protection and operations. The implementation of the security governance framework incorporates the management of network security, physical security, environmental security, business continuity measures, and user management assets. To stay up with market shifts, it is essential to maintain ongoing vigilance over the state of the technology ecosystem [23].

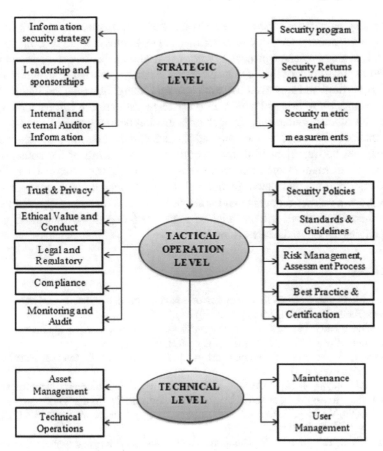

Fig. 5. Suggested ISG Framework's First Iterative Design

6 Conclusion

Information security is a major concern in a wide range of societal contexts, including the banking and financial institutes. The advancement of technology has been fast in the previous few decades. The financial system's networks are in place to make all of its services more accessible and secure connectivity. When information is sent between business partners, suppliers, and vendors through the internet, the need for high levels of information security becomes even more apparent. Unauthorized users posing a hazard to the network must be guarded against harmful assaults by taking appropriate precautions and measures. Open, accessible, affordable, and secure network connection and services are critical to banks' capacity to capitalize on new possibilities. It is beneficial for a firm to have a strong reputation for protecting information in order to retain and grow market share. Thus, comprehensive information security governance architecture is essential for financial information systems in order to provide strong protection. There are a number of broad standards and best practices, such as FFIEC, COBIT, ISO 27002, and the PCI data security standard, that may be utilized as a starting point for a bank's information security policy even if they address most of a company's particular needs. An information security governance framework with banking and IT systems in mind is being developed as part of an ongoing research project. Framework components from several sources have been used to create this design. An information security governance program must be thoroughly verified and approved by experts before being implemented in a real-world banking setting. A web-based poll was used to investigate the information security governance structure, and as long as each organization operates under separate national and international laws and regulations, additional components may be needed while others may not be important in different countries. As a result of the comprehensive study for the information security governance framework, the Control Objectives for Banking Information and related Technology (COBIT) model helps management and business process owners recognize and manage IT risks. While other types of ISG may address other organizations and financial institutes needs depend on laws and regulations in the countries they have been working in.

References

1. Kumar, M., Gupta, S.: Security perception of e-banking users in India: an analytical hierarchy process. Banks Bank Syst. **15**(1), 11 (2020)
2. Das, S.V.A., Ravi, N.: A study on the impact of e-banking service quality on customer satisfaction. Asian J. Econ. Finance Manag. **2021**, 48–56 (2021)
3. Hassana, M.K., Hassanb, A., Khodayerc, A.M., Khodayerd, O.M.: Internet security impact on e-banking users (2022)
4. Bah, C.U., Seyal, A.H., Yahya, U.: Combining PIN and biometric identifications as enhancement to user authentication in internet banking. arXiv Prepr arXiv210509496 (2021)
5. Tang, Z., Qin, Y., Jiang, Z., Krawec, W.O., Zhang, P.: Quantum-secure microgrid. IEEE Trans. Power Syst. **36**(2), 1250–1263 (2020)
6. Hayashi, V.T., Ruggiero, W.V.: Hands-free authentication for virtual assistants with trusted IoT device and machine learning. Sensors **22**(4), 1325 (2022)
7. Abass, I.A.M.: Social engineering threat and defense: a literature survey. J. Inf. Secur. **9**(04), 257 (2018)
8. Karagiannis, S., Papaioannou, T., Magkos, E., Tsohou, A.: Game-based information security/privacy education and awareness: theory and practice. In: Themistocleous, M., Papadaki, M., Kamal, M.M. (eds.) EMCIS 2020. LNBIP, vol. 402, pp. 509–525. Springer, Cham (2020). https://doi.org/10.1007/978-3-030-63396-7_34
9. Vohra, A., Bhardwaj, N.: Customer engagement in an e-commerce brand community: an empirical comparison of alternate models. J. Res. Interact. Mark. (2019)
10. Ullah, F., Qayyum, S., Thaheem, M.J., Al-Turjman, F., Sepasgozar, S.M.E.: Risk management in sustainable smart cities governance: a TOE framework. Technol. Forecast. Soc. Change **167**, 120743 (2021)
11. Kaur, J., Syan, A.S., Kaur, S., Sharma, R.R.: Understanding the factors influencing actual usage of payments banks: an empirical investigation using the extended information systems success model. FIIB Bus Rev. (2022). 23197145221099096
12. Louw, L.B., Esterhuyzen, E.: Disaster risk reduction: Integrating sustainable development goals and occupational safety and health in festival and event management. Jàmbá J. Disaster Risk Stud. **14**(1), 10 (2022)
13. Mounia Z, Bouchaib N. A new comprehensive solution to handle information security Governance in organizations. In: *Proceedings of the 2nd International Conference on Networking, Information Systems & Security.* ; 2019:1–5

14. Cindana, A., Ruldeviyani, Y.: Measuring information security awareness on employee using HAIS-Q: Case study at XYZ firm. In: 2018 International Conference on Advanced Computer Science and Information Systems (ICACSIS). IEEE, pp. 289–294 (2018)
15. Chaimaa, B., Najib, E., Rachid, H.: E-banking overview: concepts, challenges and solutions. Wirel. Pers. Commun. **117**(2), 1059–1078 (2021)
16. Daka, C.G., Phiri, J.: Factors driving the adoption of e-banking services based on the UTAUT model. Int. J. Bus. Manag. **14**(6), 43–52 (2019)
17. Aboobucker, I., Bao, Y.: What obstruct customer acceptance of internet banking? Security and privacy, risk, trust and website usability and the role of moderators. J. High Technol. Manag. Res. **29**(1), 109–123 (2018)
18. Moudoubah, L., Mansouri, K., Qbadou, M.: COBIT 5 concepts: towards the development of an ontology model. In: Maleh, Y., Alazab, M., Gherabi, N., Tawalbeh, L., Abd El-Latif, A.A. (eds.) Advances in Information, Communication and Cybersecurity. ICI2C 2021. Lecture Notes in Networks and Systems, vol. 357, pp. 247–256. Springer, Cham. https://doi.org/10. 1007/978-3-030-91738-8_24
19. AlGhamdi, S., Win, K.T., Vlahu-Gjorgievska, E.: Information security governance challenges and critical success factors: systematic review. Comput. Secur. **99**, 102030 (2020)
20. Garas, S., ElMassah, S.: Corporate governance and corporate social responsibility disclosures: the case of GCC countries. Crit. Perspect. Int. Bus. (2018)
21. Ahmad, S., Bhatti, S.H., Hwang, Y.: E-service quality and actual use of e-banking: explanation through the technology acceptance model. Inf. Dev. **36**(4), 503–519 (2020)
22. Chepkwony, F.C.: the influence of selected factors on the behaviour of share prices of commercial banks listed at the Nairobi securities exchange (2021)
23. Hijji, M., Alam, G.: A multivocal literature review on growing social engineering based cyber-attacks/threats during the COVID-19 pandemic: challenges and prospective solutions. IEEE Access **9**, 7152–7169 (2021)

Smart Autonomous Bike Hardware Safety Metrics

Menatalla Elnemr$^{(\boxtimes)}$, Hassan Soubra , and Mohamed Sabry

German University in Cairo, Cairo, Egypt
mennaelnemr@yahoo.com, {hassan.soubra,mohamed.ihab-sabry}@guc.edu.eg

Abstract. Automotive Safety Integrity Levels (ASIL) are the safety requirements set by the International Organization for Standardization (ISO 26262) protocol to ensure safety for all automotive road vehicles. As of 2018, the ISO 26262 standard has put into consideration motorcycles as a reference to any "two-wheeler system" and named the improved version of the Safety Integrity Levels (SIL) as Motorcycle Safety Integrity Levels (MSIL). There are considerable differences between ASIL and MSIL due to the differences between four wheeled and two wheeled vehicles, such as the protection level of the riders, the damage control and the different risks of their hardware components failing, as well as the weight of the four-wheeler systems in comparison to the two-wheeler systems. In this paper, an overview of the different safety requirements and standards is presented, as well as the protocols used to ensure maximum safety for both the rider and the vehicle. The metrics proposed which indicate whether the smart bike is safe to ride are also presented. The proposed approach is implemented on a real-life smart autonomous bike prototype. Furthermore, this paper provides guidelines on how to use Mean Time To Failure (MTTF) to ensure that the hardware does not fail during operation, which can cause safety hazards for riders and surrounding entities.

Keywords: Smart bike safety · Mean time to failure · Automotive safety integrity levels · ISO 26262

1 Introduction

An E-bike, also known as an electric bike, is an electric and power-assisted bike which is one of the fastest-growing technology of the bicycle industry. There are two main types of E-bikes; pedal assist and throttle assist. In a pedal-assist E-bike, as you pedal the bike, the motor gets powered and boosts the rider. A throttle-assist E-bike is similar to a normal motorbike, which operates as you accelerate the throttle.

Damage and failure of the hardware components of a vehicle can be fatal to the rider or pedestrians surrounding the vehicle. To reduce such damages and failures, the international standard for the functional safety of electrical, electronic, and programmable safety-related systems (IEC 61508) was introduced.

© The Author(s), under exclusive license to Springer Nature Switzerland AG 2023
F. P. García Márquez et al. (Eds.): ICCIDA 2022, LNNS 643, pp. 132–146, 2023.
https://doi.org/10.1007/978-3-031-27099-4_11

This standard ensures minimum risk and damages due to failures by guaranteeing that they implement the required Safety Integrity Levels-SIL. In addition, accidents and road incidents have increased rapidly in recent years and have resulted in many casualties, many of which are deadly. As a result, ASILs were introduced. ASILs are specific requirements that were implemented by ISO 26262 to minimize any risk and damages resulting from road failures.

Since cars and road vehicles are currently more common and are used more than bikes, there are significantly less safety standards and protocols in place to ensure safety for both the bike and the rider. Over the last couple of years, the use of motorcycles have increased rapidly and have gained more attention in the safety sector due to the increase in the accidents caused by the bike's behavior as well as the rider's. However, both normal and autonomous electrical bikes are still not covered by any safety standards.

This work proposes Hardware Safety Metrics for Smart Autonomous Bicycles based on measurements from a smart bike system components including, the voltages and temperatures of the motors responsible for the automated bike motion as well as the battery. The condition and safety of the hardware components of the bike were based on a range of optimum values compared to the values obtained from the tests applied on the bike. For the rest of the paper structure, Sect. 2 highlights the related work. In Sect. 3, the approach of creating bike safety integrity levels and safety requirements for smart autonomous bikes is discussed. In Sects. 4 and 5, the implementation of safety testing techniques of hardware components and the results are presented respectively. Lastly, the conclusion is presented in the last section.

2   Related Work

2.1   International Electrotechnical Commission (IEC 61508)

In [1–3], the origins of the IEC 61508 standard, its scope, and how it may be applied in real-life situations were discussed. In 1985, IEC 61508 introduced a generic standard for PESs; IEC 61508 standard. This standard was developed in 1998 to test and establish safety requirements for more complex electrical hardware systems. In 2010, the second edition and revision of the standard was published, the differences and key features were highlighted in [4].

However, even though the IEC 61508 standard has been used in several areas of the electronic world, it is limited in the sense that it does not offer safety requirements for automotive electrical systems that have been on the rise since the mid 2000s.s. Hence, the ISO 26262 standard was introduced.

Furthermore, in [5], it was proven that the placement of the IEC 61508 standard has not only made a huge difference in the electrical world, but has drastically increased awareness of functional safety. It also encouraged nonzero risk targets by implementing Safety Integrity Levels. IEC 61508 has also adopted the safety integrity levels of the UK Safety Health & Safety guidelines. They have been used by the standard since its first publication as a form of risk reduction and minimal damage and failures in electrical and electronic devices.

In addition, the protocol has proved to decrease rider injuries and accidents over the years due to its increase in functional safety awareness statistics show that most accidents and injuries are due to pedestrians, vehicles, and riders instead of component failures.

Safety Integrity Levels are classification levels used in safety-critical systems. They were used as Health and Safety Guidelines and were used by other standards such as IEC 61508 as well as other standards. There are five levels of SIL in total: SIL0, which represents no additional safety requirements, to SIL4, which represents high safety requirements demand. In SIL, architectural elements are assigned lower or equal SILs. When these SILs are combined, they fulfil the SIL of their parent functions. As a manual and time-consuming process, it has to be automated to deal with complex network architectures with multiple safety functions. In general, SILs are used to assign functional safety requirements to a system and to provide requirements for the implementation of critical functions. A function with higher criticality has to have a higher SIL. The distribution of the integrity levels can be arranged in various ways for different components. For instance, in a subsystem with SIL four, the components within this subsystem can have a SIL of three level two or a SIL of one level four. The techniques used for are based on probability using failure rates and reliability prediction models such as FMEA. Standards also consider the faults in software and hardware, which are considered systematic and introduced by humans in specification, design, manufacturing and installation.

2.2 International Organization for Standardization (ISO 26262)

ISO 26262 is an adaptation of IEC 61508 'Functional safety of electronic / electronic / programmed electronic safety related systems' initially applied to the specific requirements of light vehicles for passenger cars. This standard contains a total of 10 parts, 9 of which were published in 2011 while the last one was published in 2012 [6].

In 2018, the ISO 26262 standard was revised and republished with an addition of Motorcycle Safety Integrity Levels (MSIL). These are very similar to ASILs; however, they are mapped differently due to many differences in the architecture of the motorcycle compared to vehicles. Differences include the weight, the structure, and most importantly, the lack of protection due to the absence of a shield in the motorcycle. This was explained in [7].

In part 5 of the standard, titled 'Hardware Development', all the hardware components are designed, tested and validated. The testing and validation process is implemented using the FMEA analysis ; Failure Mode and Effects Analysis. This analysis uses failure modes to replicate real life situations and failure triggers to help deduce the causes and effects of certain failures of hardware components.

ISO 26262 has adopted SILs as a way to ensure safety and place minimal risks and damages, as these levels have been shown to aid in assessing damages and device failure as a form of failure reduction. However, since the standard is specifically for vehicle electrical devices, such as cars and road vehicles, they are now called Automotive Safety Integrity Levels (ASIL), which are determined using the HARA assessment in combination with the ASIL determination matrix shown in Table 1.

HARA's Assessment. The hazards derived during HARA are classified under three categories:Exposure (E) which is the measure of possibility of a system to fail or be in a hazardous situation. Severity (S), which is the extent of harm that may be caused to the driver and other occupants, in the event of occurrence of a hazard, and finally controllability (C), which determines the extent to which the driver of the vehicle can control the vehicle, if a safety goal is breached due to failure or malfunctioning of any automotive component. Each of these categories are subdivided into three more sub-levels; S1 (Light Injuries), S2 (Severe injuries) and S3 (Life threatening injuries) for Severity, E1 (Very low possibility), E2 (Low possibility) and E3 (Medium possibility) for Exposure and C1 (Easy to control), C2 (Moderately difficult to control) and C3 (Difficult to control) for Controllability.

The ASIL determination matrix deduced using the assessment process HARA is the most validated and safest way to allocate levels to the hardware components of road vehicles. This was highlighted and explained in [8].

Table 1. The ASIL determination matrix as shown in [8].

Severity class	Probability class	Controllability class		
		C1	C2	C3
S1	E1	QM	QM	QM
	E2	QM	QM	QM
	E3	QM	QM	A
	E4	QM	A	B
S2	E1	QM	QM	QM
	E2	QM	QM	A
	E3	QM	A	B
	E4	A	B	C
S3	E1	QM	QM	A
	E2	QM	A	B
	E3	A	B	C
	E4	B	C	D

2.3 Safety of Smart Autonomous Bikes

In [9], the dangers that may occur with smart bikes due to their different automotive nature and greater electric complexity are discussed. The study was carried out and reviewed in the Republic of Slovakia and was then compared with many countries in the EU region, such as Germany and the Netherlands, which have the highest populations of electric bike users. The speed of the bike was compared, as well as the users' opinion regarding their bikes and their road safety. In all countries, smart bike users thought that smart electric bikes were safe to ride. However, older people aged 60 and over were the users with the most problems with the smart bike. The speed of the smart bike seemed to affect and cause great discomfort for older users, this resulted in more accidents, injuries, and casualties for elder road users. In addition, older users complained about road safety in urban areas, especially where they felt the most unsafe and where the most injuries and accidents are reported. As concluded, the higher speed of the smart bike can lead to dangerous situations on the road. Second, the level of experience in riding an e-bike affects the way danger is perceived. Lastly, since some bikes lack assisted pedaling, it places older people at increased risk and higher chance of danger and accidents.

In [10], the safety effects of electric bikes and bicycles are conveyed through traffic conflicts and intersection approaches. Pedestrians crossing the lane in the presence of a bike or two electric bikes intersecting in two different directions are examples of topics discussed. To be exact, the article highlights 16 possible intersections that can cause an increase in incidents, including electric or smart bikes. The authors statistically showed that 77.5 percent of the incidents were due to reckless automobile drivers at intersections, 13.4 percent were due to the behavior of e-bike riders, and the remaining were due to the actions of regular cyclists. This concludes that users of smart bikes are in danger more often than when they are causing the danger at signalized intersections; which are proven to be present in thousands of road ways around the world. This review was conducted in China, where the population of smart bike users is leading worldwide, this aids with the study as it shows how it also affects experienced riders.

In [11], the difference between injuries, accidents, and damage to bikes was discussed in detail. The injury pattern of electric bike users resembled that of those who ride bicycles more closely than motorcyclists. Despite the fact that smart bike users were significantly more likely to wear protective gear as bike riders, smart bike accident patients were approximately 15 years older and had a greater incidence of significant head trauma than cyclist accident patients. In electric bike accidents, the frequency of pelvic injuries was double that of as in bicycle accidents, but the rate of upper injuries was higher in bicycle accidents. The injury pattern of bike riders on electric bikes is comparable to that of bikers. Higher speeds at the time of the accident, as well as various types of security and automobiles, could explain the disparities in injury patterns in motorcycle accidents. Head protection, which is considered a safety metric of the rider, similar to that offered to motorcyclists, is believed to benefit older untrained people who have a slower reaction time and less control over the electric bike.

3   Methodology

3.1   Motorcycle Safety Integrity Levels - MSIL

MSIL is a recent adaptation of ASIL implemented also by IS0 26262 that is specific to motorcycles and 'two-wheeler' systems on the road. In 2018, with the launch of a new version of the ISO 26262 standard, motorcycles were finally brought under the purview of ISO 26262 functional safety.MSILs were created due to the difference in impacts and damages that a motorcycle can cause compared to a vehicle for both road users, such as riders and pedestrians, and the actual item and its components. The same components in both a motorcycle and a vehicle can cause completely different levels of injuries and damage. For example, the difference in weight for instance as the well as difference in design. There are multiple modifications to the HARA system in MSILs due to the nature of motorcycles and because safety in a 'two-wheeler' ecosystem depends on multiple external factors such as helmets, protective gears, more training, etc. In addition, hazards that may occur due to the behavior of the motorcycle than a failure in its components are more likely to occur than with a vehicle. Hence, more emphasis was placed on rider behavior than on the machine components. The globally established level of technology in the motorcycle industry suggests that the ASIL classification is inappropriate for motorcycles. Hence, a counter-version between MSIL and ASIL classifications is established to match ISO 26262 to the worldwide capability of the motorcycle industry. The mapping of levels is the key difference in the determination of MSILs using the determination matrix, also known as the allocation table, is different for motorcycles since they are more prone to accidents and damage, mainly because they are lighter and more exposed, which is shown in Table 2.

Table 2. Mapping of MSIL to ASIL

MSIL	ASIL
QM (No safety requirements)	QM
A (The Lowest degree of safety hazard)	QM
B	A
C	B
D (The Highest degree of safety hazard)	C

The levels are different for motorcycles and 'two-wheeler' systems due to their proneness to more damage and higher risk of injuries to both the rider and the bike. This mapping table will be used to map the different levels of ASIL to Bike Safety Integrity Level (BSIL) since the smart bike is very similar to the motorcycle and its hardware components, and using the new mapped levels, a new determination matrix will be created specifically for the bike.

3.2 Mean Time to Failure - MTTF

Mean Time To Failure is a very basic measure of reliability used for non-repairable systems. It represents the length of time that an item is expected to last in operation until it needs to be replaced. MTTF can be used to represent the lifetime of a product or device. Its value is calculated by looking at numerous of the same kinds of items over an extended period and tracking how long they last. MTTF is significant because it aids in the estimation of product lifespans that are not repairable.As a reliability metric, MTTF is extremely helpful. Engineers can use it to predict how long a component will last as part of a larger machine. This is especially true when the failure of the equipment in question affects the entire company process. Because faulty parts must be replaced more frequently, a shorter MTTF entails more frequent downtime.

MTTF of Drives/Motors. Mean Time To Failure of drives and motors was discussed in [12]. A table was shown with the MTTF values for various drives, as well as the period of testing. The article discussed considering the given MTTF values for the given drives with safety functionality in mind. A function called 'Safe Torque Off' was introduced; It is a function that is intended for use in safety functions. STO is the basic foundation required for drive-based functional safety assessed by the ISO series protocols and IEC 65108 standards, as it brings a drive safely to a no-torque state. By blocking electrical impulses from power electronic devices to the motor, the STO function assures that no torque is applied to the motor. STO can reduce torque-generating energy instantly as an emergency stop function, allowing the motor to come to a complete stop using natural inertia and friction of the load. STO does not apply to DC brush motors, hence, the value of 33 years for DC drives is considered. On the other hand, STO applies to servo motors; with aid from safety standards and protocols, a new MTTF value of 10 years was calculated.

MTTF of Temperature Sensors. Temperature sensors are irreparable when exposed to any damage or failure, so their MTTF can be calculated. In [13], the MTTF for a temperature sensor is, on average, 20 years. Furthermore, the MTTF value can be increased by monitoring ambient temperatures and ensuring that they are within a range that helps the sensor to work optimally. The article also highlighted that measuring and operating at temperatures above 105 °C will gradually shorten the lifetime (also the MTTF) of the sensor, as well as it's 'useful lifetime' ; which is the lifetime the sensor will operate optimally.

3.3 Bike Safety Integrity Levels - BSIL

BSIL, will be the equivalent adaptation of the MSILs used to ensure safety in the autonomous bike. There are also 5 levels; QM, A, B, C, and D, all of which have the same characteristics and levels of the ASILs and MSILs. Firstly, we need to determine the different hardware components of the bike, and which of them would be vital in the determination of the safety of the bike. The hardware components of the smart autonomous bike are divided into three main categories: Perception; Core and System Management; Movement.

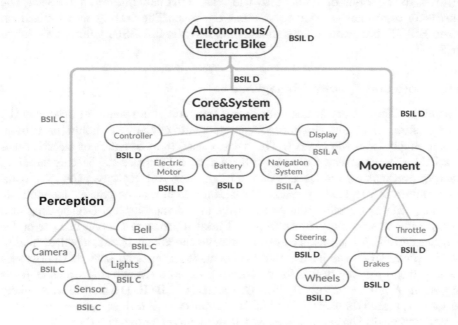

Fig. 1. Full BSIL failure tree

Each of these categories contains specific hardware components that are vital to the autonomous bike. The MSIL and ASIL mapping table in Table 2 will help with mapping the similar components found in both motorcycles and cars/vehicles to their corresponding bike safety integrity level. This mapping table is now shown in Table 3.

Table 3. Mapping of BSIL to ASIL

BSIL	ASIL
QM(No safety requirements)	QM
A (The Lowest degree of safety hazard)	QM
B	A
C	B
D (The Highest degree of safety hazard)	C

In Fig. 1, a complete BSIL tree was created to help visualize the different hardware components and to which category they belong. This will aid with determining the tree's overall BSIL if all the components fail, if only certain components fail, or even if nothing is in failure at all. The goal is to determine whether the bike is safe to ride prior to it moving, in order to minimize any injuries to the rider or damage to the bike. This can be done by testing the hardware components prior to starting the bike. The testing was focused on some BSIL-D hardware components, shown in the full BSIL failure tree shown in Fig. 1.

3.4   Core and System Management

Battery. The battery is the only component which is needed in order for the bike to start operating, the hardware components of the bike should get enough power supply from the battery. The battery used in most smart or electric bikes is a lithium-ion battery. Common failures of such batteries are battery swelling, battery overheating and leakage due to a puncture [14]. However, the most common failure is the battery pack overheating. This can be caused by multiple reasons, such as overcharging or excessive use of the battery. Overheating of a lithium-ion battery may cause fires and fatal injuries to anyone in near or far contact with it. If the battery overheats while the bike is in use, it will stop supplying power to all the hardware components it supplies to, such as the throttle for speed control, the brakes for deceleration or even the steering which controls direction. A failure in any of these components is BSIL-D and will most likely cause irreparable damage to the bike and injuries to the rider, and even vehicles or pedestrians in the premises. Excess heat also accelerates the thermal aging of the battery and decreases the MTTF, hence, shortening the lifespan of the battery greatly. The solution proposed is to measure the temperature of the battery prior to riding the bike, as well as every specified time period while the bike is in use. This will help prevent major damage and risks that may be caused due to the decreased power supply reaching the rest of the bike.

3.5   Movement

Steering. The steering is a vital component to the bike as it is the only component that controls the direction of the bike, left and right. The steering contains a DC motor for the actual movement of the steering component. If it fails, the consequences, damage, and injuries to the rider could be fatal, as seen from the full BSIL failure tree in Fig. 1 since it is level D, hence, testing these components prior to riding the bike is crucial. Over 90 percent of failures that occur in DC motors are due to overheating, which eventually causes the motor to stop working. The motor overheats as a result of two main reasons, temperature, and voltage. If the voltage going into the motor is too much, the motor may be able to handle it for a while but will slowly start overheating, and its performance will decline until it eventually stops working. If not enough voltage is going into the motor, it will be very slow and the steering may not work properly and may lag. In addition to too much voltage causing it to overheat, external factors and atmospheric temperatures may also cause overheating to the motor.

Brakes. The brakes are the only component on the bike that allows the rider to either slow down the bike or stop it completely, without it the damage and injuries to the rider will be fatal, hence it has the highest degree of safety integrity level as seen in Fig. 1. The brakes are controlled using a servo motor that helps the rider press the brakes for functionality. Similar to the throttle and steering, temperature is a huge factor that contributes to major risks and failures that could indefinitely damage the bike and severely injure the rider. Hence, testing the temperature over different periods of time may prevent and decrease these risks.

Throttle. The throttle is the component that is fully responsible for the bike's speed on the bike. it is used for speed and acceleration by the rider. The throttle is made up of a servo motor that is responsible for the movement in order to increase speed. If this component fails, the consequences are very unpredictable; meaning that the rider does know what to expect in damage to the bike or injuries that he/she may encounter, since there is no control over the speed of the bike at all. For instance, the bike may stall or misfire unpredictably. The servo motor is very similar to the DC motor in the sense that it may fail for very similar reasons; most of which are overheating. Similar to the brakes, temperature will be tested over the duration of the rider's bike ride to ensure maximum safety and be aware of any unexpected values.

4   Implementation

4.1   Voltage of DC Motor - Steering

The steering component on the smart bike uses a 20 V battery with a h-bridge to reduce the incoming voltage to 12 V for the DC motor that controls the movement ability of the steering. For optimal operation of the DC motor, 12 V to 12.5 V is the optimal and acceptable range for incoming voltage. However, if the voltage supplied is less than 12 V, the motor will operate at a much slower rate, which may cause lags and sudden stops of the motor, which would affect the entire steering component and its ability to move properly. Furthermore, an excessive amount of voltage over the suggested range will initiate overheating of the motor, which will eventually cause it to fail completely. This may occur prior to riding the bike, or during the rider's journey, in which case it may cause fatal injuries to the rider as they have no control over the movement of the steering. Excess voltage supply will, overtime, decrease the MTTF of the DC motor, shortening its life span and resulting in the quicker scheduling of a replacement. Using a resistive voltage sensor, the voltage supply to the steering's DC was measured.

4.2   Temperature of All Motors

All motors that make up the throttle, brakes, and steering have an optimal operating temperature range of 50 °C–70 °C in a nominal ambient temperature atmosphere that ranges from 25 °C to 35 °C on an average summer day, as well as a range of 7 °C–18 °C on an average winter day; these values are all considered in Cairo, Egypt. The motors will all work at a slower rate at lower temperatures, however, if the temperature is below –40 °C, the motor will not work at all. If the motor temperature exceeds 90 °C, it is a clear indication that there is excess heat; this drastically reduces the MTTF of the motor overtime. However, it will not cause the motor to stop working. On the other hand, if the temperature of one of the motors is 130 °C or higher, the motor will stop working since the internal insulation guard of the drive will not be able to handle such high temperatures and will no longer protect the motor from heat as it is destroyed, resulting in complete failure and irreparable damage to the motor. The higher temperatures will result in the constant and immediate reduction in MTTF, until the MTTF reduces to zero and the motor develops irreparable damages and stops working.

4.3   Temperature of the Battery

The battery is a sensitive hardware component and usually needs more care and supervision than the rest of the components, as it may be the most important due to its functionality; supply of power for all components to work. A lithium-ion battery operates optimally at a temperature between around 15 °C–35 °C [14]. Lower temperatures will cause a decrease and degradation in performance, and high temperatures will cause overheating. The excess heat of the battery pack will eventually cause damages, such as leakage, and will damage the battery indefinitely. The MTTF of the battery is dependent on the state of it, so, overtime, these damages will keep reducing the MTTF period and will result in an irreparable battery.

5   Results

5.1   Steering

Temperature. The DC motor used for the steering component of the bike reached the optimal temperature at minute 4, and remained within the suggested range of 50 °C–70 °C for the rest of the time the bike was used, as seen in Fig. 2. Even-though it started at a low temperature, it generated enough heat to increase its temperature to an optimum. These results suggest that the DC motor on the steering component is reliable and is within the acceptable MTTF period. This is suggested because there were no inconsistencies in the temperature values, nor was excess heat or overheating produced before, during, or after the bike ride.

Voltage. The voltage that should be supplied to the DC motor in the steering ranges from 12V–12.5 V for optimal functionality. The voltage was tested for a 10-minute bike ride, all the values were within range and full voltage is being supplied; the voltage is constant during the entire usage by the rider, meaning the DC motor is functioning without any failures or hazards. These results suggest

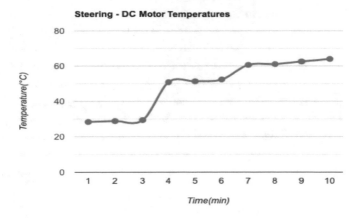

Fig. 2. Steering's DC Motor Temperature results

that both the battery from which the voltage is supplied and the DC motor in operation are in their acceptable MTTF period, which is deduced from the consistent results obtained.

5.2 Battery

The optimum temperature range for a lithium-ion battery is 15 °C–35 °C. In Fig. 3, it is evident that it started and remained in the optimal range until minute 7. Then it started to produce excess heat and exceeded the range by 5 °C. It remained at 40 °C for the remaining of the bike ride, which did not cause any damage. These results were monitored very carefully as any irregularity in temperature, unlike the other components, could immediately affect the battery, as it is significantly more sensitive and has a much smaller acceptable range of temperatures. Being more sensitive, the battery, as seen in Fig. 3, detected more heat from the surroundings than the other components. This increase in temperature is a result of both the sensitivity of the battery to its surroundings and its MTTF. The excess heat produced suggests that the battery may be closer to the end of its MTTF period, which is causing the irregularity in results and temperature values but not causing it to stop working at all, since it still hasn't reached the end of its MTTF.

5.3 Throttle

The temperature remained at 62.5 °C for the rest of the bike operation period. As shown in Fig. 4, the temperature remained in the acceptable range of 50 °C–70 °C and worked optimally from minute 7, with a temperature greater than 50 °C. It also indicated no overheating or excess heat production, as the temperatures remained below 90 °C. This suggests that the throttle's servo motor is in great condition and is reliable. With the MTTF of the servo motor being 10 years, the

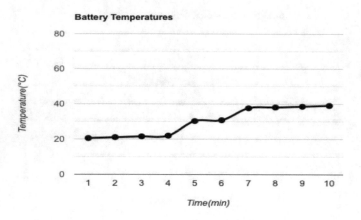

Fig. 3. Battery's Temperature results

temperature values obtained suggest that the servo motor has not been operating for that long.

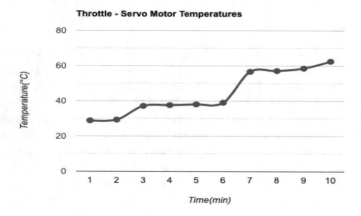

Fig. 4. Throttle's Servo Motor Temperature results

5.4   Brakes

Unlike the throttle, this servo motor did not have an abrupt and sudden increase in temperature, it was more subtle and gradual. It also reached the optimal temperature range 3 min faster than the throttle. It also remained at 63 3 °C for the rest of the bike's operation period. This suggests that overall, the brakes' servo motor has been operating for a shorter period of time than the throttle's servo motor, as well as not being as far into its MTTF period as the throttle's servo motor. This is shown in Fig. 5.

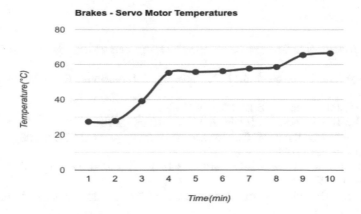

Fig. 5. Brakes' Servo Motor Temperature results

6   Conclusion

Bike Safety Integrity Levels (BSIL), an adaptation of ASILs, were introduced in this paper to ensure that functional safety requirements are provided for smart autonomous E-bikes. BSILs were created by mapping ASIL and MSIL using the HARA assessment analysis method in addition to the determination matrix. Some of the hardware components with the highest degree of automotive hazard; BSIL-D, were grouped for testing using three hardware safety insurance metrics ; Temperature, Voltage, and Mean Time to Failure (MTTF). The steering, throttle, brakes, and battery of the smart bike were analyzed for their most common causes of failure and damages. This analysis shows that these components are made up of two specific drives ; servo motors and DC motors, excluding the battery. It was also found that the main causes of damage to these drives were overheating and lack of a sufficient voltage supply. Hence, the temperature of the throttle's servo motor, the brakes' servo motor, the steering's DC motor and finally, the battery were all tested over the course of a 10-minute bike ride. The temperatures of all the motors remained in the acceptable optimal temperature range for the whole course of the bike ride. However, the temperature of the battery produced excess heat in the second half of the ride. This suggests that the battery is approaching the end of its suggested MTTF period, unlike the rest of the motors where their results conveyed that they are in a healthy condition, suggesting that they are not approaching the end of their MTTF period. Furthermore, the voltage supplied to the DC motor on the steering component was also tested and measured to ensure that the responsible battery supplied enough voltage to operate the steering component of the bike. The initial feasibility testing process showed that the required voltage was supplied for the entire 10-minute bike ride, suggesting that both the battery and the DC motor are robust and have a long MTTF period remaining. Testing over larger periods of time is still required for more accurate estimates.

References

1. MTL Instruments Group plc.: an introduction to Functional Safety and IEC 61508 (2002)
2. International Electrotechnical Commission: International IEC Standard 61508-1 (1998)
3. Ron, B.: Introduction to IEC 61508 (2006)
4. Ron, B.: Introduction & Revision of IEC 61508 (2010)
5. Foord, A.G., Gulland, W.G., Howard, C.R..: Ten years of IEC 61508; has it made any difference? (2011)
6. Road vehicles - Functional safety - Part 1 : Vocabulary. https://www.iso.org/standard/43464.html
7. Rami, D.: Overview of the 2nd Edition of ISO 26262. Functional Safety-Road Vehicles (2018)
8. Messnarz, R., et al.: IMplementing functional safety standards – experiences from the trials about required knowledge and competencies (SafEUr). In: McCaffery, F., O'Connor, R.V., Messnarz, R. (eds.) EuroSPI 2013. CCIS, vol. 364, pp. 323–332. Springer, Heidelberg (2013). https://doi.org/10.1007/978-3-642-39179-8_29
9. Gogola, M.: Are the e-bikes more dangerous than traditional bicycles?. In: 2018 XI International Science-Technical Conference Automotive Safety, pp. 1–4 (2018)
10. Bai, L., Liu, P., Chen, Y., Zhang, X., Wang, W.: Comparative analysis of the safety effects of electric bikes at signalized intersections. Transp. Res. Part D: Transp. Environ. **20**, 48–54 (2018)
11. Spörri, E., et al.: Comparison of injury patterns between electric bicycle, bicycle and motorcycle accidents. J. Clin. Med. **10**(15), 3359 (2021)
12. MTTF (Mean Time To Fail) data for Standard drives, High performance AC drives, Servo drives and DC drives. https://www.nidec-netherlands.nl/media/3911-engineering-documentatie-mttf-data-en-iss8-gen049.pdf
13. Webber, A., Haj-Omar, A.: Calculating Useful Lifetimes of Temperature Sensors (2018)
14. Ma, S., et al.: Temperature effect and thermal impact in lithium-ion batteries: a review. Prog. Nat. Sci. Mater. Int. **28**(6), 3359 (2018)

Demand Side Management Based Cost-Centric Solution for a Grid Connected Microgrid System with Pico-Hydro Storage System

Tapas Chhual Singh[1] , K. Srinivas Rao[1], Bishwajit Dey[1] ,
and Fausto Pedro Garcia Marquez[2](✉)

[1] GIET University, Gunupur, Odisha, India
[2] University of Castilla-La Mancha: Ciudad Real Campus, Rectorado UCLM, C. Altagracia, 50, 13001 Ciudad Real, Spain
faustopedro.garcia@uclm.es

Abstract. This paper uses a hybrid intelligence algorithm to reduce the producing cost of a low-voltage microgrid system equipped with a micro-turbine, fuel cell, solar system, and pumped hydro storage. Demand side management (DSM) has been implemented as the cost- management tool and the positive impacts of DSM have been analyzed in detail. Result shows that inclusion of pico hydro storage system helped in dipping the generation cost of the MG structure which was further lowered by implementing DSM strategy. Among positive impact of DSM are peak reductions and load factor improvement which is also portrayed in the results. Incorporating a pumped hydro storage system along with the RES (solar PV system) helped in minimizing the generation cost. This happened primarily because the pumping mode of pumped storage was powered by the PV system and no fuel cost was incurred. Projected hybrid WOASCA yield superior class outcome than the other algorithms compared with.

Keywords: Demand side management · Pico hydro storage · Microgrid · Optimization · Distributed generation

1 Introduction

1.1 General Overview

As per parameter optimization, every components associated with the operation don't have the same value to accommodate a same demand, and several of them are more expensive even though they contribute the same level of electricity [1, 2]. As a consequence, optimizing the allocation of a certain proportion of national consumption is rather critical for lowering fuel prices[3]. As a result, the electricity generated satisfies the load requirement [4]. Resulting in additional limitations placed by distributed generation units (DGUs) and durations, ELD has significantly more complexities [5]. The distributed generation mechanism (DGM) of a microgrid can be related to the lowering costs of highly dynamic dispatch, with venture, halted durations of DGUs, and charging

© The Author(s), under exclusive license to Springer Nature Switzerland AG 2023
F. P. García Márquez et al. (Eds.): ICCIDA 2022, LNNS 643, pp. 147–167, 2023.
https://doi.org/10.1007/978-3-031-27099-4_12

or discharging condition of batteries as many of the major technical challenges while tackling development of economic dispatch[6]. In the current situation, optimum dispatching is a critical power system optimization technique for dependable and efficient functioning, with a focus on microgrids. A collection of distributed generating units (DGUs) with a limited capacity for a given area is what is known as a microgrid. [5, 7]. The Distributed generation resources (DERs) try to make up for petroleum systems, energy storage, small turbines, and other non-renewable resources that are suited for microgrid locations[8]. Energy storage systems (ESS),that might be a Pumped storage system or flywheel or battery are also included in DGUs. The unique structure and characteristics of every microgrid necessitate specialized simulation with specific constraints, converting load flow into a challenging optimization problem for engineers and scientists to solve[9]. The three most popular types of microgrid operation are utility connected,electrically isolated and storage [10]. The effectiveness type of process is known designate, better and sustainable owing to a microgrid's right to offer energy based on huge excess electricity production of specific DGUs. The durability component is raised in utilities connected state because it can rely just on system with in case of DGU malfunction, preventing an unanticipated system termination.

1.2 Literature Review

Distributed generation is viewed as a potential electricity provider mostly in coming years, with various study papers published in the recent decade. [11, 12]. The matrix real coded genetic algorithm (MRCGA) has been employed in a grid-connected microgrid to save production costs.. Heuristic efficiency is determined using major considerations, operational limits, and varying electricity costs [13]. Kasaei [14] created an alternate power station and used an expansionist economic strategy to govern its operational and financial aspects. Basu and Chowdhury [15] applied Cuckoo Search Algorithm (CSA) was used to explore various transient and steady state optimal power flow problems, and the results were correlated with differential evolution (DE) and particle swarm optimization (PSO). The system comprises of 2 wind generators that were controlled by airspeed. Moghaddam et al. [16] conducted research in a distributed generation system with utilities attached and achieved economic-emission dispatch using a adaptive modified PSO (AMPSO), while Rabiee et al. [17] employed a novel multi tasking algorithm i.e. modified imperialist competitive algorithm (MICA). This work addresses an issue that is analogous to those discussed in this study. This research covers a period of time whenever the utility does not contributing to the micro - grid by purpose or by accident, which is not covered in the research. Six potential grid involvement situations are investigated in this study for an island network operated in low voltage (LV). References [17–19] worked on a three-unit island microgrid system with solar PV and wind systems, respectively, the interior search algorithm (ISA), modified harmony search algorithm (MHSA), and whale optimization algorithm (WOA) were used to probe fiscal load and emission dispatch and combined economic emission dispatch (CEED). On an island microgrid system, references [20–22] employed the memory-based genetic algorithm (MGA), modified personal best PSO (MPBPSO), and artificial fish swarm algorithm (AFSA).. Ramli [23] also looked into two other possibilities for the above-mentioned test system, including an energy storage unit (ESU) and a grid-linked island microgrid for unidirectional power flow, by increasing the power demands to 10% of its present value.

Guezgouz, Mohammed, et al. [24] described an effective control mechanism with combined pumped hydro-battery energy storage (BES) system with energy producers like wind and sun. The mechanism provided degree of consistency at a reasonable cost and the results compared to single battery and pumped storage systems. Naik, K. Raghavendra, et al. [25] proposed SPMS effectively runs the DC Microgrids micro hydro power plant MHPP in a manner for long-term output current may be controlled via distributed power inputs with maximal contribution of MHPP when load transient conditions while taking into account overall C-rate constraints of battery bank. Tyagi, et al. [26] employed a Small Hydro Energy Conversion SHEC system for the drawing out of active and reactive components of the load currents using an orthogonal signal generator OSG technique in the standalone mode. Asiful, et al. [27] able To minimize the outlay and burden of the system A coreless axial flux permanent magnet (AFPM) generator is worn The single stator double rotor (SSDR) method to make certain about better power density. Carravetta, et al. [28] employed a pump as turbine (PAT) which is a best electro-mechanical device for effective production of low cost and consistent electrical power for the rural area with different connection methods. Kusakana [29] developed a dynamic model using wind pumped pico hydro power generation system identified as 'hydro aeropower' using optimal scheduling of the power generation through hydro turbine. Tapia, et al. [30] identified the problems on small micro hydro power plants design considerations in isolated area, where machinery and resources are inadequate. In the proposed design Pelton wheels used for the effective generation of small hydro power and optimized through integer programming. Kamran, et al. [31] employed a grid tied small hydel power plant for the electricity generation and to supply the nearby 11 kV substation during peak hour, whilst the HOMER application software is utilised for simulation and optimization and levelized cost of energy calculation (LCOE). Kusakana [32] developed a hybrid model for pico hydro power generation in association with solar PV system to get rid from dynamic pricing of electricity and load demand which can mitigate by the use of a water storage system with optimal scheduling of power flow to the different loads from various sources. Javed, et al. [33] briefed about the various hybrid techniques for the development of pumped hydro storage system (PHS) such as wind – hydro, solar PV – hydro, wind-solar PV- hydro power plant by considering various aspects like efficient, ecological and technological aspects of combined systems. Discussed about the hybrid storage system PHS, which is a nascent choice for the lifetime storage facility.

1.3 Paper Contribution

The crucial novel contributions of the article to the state-of-art are.

- For a microgrid system, the whole cost of generating is assessed.
- For the relevant microgrid system, four distinct eventualities are examined.
- Power production costs are assessed with and without different levels of DSM input.
- A quick and simple innovative WOASCA suggested article's optimization technique.

- Events of innermost propensity with algorithm running point support the method's effectiveness also resilience. The algorithm's efficiency and resilience are supported by its propensity and system performance.

1.4   Paper Orientation

This section outlines the overall structure of the article: In addition to accounting for the equality and inequality constraints, modeling of Pico hydro storage, and DSM, Sect. 2 presents the performance index to be decreased. The optimization tool is described in full in Sect. 3. Section 4 includes case studies of the microgrid test systems, together with the key findings, the approach's validity, and an evaluation of the solutions. Section 5 concludes the entire work by summarizing its findings.

2   Formulation of an Objective Function

2.1   Micro-grid System Active Power Generation Rate Calculation

Equation (1) represents the functional form of the objective function equation for a grid-connected micro-grid system.

$$ECD_G = \sum_{t}^{24} \sum_{k=1}^{m} \left(F_k * G_{k,t} + C_{Grid,t} \cdot *G_{Grid,t} \right) \tag{1}$$

where F_k and G_k, respectively, represent the Kth DG unit's power output and fuel cost coefficient. Cgrid is the price at which the grid purchases and sells electricity.

2.2   Operating Constraint

The DERs and grid are bound to the operating constrains given by Eqs. (2) to (5)

$$\sum_{k=1}^{m} G_{k,t} = D_t \tag{2}$$

$$\sum_{k=1}^{m} G_{k,t} + P_{RES,t} = D_t \tag{3}$$

$$G_{k,min} \leq G_k \leq G_{k,max} \tag{4}$$

$$G_{grid,min} \leq G_{grid} \leq G_{grid,max} \tag{5}$$

The demand is D_t for hour t and while the RES output is $P_{RES,t}$ in terms of power.

2.3 Pumped Hydro Storage Modeling

2.3.1 Pumping Mode

Equation (6) is used to calculate the amount of power required to pump water from the lower reservoir to the upper reservoir.

$$P_{GP} = \frac{\rho_w \times g \times h \times Q_P}{\eta_{GP} \times t} \tag{6}$$

where P_{GP} the allocate of power from the grid used to supply the load (W); η_{GP} is the efficiency of the pumping system, Q_P is the pumping flow rate (m3/s), g is the gravity (9.8 m/s2), h is the useful pumping head (m);, t is the measured time.

2.3.2 Pico Hydro Power Generation During Turbine Mode

Pico Hydro Power generation during Turbine mode is given as:

$$P_{TG} = \frac{\rho \times g \times h \times Q_{TG}}{t} \tag{7}$$

where Q_{TG} is turbine flow rate (m3/s); P_{TG} is the hydro generating power; h is the water head (m).

2.3.3 Upper Reservoir

The reservoir's potential energy is determined by:

$$E_R = \rho \times V_R \times g \times h \tag{8}$$

where V_R is the size of the reservoir (m^3), E_R is potential energy (kWh);

2.3.4 Pico Hydro Constraints

The surface runoff variation across top and bottom reservoir, proportionate to the gravitational force, gross height, and fluid density, as modelled for the pumping and turbine operations, accordingly, determines the horsepower transferred between the PHS and the system (Eqs. 9 and 10). in addition (Eq. 11) limited the flow range of water.

$$P_t^{PHS,t} = \frac{g.H.\rho.q_t^{PHS,t}.\eta^{PHS}}{1000}; \forall t \in T \tag{9}$$

$$P_t^{PHS,p} = \frac{g.H.\rho.q_t^{PHS,p}}{1000.\eta^{PHS}}; \forall t \in T \tag{10}$$

$$u_t^{PHS,i}.\underline{q}^{PHS} \leq q_t^{PHS,i} \leq u_t^{PHS,i}.q^{-PHS}; \forall t \in T \wedge i \in \{p,t\} \tag{11}$$

The PHS system, like the BES, could indeed operate throughout the pumping and turbine modes simultaneously, that is assured by providing (Eq. 12). The bitwise parameters linked to the PHS operation, however, are coherent according to (Eq. 13).

$$\sum_{\forall t \in \{p,t\}} \left\{ u_t^{PHS,i} \right\} \leq 1; \forall t \in T \tag{12}$$

$$on_t^{PHS,i} - off_t^{PHS,i} = u_t^{PHS,i} - u_{t-1}^{PHS,i}; \forall t \in T \backslash t > 1 \wedge i \in \{p,t\} \tag{13}$$

Equations 14 and 15 represent the momentary quantity of water kept throughout the top and bottom reservoirs, appropriately. Additionally, the capacities of both ponds and a reasonably basic volume must be used to restrict the quantity of water that can be held in each, as presented in (Eq. 16).

$$v_t^{Upper} = v_{t-1}^{Upper} + 3600 . \Delta\tau(q_t^{PHS,p} - q_t^{PHS,t}); \forall t \in T \backslash t > 1 \tag{14}$$

$$v_t^{Lower} = v_{t-1}^{Lower} + 3600 . \Delta\tau(q_t^{PHS,t} - q_t^{PHS,p}); \forall t \in T \backslash t > 1 \tag{15}$$

$$\underline{v}^{PHS} \leq v_t^i \leq v^{-PHS}; \forall t \in T \wedge i \in \{upper, Lower\} \tag{16}$$

Comparable to the BES, the PHS has been mandated to maintain identical starting and ending power storage that is determined by (Eqs. 17 and 18). Last and not least, the scaling restriction for PHS is given by (Eq. 19)

$$v_{t=1}^{Upper} = v_{t=Size(T)}^{Upper} = v^{-PHS}; \tag{17}$$

$$v_{t=1}^{Lower} = v_{t=Size(T)}^{Lower} = \underline{v}^{-PHS}; \tag{18}$$

$$p_{t-1}^{PHS,i} - R^{PHS} \leq p_t^{PHS,i} \leq p_{t-1}^{PHS,i} + R^{PHS}; \forall t \in T \backslash t > 1 \wedge i \in \{p,t\} \tag{19}$$

2.4 Demand Side Management Strategy (DSM)

Demand Side Management (DSM) [34, 35] is a variable methodology something which operators can utilise to reduce maximum load demand while lowering costs. Users are generally given bonuses by electricity companies to motivate individuals to participate in the process. The kinds of load structuring technologies outlined in the DSM domain include valley filling, peak clipping, load shifting, strategic conservation, strategic load growth, and variable load structure. The first three are fundamental strategies, and the others are specialized forms that control quantity demanded by restricting or extending it according to the unit's management. The most recommended load control solution is load changing, that is a combination of valley filling and peak clipping. Load shifting can be accomplished with the use of controllable equipment. Final load shifting can be accomplished by employing controllable loads. With little or no difference in energy usage, the load shifting approach is utilised to shift controllable loads from peak to off-peak times.

The following are among the key characteristics of peak load modelling restructure utilising DSM methods:

1. The System's peak usage request does not surpass the initial predicted consumption after using the DSM approach.
2. The minimal requirement after using the DSM approach doesn't really surpass the device's initial lowest demand.
3. There at end of each day, the system's total and average load demand are same in both the cases i.e. with and without DSM.

The algorithm to implement DSM is declared as follows:

Algorithm 1. A scattered algorithm is implemented by each end user $n \in N$.

1: Establish the load allocation for the nth consumer (xn) and all other users excluding the nth user at random $(x - n)$
2: Revert to step 1 once more.
3: Use a linear integer programme to solve the following equation.

$$\underset{x_n = \forall n \in N}{Minimize} \sum_t q_t^{FRM} * \left(\sum_{n \in N} \sum_{q \in Q_n} x_{n,a}^t \right) \tag{20}$$

'n' is the index of user, while N is the total number of residential users who are actively registered. A is the collection of home appliances, and q_t^{FRM} denotes the tariff model constant at time interval t.
4: While X_n changes,
5: Update X_n 's value and make the information available to other users.
6: end if.
7: Change the value of x–n in accordance with any additional proclamation updates received from the consumers
8: end if.
9: Continue until no consumer update announcement is received (Fig. 1).

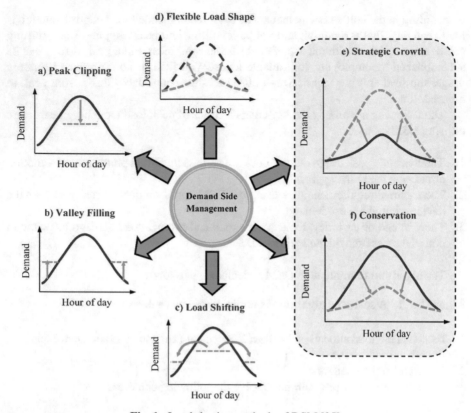

Fig. 1. Load shaping methods of DSM [35].

3   Whale Optimization Algorithm (WOA) Hybrid Approach

This programme was created by modelling the way that humpback whales hunt with bubbles net and how they catch prey. [36, 37].

3.1   Looking for the Prey Phase of Exploration

Rather than using the location of the finest whale found, during this phase, the location of a whale is altered using the position of a whale that was chosen at random. Mathematically is given by (21)–(22):

$$\vec{G} = \left| \vec{C} \cdot \vec{Y}_R - \vec{Y}_W \right| \tag{21}$$

$$\vec{Y}_W(iter + 1) = \vec{Y}_R - \vec{A}\,\vec{G} \tag{22}$$

The whale's arbitrary location vector is Y_R.

$$\vec{A} = 2.\vec{a}.\vec{r}_1 - \vec{a}$$
$$\vec{C} = 2.\vec{r}_2 \tag{23}$$

where $\vec{r}_1, \vec{r}_2 \in [0, 1]$ $[0, 1]$

In the Eq. (25), iteration-wise Vector 'a' changes from 2 to 0.

$$a = 2 * (1 - \frac{iter}{Max\_iter}) \tag{24}$$

When preys are discovered, whales encircle them, given by (25)–(26):

$$\vec{G}' = \left| \vec{C} \times \vec{Y}_p(iter) - \overrightarrow{Y_W}(iter) \right| \tag{25}$$

$$\overrightarrow{Y_W}(iter + 1) = \vec{Y}_p(iter) - \vec{A} \times \vec{G}' \tag{26}$$

where iter: the iteration,

YP: location vector of the prey.

YW: the location vector of the Whale.

3.2 Attacking Strategy Using Bubble-Net

The invading approach by bubble-net provides the stage of penetration and escalation, which is carried out by reducing the size of the encircling technique. The first approach relies on lowering the value of the 'a' stated above. A is also lowered by a [–a,a], when A is taken into account as the probability variable within the period variance range. The updated coil locations are provided by (27)

$$\overrightarrow{Y_W}(iter + 1) = \vec{G}''.e^{bl}.\cos(2\pi l) + \overrightarrow{Y_W}(iter) \tag{27}$$

where,

$$\vec{G}'' = \left| \vec{Y}_p(iter) - \overrightarrow{Y_W}(iter) \right| \tag{28}$$

Here, l is a random number lies between $[-1,1]$, is the gap between ith whale to its prey, where as the shape of the logarithmic spiral is defined by the constant b.

In this work, the concepts of the circular or diminishing encircling mechanisms are taken into consideration with a possibility of 50%. The functional form is (29):

$$\vec{Y}_p(iter + 1) = \begin{cases} \vec{Y}_p(iter) - \vec{A}.\vec{G}' & if \quad p < 0.5 \\ \vec{G}''.e^{bl}.\cos(2\pi l) + \vec{Y}_p(iter) & if \quad p \geq 0.5 \end{cases} \tag{29}$$

where p is a random number in between [0,1].

3.3 Hybrid WOASCA

This portion integrates the aforementioned WOA with a SCA, defined as WOASCA [37], in order to lower the computational price and boost the robustness of the concern method. The formulation of proposed hybrid WOASCA after incorporating necessary modifications by the involvement of SCA is explained in detail in Table 1 below.

Table 1. Formulation of proposed hybrid WOASCA

SL NO	Mathematical Equations	Traditional WOA	Hybrid WOASCA
1	Tuning Parameter	$a = 2 * (1 - \frac{iter}{Max\_iter})$ (24)	$a = 1 + \cos\left(\pi * \frac{iter}{Max\_iter}\right)$ (30)
2	Exploration: Searching the preys	$\vec{G} = \left\| \vec{C} . \vec{Y}_R - \vec{Y_W} \right\|$ (21)	$\vec{G} = rand_1 * \sin(rand_2)* \left\| \vec{C} . \vec{Y}_R - \vec{Y_W} \right\|$ if rand < 0.5 $\vec{G} = rand_1 * \cos(rand_2)* \left\| \vec{C} . \vec{Y}_R - \vec{Y_W} \right\|$ Otherwise (31)
3	Exploration: Encircle the preys	$\vec{G}' = \left\| \vec{C} \times \vec{Y}_p(iter) - \vec{Y_W}(iter) \right\|$ (25)	$\vec{G} = rand_3 * \sin(rand_4)* \left\| \vec{C} \times \vec{Y}_p(iter) - \vec{Y}_W(iter) \right\|$ if rand < 0.5 $\vec{G}' = rand_3 * \cos(rand_4)* \left\| \vec{C} \times \vec{Y}_p(iter) - \vec{Y}_W(iter) \right\|$ Otherwise (32)
4	Exploitation stage	$\vec{G}'' = \left\| \vec{Y}_p(iter) - \vec{Y}_W(iter) \right\|$ (28)	$\vec{G}'' = rand_5 * \sin(rand_6)* \left\| \vec{Y}_p(iter) - \vec{Y_W}(iter) \right\|$ $\vec{G}'' = rand_5 * \cos(rand_6)* \left\| \vec{Y}_p(iter) - \vec{Y}_W(iter) \right\|$ if rand < 0.5 Otherwise (33)

3.4 Comparison of the Hybridized WOASCA and the Issue of Electricity Consumption

In the case where D is the number of DERs involved in feeding the microgrid, T is the time period for the most advantageous configurations, and N is the number of particles in the resident's matrix, represented as (34). Constraints (2) through (11), which relate to the population's element, state that each particle's size and dimension are (D*T). Location of the whales, or the density matrix's constituent elements, provides the possible alternatives set. The target system gives the range between the whale and its prey.

$$POP = \begin{bmatrix} pop^1_{1,DER1}, & pop^2_{1,DER1}, & \cdots pop^T_{1,DER1}, & pop^1_{1,DER2}, & pop^2_{1,DER2}, & \cdots pop^T_{1,DER2} \cdots pop^1_{1,DER\,D}, & pop^2_{1,DER\,D}, & \cdots pop^T_{1,DER\,D} \\ pop^1_{2,DER1}, & pop^2_{2,DER1}, & \cdots pop^T_{2,DER1}, & pop^1_{2,DER2}, & pop^2_{2,DER2}, & \cdots pop^T_{2,DER2} \cdots pop^1_{2,DER\,D}, & pop^2_{2,DER\,D}, & \cdots pop^T_{2,DER\,D} \\ pop^1_{3,DER1}, & pop^2_{3,DER1}, & \cdots pop^T_{3,DER1}, & pop^1_{3,DER2}, & pop^2_{3,DER2}, & \cdots pop^T_{3,DER2} \cdots pop^1_{3,DER\,D}, & pop^2_{3,DER\,D}, & \cdots pop^T_{3,DER\,D} \\ \cdots\cdots\cdots & & & & & \\ \cdots\cdots\cdots & & & & & \\ pop^1_{N,DER1}, & pop^2_{N,DER1}, & \cdots pop^T_{N,DER1}, & pop^1_{N,DER2}, & pop^2_{N,DER2}, & \cdots pop^T_{N,DER2} \cdots pop^1_{N,DER\,D}, & pop^2_{N,DER\,D}, & \cdots pop^T_{N,DER\,D} \end{bmatrix} \quad (34)$$

4 Results and Discussions

4.1 A General Description of the Test System and Work Environment

The subject low voltage grid connected microgrid system is showed in Fig. 2. It comprises of micro-turbine system (MT), fuel cell (FC), PV system and pico-hydro (pumped) energy storage system. The operating ranges and bid prices of MT and FC are displayed in Table 2. Since the pumping mode of the pico-hydro system was supported by the PV system, hence no price bid is associated with it. The system's grid pricing and hourly load demand are shown in Fig. 3. Figure 4 depicts the hourly electricity produced by a PV system on a typical day with poor weather (cloudy). Four optimization algorithms were implemented along with the proposed hybrid WOASCA for evaluating the lowest production price for four different instances. Using a computer with an Intel Core i5 processor, the codes got executed.8th Generation in MATLAB environment. The population size was considered as 80 and 500 iterations was the stopping criteria for the algorithms.

Table 2. DER parameters

DER	MT	FC	Grid
Operating range (kW)	6/30	3/30	30 kW↑, 80 kW↓
Bid ($/kW)	0.457	0.294	Refer to Fig. 3

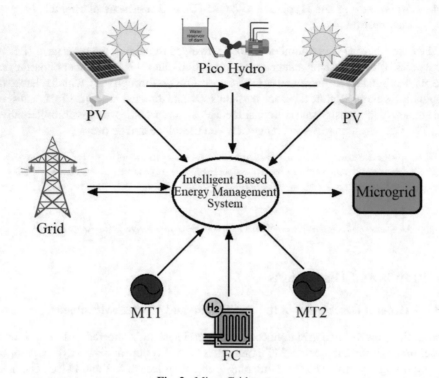

Fig. 2. Micro Grid system

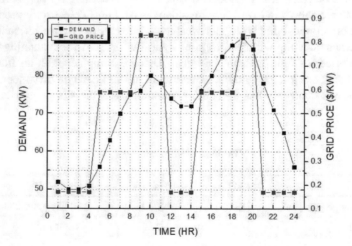

Fig. 3. Electricity market price and load demand

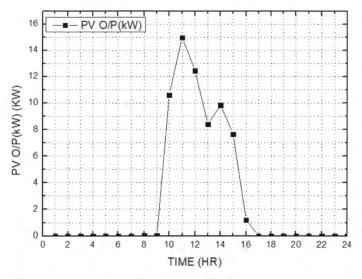

Fig. 4. PV output

4.2 DSM Evaluation for Load Profile Improvement with Respect to Generation Cost

Initially the optimization algorithm was utilized to restructure the load demand for 15% and 20% DSM participation as discussed in Sect. 2.4. As mentioned above, 15% and 20% DSM participation means 15/20% of the total loads every hour was elastic in nature and could be optimally shifted to hours when electricity price was less. Table 3 shows the positive effects of DSM implementation on load profile which are as follows:

a. The load demand was reduced to 84.8 kW and 84.05 kW from 90 kW when 15% and 20% DSM was considered respectively. This corresponded to a peak load reduction of 5.76% and 6.61%.
b. The microgrid system's overall consumption and average load demands remained constant. Which is one of the benefits of restructuring load demand using DSM.
c. The load factor was improved from 0.78 to 0.83 with 15% DSM and 0.84 with 20% DSM.

Table 3. Effects of DSM implementation

	Without DSM	With 15% DSM	With 20% DSM
Peak	90	84.81	84.05
Total demand	1695	1695	1695
Avg demand	70.625	70.62	70.62
Reduction in peak		5.76%	6.61%
Load factor	0.784722222	0.832684825	0.840214158

Figure 5 depicts the load requirement for the microgrid by means of 15% and 20% DSM and without DSM implementation.

4.3 Discussion on Generation Costs for Various Cases

Initially the generation cost was evaluated without considering DERs i.e. only grid was utilized to deliver the power. The generation cost during this case was found to be **$808.47**.

Thereafter, in *Case 1*, the generation cost was evaluated using MT, FC, PV and grid. The best values of minimum generation cost during this case was $562 using GWO and WOA, $560.9 using SCA and **$560.5** using proposed WOASCA. Figure 6 shows the hourly output of DERs and grid when the fuel cost was minimized using proposed algorithm for Case 2.

Thereafter, in *Case 2*, a pumped hydro storage system was incorporated along with the PV system. During peak hours of solar power output, the water is pumped up to the reservoir using solar power. When compared to using the other DERs, this way of lifting water helps to cut down on fuel costs and pollution emissions. The best values of minimum generation cost during this case was $555 using GWO and WOA, $554.7 using SCA and **$553.3** using proposed WOASCA. Figure 7 shows the pumping and generation mode of the pumped hydro storage and the water level in the reservoir. The maximum power generated and consumed by the pumped hydro storage system lies between $\pm$ 5 kW. Figure 8 shows the hourly output of DERs and grid when the fuel cost was minimized using proposed algorithm for Case 2.

Cases 3 and 4 are nothing but Case 2 with restructured load demand considering 15% and 20% DSM participation level respectively. As shown in Table 4, the best fuel cost obtained using the proposed WOASCA are $529 and $520 for 15% and 20% DSM participation respectively. This demonstrates unmistakably the drop in production cost when the DSM method has been taken into account to alter and reconstruct the current anticipated load demand of the LV subject MG system.. Figures 9 and 10 shows the hourly load distribution of the system when minimum fuel cost was obtained using proposed WOASCA as the optimization tool.

The overall decrease in fuel cost of the system can be easily realized from Fig. 11. There was a meagre decrease from $560 to $553 when pumped hydro storage was

Table 4. Cost analysis for various cases using different algorithms

Algorithm		CASE-1	CASE-2	CASE-3	CASE-4
	Without DER	Without Pico hydro	With Pico	With Pico + 15% DSM	With Pico + 20% DSM
GWO	808.47	562.2706	555.4162	530.8658	522.8753
WOA		562.0728	555.1672	530.8358	521.8291
SCA		560.9114	554.7697	529.1947	521.0725
WOASCA		560.5641	553.3243	528.8588	520.5477

incorporated in the system. This fuel cost further dropped to about $25-$30 when the load demand was restructured using DSM strategy for 15% and 20% elastic loads.

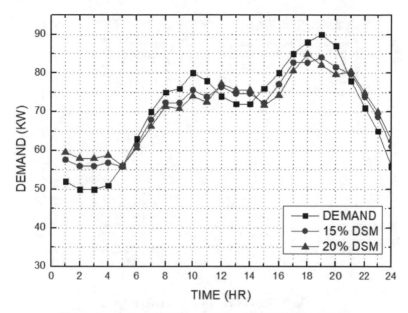

Fig. 5. Load demand with and without DSM implementation

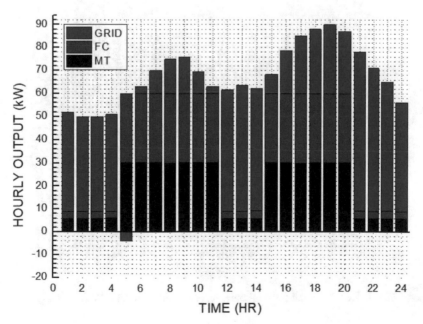

Fig. 6. Hourly output for minimum cost obtained with WOASCA for Case 1

Figure 12 shows the convergence curve of proposed WOASCA when minimum generation cost was obtained for all the cases.

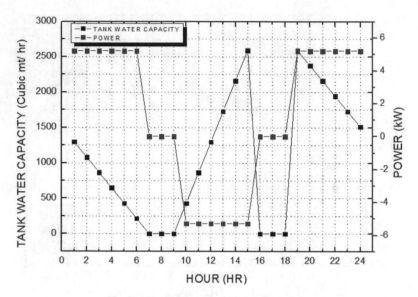

Fig. 7. Pico hydro generation and pumping

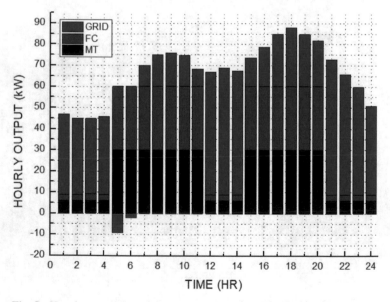

Fig. 8. Hourly output for minimum cost obtained with WOASCA for Case 2

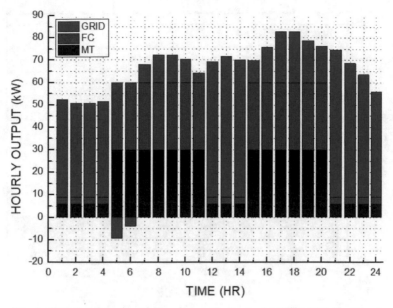

Fig. 9. Hourly output for minimum cost obtained with WOASCA for Case 3

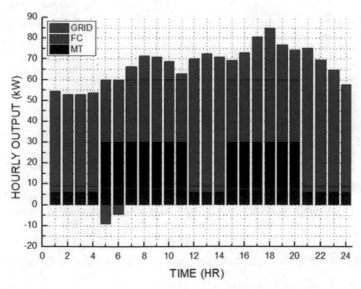

Fig. 10. Hourly output for minimum cost obtained with WOASCA for Case 4

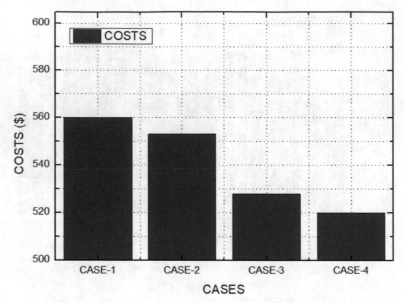

Fig. 11. Cost comparison for various cases using WOASCA

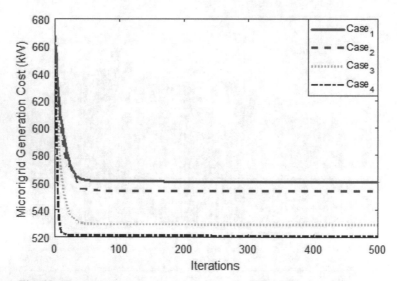

Fig. 12. Cost convergence characteristics using WOASCA for all the Cases

5 Conclusion

The results obtained in this study lead to the following concluding remarks:

I. Implementation of DSM to restructure or modify the load demand of the distribution system has major benefits like reduction of peak, improving load factor and minimizing the production cost. Most importantly all of these can be achieved without affecting the final day's overall load demand.

II. Incorporating a pumped hydro storage system along with the RES (solar PV system) helped in minimizing the generation cost. This happened primarily because the pumping mode of pumped storage was powered by the PV system and no fuel cost was incurred.

III. Proposed WOASCA proved consistent in delivering a better quality solution for all the cases studied.

As a future scope of work, emission constraint can be considered and combined economic emission dispatch could be studied for the same types of system.

Acknowledgement. Authors would like to acknowledge the support of the Department of Science and Technology (DST), Govt. of India, and GIET University, Gunupur, Odisha, India for the financial & technical support for this work.

Declaration of Conflicts of Interest. The authors affirm that they have no known financial or interpersonal conflicts that would have seemed to have an impact on the research presented in this study.

References

1. Sasaki, Y., Yorino, N., Zoka, Y., Wahyudi, F.I.: Robust stochastic dynamic load dispatch against uncertainties. IEEE Trans. Smart Grid **9**, 5535–5542 (2017)
2. Nayak, A., Maulik, A., Das, D.: An integrated optimal operating strategy for a grid-connected AC microgrid under load and renewable generation uncertainty considering demand response. Sustain. Energy Technol. Assess. **45**, 101169 (2021)
3. Kayalvizhi, S., DM, V.K.: Optimal planning of active distribution networks with hybrid distributed energy resources using grid-based multi-objective harmony search algorithm. Appl. Soft Comput. **67**, 387–398 (2018)
4. Wood, A.J., Wollenberg, B.F., Sheblé, G.B.: Power Generation, Operation, and Control. Wiley, Hoboken (2013)
5. Das, B.K., Hasan, M., Rashid, F.: Optimal sizing of a grid-independent PV/diesel/pump-hydro hybrid system: a case study in Bangladesh. Sustain. Energy Technol. Assess. **44**, 100997 (2021)
6. Kiran, P., Chandrakala, K.V.: New interactive agent based reinforcement learning approach towards smart generator bidding in electricity market with micro grid integration. Appl. Soft Comput. **97**, 106762 (2020)
7. Hatziargyriou, N.: Microgrids: Architectures and Control. Wiley, Hoboken (2014)
8. Ray, P.K., Mohanty, A.: A robust firefly–swarm hybrid optimization for frequency control in wind/PV/FC based microgrid. Appl. Soft Comput. **85**, 105823 (2019)
9. Sanjari, M.J., Karami, H., Yatim, A., Gharehpetian, G.B.: Application of Hyper-Spherical Search algorithm for optimal energy resources dispatch in residential microgrids. Appl. Soft Comput. **37**, 15–23 (2015)
10. Dong, W., Yang, Q., Fang, X., Ruan, W.: Adaptive optimal fuzzy logic based energy management in multi-energy microgrid considering operational uncertainties. Appl. Soft Comput. **98**, 106882 (2021)

11. Aprilia, E., Meng, K., Al Hosani, M., Zeineldin, H.H., Dong, Z.Y.: Unified power flow algorithm for standalone AC/DC hybrid microgrids,. IEEE Trans. Smart Grid **10**, 639–649 (2017)
12. Tang, Z., Hill, D.J., Liu, T.: A novel consensus-based economic dispatch for microgrids. IEEE Trans. Smart Grid **9**, 3920–3922 (2018)
13. Chen, C., Duan, S., Cai, T., Liu, B., Hu, G.: Smart energy management system for optimal microgrid economic operation. IET Renew. Power Gener. **5**, 258–267 (2011)
14. Kasaei, M.J.: Energy and operational management of virtual power plant using imperialist competitive algorithm. Int. Trans. Electr. Energy Syst. **28**, e2617 (2018)
15. Basu, M., Chowdhury, A.: Cuckoo search algorithm for economic dispatch. Energy **60**, 99–108 (2013)
16. Moghaddam, A.A., Seifi, A., Niknam, T., Pahlavani, M.R.A.: Multi-objective operation management of a renewable MG (micro-grid) with back-up micro-turbine/fuel cell/battery hybrid power source. Energy **36**, 6490–6507 (2011)
17. Rabiee, A., Sadeghi, M., Aghaei, J.: Modified imperialist competitive algorithm for environmental constrained energy management of microgrids. J. Clean. Prod. **202**, 273–292 (2018)
18. Trivedi, I.N., Jangir, P., Bhoye, M., Jangir, N.: An economic load dispatch and multiple environmental dispatch problem solution with microgrids using interior search algorithm. Neural Comput. Appl. **30**(7), 2173–2189 (2016). https://doi.org/10.1007/s00521-016-2795-5
19. Elattar, E.E.: Modified harmony search algorithm for combined economic emission dispatch of microgrid incorporating renewable sources. Energy **159**, 496–507 (2018)
20. Dey, B., Roy, S.K., Bhattacharyya, B.: Solving multi-objective economic emission dispatch of a renewable integrated microgrid using latest bio-inspired algorithms. Eng. Sci. Technol. Int. J. **22**, 55–66 (2019)
21. Kumar, K.P., Saravanan, B., Swarup, K.: Optimization of renewable energy sources in a microgrid using artificial fish swarm algorithm. Energy Procedia **90**, 107–113 (2016)
22. Gholami, K., Dehnavi, E.: A modified particle swarm optimization algorithm for scheduling renewable generation in a micro-grid under load uncertainty. Appl. Soft Comput. **78**, 496–514 (2019)
23. Ramli, M.A., Bouchekara, H., Alghamdi, A.S.: Efficient energy management in a microgrid with intermittent renewable energy and storage sources. Sustainability **11**, 3839 (2019)
24. Guezgouz, M., et al.: Optimal hybrid pumped hydro-battery storage scheme for off-grid renewable energy systems. Energy Conv. Manag. **199**, 112046 (2019)
25. Naik, K.R., et al.: Power management scheme of DC micro-grid integrated with photovoltaic-Battery-Micro hydro power plant. J. Power Sources **525**, 230988 (2022)
26. Tyagi, S., Singh, B., Das, S.: ELD-OSG control of battery based electronic load controller for small hydro energy conversion system. IEEE Trans. Ind. Appl. (2022)
27. Asiful, H., Che, H.S.: Design and simulation of axial flux permanent magnet generator for residential pico-hydro power generation. In: 2018 International Conference on Intelligent and Advanced System (ICIAS). IEEE (2018)
28. Carravetta, A., Fecarotta, O., Ramos, H.M.: A new low-cost installation scheme of PATs for pico-hydropower to recover energy in residential areas. Renew. Energy **125**, 1003–1014 (2018)
29. Kusakana, K.: Hydro aeropower for sustainable electricity cost reduction in South African farming applications. Energy Rep. **5**, 1645–1650 (2019)
30. Tapia, A., Millán, P., Gómez-Estern, F.: Integer programming to optimize Micro-Hydro Power Plants for generic river profiles. Renew. Energy **126**, 905-914 (2018)
31. Kamran, M., et al.: Designing and economic aspects of run-of-canal based micro-hydro system on Balloki-Sulaimanki Link Canal-I for remote villages in Punjab, Pakistan. Renew. Energy **141**, 76–87 (2019)

32. Kusakana, K.: Optimal operation scheduling of grid-connected PV with ground pumped hydro storage system for cost reduction in small farming activities. J. Energy Storage **16**, 133–138 (2018)

33. Javed, M.S., et al.: Solar-wind-pumped hydro energy storage systems: review and future perspective. Renew. Energy (2019)

34. Dey, B., Basak, S., Pal, A.: Demand-side management based optimal scheduling of distributed generators for clean and economic operation of a microgrid system. Int. J. Energy Res. **46**(7), 8817–8837 (2022)

35. Basak, S., Bishwajit, D., Biplab, B.: Demand side management for solving environment constrained economic dispatch of a microgrid system using hybrid MGWOSCACSA algorithm. CAAI Trans. Intell. Technol. **7**, 256–267 (2022)

36. Dey, B., Bhattacharyya, B.: Comparison of various electricity market pricing strategies to reduce generation cost of a microgrid system using hybrid WOA-SCA. Evol. Intell. 1–18 (2021). https://doi.org/10.1007/s12065-021-00569-y

37. Devarapalli, R., et al.: An approach to solve OPF problems using a novel hybrid whale and sine cosine optimization algorithm. J. Intell. Fuzzy Syst. **42**(2), 957–967 (2022)

Impact of Multimodal Model Complexity on Classification of Diabetic Retinopathy Level

Maksym Shulha⑩, Yuri Gordienko$^{(\boxtimes)}$ ⑩, and Sergii Stirenko⑩

National Technical University of Ukraine "Igor Sikorsky Kyiv Polytechnic Institute",
Kyiv, Ukraine
yuri.gordienko@gmail.com

Abstract. The diabetic retinopathy (DR) severity classification problem was considered on the basis of RetinaMNIST dataset by means of the single modality (with image input) models and multi modality (with image and text inputs) deep neural network (DNN) models. In practice, the additional data such as subjective "patient" opinion even about the patient health status (that offer "data leakage" on specific classes) can be useful. Additional (augmented) metadata from the simulated surveys was used to imitate these opinions. As a result, multi modality models (MP) allowed us to reach the statistically significant improvements of classification performance by AUC value for some classes in the range from 15% to 26% (depend on DNN model complexity). The improvements are rather beyond the limits of the standard deviation of 3–8% (depend on DNN model complexity) measured by cross-validation and can be estimated as significant ones. In general, this approach could be useful strategy for the classification performance improvement and, especially, effective for the small models (like MobileNetV2) for Edge Computing devices with limited computational resources to classify DR severities and in the more general medical and other contexts.

Keywords: Multi-class classification · Neural networks · Deep learning · Metadata augmentation · Multimodal model · Retina

1 Introduction

The most common outcome in diabetes and the leading global cause of vision impairment in adults is diabetic retinopathy (DR) [1–3]. For an early diagnosis and prompt treatment of DR to prevent blindness, periodic eye screening is required. The creation and implementation of efficient computer-aided screening applications can achieve this goal. The potential of artificial intelligence (AI) approaches in medicine is actively being used by the medical community [4–6].

Supported by "Knowledge At the Tip of Your fingers: Clinical Knowledge for Humanity" (KATY) project funded from the European Union's Horizon 2020 research and innovation program under grant agreement No. 101017453.

© The Author(s), under exclusive license to Springer Nature Switzerland AG 2023
F. P. García Márquez et al. (Eds.): ICCIDA 2022, LNNS 643, pp. 168–180, 2023.
https://doi.org/10.1007/978-3-031-27099-4_13

Deep neural networks (DNNs) have recently shown their effectiveness and suitability for Computer-Aided Detection (CADe) and Computer-Aided Diagnosis (CADx). By autonomously processing medical data without the assistance of medical personnel during the screening stage and making it regularly available worldwide, AI-related approaches can automate and greatly speed up screening applications. In this work, DR severity classification problem is considered on the basis of RetinaMNIST dataset by means of the DNN models. Section 2 contains description of the state of the art, Sect. 3 describes dataset, models, experiments, and the whole workflow, Sect. 4 gives the results obtained during the experiments, Sect. 5 contains discussions of results and resumes them.

2   Background and Related Works

It is well known that more than 10 million of people worldwide suffer from various vision impairments caused by glaucoma, trachoma, and DR [7]. As to DR, it causes more than 2.5% cases of blindness worldwide [6,7].

Several DL-based on DNNs recently were efficiently proposed for different medical imaging purposes [5,8] and, especially, for diabetic retinopathy [9,10]. The current state of the art was many times summarized in thorough reviews of such attempts for diabetic retinopathy classification [6,11,12]. During the last years many intriguing and promising studies appeared [9,10,13–15] that are worth of a separate consideration.

In the context of this paper, some promising approaches should be noted like a novel diagnostic tool for automated DR detection that processed color fundus images and classified them as healthy (no DR) or having DR, identifying relevant cases for medical referral [9]. Several computer vision and DL models were used and compared for quantifying the body state features into different classes [13]. The multistage transfer learning was used as a screening method for early detection of DR with high sensitivity and specificity [14]. Also some versions of a DR prediction system were prepared where the two input versions were either a set of three-field or one-field colour fundus photographs [15].

The progress of these studies was significantly fostered by appearance of the pertinent medical datasets that were proposed in an open science manner. For these and other purposes recently, a large retinal image dataset, DeepDR (Deep Diabetic Retinopathy), was proposed in IEEE DeepDR Diabetic Retinopathy Image Dataset (DeepDRiD) competition [16] with 5 classes: a) no apparent proliferative - Class 0, b) mild proliferative - Class 1, c) moderate proliferative - Class 2, d) severe proliferative - Class 3, and e) proliferative diabetic retinopathy - Class 4. On its basis, many other datasets appeared that were actively used to develop and test the new DL methods. For example, MedMNIST v2, a large-scale MNIST-like dataset collection of standardized biomedical images, was recently proposed for research and educational purposes (Fig. 1) [17,18].

3 Methodology

3.1 Dataset

RetinaMNIST dataset (a component of MedMNIST) on the basis of DeepDRiD dataset [16] was used here. In RetinaMNIST $3 \times 1736 \times 1824$ source images were center-cropped and downsized to $3 \times 28 \times 28$ (Fig. 1). Here due for usage with the standard DNNs (see below), the train (1080 images), validation (120 images), and test (400 images) parts were pre-processed from $3 \times 28 \times 28$ to $3 \times 32 \times 32$ format.

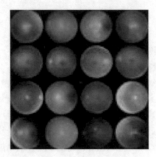

Fig. 1. Images from MedMNIST dataset for the DR severity classification problem.

In some real-world circumstances, the influence of extra data, such as the subjective patient evaluation of the patient's health status, can be useful. This opinion was simulated by additional (augmented) metadata from simulated questionnaires to get: "patient" opinion as an answer to the question "do you feel healthy?" with 2 possible answers: "healthy" (for Class 0) or "ill" (for classes from 1 to 4), where any patient discomfort is roughly classified as an "ill" state [19]. In general, the patient's opinion might be far more complicated and include words that are comparable yet have different semantic connotations. In this research, it was made very simple to understand how extra modalities affected the metrics of the DL procedure. As a result the following variants of input values (column "Model") were prepared and shown in Table 1:

- Single modality model (SM) - input consists of image only (input_image),
- Multi model with patient opinion (MP) - input consists of image and patient opinion text, namely, input_image + input_text as "healthy" *or* "ill".

Table 1. Input values for the models in Table 2

Model	Inputs	Class 0	Class 1–4
SM	Input_image	Image	Image
MP	Input_image + Input_text	Image + healthy	Image + ill

3.2 Models

Here the several DNN architectures were used to work with input data of various modalities (Table 1) on the basis of the standard convolutional neural network (CNN) (Table 2) to process image inputs and recurrent neural network (RNN) to process text inputs:

- single modality (SM models): different kinds of CNN architecture (Table 2),
- multi modalities (MP models): different kinds of CNN architecture and RNN with LSTM architecture (Table 2).

Table 2. DNN architectures

Model	CNN		RNN
SM	ResNet50, MobileNetV2, DenseNet201, NASNetLarge, EfficientNetB7, VGG19, EfficientNetV2L, NASNetMobile, EfficientNetB0, EfficientNetV2S	+	−
MP			LSTM

To compare the approaches proposed here with the benchmarked results, especially on ResNet50 [19], the different CNN architectures was used in the work. Several classic DNNs were used for CNN component of the DNN architectures (Table 2) used here: VGG [20], ResNet50 [21], MobileNetV2 [22], DenseNet201 [23], NASNet [24], and EfficientNet [25]. The general architecture of CNN model (Table 2) used in [19] was applied to other kinds of CNN model that were used in this work. The area under curve (AUC) and number of parameters in CNN components were chosen as the main metrics for analysis of the results obtained.

The whole workflow was implemented as a Cross-Validation (CV) [26] regime and the further Out-of-Cross-Validation (OoCV) regime (by training on the whole train part (1080 images) of the original RetinaMNIST dataset, by selecting the best model by AUC value after validating on the validation part (120 images) of the original RetinaMNIST dataset, and the final testing on the test part (400 images) of the original RetinaMNIST dataset) which were initially used and described in details in [19].

4 Results

4.1 Single Modality Model

Cross-Validation Study. Various versions of SM model, actually the different kinds of CNN architectures (Table 2), were trained for 100 epochs in 6-fold CV regime. The configuration parameter for k-fold cross-validation, k, defines the number folds in which the dataset was split. The most popular value used in machine learning purposes is $k = 10$, but here $k = 6$ to provide the more samples for validation parts. Then the AUC values per class (see Table 5 below in the next sections) were calculated at each fold with determination of their mean and standard deviation. Also the macro and micro AUC values were determined at each fold (Table 3) and visualized as a bar chart (Fig. 2).

Table 3. AUC values after 6-fold CV and OoCV for SM model

Model	6-fold CV		OoCV	
	AUC_{macro}(mean ± std) std	AUC_{micro}(mean ± std)	AUC_{macro}	AUC_{micro}
ResNet50	0.697 ± 0.074	0.796 ± 0.006	0.667	0.759
MobileNetV2	0.503 ± 0.007	0.707 ± 0.015	0.500	0.708
NASNetLarge	0.556 ± 0.027	0.688 ± 0.043	0.500	0.647
EfficientNetB7	0.658 ± 0.092	0.757 ± 0.009	0.680	0.782
DenseNet201	**0.704 ± 0.073**	**0.798 ± 0.010**	**0.722**	**0.794**
VGG19	0.610 ± 0.069	0.745 ± 0.024	0.499	0.721
EfficientNetV2L	0.613 ± 0.068	0.738 ± 0.023	0.659	0.776
NASNetMobile	0.619 ± 0.074	0.738 ± 0.021	0.617	0.744
EfficientNetB0	0.693 ± 0.079	0.781 ± 0.022	0.669	0.752
EfficientNetV2S	0.647 ± 0.095	0.765 ± 0.007	0.690	0.767

Out-of-Cross-Validation Study. Also these versions of SM model (again, the different kinds of CNN architectures from Table 2) were trained for 100 epochs in OoCV regime. Then the AUC values per class (see below in the next sections) and the macro and micro AUC values were determined (Table 3) and visualized as a bar chart (Fig. 2).

4.2   Multi Modality with Patient Opinion

In the similar way as with the SM models above, various versions of the Multi modality with Patient opinion (MP) model, actually different kinds of CNN architecture and RNN with LSTM architecture, were trained for 100 epochs in 6-fold CV regime and in OoCV regime. Then the AUC values per class (see Table 5 below in the next sections) and the macro and micro AUC values were determined (Table 4) and visualized as a bar chart (Fig. 3).

5   Discussion

Macro AUC Values Vs Model Complexity. In the cross-validation (CV) study, it was found that few DNN architectures allowed us to reach the highest AUC macro values (AUC_{macro} in Fig. 4) among all architectures after 6-fold CV for SM model. The highest AUC value lies in range 0.693–0.704 for ResNet50, DenseNet201 and EfficientNetB0. As to the performance observed for MP model, it was found that the same architectures as for SM model allowed us to reach the highest AUC value among all architectures after 6-fold CV. The highest AUC

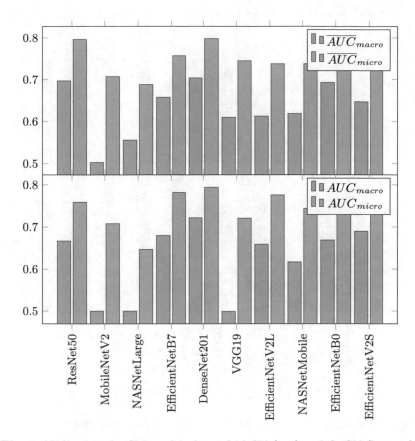

Fig. 2. AUC values for SM model after 6-fold CV (top) and OoCV (bottom).

Table 4. AUC values after 6-fold CV and OoCV for MP model

Model	6-fold CV		OoCV	
	AUC_{macro}(mean ± std)	AUC_{micro}(mean ± std)	AUC_{macro}	AUC_{micro}
ResNet50	0.815 ± 0.089	0.911 ± 0.007	0.799	0.871
MobileNetV2	0.803 ± 0.098	0.918 ± 0.004	0.805	0.923
NASNetLarge	0.656 ± 0.082	0.742 ± 0.088	0.500	0.647
EfficientNetB7	0.796 ± 0.103	0.915 ± 0.009	**0.854**	**0.930**
DenseNet201	**0.826** ± 0.085	0.915 ± 0.006	0.808	0.890
VGG19	0.811 ± 0.093	0.919 ± 0.005	0.801	0.922
EfficientNetV2L	0.807 ± 0.096	0.916 ± 0.003	0.846	0.924
NASNetMobile	0.771 ± 0.119	0.864 ± 0.036	0.713	0.787
EfficientNetB0	0.817 ± 0.094	**0.922** ± 0.006	0.834	**0.930**
EfficientNetV2S	0.808 ± 0.100	0.917 ± 0.008	0.793	0.921

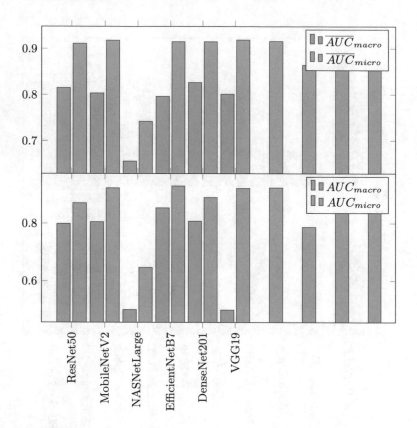

Fig. 3. AUC values for MP model after 6-fold CV (top) and OoCV (bottom).

value lies in range 0.815–0.826 for ResNet50, DenseNet201 and EfficientNetB0. Comparing the size of architectures with the best AUC value, it can be found that two of them are of medium size (ResNet50 and DenseNet201) and one of small size (EfficientNetB0).

In the out-of-cross-validation (OoCV) study, as to the performance obtained after OoCV for SM model, it was found that only DenseNet201 architecture allowed to reach the highest AUC value. This value is 0.722, the second highest value is only 0.690. However, there are two best architectures for MP model. They are EfficientNetB7 and EfficientNetV2L with AUC values 0.854 and 0.846 respectively. Comparing the size of architectures with the best AUC value, it can be found that two of them are of large size (EfficientNetB7 and EfficientNetV2L) and one of medium size (DenseNet201).

In general, among all DNN models three groups of their CNN components can be categorized by their size: a) small: MobileNetV2, EfficientNetB0, NASNet-Mobile; b) medium: VGG19, ResNet50, DenseNet201, EfficientNetV2S; c) large:

Table 5. AUC values after 6-fold CV

CNN	Model	Class 0	Class 1	Class 2	Class 3	Class 4
ResNet50	SM	0.818 ± 0.01	0.600 ± 0.04	0.668 ± 0.01	0.734 ± 0.02	0.663 ± 0.04
	MP	0.986 ± 0.01	0.784 ± 0.04	0.775 ± 0.02	0.803 ± 0.01	0.725 ± 0.06
MobileNetV2	SM	0.502 ± 0.00	0.499 ± 0.00	0.496 ± 0.01	0.502 ± 0.02	0.517 ± 0.03
	MP	0.995 ± 0.00	0.745 ± 0.01	0.783 ± 0.00	0.758 ± 0.01	0.733 ± 0.02
NASNetLarge	SM	0.572 ± 0.10	0.505 ± 0.02	0.551 ± 0.05	0.581 ± 0.09	0.570 ± 0.08
	MP	0.807 ± 0.22	0.584 ± 0.10	0.677 ± 0.13	0.618 ± 0.12	0.596 ± 0.07
EfficientNetB7	SM	0.783 ± 0.03	0.536 ± 0.02	0.671 ± 0.02	0.725 ± 0.03	0.574 ± 0.03
	MP	0.995 ± 0.00	0.755 ± 0.04	0.789 ± 0.01	0.741 ± 0.04	0.702 ± 0.03
DenseNet201	SM	0.825 ± 0.01	0.625 ± 0.02	0.676 ± 0.01	0.745 ± 0.02	0.647 ± 0.03
	MP	0.989 ± 0.01	0.800 ± 0.02	0.785 ± 0.02	0.820 ± 0.03	0.738 ± 0.03
VGG19	SM	0.726 ± 0.12	0.520 ± 0.03	0.565 ± 0.12	0.617 ± 0.17	0.621 ± 0.07
	MP	0.995 ± 0.00	0.764 ± 0.03	0.787 ± 0.01	0.763 ± 0.03	0.746 ± 0.04
EfficientNetV2L	SM	0.715 ± 0.03	0.513 ± 0.02	0.608 ± 0.06	0.654 ± 0.08	0.578 ± 0.06
	MP	0.995 ± 0.00	0.765 ± 0.03	0.783 ± 0.01	0.771 ± 0.02	0.721 ± 0.02
NASNetMobile	SM	0.715 ± 0.13	0.522 ± 0.03	0.607 ± 0.07	0.692 ± 0.10	0.560 ± 0.09
	MP	0.995 ± 0.00	0.670 ± 0.07	0.765 ± 0.02	0.757 ± 0.04	0.669 ± 0.04
EfficientNetB0	SM	0.811 ± 0.01	0.581 ± 0.03	0.663 ± 0.05	0.747 ± 0.02	0.662 ± 0.02
	MP	0.995 ± 0.00	0.791 ± 0.02	0.791 ± 0.01	0.793 ± 0.04	0.714 ± 0.04
EfficientNetV2S	SM	0.779 ± 0.03	0.552 ± 0.04	0.624 ± 0.04	0.735 ± 0.03	0.545 ± 0.07
	MP	0.995 ± 0.00	0.781 ± 0.02	0.783 ± 0.01	0.787 ± 0.03	0.693 ± 0.04

EfficientNetB7, EfficientNetV2L, NASNetLarge (Fig. 4). From Table 3, Table 4, and Fig. 4 one can see that application of additional modality in MP models allow to get the higher performance with regard to $(\overline{AUC}_{macro})$ values. Actually, $\overline{AUC}_{macro}$ values for MP models are much bigger than $\overline{AUC}_{macro})$ values for SP models in comparison to the observed standard deviations. It is interesting that the performance increase can quite different for various DNN models, for example, for NASNetLarge the increase of performance is not very pronounced, while for MobileNetV2 it is greatest among all.

Per Class AUC Values Vs Model Complexity. Due to some visualization (Fig. 5) for the per class AUC values (Table 5) some differences of the performance improvement $(\Delta = \overline{AUC}^{MP} - \overline{AUC}^{SM})$ can be observed among the categories of models by size for various classes. For example, the hierarchy of the performance improvement Δ can be observed as follows: Δ(Class 1) (red, Fig. 5) $> \Delta$(Class 2) (green, Fig. 5) $> \Delta$(Class 3) (blue, Fig. 5) $\geq \Delta$(Class 4) (black, Fig. 5). In this hierarchy all of the small models for Class 1 and 2 demonstrate the best performance improvements Δ which are much higher than the standard

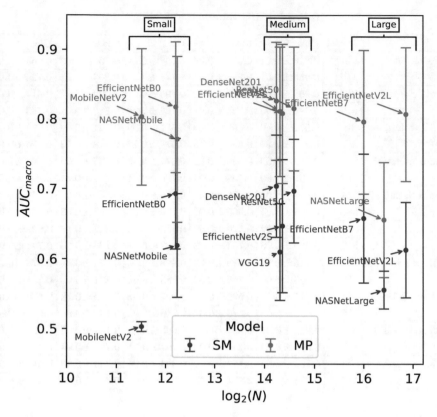

Fig. 4. The mean AUC values ($\overline{AUC}_{macro}$) with their standard deviations (shown as error bars) versus the number of model parameters N (in thousands) shown on the logarithmic scale as $\log_2(N)$.

deviations ("std" in Fig. 5) as the averaged ("aver") or maximal ("max") values estimated over all standard deviation values for all models from Table 5. Also, the small and medium models for Class 3 and 4 demonstrate just little better performance improvement in comparison to the large models. This means that addition of "data leakage" on Class 0 in MP models allowed us to get the higher performance improvement Δ for the nearest Classes 1 and 2 (that are similar to Class 0), but Δ was not so high for the farther Classes 3 and 4 (that are more different from Class 0).

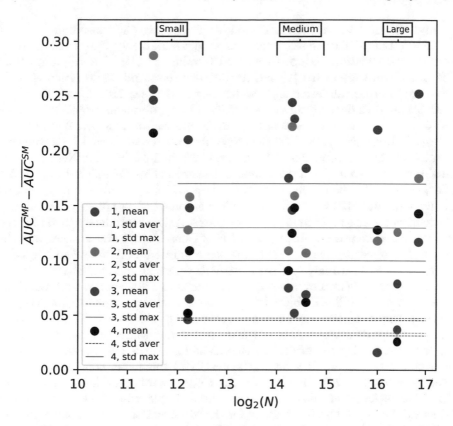

Fig. 5. Increase of classification performance as a difference of the mean AUC values $(\overline{AUC}^{MP} - \overline{AUC}^{SM})$ the number of model parameters N (in thousands) shown on the logarithmic scale as $\log_2(N)$ for each class mentioned in the legend. The legend contains the class number, type of value ("mean" or "std"), where "std" can be averaged ("aver") or maximal value ("max") over all standard deviation values (from Table 5).

6   Conclusions

In conclusion, the DR level classification problem was considered for single modality (with image input) model and multi modality (with image and text input) model are considered on the basis of the "pedagogical" dataset RetinaMNIST. The influence of additional data like subjective "patient" opinion about patient health state (that provide "data leakage" on one of the classes) can be useful in some practical situations. The patient opinions were imitated by the additional (augmented) data generated from simulated questionnaires. As a result the single modality model (SM) with input image only and the multi modality model with input image and patient opinion text under the title Multi modality model with Patient opinion (MP) were investigated. Several different classic CNN models from small (MobileNetV2, EfficientNetB0, NASNetMobile), medium (VGG19, ResNet50, DenseNet201, EfficientNetV2S), and to large (EfficientNetB7, EfficientNetV2L, NASNetLarge) were used as input image processing components for

SM and MP models. All these MP models with various CNN backbones (except for NASNetLarge) allowed us to reach the various statistically significant improvements of classification performance by AUC value for Class 1 in the range from 15% to 26% that are rather beyond the limits of the standard deviation of 3–4% measured by cross-validation and can be estimated as significant ones. For Class 2, all MP models demonstrated the significant improvements from 11% to 28% that are also beyond the limits of the standard deviation of 3–4%. As to Class 3 and 4 only MobileNetV2 and EfficientNetV2L show the significant improvements from 22% to 26% (MobileNetV2) and from 12% to 14% (EfficientNetV2L) that are also beyond the limits of the standard deviation of 3% (MobileNetV2) and 8% (EfficientNetV2L). It was found that the best $\overline{AUC}_{macro}$ values among models (both for SM and MP) with different CNN components does not vary significantly on architecture size and number of parameters. Also, it was determined that the best performance was obtained with CNN of DenseNet201 architecture, and the best performance improvement Δ can be observed for MobileNetV2 architecture. But the more careful analysis of per class AUC values brought to light the slightly higher performance improvement Δ (in the limits of the standard deviations) for the smaller models. The inability of the bigger models to reach the same progress can be explained by the very small size of the pedagogical dataset RetinaMNIST which is the evident weaknesses of the current study, which can be solved in the future by the further investigations that should be carried on the full version of its original version [16]. In general, the novelty of this paper consist in the approach based on metadata augmentation, namely, usage of the additional modalities with "data leakage" on the extreme class, for example, with the lowest (Class 0) DR severity could be useful strategy for the classification performance improvement of the models. The additional novelty relates to model complexity analysis. It is especially important and effective for selection of small DNN models (like MobileNetV2) for usage on Edge Computing devices [27,28] with limited computational resources used to classify DR severities like Classes 1–4 here and in the more general medical and other contexts. In addition to the previous attempts on DNN performance improvement by tuning the input size [29,30], types of components used (like various activation functions [31]), batch size [32], and other parameters, this work demonstrate that additional novel improvement can be obtained by including additional modality which can be contextually relevant to the targeted objects. It is especially important in the view of appearance of the new approaches for detection of DR by means of the new DNN architectures and DL systems [33–35].

References

1. Kertes, P.J., Johnson, T.M.: Evidence-Based Eye Care. Lippincott Williams Wilkins, Philadelphia (2007)
2. Cunha-Vaz, J.G.: Diabetic Retinopathy. World Scientific (2011)
3. Scanlon, P.H., Sallam, A., Wijngaarden, P.V.: A Practical Manual of Diabetic Retinopathy Management. Wiley, New York (2017)

4. Esteva, A., et al.: A guide to deep learning in healthcare. Nature Med. **25**(1), 24–29 (2019)
5. Chen, Y.-W., Jain, L.C. (eds.): Deep Learning in Healthcare. ISRL, vol. 171. Springer (2020). https://doi.org/10.1007/978-3-030-32606-7
6. Atwany, M.Z., Sahyoun, A.H., Yaqub, M.: Deep learning techniques for diabetic retinopathy classification: a survey. IEEE Access **10**, 28642–28655 (2022)
7. Who Team. World Report on Vision. World Health Organization (2019)
8. Alienin, O., Rokovyi, O., Gordienko, Y., Kochura, Y., Taran, V., Stirenko, S.: Artificial Intelligence Platform for Distant Computer-Aided Detection (CADe) and Computer-Aided Diagnosis (CADx) of Human Diseases. In: Hu, Z., Zhang, Q., Petoukhov, S., He, M. (eds.) Advances in Artificial Systems for Logistics Engineering. ICAILE 2022. Lecture Notes on Data Engineering and Communications Technologies, vol 135. Springer, Cham (2022). https://doi.org/10.1007/978-3-031-04809-8_8
9. Gargeya, R., Leng, T.: Automated identification of diabetic retinopathy using deep learning. Ophthalmology **124**(7), 962–969 (2017)
10. Grauslund, J.: Diabetic retinopathy screening in the emerging era of artificial intelligence. Diabetologia 1–9 (2022)
11. Asiri, N., Hussain, M., Adel, F.A., Alzaidi, N.: Deep learning based computer-aided diagnosis systems for diabetic retinopathy: a survey. Artifi. Intell. Med. **99**, 101701 (2019)
12. Alyoubi, W.L., Shalash, W.M., Abulkhair, M.F.: Diabetic retinopathy detection through deep learning techniques: a review. Inf. Med. Unlocked **20**, 100377 (2020)
13. Dutta, S., Manideep, B., Basha, S.M., Caytiles, R.D., Iyengar, N.: Classification of diabetic retinopathy images by using deep learning models. Int. J. Grid Distrib. Comput. **11**(1), 89–106 (2018)
14. Tymchenko, B., Marchenko, P., Spodarets, D.: Deep learning approach to diabetic retinopathy detection. arXiv preprint arXiv:2003.02261 (2020)
15. Bora, A., et al.: Predicting the risk of developing diabetic retinopathy using deep learning. Lancet Digital Health **3**(1), e10–e19 (2021)
16. IEEE. The 2nd diabetic retinopathy - grading and image quality estimation, Challenge (2020). https://isbi.deepdr.org/data.html. Accessed 30 Jul 2022
17. Yang, J., Shi, R., Ni, B.: MedMNIST classification decathlon: a lightweight automl benchmark for medical image analysis. In: IEEE 18th International Symposium on Biomedical Imaging (ISBI), pp. 191–195 (2021)
18. Yang, J., et al.: MedMNIST v2: a large-scale lightweight benchmark for 2D and 3D biomedical image classification. arXiv preprint arXiv:2110.14795 (2021)
19. Shulha, M., Gordienko, Y., Stirenko, S.: Deep learning with metadata augmentation for classification of diabetic retinopathy level. In: 3rd International Conference on Sustainable Expert Systems (ICSES) (2022)
20. Karen, S., Andrew, Z.: Very deep convolutional networks for large-scale image recognition. arXiv preprint arXiv:1409.1556 (2014)
21. Kaiming, H., Xiangyu, Z., Shaoqing, R., Jian, S.: Deep residual learning for image recognition. arXiv preprint arXiv:2202.13981 (2022)
22. Mark, S., Andrew, H., Menglong, Z., Andrey, Z., Chieh, C.L.: MobileNETv2: inverted residuals and linear bottlenecks. arXiv preprint arXiv:1801.04381 (2018)
23. Gao, H., Zhuang, L., van der Maaten, L., Weinberger Kilian, Q.: Densely connected convolutional networks. arXiv preprint arXiv:1608.06993 (2016)
24. Barret, Z., Vijay, V., Jonathon, S., Le Quoc, V.: Learning transferable architectures for scalable image recognition. arXiv preprint arXiv:1707.07012 (2017)

25. Mingxing, T., Le Quoc, V.: EfficientNet: rethinking model scaling for convolutional neural networks. arXiv preprint arXiv:1905.11946 (2019)
26. Refaeilzadeh, P., Tang, L., Liu, H.: Cross-validation. Encycl. Database Syst. **5**, 532–538 (2009)
27. Gordienko, Y., et al.: Scaling analysis of specialized tensor processing architectures for deep learning models. In: Pedrycz, W., Chen, S.-M. (eds.) Deep Learning: Concepts and Architectures. SCI, vol. 866, pp. 65–99. Springer, Cham (2020). https://doi.org/10.1007/978-3-030-31756-0_3
28. Gordienko, Y., et al.: "last mile" optimization of edge computing ecosystem with deep learning models and specialized tensor processing architectures. In: Advances in Computers, vol. 122, pp. 303–341. Elsevier (2021)
29. Gordienko, Y., Kochura, Y., Taran, V., Gordienko, N., Bugaiov, A., Stirenko, S.: Adaptive iterative pruning for accelerating deep neural networks. In: 2019 XIth International Scientific and Practical Conference on Electronics and Information Technologies (ELIT), pp. 173–178. IEEE (2019)
30. Doms, V., Gordienko, Y., Kochura, Y., Rokovyi, O., Alienin, O., Stirenko, S.: Deep learning for melanoma detection with testing time data augmentation. In: Hu, Z., Zhang, Q., Petoukhov, S., He, M. (eds.) ICAILE 2021. LNDECT, vol. 82, pp. 131–140. Springer, Cham (2021). https://doi.org/10.1007/978-3-030-80475-6_13
31. Kochura, Y., Stirenko, S., Gordienko, Y.: Comparative performance analysis of neural networks architectures on H2O platform for various activation functions. In: 2017 IEEE International Young Scientists Forum on Applied Physics and Engineering (YSF), pp. 70–73. IEEE (2017)
32. Kochura, Y.: Batch size influence on performance of graphic and tensor processing units during training and inference phases. In: Hu, Z., Petoukhov, S., Dychka, I., He, M. (eds.) ICCSEEA 2019. AISC, vol. 938, pp. 658–668. Springer, Cham (2020). https://doi.org/10.1007/978-3-030-16621-2_61
33. Dai, L., et al.: A deep learning system for detecting diabetic retinopathy across the disease spectrum. Nature Commun. **12**(1), 1–11 (2021)
34. Feng, C., Hung, J.P., Li, A., Yang, J., Zhang, X.: MTCSNN: multi-task clinical siamese neural network for diabetic retinopathy severity prediction. arXiv preprint arXiv:2208.06917 (2022)
35. Nasir, N., Oswald, P., Alshaltone, O., Barneih, F., Shabi, M.A., Shammaa, A.A.: Deep DR: detection of diabetic retinopathy using a convolutional neural network. In: 2022 Advances in Science and Engineering Technology International Conferences (ASET), pp. 1–5. IEEE (2022)

Secure Image Data Encryption Scheme for 5G Internet of Things Applications

Renjith V. Ravi[1]([✉]) [iD], S. B. Goyal[2] [iD], Chawki Djeddi[3,4] [iD],
and Vladimir Kustov[5] [iD]

[1] Department of Electronics and Communication Engineering,
M.E.A Engineering College, Kerala, India
renjithravi@meaec.edu.in

[2] Faculty of Information Technology, City University, Petaling Jaya 46100, Malaysia
sb.goyal@city.edu.my

[3] Department of Mathematics and Computer Science, Larbi Tebessi University,
Tebessa, Algeria
c.djeddi@univ-tebessa.dz

[4] LITIS Lab, Rouen University, Rouen, France

[5] Saint Petersburg Railway Transport University of Emperor Alexander I,
Saint Petersburg, Russia
kvnvika@mail.ru

Abstract. The foundation of telecommunications for linking things in the Internet of Things (IoT) ecosystem is fifth-generation (5G) communications. The 5G network is being developed to deliver massive capacity, large bandwidth, and reduced latency. The creation and invention of new 5G-IoT approaches will undoubtedly result in massive new security and privacy issues. 5G-IoT technologies will need secure data transfer mechanisms as the foundation to solve these issues. Chaos theory-based image encryption and DNA sequence computations are combined in the suggested encryption approach. The encryption process is performed independently on three colour image channels (R, G, and B) with an appropriate chaotic map. Here the selection algorithm chooses an appropriate chaotic map according to the entropy values of image attributes and few other factors. Following chaotic encryption, a DNA encoding approach is used. According to simulation and security analysis, the suggested image encryption technique performs very well in terms of security.

Keywords: 5G IoT Security · Chaotic map selection · DNA Complementary Operation · Image encryption · Image cryptography

1 Introduction

The Internet of Things (IoT) has become a popular component of fifth-generation (5G) communication networks [2, 7]. It is imagined as a linked set of everything, anywhere, anytime, and any service. Like intelligent objects/things, 5G-IoT technology is becoming a more active actor in IoT-enabled environments, with the

F. P. García Márquez et al. (Eds.): ICCIDA 2022, LNNS 643, pp. 181–192, 2023.
https://doi.org/10.1007/978-3-031-27099-4_14

ability to connect and deal with relationships across various apparatuses, sensors, services, applications, and data sources. 5G-IoT intends to integrate the real world with the virtual world by using 5G networks as a communication and information exchange channel [1]. The IoT has become a popular component of 5G networks. It is considered as a linked set of everything, anywhere, anytime, and any service. Like intelligent objects/things, 5G-IoT technology has become a much more active actor in IoT-enabled ecosystems, with the ability to connect and interact with relationships across various apparatuses, sensors, services, applications, and data sources. 5G-IoT intends to integrate the real world with the virtual world by using 5G communication networks as a communication and information interchange channel.

Because of its huge storage capability, parallelism, and low energy consumption, DNA computation is often used in cryptography systems. Numerous DNA series encryption techniques have been developed in recent years.

DNA computing is often utilised in encryption systems due to its enormous storage capacity, scalability, and low energy usage [4, 15]. Recently, a variety of DNA encryption methods have been established [10].

Hu et al. [5] used a three-dimensional chaotic map with the cyclic operations of DNA sequences to operate a simple image. They encode the original plaintext image as a DNA matrix using Chen's approach for producing DNA encoding criteria. The DNA matrix was then put through a series of related DNA sequencing processes. An assortment of neighbouring and non-neighbouring connected mapping lattices were produced by Zhang et al. in [14]. They built a linked map lattice and used DNA computation to encrypt the image. They created a connected map lattice and then enciphered the image using DNA computing technique.

We introduce an unique encryption scheme that combines CML and DNA sequences that includes permutation using Arnold Cat Map, DNA encoding, and diffusion using a chaotic sine map, considering the aforesaid analytical aspects.

2   Materials and Methods

Image encryption may be accomplished using either the scrambling encryption technique or the pixel gray value encryption method. The process of applying algorithms to modify the location of pixels is referred to as scrambling (Shuffling) encryption. The association between pixels may be lessened by scrambling encryption, but the values of the pixels cannot be altered.

2.1   Chaotic Maps

It is more crucial to specify the attributes of many chaotic maps and select the optimal one for the algorithm before presenting the recommended technique. Chaotic systems are mathematical tools that exhibit chaotic behaviour. Choosing the map to be utilised in conjunction with the encryption-decryption technique was one of the most challenging issues encountered throughout this project. There are several chaotic maps, each of which has its unique set of pros and downsides. The best map from among the most often used ones is then chosen by performing a basic encipherment on an example image using each of the maps. Then, for each map, we look at the unpredictability of its generated values, such as the chaotic ranges and evaluation parameter values. The chaotic Maps [6] Used here are Henon Map, Logistic Map, Duffing Map, Arnold Cat Map (ACM), Sine Map and Tent Map.

2.2   Selection of Chaotic Maps

One of the most critical aspects of the encryption system is map selection. The chaotic character of the map's sequences adds to the security. It stops attackers from disclosing or breaching the encrypted images. The encryption quality is influenced by the maps used. To better disguise the image information pattern, we must choose the finest maps.

Entropy was employed in the method [9] to choose the best maps from a batch of N maps. We chose it at random after considering its chaotic actions trait. An encrypted image's entropy value displays a random values of pixel from 0 to 255, covering the whole image area, and obscuring the image information. Entropy has a theoretical value of eight. Because a successful encryption scheme should cause the information entropy to gravitate to 8 [13], entropy was chosen as the criterion. As a consequence, the information pattern of the image is hidden. The selection of an appropriate map that yields increased entropy is very desired.

We analyzed the entropy levels of the associated enciphered images using a simple Map Selecting() algorithm [9] on a test image named *Lena* with a size of 256×256 for each of the selected maps as inputs. The Arnold, Logistic, Tent, Henon, and Duffing maps are used to form a chaotic map collection in our research $(N = 5)$. As a result, we have five encrypted Lena images to evaluate their entropy levels. The maps, beginning values, as well as other factors employed in our experiment are listed in [6]. After applying the Map Selection technique to the enciphered image, the maps with the highest entropy value are chosen for further computations.

The preceding formula in Eq. 1 is used to determine the value of entropy.

$$E(m) = -\sum_{j=0}^{255} P\left(p_j\right) \cdot \log\left[P\left(p_j\right)\right] \tag{1}$$

Here p_j is the jth gray value and $P(p_j)$ is the probability that p_j will emerge. When paired with a normal Lena image, the Arnold map produces a high value of entropy. As a consequence, the Arnold map is used for cryptographic operations in our system.

2.3   DNA Encoding and Decoding

Four different types of de oxyribo nucleic acids are used in a DNA sequence. The name of these DNAs are Adenine (A), Guanine (G), Cytosine (C), and Thymine (T). Due to the complimentary nature of the binary values 0 and 1, the two-digit

binary numbers 00 and 11, as well as 01 and 10, are likewise complementary. Therefore, if we encode A, T, G, and C as 00, 11, 01, and 10, we can say that A & T and
G & C are complement each other. In DNA encoding, there are 24 different types
of coding rules are available. Among these only eight of them only meets Watson-
Crick complement rule. This work uses these eight rules for encrypting each pixel
during the DNA encoding process. For example, if suppose the grey values 55 and
100 with binary values "00, 110, 111" and "01, 100, 100" respectively. Using Rule 3
for DNA encoding, the first pixel will be encoded as $GCAC$ and the second pixel
using Rule 8 is $TATC$. We may decode the appropriate binary sequences during
the inverse process using the DNA decoding methods in [12]. Moreover, the DNA
Complementary Operation [8,12] is also used to achieve the encryption.

2.4 Proposed Cryptographic Algorithm

Encryption. In the proposed encryption algorithm, the plaintext image in
colour format will be separated into its three color planes: R, G, and B. Afterwards, the entropy of these three channels will be calculated for applying the map
selecting procedure for permutations. The maps will be selected according to the
entropy values, and these three channels will undergo initial permutation. Again
the DNA encoding operations with Rule 2,4 and 8 will be applied to R, G, and
B channels, respectively. Further, these three channels will undergo DNA complement operation before the DNA decoding operation. At this point, the DNA
operations are over. Again, to ensure security and make the algorithm against
histogram arracks, we will perform a diffusion operation using a chaotic map
selected from the same ap selection procedure. Finally, these three encrypted
colour channels will be combined to produce the ciphertext image.

The step-by-step processes in the encryption process are as follows.

1. Decompose the plaintext image into its three colour channels, R, G, and B.
2. Calculation of Entropy for Chaotic Map Selection
3. Chaotic Map Selection for Permutation and Diffusion
4. Perform the DNA encoding operations using Encode rules 2, 4, and 8 for the
   R, G, and B channels, respectively.
5. DNA complement operation on these three channels individually.
6. DNA decoding using the three rules 2, 4 and 8 for the R, G, and B channels,
   respectively.
7. Chaotic diffusion individually on the three channels.
8. Integration of R, G, and B planes (channels) to produce the ciphertext.

Figure 1a depicts the block diagram perspective of the suggested encryption approach.

Decryption. Decryption is the exact inverse action of encryption. The actions
taken during encryption will be reversed during the decryption phase. The R, G,
and B color planes of the ciphertext image in color format will be split into its
R, G, and B color planes. Further, the entropy will be calculated for the doing
the chaotic map selection process. (These selected maps will be used for inverse

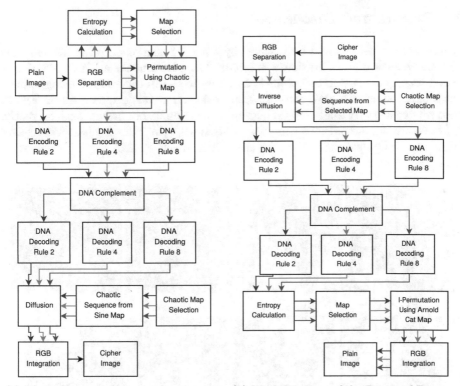

(a) Block diagram of proposed encryption algorithm (b) Block Diagram of the Proposed Decryption Algorithm

Fig. 1. (a) Block diagram of proposed encryption algorithm (b) Block diagram of the proposed decryption algorithm

diffusion and permutation). The next step is inverse diffusion using the same chaotic map. These three channels will be encoded using DNA encoding rules 2, 4 and 8 for DNA complement operation. After the DNA complement operation, the DNA decoding operations will be performed on these three channels using DNA decoding rules 2,4 and 8, respectively. Further, the inverse permutation operation will be done on these three channels using the selected chaotic map. The step by step processes are listed as follows, and the block diagram is depicted in Fig. 1b

1. Separate the colour channels R, G and B in the ciphertext image.
2. Calculate the entropy for map selection.
3. Inverse diffusion using chaotic maps.
4. Perform DNA encoding operations with Rules 2, 4 and 8 in R, G, and B channels.
5. Perform DNA complement operation.
6. Perform DNA decoding operations using the same rules applied in encoding.
7. Inverse permutation on the three channels.
8. Integration of R, G and B (planes) channels to get the plaintext image.

3   Results and Discussion

Statistical assaults should be resisted by a good image encryption system that uses permutation and diffusion operations. The performance analysis of the proposed image encryption technique is carefully tested in this section using numerous statistical studies. We chose several basic colour images, as given in Fig. 2, each having a size of 256×256, to highlight the benefits of our approach correctly. All tests are carried out on a computer system with the following hardware setup: Windows 10 Operating system, 1.8 GHz CPU, 6 GB RAM.

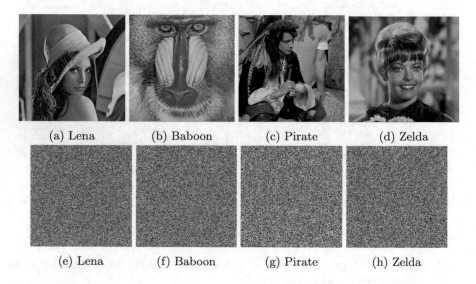

(a) Lena	(b) Baboon	(c) Pirate	(d) Zelda
(e) Lena	(f) Baboon	(g) Pirate	(h) Zelda

Fig. 2. Test images

3.1   Key Sensitivity

The encryption method must take the map's beginning and parameter costs into consideration. For a little change in the keys, the cryptosystem should return a different result. The examination of key sensitivities is shown in Fig. 3. We formed a wrong key by slightly altering one among the initial parameters of the ACM (the value $X_0 = 0.1057950190$ is transformed to $X_0 = 00.1057950200$), then applied encryption to the test images using the exact keys, as shown in Fig. 3a. We then attempted to perform decryption of the images using the modified key. The resulting decrypted images are shown in Fig. 3b. It is evident that, despite a minor modification to the keys, the encrypted images acquired with the modified keys are very different from the original images. This demonstrates that the suggested cryptosystem has complete key sensitivity, shielding it against all types of assaults.

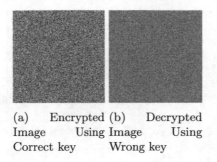

(a)   Encrypted (b)   Decrypted
Image   Using Image   Using
Correct key   Wrong key

Fig. 3. Key sensitivity test on the test image *Lena*

3.2   Histogram Analysis

An image histogram gives some statistical data about the image and displays the pattern of pixel intensity levels. To stave against statistical assaults, a reliable image encryption algorithm may give the ciphertext image a constant histogram. The histogram of plaintext and encrypted images are shown in Fig. 4. The plaintext image distribution varies dramatically from the ciphertext distribution. An image histogram depicts the pixel intensity value distribution and gives the statistical information about the picture. A powerful image encryption technique could provide the encrypted image a consistent histogram in order to defend against statistical attacks. The histogram of plaintext and encrypted images is shown in Fig. 4. The plaintext image distribution varies significantly from the cypher image distribution. As a consequence, our approach is designed to apply an uniform pixel distribution to the encrypted image, disguising the true image pattern. As a result, no sequences/patterns are apparent in the associated encrypted images, as shown in the Fig. 4.

3.3   Correlation Analysis

Every pair of neighboring pixels in every image maintains some amount of association. To protect data from various assaults, suitable encryption methods should prevent or obscure such correlations among pixels. To identify the correlations between pairs of pixels, choose some neighboring pixels from the input image in the three orientations (horizontal (H), vertical (V), and diagonal (D)).

The following equations in Eq. 2 Eq. 3 and Eq. 4 are the formulas used to compute the correlation between pixel pairs:

$$r_{cd} = \frac{S^2 \cdot \text{cov}(c, d)}{\sum_{i=1}^{S} (c_i - E_c)^2 \cdot \sum_{i=1}^{S} (d_i - E_d)^2} \quad (2)$$

$$E_c = \frac{\sum_{i=1}^{S} c_i}{S} \quad (3)$$

$$\text{cov}(c, d) = E\left((c - E_c)(d - E_d)\right) \quad (4)$$

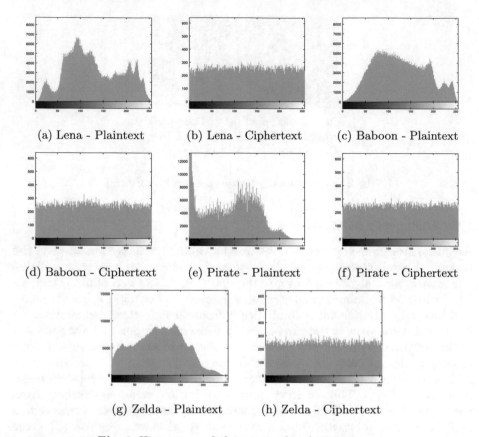

(a) Lena - Plaintext   (b) Lena - Ciphertext   (c) Baboon - Plaintext

(d) Baboon - Ciphertext   (e) Pirate - Plaintext   (f) Pirate - Ciphertext

(g) Zelda - Plaintext   (h) Zelda - Ciphertext

Fig. 4. Histograms of plaintext and ciphertext images

where (c, d) are the two adjacent horizontals, vertical, or diagonal pixels sequences, and S is the image size.

Table 1 and Table 2 illustrates the distribution of correlation within each pixel in horizontal, vertical and diagonal directions for the plaintext images, as well as the dispersion of the matching ciphertext images. Table 2 displays the coefficient correlation for all of the ciphertext images in the proposed work and other works in the literature. In virtually all of the ciphertext images, the correlation coefficients between each pair of pixels are extremely less in all the three directions, as seen in Table 1 and Table 2. As a result, the ciphertext images hide all types of patterns, making them impenetrable to attackers.

3.4 Shannon's Entropy

Information entropy or Shannon's entropy is a concept that describes the congestions of an image or the amount of data or information that may be concealed in an image using such a technique. Shannon Entropy is the measure of

Table 1. Correlation coefficients in three directions

Images	Plaintext			Ciphertext		
	H	V	D	H	V	D
Lena	00.9572	00.9789	00.9339	00.0068	00.0015	−00.0041
Baboon	00.9474	00.9208	00.9034	00.0080	00.0057	−00.0047
Pirate	00.9404	00.9539	00.9098	00.0024	−00.0041	00.0023
Zelda	00.9696	00.9814	00.9542	00.0027	00.0018	−00.0001

Table 2. Comparison of correlation coefficients

Direction	Plaintext	Ciphertext	Ref [11]	Ref [3]	Ref [6]	Ref [12]
Horizontal	00.9572	00.0068	00.0013	−00.00522	−00.0047	−00.0021
Vertical	00.9789	00.0015	−00.0009	−00.012447	00.0022	00.0009
Diagonal	00.9339	−00.0041	00.0012	00.002597	−00.0061	00.0003

unpredictability in an image. For example, the Shannon entropy for an image in 8-bit format is calculated as in Eq 5:

$$E(m) = -\sum_{j=0}^{255} P(p_j) \times \log P(p_j) \tag{5}$$

$P(p_j)$ is the probability of p_j in an image, where p_j is the jth grey value. The entropy value of a suitable encryption method should be close to 8. For different images, our encryption approach provides the maximum entropy value. The entropy values of several images are shown in Table 3 and Table 4.

Table 3. Entropy values obtained

Test image	Plaintext image	Ciphertext image
Lena	07.2417	07.9975
Baboon	07.6058	07.9972
Pirate	07.5361	07.9963
Zelda	07.7784	07.9968

Table 4. Comparison of entropy values obtained for the ciphertext image of *Lena*

Reference	Entropy
Propose work	07.9975
Wang et al. [11]	07.9974
Elkandoz et al. [3]	07.9734
Jithin et al. [6]	07.99924
Wang et al. [12]	07.9971656

3.5 NPCR and UACI

In a differential attack, a particular pixel in the plaintext image is changed, and a meaningful relationship is found by tracing the alterations to the equivalent encrypted image. A selected plaintext image attack is another name for this.

A good encrypted image should be sensitive to slight adjustments, and even a single bit alteration in the plaintext image should cause a variety of adjustments in the cypher image.

The NPCR calculates how many pixels change in a ciphertext image for every one bit that is altered in the plaintext image. According to Eq. 6's calculation, this value should be 1 for an ideal encryption method. Consider

$$\text{NPCR} = \frac{1}{p \times q} \sum_{k=1}^{p} \sum_{l=1}^{q} f(k,l) \times 100 \tag{6}$$

$$f(k,l) = \begin{cases} 0, & \text{if } c_1(k,l) = c_2(k,l) \\ 1, & \text{if } c_1(k,l) \neq c_2(k,l) \end{cases} \tag{7}$$

where c_1 and c_2 are generated by encrypting two plaintext images with a size of p by q and one random bit difference.

The unified average changing intensity (UACI) among two ciphertext images with a change of just one bit in the respective plaintext images. A formula for calculating the UACI is in Eq. 8:

$$\text{UACI} = \frac{1}{p \times q} \sum_{k=1}^{p} \sum_{l=1}^{q} \frac{|c_1(k,l) - c_2(k,l)|}{255} \times 100 \tag{8}$$

The Table 5 shows the NPCR and UACI obtained for the proposed work and Table 6 shows its comparison with other works in the literature.

Table 5. NPCR and UACI Values obtained

Input image	NPCR	UACI
Lena	99.63%	33.10%
Baboon	99.85%	33.23%
Pirate	99.56%	33.61%
Zelda	99.68%	33.81%

Table 6. Comparison of NPCR and UACI values obtained for *Lena* image

Reference	NPCR	UACI
Proposed work	99.63%	33.10%
Wang et al. [11]	99.55	33.41
Jithin et al. [6]	99.57	33.8
Wang et al. [12]	99.5956	33.4588

4   Conclusion

This study seeks to enable secure image sharing and storage in 5G networks since image sharing is one of the most used multimedia sharing services. This paper provides an enhanced image encryption approach depending on a chaotic map and DNA encoding. The Arnold map has been chosen as the most refined map to utilize with our encryption technology. An appropriate and transparent map selection approach is devised to complete the map choosing. To improve security, encryption is done independently to each of the three color planes of the RGB image. The security of transferring image data has been enhanced thanks to the chaos in image encryption. Because our technique encrypts using pseudo randomly created sequences from chaotic maps. The attacker never deciphers the cipher image created by our algorithm, and the vast keyspace prevents brute-force assaults. Entropy, Correlation analysis, Histogram analysis, NPCR, UACI, and other analytic techniques are used in this project. When these assessment metrics are used, it is found that the results are more significant than earlier studies. It shows that the algorithm provides a secure transmission of image data.

References

1. Abd El-Latif, A.A., Abd-El-Atty, B., Mazurczyk, W., Fung, C., Venegas-Andraca, S.E.: Secure data encryption based on quantum walks for 5g internet of things scenario. IEEE Trans. Netw. Serv. Manage. **17**(1), 118–131 (2020). https://doi.org/10.1109/tnsm.2020.2969863
2. Akpakwu, G.A., Silva, B.J., Hancke, G.P., Abu-Mahfouz, A.M.: A survey on 5g networks for the internet of things: communication technologies and challenges. IEEE access **6**, 3619–3647 (2017). https://doi.org/10.1109/access.2017.2779844
3. Elkandoz, M.T., Alexan, W., Hussein, H.H.: Logistic sine map based image encryption. In: 2019 Signal Processing: Algorithms, Architectures, Arrangements, and Applications (SPA), pp. 290–295. IEEE (2019). https://doi.org/10.23919/SPA.2019.8936718
4. Head, T., Rozenberg, G., Bladergroen, R.S., Breek, C., Lommerse, P., Spaink, H.P.: Computing with dna by operating on plasmids. Biosystems **57**(2), 87–93 (2000). https://doi.org/10.1016/s0303-2647(00)00091-5
5. Hu, T., Liu, Y., Gong, L.H., Ouyang, C.J.: An image encryption scheme combining chaos with cycle operation for dna sequences. Nonlinear Dyn. **87**(1), 51–66 (2017). https://doi.org/10.1007/s11071-016-3024-6
6. Jithin, K., Sankar, S.: Colour image encryption algorithm combining arnold map, dna sequence operation, and a mandelbrot set. J. Inf. Secur. Appl. **50**, 102428 (2020). https://doi.org/10.1016/j.jisa.2019.102428
7. Li, S., Da Xu, L., Zhao, S.: 5g internet of things: a survey. J. Ind. Inf. Integr. **10**, 1–9 (2018). https://doi.org/10.1016/j.jii.2018.01.005
8. Liao, X., Kulsoom, A., Abbas, S.A., et al.: Selective encryption for gray images based on chaos and dna complementary rules. Multimed. Tools Appl. **74**(13), 4655–4677 (2015). https://doi.org/10.1007/s11042-013-1828-7
9. Sneha, P., Sankar, S., Kumar, A.S.: A chaotic colour image encryption scheme combining walsh-hadamard transform and arnold-tent maps. J. Ambient. Intell. Humaniz. Comput. **11**(3), 1289–1308 (2020). https://doi.org/10.1007/s12652-019-01385-0
10. Wang, X.Y., Zhang, Y.Q., Bao, X.M.: A novel chaotic image encryption scheme using dna sequence operations. Opt. Lasers Eng. **73**, 53–61 (2015). https://doi.org/10.1016/j.optlaseng.2015.03.022
11. Wang, X., Hou, Y., Wang, S., Li, R.: A new image encryption algorithm based on cml and dna sequence. IEEE Access **6**, 62272–62285 (2018). https://doi.org/10.1109/access.2018.2875676
12. Wang, X., Wang, Y., Zhu, X., Luo, C.: A novel chaotic algorithm for image encryption utilizing one-time pad based on pixel level and dna level. Opt. Lasers Eng. **125**, 105851 (2020). https://doi.org/10.1016/j.optlaseng.2019.105851

13. Zhang, X., Wang, X.: Multiple-image encryption algorithm based on dna encoding and chaotic system. Multimed. Tools Appl. **78**(6), 7841–7869 (2019). https://doi.org/10.1007/s11042-018-6496-1
14. Zhang, Y.Q., Wang, X.Y., Liu, J., Chi, Z.L.: An image encryption scheme based on the mlncml system using dna sequences. Opt. Lasers Eng. **82**, 95–103 (2016). https://doi.org/10.1016/j.optlaseng.2016.02.002
15. Zheng, X., Xu, J., Li, W.: Parallel dna arithmetic operation based on n-moduli set. Appl. Math. Comput. **212**(1), 177–184 (2009). https://doi.org/10.1016/j.amc.2009.02.011

Symptom Based Health Status Prediction via Decision Tree, KNN, XGBoost, LDA, SVM, and Random Forest

Elif Meriç[1](✉)[iD] and Çağdaş Özer[2][iD]

[1] Istanbul Kultur University, Istanbul, Turkey
eelifmeric@gmail.com
[2] Istanbul Bilgi University, İstanbul, Turkey
cagdas.ozer@bilgi.edu.tr

Abstract. Machine learning applications in health science become more important and necessary every day. With the help of these systems, the load of the medical staff will be lessened and faults because of a missing point, or tiredness will decrease. It should not be forgotten that the last decision lies with the professionals, and these systems will only help in decision-making. Predicting diseases with the help of machine learning algorithm can lessen the load of the medical staff. This paper proposes a machine learning model that analyzes healthcare data from a variety of diseases and shows the result from the best resulting algorithm in the model. It is aimed to have a system that facilitates the diagnosis of diseases caused by the density of data in the health field by using these algorithms of previously diagnosed symptoms, thus resulting in doctors going a faster way while diagnosing the disease and have a prediction about the diseases of people who do not have the condition to go to the hospital. In this way, it can ease the burden on health systems. The disease outcome corresponding to the 11 symptoms found in the data set used is previously experienced results. During the study, different ML algorithms such as Decision Tree, Random Forest, KNN, XGBoost, SVM, LDA were tried and compatibility/performance comparisons were made on the dataset used. The results are presented in a table. As a result of these comparisons and evaluations, it was seen that Random Forest Algorithm gave the best performance. While data was being processed, input parameters were provided to each model, and disease was taken as output. Within this limited resource, our model has reached an accuracy rate of 98%.

Keywords: Machine Learning (ML) · K-Nearest Neighbors (KNN) · Random Forest (RF) · GridSearchCV · Extreme Gradient Boosting (XGBoost) · Decision Tree (DT) · Mean Absolute Error (MAE) · Linear Discriminant Analysis (LDA) · Support Vector Machine (SVM)

1 Introduction

Machine learning applications are used in many different fields today, and especially the importance of machine learning in the medical field has grown enor-

F. P. García Márquez et al. (Eds.): ICCIDA 2022, LNNS 643, pp. 193–207, 2023.
https://doi.org/10.1007/978-3-031-27099-4_15

mously at a daily rate. Also, it is of great importance that the diagnoses made are correct. Today, where many different diseases are common in humans, the importance of machine learning algorithms for an accurate diagnosis is increasing. With machine learning models created using datasets that contain a wide range of data, high accuracy rates can be obtained and mistakes can be prevented. In this direction, in this study, which was conducted with a data set of various diseases and their symptoms, a machine learning model was created by trying to get the highest possible accuracy in order to provide doctors with an additional view in the field of medicine, where accurate diagnosis is very important.

This paper presents the application of health status prediction in Python programming language using machine learning models. The machine learning model was trained using a symptom-based disease dataset. Throughout the study, KNN, Decision Tree, XGBoost, SVM, LDA and Random Forest algorithms were compared, which algorithm worked better with the imbalanced multi-class dataset used was analyzed, and a model was created with Random Forest, which is the best performing algorithm in this direction. Label Encoding was done during the preprocessing stage and GridSearchCV was used to determine the parameters that showed the best performance with the algorithm.

By applying machine learning algorithms to the dataset of the disease and various symptoms, since the correct diagnosis is of great importance in the field of health, it is aimed to determine the algorithm that gives the best result among these algorithms and to provide a realistic result by obtaining the best accuracy rate, by using machine learning algorithms in the field of health. Pre-estimation is provided by the machine learning algorithm, and the most realistic results are obtained with the Random Forest Classifier, which gives the best accuracy rate.

The paper is divided into 7 different sections. Section 2 includes different studies that have been done on the same, or similar subjects and the methods they recommend. Section 3 provides the methods followed during this study and a detailed explanation of these methods. Section 4 contains proposed work for the paths followed throughout the study. Section 5 includes the results and findings from the study and Sect. 6 is the discussions on these findings. Finally, the concluding remarks are in Sect. 7.

2   Literature Review

A wide variety of studies have been conducted in the past years on symptom-based disease prediction. In some of these studies, a dataset containing the symptoms of a specific disease was used [1,2,4], while in others, a dataset containing the symptoms of various diseases [3] was used. In a similar study done before, a web-based and machine learning integrated disease diagnosis prediction [5] was presented using KNN, Naïve Bayes, and Random Forest algorithms. Ahmed, Maruf [4] et al. proposed a machine learning approach for heart disease prediction based on external factors. They did their work by trying various algorithms such as Naïve Bayes, Quadratic Discriminant Analysis, Logistic Regression, Support Vector Machine, Decision Tree, Random Forest for the best result, and got the highest accuracy value of 95% with the Support Vector Machine, but Random Forest Classifier gave an accuracy rate of 92%. In the study of John, Rinehart [6], linear and tree-based classifiers were compared and it was observed that tree-based classifiers gave a more accurate result with the disease dataset used. Classifiers such as GLM, LDA, QDA, Random Forest were used as algorithms. An accuracy rate of 74% was achieved with Random Forest. But this is not the highest rate among algorithms. They had the highest accuracy rate of 75% with bagging. One of the comparative studies on disease prediction by trying various algorithms is [7] focused on Artificial Neural Networks, Logistic Regression, KNearest Neighbors, Decision Tree, and Random Forest algorithm. In this study, in addition to other general machine learning algorithms such as KNN, Decision Tree, Random Forest, etc. artificial neural networks were also used and the highest accuracy rate was obtained with neural networks. This accuracy for the test dataset is 80.2% and for the training dataset is 82.3%. The algorithm that gives the highest accuracy rate after neural networks are Random Forest with 78.6%. Vijaya Shetty S. et al. [8] among other studies mentioned, proposed a model that uses the capabilities of different ML algorithms combined with text processing. Text processing is done using tokenization and similarities and outputs are combined with various algorithms. These algorithms are Decision Tree, Random Forest, and Naïve Bayes. Among them, Naïve Bayes gave the highest accuracy with a rate of 98.55%. While Decision Tree gave an accuracy rate of 98.18%, Random Forest gave the lowest accuracy at 98.05%. Joshi, Tejas et al. [9] aimed to predict diabetes via different supervised machine learning methods such as Support Vector Machine(SVM) and Logistic Regression. They also aimed to propose a technique for earlier detection of diabetes. They achieved 78.645% accuracy via SVM, and 78.125% accuracy via Logistic Regression. There is little difference between these two machine learning algorithms in this study, and the accuracy of SVM is higher than Logistic Regression. Muhammad Daniyal Baig et al. [11] used logistic regression, decision tree, and random forest algorithms to predict diabetes. They applied data preprocessing techniques before processing the data. The model has achieved 98% accuracy with Random Forest Algorithm, and logistic regression gave the lowest accuracy with 84%. In some studies [11,13], the diabetes dataset was used and various machine learning algorithms were implemented to detect diabetes. Baig, Muhammad et al. [11] evaluated the results on four different machine learning algorithms such as Logistic Regression, KNN, Random Forest, Gradient Boosting. They achieved the highest accuracy with 98% via Random Forest algorithm, and Logistic Regression gave the lowest accuracy with 84%. Moreover, Random Forest Algorithm obtained the best ROC with 99%. Mujumdara, Aishwarya et al. [13] implemented Support Vector Classifier, Decision Tree Classifier, Random Forest Classifier, Extra Tree Classifier, Ada Boost Algorithm, Perceptron, Linear Discriminant Analysis Algorithm, Logistic Regression, K-Nearest Neighbor, Gaussian Naïve Bayes, Bagging algorithm, Gradient Boost Classifier algorithms and applied it on the diabetes dataset they used in the paper. According to their evaluation, Logistic Regression gave the highest accuracy with a rate of

96%, and they obtained 91% accuracy with Random Forest Algorithm. Among these algorithms, SVM gave the lowest accuracy with 60%. Furthermore, they got the highest accuracy of 97.2% for Logistic Regression. AdaBoost Classifier gave the best result with 98.8% accuracy with the pipeline.

3   Methodology

First of all, non-numeric data in the dataset in the preprocessing stage are converted into numeric data with the Label Encoder. The purpose of Label Encoder is to convert the data into numerical form that the model expects and will perform better. Subsequently, data reduction and data cleaning processes were carried out in order to obtain a better dataset. Afterwards, 70% of the data was reserved for training and 30% for testing, and training was carried out with various machine learning algorithms. GridSearchCV, whose mathematical formula is calculated by the estimator, is a function that helps to process the specified hyperparameters and fit the model to the training data, choosing the best parameters from the list. With this function, the best parameters are determined for the algorithms detailed below. Additionally, this study was carried out using a computer with 8 GB RAM, 8th Generation Intel Core i7 CPU, 1 TB SSD hardware. Also, Jupyter Notebook, Google Colab and Visual Studio were used in the project.

3.1   Decision Tree Classifier

This algorithm is supervised and non-parametric. It begins with a node and is in a tree structure. There are multiple criteria that are effective in making the decision to partition a node into child nodes. Thus, the node purity increases according to the target variables. Criteria selection is made according to the type of target variables. Two of these criteria are "Entropy" and "Gini". Entropy is a measure of the uncertainty associated with the data. Therefore, minimizing entropy requires dividing the data well because the prediction is directly proportional to the division. The better the splits, the better the forecast. In the entropy formula $Entropy = -\sum_{i=1}^{n} p_i \log p_i$, p_i is the percentage of the group belonging to a particular class. In this project, "entropy" was used as a criterion in the Decision Tree Algorithm. In order to minimize the entropy value, it is necessary to determine the best splitting. For this, the formula "information gain" $InfoGain(S, D) = H(S) - \sum_{V \in D} \frac{|V|}{|S|} H(V)$ is used. S is the original dataset and D represents a split part of the set. Each V is a subset of S. All of V is discrete and form S. The next formula, $GiniIndex = 1 - \sum_{i=1}^{n} p_i^2$, is the measure of inequality in the sample.

3.2   K-Nearest Neighbors (KNN)

In this algorithm, the parameter k, which is the number of nearest neighbors to the given point, is determined. For instance, when k = 3 is given, the classification is made according to the closest 3 neighbors. Then, the Euclidean distances of other points from the target point are calculated. The nearest neighbors are found according to these calculated distances. Then, the nearest neighbor categories are summed up. After that, the most appropriate neighbor category is selected. The parameter n_neighbors is assigned as 5 to make the classification according to the 5 closest neighbors. The Euclidean distance formula is used to calculate the distance between two points.

$$d(p,q) = d(q,p) = \sqrt{\sum_{i=1}^{n}(q_i - p_i)^2}$$

3.3   Extreme Gradient Boosting (XGBoost)

Its working principle is based on decision trees. First, a base score is determined. Because the proper result will be reached by converging with the operations in the other phases, the base score may be any value. The estimate is analyzed with residual. The predicted value is subtracted from the observed value, and errors are detected in this way. Then, a decision tree that predicts errors is set up. The aim is to know the errors and get closer to an accurate prediction. A similarity score is generated for each of the tree's branches to determine how well the data is grouped into branches. In the formula $SimilarityScore = \frac{(S.R.)^2}{N+\lambda}$, Lambda is the regularization parameter. To answer whether a better prediction can be made after similarity scores are calculated, trees of all possible possibilities are built. Similarity scores are calculated for all. Then, the information gain $(IG = \frac{1}{2}[\frac{G_L^2}{H_L+\lambda} + \frac{G_R^2}{H_R+\lambda} - \frac{(G_L+G_R)^2}{H_L+H_R+\lambda}] - \gamma)$ is calculated to determine a better tree.

$$SimilarityScore = \frac{(S.R.)^2}{N+\lambda}$$

$$IG = \frac{1}{2}[\frac{G_L^2}{H_L+\lambda} + \frac{G_R^2}{H_R+\lambda} - \frac{(G_L+G_R)^2}{H_L+H_R+\lambda}] - \gamma$$

The γ value is chosen for pruning. If a branch has a gain score lower than its gamma score, it is pruned. Accordingly, increasing the γ prevents overfitting because increasing γ helps retain only valuable branches. Pruning proceeds upwards from the final branch. There is no need to investigate the branches above the lowest branch if it is determined not to prune it. Based on the similarity score formula, it can be said that when lambda increases, the calculated similarity score decreases and thus the gain score decreases. Consequently, not only branches with high scores are pruned. Also, as lambda increases, learning becomes difficult and overfitting is prevented. The more values in the branch, the less it will lower the lambda similarity score. This feature also helps prevent overfitting. In iteration, the objective function (loss function and regularization) t is to minimize. In this project, the base score was set at 0.5, gamma was set at 0, and the learning rate(eta) was set as 0.300000012. The second following formula is used for this:

$$Output = \frac{Sum\,of\,the\,Values\,in\,the\,Branch}{Number\,of\,Values\,in\,the\,Branch + \lambda}$$

$$L^t = \sum_{i=1}^{n} l(y_i, \overset{y^{(t-1)}}{\underset{i}{}} + f_t(X_i))\Omega(f_t)$$

3.4   Support Vector Machine

In this algorithm, a linear line is drawn to separate the classes, and it aims to maximize the distance between the classes, called the margin, in order to distinguish the classes in the best way. In fact, this approach, which is suitable for binary classification problems, is also suitable for multi-class classification problems with the One vs Rest method. In this method, one of the multiple classes is taken and compared with the others by drawing a linear line with the SVM approach. In the figure (Fig. 1) below, there are two different classes, black and white. Since this is a classification problem, the goal is to determine which class the data is in. For this, a line is drawn that separates the two classes in the Support Vector Classifier algorithm, and the region between this line is the margin. The wider the margin, the better the classes are separated.

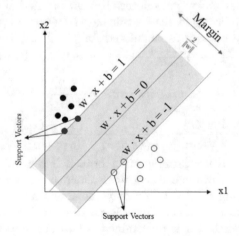

Fig. 1. Support vector machine

3.5   Linear Discriminant Analysis

It is a supervised algorithm. To distinguish between classes, it examines the distribution of classes and uses the difference between their mean values. It performs size reduction in the dataset by maximizing the distance between classes. It performs well with datasets where there are more than two classes and the output is categorically independent. LDA performs well on disease prediction.

Since the data are categorical in these datasets consisting of data such as disease and symptom, classification with LDA gives a good performance. In this algorithm, the distance between the averages of the classes is measured with the formula $S_b = \sum_{i=1}^{g} N_i(\overline{x_i} - \overline{x})(\overline{x_i} - \overline{x})^T$. Then with the following formula $S_w = \sum_{i=1}^{g}(N_i-1)S_i = \sum_{i=1}^{g}\sum_{j=1}^{N_i}(X_{i,j}-\overline{x_i})(X_{i,j}-\overline{x_i})^T$, the distance between the average of the classes and the sample is calculated. With the last formula $P_{lda} = argmax\frac{|P^T S_b P|}{|P^T S_w P|}$, sub-dimensional space is created to maximize the distance between classes and to minimize the distance between classes.

3.6   Random Forest Classifier

Random forests generate many decision trees during training. Predictions from all various trees are aggregated to produce an estimate. In the classification, among the predictions, the highest rated prediction is selected. Since training takes place on different datasets in the random forest model, the variance, or overfitting is reduced. Random forest should be used to model large datasets, due to bootstrap resampling, training datasets on samples would not give the best result in manners of accuracy and precision, so it gives out the best performance with large datasets. To calculate the node importance for each decision tree, Gini Importance is used. ni_j is the node importance of node i in its formula W_i
$ni_i = W_i C_i - W_{left(i)} C_{left(i)} - W_{right(i)} C_{right(i)}$.

fi_i is the importance of feature i, and ni_j represents the importance of node j. The sum of the feature importances is divided to normalize the values between 0 and 1. The mean of all trees determines the final feature relevance in Random Forest. After calculating the sums of the importance of the features in the trees, it is calculated by dividing the total number of trees. $RFfi_i$ represents the importance of feature i computed from all trees in the Random Forest model. $normfi_{ij}$ indicates the normalized feature importance of i in tree j, and T indicates the number of trees in total.

fi_i represents the importance of feature i and it is calculated with the formula $fi_i = \frac{\sum_{i:node\,j\,splits\,on\,feature\,i} ni_j}{\sum_{k\in all\,nodes} ni_k}$, s_j is the number of instances that arrive at node j, and C_j is the impurity of j. Each tree's feature importance is normalized, and the final importance is computed. In the formula $fi_i = \sum j : nodes\,j\,splits\,on\,feature\,i\, S_j C_j$, $normalizedfi_i$ is the normalized importance of feature i and it is calculated as $normalizedfi_i = \frac{fi_i}{\sum_{j\in all\,features} fi_j}$, and the feature importance of i is fi_i. Finally, feature importance values from all trees are summed and normalized. $RFfi_i$ indicates the feature importance of i as computed across all trees in the Random Forest model and its formula is $RFfi_i = \frac{\sum_{j\in all\,trees} normfi_{ij}}{T}$, and $normalizedfi_{ij}$ indicates the normalized feature importance for i in tree j, and it is calculated as $normalizedfi_{ij} = \frac{fi_i}{\sum_{j\in all\,features} fi_j}$. In this model, Random Forest Classifier was used for the disease prediction with the parameter n_estimators=26, which indicates the quantity of trees. Thus, the number of trees was set to 26.

$$RFfi_i = \frac{\sum jnormfi_{ij}}{\sum j \in allfeatures, k \in alltrees^{normfi_{jk}}}$$

3.7   Dataset

The dataset used in this study is consisting of 38 unique diseases and 41 unique symptoms collected from different patients by Pranay Patil & Pratik Rathod [12], and taken from Kaggle. It consists of 254 rows and 12 columns. 11 of these columns include symptoms, while 1 includes diseases. In the figure (Fig. 2) below, a part of the dataset taken from Kaggle can be seen. As seen below, there are 11 symptom columns and 1 disease column in total. The dataset consists of different diseases, not specific diseases. The data in the symptom columns consist of different symptoms according to the diseases. There are 11 symptom columns from Symptom_1 to Symptom_11. The bar graphs below (Fig. 3) show the disease counts and symptom distribution. The first bar graph below illustrates the disease counts. According to the graph, it can be said that the most important features are Hepatitis D and Migraine. In the second bar graph below, the symptom distribution is shown. According to these, the most important symptom is *fatigue* with a rate of 40. In the 11 symptom columns, the symptoms are mixed according to the diseases.

Disease	Symptom_1	Symptom_2	Symptom_3	Symptom_4	Symptom_5
fungal infection	itching	skin rash	nodal skin erupti	dischromic patch	irritation
fungal infection	skin rash	nodal skin erupti	dischromic patch	itching	irritation
fungal infection	itching	nodal skin erupti	dischromic patc	skin rash	irritation
fungal infection	itching	skin rash	dischromic patc	nodal skin erupti	irritation
fungal infection	itching	skin rash	nodal skin erupti	dischromic patch	irritation
fungal infection	skin rash	itching	dischromic patc	nodal skin erupti	irritation
fungal infection	skin rash	itching	nodal skin erupti	dischromic patc	irritation
fungal infection	itching	skin rash	nodal skin erupti	dischromic patc	irritation
allergy	continuous snee	shivering	chills	watering from ey	hives
allergy	shivering	chills	watering from ey	continuous snee	hives
allergy	continuous snee	chills	watering from ey	shivering	hives

Fig. 2. The dataset used in the study

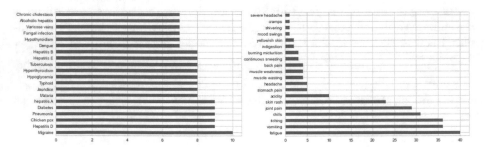

Fig. 3. Feature counts for diseases and symptoms

4   Proposed Work

In the data preprocessing process, label encoding was used to make the labels machine-readable and unique numbers were assigned to categorical variables. Before deciding on the Random Forest algorithm, SVM, LDA, XGBoost, Decision Tree, and KNN machine learning algorithms were applied and comparisons were made. The proposed architecture can be seen in the figure (Fig. 4) below.

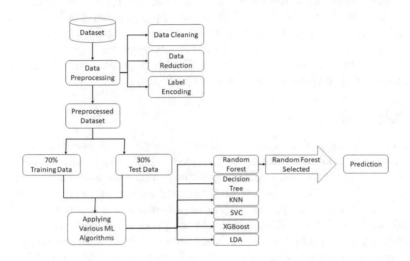

Fig. 4. The proposed architecture

4.1   Decision Tree Classifier

Decision Tree Classifier was chosen because it is non-distributional, does not depend on probability distribution assumptions, and therefore can work well with high-dimensional data. Before using this algorithm in the model, label encoding was done during the preprocessing phase. In this project, the ID3 decision tree algorithm is used. Since this algorithm is based on entropy while classifying, the criterion parameter is given as "entropy". Entropy increases with the increasing number of classes and decreases with the decrease. In this algorithm, first, the overall entropy is calculated and then the entropy of each attribute is calculated. By subtracting the calculated entropy values from the general entropy, it is determined which feature decreases the entropy value the most. A decrease in entropy is gain. In this algorithm, other hyperparameters other than criterion are default values. However, this algorithm has been abandoned because it cannot prevent overfitting and does not give a good accuracy rate as well as Random Forest Classifier.

4.2   K-Nearest Neighbor

It calculates the best vectors or separations of the values according to their distance and proximity to each other. For this reason, it is necessary to normalize

or standardize the data beforehand. Therefore, the Standard Scaler was used
in the preprocessing stage. After that, the default value of 5 is assigned to the
neighbor number parameter of k-Nearest-Neighbor. In the K-Nearest Neighbor
algorithm, the k value represents the number of nearest neighbors. Since this
is a classification problem, giving the k value higher will result in more nearest
neighbors being looked at and will yield better results. First of all, the k value
was given as 3 and it was seen that the result obtained was not sufficient. Subsequently, sufficient final results were obtained by giving 5. The reason why the
KNN is tried in the study is that it does not require training before making the
estimation and therefore it is a simple algorithm that works fast in the training
phase. The reason for not continuing the project with this algorithm is that the
test phase is slow, it does not perform well with large datasets, it takes up too
much memory, and gives a low accuracy rate.

4.3   XGBoost Classifier

The XGBoost algorithm was used because it prevents overfitting and is a
fast-running algorithm. DMatrix and GridSearchCV were used with XGBoost
Classifier in the study and optimal hyperparameters were adjusted with Grid-
SearchCV. The parameter values were assigned as learning rate 0.1, maximum
depth 3, eta 0.2, gamma 0, max depth step 3, estimator number 100, silent 0,
and subsample 0.8. Since XGBoost could not process categorical data by itself,
XGBoost algorithm was used after label encoding in the preprocessing phase.
The parameter values used for GridSearchCV are −1 for the number of jobs, 3
for cross-validation splitting (cv), and verbose 2.

4.4   Support Vector Classifier

Support Vector Machine, another algorithm whose performance has been tested
in this study, is a supervised machine learning algorithm that can be used for classification or regression problems. This study is a classification problem. According
to the working principle of the algorithm, classification is done using a hyperplane.
Maximum margin gives the best classification result. Hyperparameters were determined with the help of GridsearchCV. As the best parameter values, the value of
10 is assigned to C, which is the regularization parameter, and the value of 'rbf' is
assigned to the kernel, which indicates the kernel type. Here rbf is a method that
calculates and classifies how similar each point is to a particular point with a normal distribution. Other parameters are assigned by default.

4.5   Linear Discriminant Analysis

This supervised algorithm, which was used in classification problems and performed well with two or more classes and categorical independent data, was
applied in this study because it was predicted to perform well with the dataset
containing the diseases and symptoms used in the study. First of all, GridsearchCV was used to determine hyperparameters and the best parameter values were determined. The number of components is set to 10 and the solver

parameter is set to singular value decomposition (svd). Since there are too many features in the dataset used, the value 'svd' is given, which does not calculate the covariance. Other parameters are assigned by default.

4.6   Random Forest Classifier

The final ML model in this study was created with Random Forest Classifier. The reason for deciding on this algorithm is that each of the hundreds of decision trees created in Random Forest makes an individual prediction, since the study is a classification problem, it is more successful than other algorithms in preventing overfitting by choosing the most voted among these predictions (especially when compared to Decision Tree Classifier) and can perform well with large datasets. Before applying the algorithm, label encoding was done during the preprocessing phase and the algorithm was used with GridSearchCV to set its hyperparameters. The pseudo-code of the Random Forest algorithm that gives the best performance is as follows, and the following figure (Fig. 5) shows the Random Forest algorithm. The figure below (Fig. 6) represents the working principle of the Random Forest Classifier used in this study.

Algorithm 1 Random Forest for Classification (RFC)

1: To create n classifiers
2: **for** $i = 1 \to n$ **do**
3:     Select random samples to create a bootstrapped dataset from the Disease Dataset.
4:     Create the tree.
5:         Select variables randomly.
6:         Pick the best split point among the selected variables.
7:         Split the node.
8: **end for**
9: Run test data, and get a prediction from each Decision Tree
10: Store the prediction
11: Take and calculate the votes from targets.
12: Output the highest voted prediction

Fig. 5. Random forest classifier

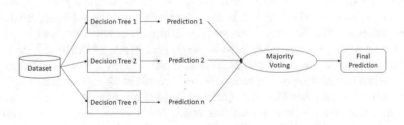

Fig. 6. Random forest classifier working principle

5   Results

Due to the large dataset that was fed to the model, and numerous input data that decides the output, Random Forest Classifier(RFC) was the best option to go on with, as multiple decision trees were constructed and the highest rated prediction was selected among the predictions. Also, from lots of models that work very well with a large dataset, RFC was chosen due to the Confusion Matrix(CM) results. With each appropriate model, a CM was created and compared to one other, so the best option was RFC. In addition, 90.20% accuracy was obtained with the Linear Discriminant Analysis (LDA) algorithm. After Decision Tree, the best result was obtained with LDA. After LDA, Support Vector Classifier gave the highest accuracy with 86.27% accuracy. The classification report of the Random Forest algorithm can be seen in Table 1 below. In this table, only the result of some classes is available as an example. According to this table, the precision indicates the number of actually positive values forecasted as positive. F1 Score value shows the harmonic mean of precision and recall values. The Recall value indicates the number of positively forecasted values that should be forecasted positively. The support value indicates how many of a class there are.

Table 1. Classification report for random forest

	Precision	Recall	F1-score	Support
Vertigo	1.00	1.00	1.00	2
AIDS	1.00	1.00	1.00	1
Acne	1.00	1.00	1.00	1
Allergy	1.00	1.00	1.00	2
Arthritis	1.00	1.00	1.00	1
Bronchial Asthma	1.00	1.00	1.00	1
Cervical spondylosis	1.00	1.00	1.00	1
Chicken pox	0.67	1.00	0.80	2
Chronic cholestasis	1.00	1.00	1.00	2
Dengue	1.00	0.67	0.80	3
Accuracy			0.99	76
Macro avg	0.99	0.99	0.99	76
Weighted avg	0.99	0.99	0.99	76

The table of the model evaluation results based on accuracy, MAE, and F1-score can be seen in the Table 2 below. According to the F1-score evaluation results obtained, the Random Forest Algorithm gives the best result with an F1 Score of 0.960, an accuracy rate of 98%, and MAE of 0.210. Since the F1 score is a harmonic mean of precision and recall values, extreme cases are not ignored when calculating and a more accurate result is seen. The final model was created with the Random Forest Algorithm. This is because Random Forest gives higher accuracy than other algorithms tried, has a lower mean absolute error, and prevents overfitting. When Random Forest and Decision Tree algorithms are compared with each other, Random Forest's ability to prevent overfitting is better than Decision Tree's. This is because Decision Tree makes predictions on a single tree, while Random Forest creates a forest structure from multiple trees and selects the prediction with the most votes from these trees. Additionally, KNN gave the worst result with the lowest accuracy of 70%, and a higher MAE of 1.57. After Random Forest Algorithm, XGBoost algorithm gave the highest accuracy rate of 93.98%. The MAE value of this algorithm was observed as 0.83 and the F1 score as 94%. Subsequently, 91.07% accuracy was obtained with the Decision Tree Algorithm. With this algorithm, the MAE value was observed as 1.25 and the F1 score as 0.910. After the Decision Tree Algorithm, the highest accuracy value was obtained with LDA with 90.20%. LDA has an MAE of 1.05 and an F1 score of 0.9019. Subsequently, 86.27% accuracy, 2.13 MAE and 0.8627 F1 score values were obtained with SVM. The following table (Table 2) shows the results obtained from the algorithms applied in this study.

Table 2. Model evaluation results

Algorithm	Accuracy	MAE	F1 Score
Decision Tree	91.07%	1.25	0.910
KNN	70%	1.57	0.7
XGBoost	93.98%	0.83	0.94
Random Forest	98%	0.210	0.960
SVM	86.27%	2.13	0.8627
LDA	90.20%	1.05	0.9019

6   Discussion

The result of this study supports the finding that a single algorithm is not enough on a specific dataset to get good performance from the model, and different algorithms should be tested on the dataset to achieve maximum performance. When the studies on different datasets using various algorithms applied for similar purposes were examined, it was seen that different accuracy results were obtained from each of them. Among the studies using independent datasets such as Heart Disease and Diabetes, the study [10] with the best accuracy of 98% was conducted using Random Forest and the Heart Disease dataset was used. With this 98% accuracy rate, they obtained the same rate as obtained in this study. In addition, in the study by Joshi, Tejas et al. [9], a lower accuracy of 79% was obtained compared to other studies. SVM was used in this study.

In the table below (Table 3), the evaluation results of two SVMs, two Random Forests (one used in the current study), one Logistic Regression, and one LDA algorithm used in different studies and with different datasets are given. By comparison, the best performing among these evaluations was the study using Random Forest [10], which gave 98% accuracy. This study gave the same result as the current study in terms of accuracy. According to the table, the algorithm that gives the second best accuracy result with 96% is Logistic Regression [11]. It has been observed that 80% accuracy is obtained in the study using the LDA algorithm [6] in the table below and on a similar subject. In this study, 90.20% accuracy was obtained with LDA. This accuracy rate obtained with LDA in another study is lower than the rate obtained with LDA in this study. In the first of the studies using SVM [4], 95% accuracy was obtained. In this study, 86.27% accuracy was obtained with SVM. Here, better performance was observed with SVM in the other study [4]. In the second study with SVM [9], a lower accuracy was obtained with 79% accuracy compared to this study.

Table 3. Comparison with other studies

Reference	Algorithm	Accuracy	MAE	F1 Score
[4]	SVM	95%	–	0.94
[9]	SVM	79%	–	–
[10]	Random Forest	98%	–	–
[11]	Logistic Regression	96%	–	–
[6]	LDA	80%	–	–
This study	Random Forest	98%	0.210	0.960

7   Conclusion and Future Work

In this paper, a disease prediction machine learning project developed to help physicians is presented. Considering the possibility of physicians being mistaken when making a diagnosis, it can be realized that this is a very vital point for the medical field. For this reason, making use of machine learning methods is a good choice. In the project, a dataset of various diseases and their symptoms was used, and the user was asked to enter their symptoms, and then a disease prediction was made to the model. It is important to get the best accuracy rate, especially in medicine and disease diagnosis prediction. Different machine learning algorithms have been tried to get the best accuracy from the model. Among the tried LDA, KNN, SVM, Decision Tree, XGBoost and Random Forest algorithms, Random Forest has proven to be the best algorithm in terms of preventing overfitting, giving the highest accuracy with 98%, the most compatible with the dataset, and the project was carried out with Random Forest algorithm. In future works, artificial neural networks, and various other improved algorithms besides the traditional machine learning algorithms used will be tried on a dataset containing more comprehensive details, to determine the algorithm that works best with this more comprehensive dataset, and thus create a model that can be more useful and can give more realistic results has been planned.

References

1. Akhtar, N.: Heart Disease Prediction (2021)
2. Jany Shabu, S.L., Nithin, M.S., Santhosh, M., Roobini, M.S., Mohana Prasad, K., Joshila Grace, L.K.: Skin disease prediction. J. Comput. Theor. Nanosci. **17**(8), 3458–3462 (2020)

3. Shilimkar, G., Shivam, P.: Disease prediction using machine learning. Int. J. Sci. Res. Sci. Technol. **8**(3), 551–555 (2021)
4. Tamal, M.A., Islam, M.S., Ahmmed, M.J., Aziz, M.A., Miah, P., Karim, M.R.: Heart disease prediction based on external factors: a machine learning approach. Int. J. Adv. Comput. Sci. Appl. **10** (2019) https://doi.org/10.14569/IJACSA.2019.0101260
5. Rajora, H., Punn, N.S., Sonbhadra, S.K., Agarwal, S.: Web based disease prediction and recommender system (2021)
6. John, R.: An application of machine learning in IVF: comparing the accuracy of classification algorithms for the prediction of twins. Gynecol. Obstet. **9**(497), 0932–2161 (2019). https://doi.org/10.4172/2161-0932.1000497
7. Lee, R., Chitnis, C.: Improving health-care systems by disease prediction. In: 2018 International Conference on Computational Science and Computational Intelligence (CSCI), pp. 726–731 (2018). https://doi.org/10.1109/CSCI46756.2018.00145
8. Shetty, S.V., Karthik, G.A., Ashwin, M.: Symptom based health prediction using data mining. In: 2019 International Conference on Communication and Electronics Systems (ICCES), pp. 744–749 (2019). https://doi.org/10.1109/ICCES45898.2019.9002132
9. Joshi, T.N., Chawan, P.M.: Logistic regression and SVM based diabetes prediction system. Int. J. Technol. Res. Eng. **5**, 4347–4350 (2018)
10. Lafta, R., Zhang, J., Tao, X., Li, Y., Tseng, V.S.: An intelligent recommender system based on short-term risk prediction for heart disease patients. In: 2015 IEEE/WIC/ACM International Conference on Web Intelligence and Intelligent Agent Technology (WI-IAT), vol. 3, pp. 102–105. IEEE (2015). https://doi.org/10.1109/WI-IAT.2015.47
11. Baig, M., Nadeem, M.: Diabetes prediction using machine learning algorithms (2020). https://doi.org/10.13140/RG.2.2.18158.64328
12. https://www.kaggle.com/itachi9604/disease-symptom-description-dataset
13. Mujumdar, A., Vaidehi, V.: Diabetes prediction using machine learning algorithms. Int. Conf. Recent Trends Adv. Comput. ICRTAC **165**, 292–299 (2019)

Multiband Microstrip Elliptical Monopole Antenna Design with Mushroom-Like Loadings for DCS, 5G and Ku-Band Applications

Gürtay Sezay Gürsoy[1](✉) and Mustafa Hikmet Bilgehan Uçar[2]

[1] İstanbul Arel University, İstanbul, Turkey
gurtaysezaygursoy@arel.edu.tr
[2] Kocaeli University, Kocaeli, Turkey

Abstract. In this study, a compact multiband, low-profile and low-cost microstrip antenna design covering DCS (Digital Cellular System, 1.7–1.8 GHz), mid-band 5G and Ku-band has been realized. The proposed monopole antenna consists of an elliptical patch element and mushroom-shaped metallic loadings placed on both sides of this patch. The elliptical patch element operates in the 3.3–3.8 GHz range for the mid-band 5G frequency band alone. In addition, the mushroom-shaped structures used on the front surface of the design enables the antenna to operate in the frequency ranges of 1.72–1.8 GHz DCS and 10–15 GHz Ku-band, as well as the 5G operating frequency. The proposed microstrip line fed antenna is 50 × 40 mm^2 in size. The antenna has a bandwidth of 1.71 GHz and a maximum gain of 5.73 dBi at the 5G operating frequency. In addition, the designed antenna has a bandwidth of 80 MHz for the DCS band and 5 GHz for the Ku-band. The simulations of the proposed antenna were carried out in the CST Microwave Studio and the obtained performance parameters are presented in detail.

Keywords: Microstrip monopole antenna · DCS · 5G · Ku-band · Mushroom-shaped structures

1 Introduction

Due to the increase in data transfer traffic, high data rates are needed for a well communication. In order to meet this need, high-bandwidth 5G technology has been developed. many countries use the 3.3–3.8 GHz frequency band for 5G technology. In this context, various antenna designs have been carried out in the literature for the last few years in order to support 5G technology:

Ciydem et al. designed a 5G antenna operating in the 3.3–3.8 GHz frequency bands by using a patch on the dielectric layer. In the antenna, FR4 was preferred as the substrate material and probe feeding was preferred as the feeding method. The resulting antenna; It has a size of 75 × 75 mm and a maximum gain of 8.95 dBi [1].

Kim et al. by designing a liquid crystal-based antenna, it designed a 5G antenna operating in the 27–28 GHz frequency bands. Taconic TLY was preferred as the substrate material in the antenna. The resulting antenna; It has a size of 18 × 14 mm and a maximum gain of 3.9 dBi [2].

Puskely et al. designed an antenna with a 28 GHz bandwidth operating in the 5G frequency band by utilizing the substrate integrated waveguide (SIW) technology [3].

Xiao et al. by designing a structure-shared antenna, it designed a 5G antenna operating in the 27.3–28.8 GHz frequency bands. TLY-5–0200 was preferred as the substrate material and microstrip line feeding was preferred as the feeding method. The resulting antenna; It has a size of 44 × 30 mm and a maximum gain of 3.4 dBi [4].

Abdalrazik et al. using slits in the patch formed on the dielectric layer; it has designed a multiband 5G antenna operating in 3.5/1.17/1.57/2.4/0.7 GHz frequency bands. Microstrip line feeding is preferred as the feeding method in the antenna. The resulting antenna; it has a size of 150 × 100 mm and a maximum gain of 9 dBi [5].

One of the methods used in 5G microstrip antenna design is MIMO antenna design [6–9]. Cheng et al. By performing MIMO antenna design, it has designed a 5G antenna operating in the 4.4–5.5 GHz frequency bands. FR4-epoxy was preferred as the substrate material in the antenna. The resulting antenna; It has a size of 36 × 36 mm and a maximum gain of 2.28 dBi [10].

He et al. designed a millimeter-wave patch antenna and designed a 5G antenna operating in 24–28 GHz frequency bands. Rogers RO4450F was preferred as the substrate material and capacitive feed was preferred as the feeding method. The resulting antenna; It has a size of 18 × 30.6 mm and a maximum gain of 5.8 dBi [11].

One of the studies done in 5G applications in recent years is to create endfire arrays [12–14] and millimeter-wave patch antenna [15].

Within the scope of this study, in addition to the 5G frequency band (3.3–3.8 GHz), by using the mushroom-shaped structure (MSS) method in the microstrip antenna; A structure operating in the frequency bands of 1.72–1.8 and 10–15 GHz has been developed. Arlon DiClad 870 was preferred as the substrate material and microstrip line feeding was preferred as the feeding method. The resulting antenna; It has a size of 50 × 40 mm and a maximum gain of 5.73 dBi.

2   Antenna Design

The proposed antenna design has elliptical patch geometry and DGS structure. Mushroom-shaped structures (MSSs) and DGS structure used in the design enabled the antenna to radiate in three frequency bands covering DCS, 5G and Ku frequency band. In the first step, a structure with an elliptical patch and DGS ground plane is designed and the antenna was provided to operate in the 5G (3.3–3.8 GHz) frequency band. In the second step, MSSs are added to the sides of the elliptical patch plane and the antenna is enabled to operate in the frequency bands of 1.72–1.8 GHz and 10–15 GHz. In the third step, MSSs are placed on the right side of the patch plane, increasing the bandwidth of the antenna in the 5G frequency band. Finally, in the fourth step, one more MSS is placed on the right side of the patch plane to improve the radiation pattern of the antenna.

The dimensions of the antenna are shown in Fig. 1. The width of the microstrip line feed is determined as 5.2 mm to ensure 50-Ω impedance matching. As indicated in

Table 1, the dimensions of the antenna are 50×40 mm^2. The substrate material used in antenna design; It is Arlon DiClad 870 material with a dielectric constant of 2.33 and a thickness of 1.575 mm. The ground plane with DGS structure, positioned on the rear surface of the antenna, has a height of 10 mm. Antenna size parameters and their descriptions are as shown in Table 1.

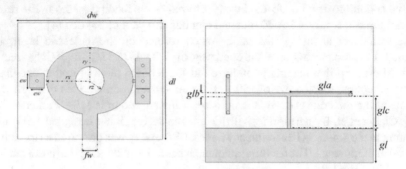

Fig. 1. The dimensions of the antenna

Table 1. Antenna size parameters

Parameters	Description	Size (mm)
dw	Substrate width	50
dl	Substrate length	40
gl	Ground length	10
fw	Feed line width	5.2
rx	Outer patch Radius-1	30
ry	Outer patch Radius-2	23
rz	Inner patch radius	10
ew	MSSs width	5
gla	MSSs via ground distance-1	18.25
glb	MSSs via ground distance-2	1
glc	MSSs via ground distance-3	9.5

3   Numerical Results

The simulations of the proposed antenna are carried out in Computer Simulation Technology (CST). Under this title, in terms of examining the performance parameters of the designed structure; S_{11} parameter, gain, group delay, surface currents and radiation pattern graphs are given. In addition, parametric analyzes of antenna dimensions were performed and optimum values of these dimensions are determined for numerical results.

Antenna design was carried out in four steps. The configurations of the designs performed in each step are as shown in Fig. 2.

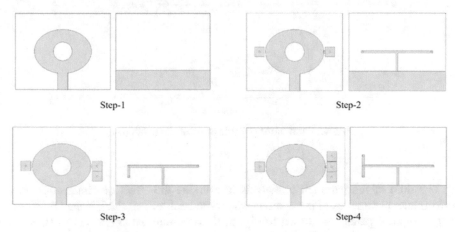

Step-1   Step-2

Step-3   Step-4

Fig. 2. The configurations of the designs performed step by step

The first of the parametric analyzes carried out in the antenna design is carried out for the length parameter of the antenna, *dl*. As shown in Fig. 3, the optimum value appears to be 40 mm when the *dl* length is simulated in four different dimensions for the antenna to radiate optimally in the 5G frequency band.

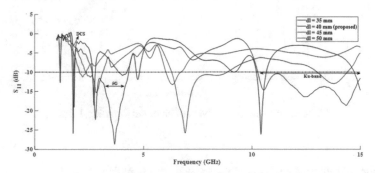

Fig. 3. Parametric analysis of substrate length (*dl*)

The second of the parametric analysis carried out in the antenna design is carried out for *dw*, which is the width parameter of the antenna. As shown in Fig. 4, it is seen that the optimum value is 50 mm when the *dw* length is simulated in four different dimensions for the antenna to radiate optimally in both the 5G frequency band and the 10–15 GHz frequency band.

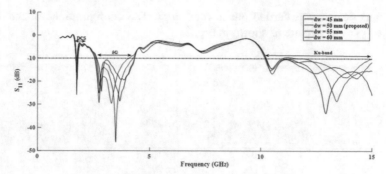

Fig. 4. Parametric analysis of substrate width (*dw*)

The third of the parametric analysis to examine the effect of dimensions on the performance parameters of the antenna was carried out for *rx*, which is the width radius of the elliptical patch. As shown in Fig. 5, the optimum value appears to be 30 mm when the *rx* length is simulated in four different dimensions for the antenna to radiate optimally in the 5G frequency band.

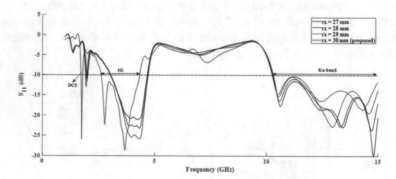

Fig. 5. Parametric analysis of outer patch radius-1 (*rx*)

Another parametric analysis is performed for *ry*, which is the length radius of the elliptical patch. As shown in Fig. 6, the optimum value appears to be 23 mm when the *ry* length is simulated in four different dimensions for the antenna to radiate optimally in the 5G frequency band.

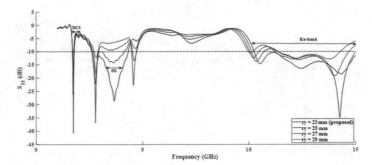

Fig. 6. Parametric analysis of outer patch radius-2 (*ry*)

Gain and group delay graphs of the designed antenna are as shown in Fig. 7 and Fig. 8 respectively. Accordingly, the antenna has a maximum gain of 5.73 dBi in the 5–10 GHz frequency range and a minimum group delay of 4.5 ns in the 1–2 GHz frequency range.

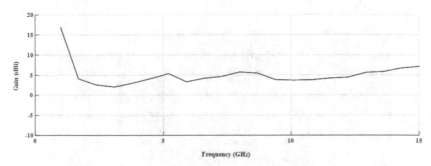

Fig. 7. Antenna gain performance

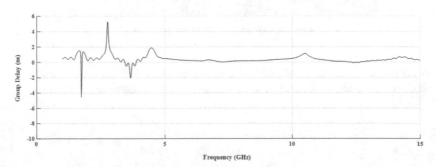

Fig. 8. Antenna group delay graph

The design of the proposed antenna is carried out in 4 steps and the S_{11} parameter values obtained in each step are as shown in Fig. 9.

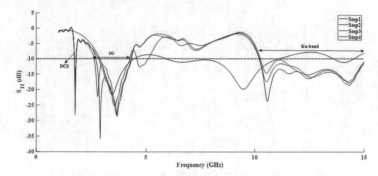

Fig. 9. Antenna S_{11} parameters result step by step

Other performance parameters of the antenna measured in the simulation program are the surface current distribution and the radiation pattern graphs. These are as shown in Fig. 10 and Fig. 11, respectively.

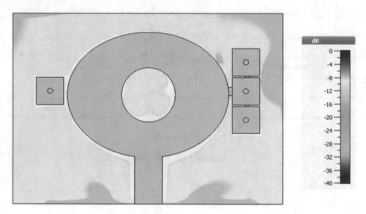

Fig. 10. Antenna surface current graph

After the final design is created and the findings are obtained, the comparison with the studies conducted in the literature in recent years is as shown in Table 2.

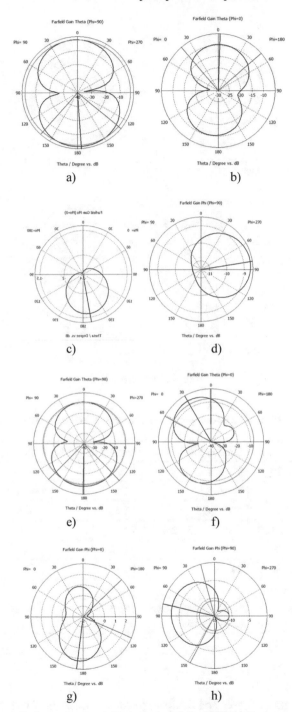

Fig. 11. Antenna radiation patterns; a) E-plane co-pol for DCS, b) E-plane cross-pol for DCS, c) H-plane co-pol for DCS, d) H-plane cross-pol for DCS, e) E-plane co-pol for mid-band 5G, f) E-plane cross-pol for mid-band 5G, g) H-plane co-pol for mid-band 5G, h) H-plane cross-pol for mid-band 5G, i) E-plane co-pol for Ku-band, j) E-plane cross-pol for Ku-band, k) H-plane co-pol for Ku-band, l) H-plane cross-pol for Ku-band

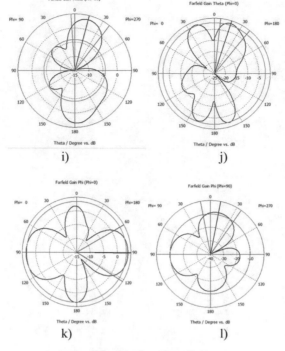

Fig. 11. (*continued*)

Table 2. The comparison of the studies in the literature

Reference	Size (mm)	Bandwidth (GHz)	Peak Gain (dBi)	Min. Return Loss (dB)
[1]	75 × 75	3.3–3.8	8.95	−50
[2]	18 × 14	27–28	3.9	−25
[3]	44 × 30	27.3–28.8	3.4	−45
[4]	150 × 100	3.5/1.17/1.57/2.4/0.7	9	−30
[5]	36 × 36	4.4–5.5	2.28	−20
[6]	18 × 30.6	24–28	5.8	−32
[7]	4.4 × 19	28, 39	10	−30
This study	**50 × 40**	**1.72–1.8/2.61–4.32 (5G)/10–15**	**5.73**	**−28.61**

4   Conclusion

This scope of work; A simple and compact microstrip antenna design operating in the DCS, mid-band 5G and Ku-band frequency band has been realized. By using mushroom-shaped metallic loadings placed on both sides of this patch and DGS methods in the designed antenna, the antenna radiates in the frequency ranges of 1.72–1.8 GHz and 10–15 GHz besides the 5G frequency band. The antenna has a bandwidth of 80 MHz, 1.71 GHz and 5 GHz in the DCS, mid-band 5G and Ku-band frequency band, respectively. When compared with the studies in the literature; it has been seen that good results are obtained in terms of size, bandwidth and gain. Since the designed antenna works in the Ku-band frequency band as well as the 5G frequency band, it will be possible to use it in satellite communication systems in the future. Moreover, the array version of the proposed antenna can be designed within the scope of future studies, and it can be evaluated in the use of base stations as well as the use of mobile terminals.

References

1. Ciydem, M., Miran, E.A.: Dual-polarization wideband sub-6 GHz suspended patch antenna for 5G base station. IEEE Antennas Wirel. Propag. Lett. **19**(7), 1142–1146 (2020)
2. Kim, J., Oh, J.: Liquid-crystal-embedded aperture-coupled microstrip antenna for 5G applications. IEEE Antennas Wirel. Propag. Lett. **19**(11), 1958–1962 (2020)
3. Puskely, J., Mikulasek, T., Aslan, Y., Roederer, A., Yarovoy, A.: 5G SIW-based phased antenna array with cosecant-squared shaped pattern. IEEE Trans. Antennas Propag. **70**(1), 250–259 (2021)
4. Xiao, F., Lin, X., Su, Y.: Dual-band structure-shared antenna with large frequency ratio for 5G communication applications. IEEE Antennas Wirel. Propag. Lett. **19**(12), 2339–2343 (2020)
5. Abdalrazik, A., Gomaa, A., Kishk, A.A.: A hexaband quad-circular-polarization slotted patch antenna for 5G, GPS, WLAN, LTE, and radio navigation applications. IEEE Antennas Wirel. Propag. Lett. **20**(8), 1438–1442 (2021)
6. Zhang, H.H., et al.: Low-SAR MIMO antenna array design using characteristic modes for 5G mobile phones. IEEE Trans. Antennas Propag. **70**(4), 3052–3057 (2021)
7. Yuan, X.T., Chen, Z., Gu, T., Yuan, T.: A wideband PIFA-pair-based MIMO antenna for 5G smartphones. IEEE Antennas Wirel. Propag. Lett. **20**(3), 371–375 (2021)
8. Hu, W., et al.: Wideband back-cover antenna design using dual characteristic modes with high isolation for 5G MIMO smartphone. IEEE Trans. Antennas and Propag. (2022)
9. Chang, L., Zhang, G., Wang, H.: Triple-band microstrip patch antenna and its four-antenna module based on half-mode patch for 5G 4 × 4 MIMO operation. IEEE Trans. Antennas Propag. **70**(1), 67–74 (2021)
10. Cheng, B., Du, Z.: Dual polarization MIMO antenna for 5G mobile phone applications. IEEE Trans. Antennas Propag. **69**(7), 4160–4165 (2020)
11. He, Y., Lv, S., Zhao, L., Huang, G.L., Chen, X., Lin, W.: A compact dual-band and dual-polarized millimeter-wave beam scanning antenna array for 5G mobile terminals. IEEE Access **9**, 109042–109052 (2021)
12. Parchin, N.O., Zhang, J., Abd-Alhameed, R.A., Pedersen, G.F., Zhang, S.: A planar dual-polarized phased array with broad bandwidth and quasi-endfire radiation for 5G mobile handsets. IEEE Trans. Antennas Propag. **69**(10), 6410–6419 (2021)
13. Seo, J., et al.: Miniaturized dual-band broadside/endfire antenna-in-package for 5G smartphone. IEEE Trans. Antennas Propag. **69**(12), 8100–8114 (2021)
14. Khajeim, M.F., Moradi, G., Shirazi, R.S., Zhang, S.: Broadband dual-polarized antenna array with endfire radiation for 5G mobile phone applications. IEEE Antennas Wirel. Propag. Lett. **20**(12), 2427–2431 (2021)
15. Gu, X., et al.: Antenna-in-package integration for a wideband scalable 5G millimeter-wave phased-array module. IEEE Microwave Wirel. Compon. Lett. **31**(6), 682–684 (2021)

An IoT-Based Modular Avionics and Electrical System for Nanosatellite Systems

Murat Bakirci$^{(\boxtimes)}$ ⓘ and Muhammed Mirac Özer ⓘ

Faculty of Aeronautics and Astronautics, Tarsus University, 33400 Mersin, Turkey
{muratbakirci,muhammed_mirac}@tarsus.edu.tr

Abstract. This study presents an avionics and electrical system design using reliable, high-performance hardware and sensors for advanced scientific experimentation missions compatible with air-land-sea vehicle platforms, particularly nanosatellite platforms. The nanosatellite avionics, which has a real-time operating system that supports frequently used interfaces, processes and manages sensory and physical data based on a central processor. It executes all operations by defining IoT requirements and computing connection parameters for IoT applications. The modular design brought into the system provides both ease of access and integration into the target platform, and also provides reliable storage for telemetry and flight data. Through the IoT station, it reliably receives information from the satellite and transmits it to smart devices while maintaining the desired signal quality. Moreover, through processing the data obtained from the sensors, critical information such as instant detection and tracking of systems errors are transmitted to the cloud, and as a result, proper control can be provided regardless of location. This critical data obtained from the cloud is straightforwardly tracked by the software platform. This design will provide the space technologies inventory as the basis for a new satellite platform and a system design for researchers to further develop.

Keywords: Nanosatellite · Avionics · IoT · Flight computer · MEMS sensors

1 Introduction

With the popularity of space studies in recent years, nanosatellites with increasing usage areas, capabilities and development potentials, have an undeniable importance in today's space technology [1, 2]. Compared to common satellite types, their small size, simple design, fast production and low cost have put these systems at the heart of many space studies [3]. On the other hand, people use internet-connected devices or objects by creating the main communication link with each other. The basis of this is the communication of objects over the internet, that is, the Internet of Things (IoT), which is the interpretation of the information obtained by artificial intelligence with its own computation procedures [4]. The link between these devices and objects creates IoT applications [5]. Activities through IoT applications create smart devices that make every aspect of human life easier [6]. In this context, the phenomenon that enables intelligent systems and items

F. P. García Márquez et al. (Eds.): ICCIDA 2022, LNNS 643, pp. 218–229, 2023.
https://doi.org/10.1007/978-3-031-27099-4_17

identified as entities to be transferred through the network without any human action, is smart devices and sensors [7]. Large amounts of data are transferred via satellite links more efficiently than terrestrial link as communication protocol with Wi-Fi, Bluetooth [8], RFID, NFC, UWB and LTE technology [9]. In this way, the use of satellites increases the efficiency of IoT applications [10]. Since IoT applications can communicate at the speed of light, a speed that cannot be achieved with wired communication between smart systems, faster data transmission is achieved [11].

Nanosatellites performing advanced scientific experiments with high resolution can be designed for various purposes such as real-time acquisition of tactical imagery, environmental impact monitoring, various observations on multiple scenarios, meteorological research, monitoring forest and vegetation, general mapping applications, surveying agricultural areas, climate changes and disasters [12]. A nanosatellite constellation, which can be used in this or many other different areas, can transfer the data received from the ground from one end of the world to the other using a safe data communication network between the receiver/transmitter and the nanosatellite [13]. When it leaves the coverage area, it can provide it through another satellite in the coverage area of the same network. Transmitting the data between other satellites to the satellite that is within the coverage area of the desired region and the data needed by that satellite can be downloaded by the user.

Although some commercial avionics and electrical systems have been developed for nanosatellites, these systems do not differ greatly. While this means faster expansion of technical and scientific knowledge and greater involvement of local industry, the avionics and electrical system development procedure is mandated by legislations and advanced for safety, as poor avionics and electrical system design has resulted in many failed missions and causalities. These designed avionic systems are devoid of the functionality of professional avionic systems, and they cannot provide much output for flight except for some basic information. Some configurations are not a good option for engineers unfamiliar with such devices. For this reason, it may not be a very efficient method to bring together different instruments regardless of the purpose of the mission. With the rapid development of MEMS technology and reducing production costs, some designers have designed the avionics system for missions with limited budgets [14–16]. Assuming that nanosatellite subsystems are not subject to the same communication bus and a single mission design, system requirements such as electric power source (EPS) [17], command and data handling (C&DH) [18], communication (COMM) [19], attitude determination and control system (ADCS) [20] and security criteria are expensive operations, which is also emerged in the high cost of electrical systems and avionics.

This study has shown that it is possible with today's technologies to design a long-lasting, low cost but reliable, IoT-based modular avionics and electrical system that meets the flight requirements of nanosatellites. To finalize a nanosatellite avionics and electrical design for IoT application, general information is given, parameters are evaluated, and best fits are made. In addition to altitude, temperature and time, other key parameters such as pressure, air velocity, particle detector, air/oxygen ratio, attitude, acceleration and inclination of the Earth's magnetic field can also be detected, thus, they are stored for post-flight analysis. Moreover, entire avionics system can be remotely controlled through a wireless network. Recovery, malfunction, power consumption status, sensor data and location information obtained were recorded and stored through IoT.

The rest of the article is organized as follows: Sect. 2 provides an overview of the proposed system. The details of the measurement subsystems that the nanosatellite will utilize during its mission are examined in the Sect. 3. The required electrical power unit and system components are described in the Sect. 4. In the Sect. 5, all the details of the communication subsystem, which is the most important component, are given. In the Sect. 6, the features of the flight software were mentioned, and in the Sect. 7, necessary discussions were made under the title of Conclusions.

2   System Overview

As shown in the system operation flowchart given in Fig. 1, the nanosatellite system consisting of a carrier and a payload is launched to be deployed near the apogee. During the launch and the entire flight, sensors within the nanosatellite begin to collect data and transmit it in real time to the ground control station and other IoT-based stations.

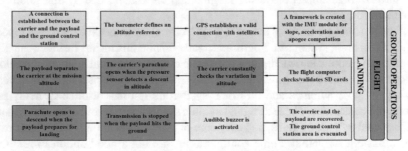

Fig. 1. Avionics system operation flowchart.

The carrier protects the mission payload from damage during launch and deployment and releases it at the mission's altitude. When the payload is released at the apogee, the carrier descends with a rescue parachute system, while the payload collects sensor data for the mission from release to landing. When it prepares for landing, it opens the parachute and descends smoothly, and when it lands, all systems are turned off and the telemetry system is stopped. Meanwhile, the ground control station designs and implements the electronic system and a portable ground control mechanism for receiving, processing, displaying, storing and transmitting data from the nanosatellite.

3   Measurement Systems

The collection of flight data is an integral part of design optimization as it allows the nanosatellite to be studied under different conditions. Some of the most important flight parameters include position, altitude, velocity, pressure, acceleration, imagery, and airborne particles. The avionics system is a computer that contains sensors, microcontroller and sensor boards containing external peripherals. Various physical quantities measured by sensors are converted into electrical signals that can be read and processed by a micro-controller. The obtained data is stored for use both in real time and after the mission flight.

3.1 Sensor Subsystem

Carrier and payload avionics systems, which include the sensor subsystem, are designed as shown in Fig. 2 within the scope of recovery requirements. The carrier flight computer uses the high accuracy (± 8 Pa/$\pm 0,5$ °C), low power consumption (3.3 V) and small size (21.6 × 16.6 × 3.0 mm) BMP388 digital pressure and temperature sensor, which can be used for 300–1250 hPa to detect where to drop the payload.

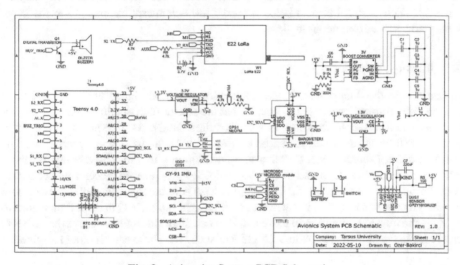

Fig. 2. Aviyonics System PCB Schematic.

Payload flight computer uses a low power consumption (3.3 V), small size, 10 degrees of freedom GY91 (MPU-9250+BMP280) module that houses the pressure sensor along with the 3-axis gyroscope, accelerometer, magnetometer to activate the recovery system that will ensure a safe landing. The payload avionics system also uses the Sharp GP2Y10 optical dust sensor to detect airborne particles, which can be used to detect very fine particulate dust grains such as cigarette smoke, and to measure air pollution. The Sharp GP2Y10 is a sensor with low current consumption (max 20 mA), capable of operating with a supply voltage of up to 7 V, and giving an analog voltage output proportional to the dust density with a precision of 0.5 V/1 mg/m^3, which fully meets this requirement. Moreover, after the payload is released, it will record the flight with the SQ11 Mini Dvr camera, which is positioned to see the ground, has a 140° wide angle, contains 6 IR LEDs, has smart motion detection, battery and micro SD card slot.

Avionics systems will provide instantaneous speed-position data with NEO-7M GPS module, which has GPS serial output of raw data that can be accessed to a 3.3 V single-chip serial solution, allows changing the baud rate and recording without power thanks to the EEPROM, as a packet with 56 channels and 500 m/s speed limit. E22 900T22D communication module is used to transmit instantaneous sensor data from electronic systems during flight to the ground control station, which provides low energy consumption, long range wireless communication, and easy use with only 2 cables due to its UART structure.

Since the data collected but not retained is not useful to the ground control station, the avionics system must store the data it collects in an easily accessible format. For this purpose, flight data is stored with the SD card module. Thus, in case of system failure, the SD card can be read with ordinary memory card readers. This not only makes data retrieval much easier, but also helps preserve the ephemeris by preventing the same systematic errors from being made.

3.2 Backup Altimeter

In addition to the data obtained from the sensors in the designed avionics systems, a commercial altimeter was used to ensure that the sensor data obtained from the carrier and payload throughout the mission were acquired safely. As a commercial altimeter, AltimeterThree is used, which can be controlled by a smart phone or tablet with its wireless control feature, enables recording and stopping by pressing a button on these devices, and can download and display the recorded pressure and accelerometer sensor data through interactive graphing. Each flight phase is stamped on the mobile device with the GPS location. Moreover, it can record accelerations up to 24 G in three dimensions to analyze the mission operation in more detail. It can also add flight notes for the mission, create an Excel spreadsheet with all flight statistics, and share it with other recipients.

4 Electrical Power Subsystem

As shown in Fig. 3, two Sony VTC6 Lithium-ion 3.7 V batteries, connected in parallel were used to obtain 6260 mAh for the payload flight computer power system. For the carrier flight computer power system, a single VTC6 was used to obtain 900 mAh as shown in Fig. 4. Umblical power supply was used for an external power supply. The Internal Real Time Clock is powered by the microcontroller when the microcontroller is on, and an external 3 V battery when it is closed or reset. The reset button is used to reset the mission computers.

Two voltage regulators are used to regulate stable 3.3 V and 1.8 V voltages and a synchronous boost converter to provide 5 V. A power switch (external on/off switch) is used to control the power of the systems. Battery voltage was measured using the voltage divider method. A 92 dB buzzer was used in order to easily detect where the carrier and the payload landed. The SQ11 Mini Dvr camera, SD card module and optical dust sensor are powered by 5 V. The BMP388 sensor, the 10-Dof inertial measurement unit, and the GPS module are powered by 3.3 V.

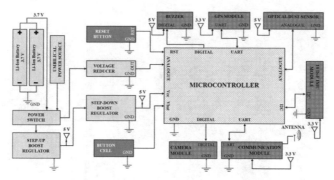

Fig. 3. Block diagram of payload electrical system components.

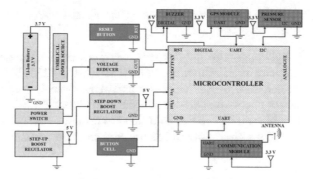

Fig. 4. Block diagram of carrier electrical system components.

5 Communication Subsystem

5.1 Ground Control Station

For the microcontroller responsible for taking all decisions, Teensy 4.0, which supports development in different operating systems, has 1 Mb SRAM, 2048 Kb flash memory and 64 Kb high-capacity EEPROM, works with 600 MHz high processor speed, is used. Although information about the boot time that takes for a device to be ready for operation after power-on is not available from the resources, it has been measured experimentally for about one second. The interfaces that the microcontroller in carrier and payload avionics systems communicate with the sensors are given in Fig. 5.

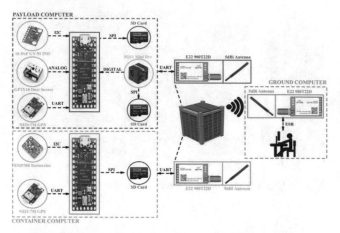

Fig. 5. Communication scheme of the nanosatellite avionics system components.

The BMP388 digital pressure and temperature sensor provides pressure, temperature and relative air velocity measurements. The GY91(MPU-9250+BMP280) module provides slope, acceleration, angular velocities and magnetic field measurements. The Sharp GP2Y10 sensor counts particles, while the NEO-7M GPS module determines the nanosatellite's geolocation. The E22 900T22D transceiver communication module sends telemetry data to the ground control station and receives commands from it.

Real Time Clock (RTC)
A hardware Real Time Clock (RTC) is used with a built-in power detection circuit that perceives outages and accordingly switches to a recovery source. RTC is a powerful module which stays operational in low power modes and is powered by the battery power supply. The battery power supply will count the time at the start of the mission and save it in the EEPROM in the microcontroller. In case of CPU reset, the mission start time in the EEPROM will be used as a reference. In this way, every time the microcontroller is reset or stopped working, it will continue to work and the data will not be reset.

Radio Configuration
E22 900T22D was chosen as the radio model. First of all, parameter settings are made by means of code blocks. After the code is loading is completed, the parameter settings of the LoRa module are changed via RF setting program. However, the parameter settings of the LoRa module can be changed remotely. As shown in the block diagram in Fig. 6, the avionics system microcontrollers collect the sensor data and send it to the serial. A 5 dBi antenna is used to minimize data loss during the mission flight. This way, it provides higher gain and suitable beam width during signal transmission for the targeted direction. The receiving communication module sends and receives packets. Connection from communication module to USB with USB stick is enabled. The computer processes the obtained data and saves it in a '.csv' file.

Fig. 6. Ground kontrol station design flowchart.

5.2 IoT-Based Station

By developing an IoT-based nanosatellite, remote control can be achieved by ground control computer and smart devices. As shown in Fig. 7, data such as pressure, air velocity, airborne particle rate, and acceleration from carrier and payload sensors are transferred to the cloud with LoRa. The status of the nanosatellite mission provided by the avionics system computers can be read and monitored in real time via the ground control station and IoT-based computer and smart devices.

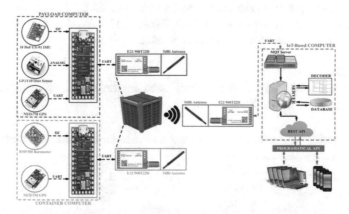

Fig. 7. Communication scheme of the IoT-based nanosatellite system components.

LoRa

LoRa, a wireless communication technology, uses a new wireless sensor network with IoT. Through this technology, the nanosatellite utilizes E22 900T22D wireless communication module with 30 dBm transmission power, low power consumption, long range use, working with transceiver logic and 30 dBm transmission power. 4.7k resistors are used on the AUX and TXD pins to prevent any exhaustion problems that may occur in Node MCU and Teensy 4.0 microcontrollers.

Sensor data is collected from avionics system microcontrollers. There are communication layers in the form of physical layer, data link layer and application layer to receive the sensor data from the nanosatellite and send it to the IoT-based station in the ground station. Figure 8 shows the communication layers for the IoT-Based station to communicate with the satellite sensor data.

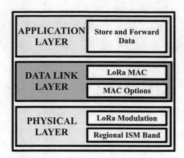

Fig. 8. IoT communication layers.

Wi-Fi Module

The data from the sensors are transferred to the server wirelessly with a Wi-Fi module. The ESP8266 Wi-Fi module, which is frequently preferred in IoT applications, communicates with RX/TX pins and UART serial, and can connect to the internet wirelessly, is used on a development board integrated into NodeMCU V3. Since it was developed using the ESP8266 SDK, it supports GPIO, PWM, I2C, 1-Wire and ADC connections without the need for an additional microcontroller.

GSM-GPRS

The connection schematic of the sensors used in the IoT-based computer on the NodeMCU is shown in Fig. 9.

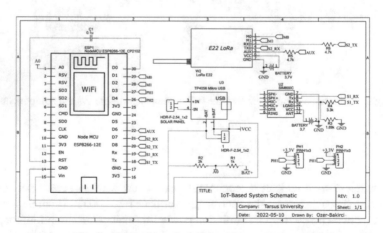

Fig. 9. PCB schematic IoT-based system sensors.

Since locating the nanosatellite and receiving the sensor data are within the scope of the mission requirements, in case of a problem with ESP8266 Wi-Fi module in the IoT-based computer, the ground station tracks the payload via mobile device through the GPRS/GSM SIM800C, and SMS can be sent and received. This allows the IoT system to be monitored not only with smart devices but also with standard devices. It can also send and receive MMS through internet connection. It can also be used for long range missions with low cost, low power consumption and small footprint and quad-band frequency support. Once the power systems are powered up, it searches for the cellular network and logs in automatically. In general, it receives and stores SMS/MMS/data from one location for the task of storing and sending and then downloading it to the ground control station at another location.

6 Flight Software

As part of the flight software design tasks given in Fig. 10, the electronic system is activated with the power (on/off) button. When the sensors receive the command "calibrate the system", the reference altitude is determined by resetting the EEPROM and the system is calibrated.

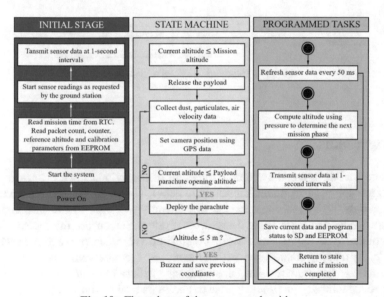

Fig. 10. Flow chart of the recovery algorithm.

The nanosatellite is launched with the rocket and is released at the apogee when the separation requirements are met based on the data from the IMU module after the rocket ascent is complete. After the payload leaves the carrier, the camera starts recording video relative to the reference point. Shows the coordinates provided by the GPS during the glide time. As the payload descends, the parachute opens when the specified requirements are met by data from the barometric sensor module. When the altitude drops below 5 m, the audible warning is activated and data transfer is stopped. The buzzer continues to sound until the electronic system is turned off with the power (on/off) button. The ground control station finds the systems, the components are turned off, and the mission is complete.

The transfer from the ground control station to the payload starts at power-up, and they communicate in unicast mode and remain in contact. The data is transmitted back to the ground control station with a frequency of 1 Hz. Transmission control is handled by the ground station when the payload is stationary at the launch pad and by the FSW in flight. At 5 m from the ground, the audible warning is activated via FSW and data transfer is stopped. Additionally, the csv file containing all locally received telemetry data is saved in nonvolatile memory.

The microcontroller resets when a temporary power outage occurs. Packet count, reference altitude, calibration status and counter data are saved in the internal EEPROM in the microcontroller. In this way, data loss will be prevented in case of resetting the microcontroller. Mission flight time is recorded in the internal EEPROM of the RTC module. Through this, data loss is prevented if an instant power outage occurs. Figure 11 shows the desired data to be retrieved from the EEPROM in case the microcontroller is reset.

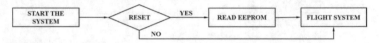

Fig. 11. Recovering avionics system data.

7 Conclusions

The conceptual design and system architecture of the avionics and electrical systems of the nanosatellite platform are presented in detail. The designed systems are stated as clearly as possible by other researchers to validate the design, ensure safe flight and a successful payload mission. In the design of the system, the sensors, microcontrollers and PCBs required for avionic systems were brought into a complete package and each sensor was coded by making connections within the scope of requirements. Telemetry systems have been tested to ensure real-time data exchange with satellite mission computers. The developed modular design provides ease of access as well as simple integration into the target platform. It also ensures reliable storage of mission-critical telemetry and flight data. It is ensured that the desired signal quality is maintained while reliably receiving information from the satellite via the IoT station and transmitting it to smart devices. Moreover, through the proposed avionics architecture, mission-critical data obtained from the sensors can be transmitted to the cloud, thus, proper control can be provided regardless of location. It has been verified that the data obtained from the cloud can simply be followed via software platform.

References

1. Sweeting, M.N.: Modern small satellites-changing the economics of space. Proc. IEEE **106**(3), 343–361 (2018)

2. Liddle, J.D., Holt, A.P., Jason, S.J., O'Donnell, K.A., Stevens, E.J.: Space science with CubeSats and nanosatellites. Nat. Astron. **4**, 1026–1030 (2020)
3. Almonacid, V., Laurent, F.: Extending the coverage of the internet of things with low-cost nanosatellite networks. Acta Astronaut. **138**, 95–101 (2017)
4. Ray, P.P.: A survey on internet of things architectures. J. King Saud Univ. Comput. Inf. Sci. **30**(3), 291–319 (2018)
5. Bandyopadhyay, D., Sen, J.: Internet of things: applications and challenges in technology and standardization. Wirel. Pers. Commun. **58**, 49–69 (2011)
6. Krco, S., Pokric, B., Carrez, F.: Designing IoT architecture(s): a European perspective. In: IEEE World Forum on Internet of Things, South Korea, pp. 79–84. IEEE (2014)
7. Sha, K., Wei, W., Yang, T.A., Wang, Z., Shi, W.: On security challenges and open issues in Internet of Things. Future Gener. Comput. Syst. **83**, 326–337 (2018)
8. Mackensen, M., Lai, M., Wendt, T.M.: Bluetooth low energy (BLE) based wireless sensors. In: IEEE Sensors, pp. 1–4, Taiwan. IEEE (2013)
9. Ghaffari, K., Lagzian, M., Kazemi, M., Malekzadeh, G.: A socio-technical analysis of internet of things development: an interplay of technologies, tasks, structures and actors. Foresight **21**(6), 640–653 (2019)
10. Gubbi, J., Buyya, R., Marusic, S., Palaniswami, M.: InternetofThings (IoT): a vision, architectural elements, and future directions. Future Gener. Comput. Syst. **29**(7), 1645–1660 (2013)
11. Narayanasamy, A., Ahmad, Y.A., Othman, M.: Nanosatellites constellation as an IoT communication platform for near equatorial countries. IOP Conf. Ser. Mater. Sci. Eng. **260**(1), 12–28 (2017)
12. Bacco, M., et al.: Iot applications and services in space information networks. IEEE Wirel. Commun. **26**(2), 31–37 (2019)
13. Fraire, J.A., Céspedes, S., Accettura, N.: Direct-to-satellite IoT - a survey of the state of the art and future research perspectives. In: Palattella, M.R., Scanzio, S., Coleri Ergen, S. (eds.) ADHOC-NOW 2019. LNCS, vol. 11803, pp. 241–258. Springer, Cham (2019). https://doi.org/10.1007/978-3-030-31831-4_17
14. Kok, M., Hol, J.D., Schön, T.B.: Using inertial sensors for position and orientation estimation. Found. Trends Signal Process. **11**(1–2), 1–153 (2017)
15. Bijjahalli, S., Sabatini, R.: A high-integrity and low-cost navigation system for autonomous vehicles. IEEE Trans. Intell. Transp. Syst. **22**(1), 356–369 (2021)
16. Akiyama, M., Saito, T.: Influence of radio waves generated by XBee module on GPS positioning performance. In: 2020 IEEE International Conference on Consumer Electronics, pp. 1–2, Taiwan. IEEE (2020)
17. Slongo, L.K., Martinez, S.V., Eiterer, B.V.B., Bezerra, E.A.: Nanosatellite electrical power system architectures: models, simulations and tests. Int. J. Circuit Theory Appl. **48**(12), 2153–2189 (2020)
18. Zhang, R.Y., Zhan, Y.F., Lu, J.H.: A new algorithm for main carrier acquisition in deep space communication. J. Electron. **28**, 169–173 (2011)
19. Kodheli, O., Lagunas, E., Maturo, N., Sharma, S.K., Shankar, B.: Satellite communications in the new space era: a survey and future challenges. IEEE Commun. Surv. Tutor. **23**, 70–109 (2021)
20. Akiyama, M., Saito, T.: A novel CanSat-based implementation of the guidance control mechanism using goal-image recognition. In: IEEE 9th Global Conference on Consumer Electronics, Japan, pp. 580–581. IEEE (2020)

Reinforcement Learning for Intrusion Detection

Ahmed Mohamed Saad Emam Saad[1]([✉]) [iD] and Beytullah Yildiz[2] [iD]

[1] Texas A&M University-Corpus Christi, Corpus Christi, TX 78412, USA
asaad1@islander.tamucc.edu
[2] Atilim University, Ankara, Turkey
beytullah.yildiz@atilim.edu.tr

Abstract. Network-based technologies such as cloud computing, web services, and Internet of Things systems are becoming widely used due to their flexibility and preeminence. On the other hand, the exponential proliferation of network-based technologies exacerbated network security concerns. Intrusion takes an important share in the security concerns surrounding network-based technologies. Developing a robust intrusion detection system is crucial to solving the intrusion problem and ensuring the secure delivery of network-based technologies and services. In this paper, we propose a novel approach using deep reinforcement learning to detect intrusions to make network applications more secure, reliable, and efficient. As for the reinforcement learning approach, Deep Q-learning is used alongside a custom-built Gym environment that mimics network attacks and guides the learning process. The NSL-KDD dataset is used to create the reinforcement learning environment to train and evaluate the proposed model. The experimental results show that our proposed reinforcement learning approach outperforms other related solutions in the literature, achieving an accuracy that exceeds 93%.

Keywords: Reinforcement learning · Deep Q-learning · OpenAI Gym · Network security · Machine learning · Intrusion detection system

1 Introduction

Network based computer systems and technologies like web services, cloud computing, and Internet of Things (IoT) systems are becoming more popular. These technologies are prone to intrusion, and the growing popularity of network-based systems made the intrusion issue worse. We can get an estimate on the expansion magnitude of network-based technologies and in return the intrusion problem by examining the market of specific network-based technologies such as cloud computing services. There has been a massive increase in the market and the revenue of cloud services at around 54.9 [1] and 129 [2] billion US dollars from 2017 to 2020, respectively. It is evident from such an increase that network-based technologies are getting a lot of attraction, which increases the scale of the intrusion issue. This increasing scale comes with significant economic costs, which has

© The Author(s), under exclusive license to Springer Nature Switzerland AG 2023
F. P. García Márquez et al. (Eds.): ICCIDA 2022, LNNS 643, pp. 230–243, 2023.
https://doi.org/10.1007/978-3-031-27099-4_18

been confirmed by two studies carried out by McAfee cyber-security firm. The two studies were conducted over a 6-year period and show the alarming increase in the economic cost of cyber-crimes. The first study shows that in 2014 the cost of cyber-attacks was around 475 billion US dollars [3], and the second study shows that in 2020 the cost of cyber-attacks was almost 1 trillion US dollars [4]. Since most of the network-based technologies and resources are obtained from a remote service provider that is not locally present through a medium, this raises the question of how to secure the medium used to obtain these services from intrusion, which in this case is the network system. The answer will significantly reduce the economical cost caused by intrusions and cyber-crimes. Securing that medium requires an absolute necessity of a modern solution to detect intrusions using a particular intrusion detection system (IDS).

Intrusion detection systems work in different ways and have many architectures. The most common two types of network intrusion detection systems are signature (misuse) based intrusion detection system and anomaly-based intrusion detection system [5]. Anomaly-based intrusion detection system attempts to deal with a novel attack by learning the normal pattern of network traffics, and any deviation from that normal pattern is considered as an intrusion [5]. The downside of this system is its high sensitivity that leads to a high false positive rate. Signature-based intrusion or misuse detection system works similarly to traditional rule-based intrusion detection systems like snort intrusion detection system [6]. Both were used to tackle and solve the intrusion problem. However, with the increase in novel attacks and the continuous change in the attack types and styles, rule-based intrusion detection systems are vulnerable. Even with rapid updates to their rules, they can not keep up with the continuous change in malicious attacks.

Therefore, creating a novel, scalable, and adaptive approach for detecting intrusions in network systems that copes with the new malicious attacks is a necessity. This is where machine learning comes into place with its adaptability and flexibility. It can provide a solution for the intrusion issue addressed before and solve the limitation of rule-based intrusion detection systems.

Although machine learning offers a solution, not all machine learning approaches are created equal. Machine learning can be categorized into three major topics: the first is reinforcement learning (RL), the second is unsupervised learning and the third is supervised learning. Many solutions were developed for intrusion detection using the three different machine learning approaches. Most of the researched and implemented approaches are based on supervised and the unsupervised learning. Supervised learning is based on the idea of recognizing attacks from captured and labeled network traffic attack data upon which it can detect relatively similar attacks. The unsupervised learning approach uses unlabeled datasets to learn and classify attacks based on their common features and patterns.

Both approaches have downsides. Most of the real traffic data are present as unlabeled data. Labeling data is an extremely cumbersome and costly process. In addition, the continuously developed attacks and changes in attack patterns render the supervised learning approaches inefficient and lead to a high false positive rate. On the other hand, the unsupervised learning approaches are inferior to the supervised ones in performance when presented with numerous features. Moreover, the feature engineering process for the large number of features in unlabeled data is a laborious task.

In this research, we propose a novel approach called OpenAI Gym Env-DQN (OGE-DQN) using reinforcement learning because, in theory, it is superior to other machine learning approaches in intrusion detection for the following reasons. First, it can go beyond the dataset by solving the labeling issue. Second, it can generalize and approximate when dealing with large observation space or features. Third, it can scale and adapt to numerous attack patterns, and it is not volatile to changes. The novel machine learning approach introduced utilizes a reinforcement learning-based algorithm called Deep Q-Network (DQN).

Reinforcement learning utilizes an environment to train an agent. We use OpenAI Gym library [7] to build the environment. The Gym environment guides the learning process through positive and negative rewards and makes use of the NSL-KDD intrusion detection dataset as a source of network traffic. In other words, the reinforcement learning agent learns from its previous actions by observing the states and rewards from the environment, so it can perform better actions in the future by maximizing the reward it gets from the environment. The reinforcement learning agent uses deep neural networks as a function approximator for Q-values associated with decision (action) making. For the NSL-KDD dataset, it is the largest and the most diverse in terms of attack types, and it fits the eleven criteria for an appropriate IDS dataset [8]. It is well-suited and serves the goal of this research. The dataset goes through a preprocessing stage before being used as the source of encoded network traffic.

The main contributions of this study are as follows:

1. We present the first attempt to create a custom-built standardized intrusion detection reinforcement learning environment using OpenAI Gym framework.
2. Our novel reinforcement learning approach offers a significant improvement in terms of metrics, such as accuracy, recall, and precision, compared to other relevant works from the literature.
3. A novel Gym environment is proven to be more expressive about intrusions and has more resolution. This is one of the key elements for the success of the reinforcement learning model.

The rest of this paper is organized as follows. Sections 2 and 3 provide discussion of the related work and a brief background, respectively. The research methodology is explained in Sect. 4. In Sect. 5, evaluation and benchmarking are explored. Finally, we conclude with the outcomes in Sect. 6.

2   Related Work

Numerous intrusion detection solutions using machine learning approaches and techniques were developed.

Liang et al. [9] suggested a hybrid approach of a multi-agent reinforcement learning model consisting of three parts: data management, analysis and response modules, and data collection. The analysis modules are based on deep learning to detect anomalies from the transport layer in the network. The dataset used for evaluation was the NSL-KDD dataset. The anomaly detection accuracy of 98% was achieved in an IoT environment. The ability for the proposed model to classify different types of attacks was accurate by 97%. Nevertheless, the model was only tested in an IoT environment where the types of attacks are very limited.

Koduvely [10] proposed making a Gym environment based on the OpenAI Gym environment concept to detect network intrusions using reinforcement learning and policy gradient model, which inspired us into building a Gym environment as part of the approach suggested in this paper. The proposed approach works by solving the environment, and for evaluation, a receiver operating characteristic (ROC) curve is used. The proposed solution's performance was not evaluated, and the False Positive (FP) and False Negative (FN) rates were unknown. The research also suggested implementing other techniques such as deep neural network and deep and wide neural network, but there was no continuation on this proposal.

The first approach that integrated a reinforcement learning framework as intrusion detection solution was introduced by Caminero et al. [11]. The approach was named Adversarial Environment using Reinforcement Learning (AE-RL). They created an environment that provides network traffic samples to the agent and also act as a second adversarial agent by increasing the classifier's incorrect predictions. Moreover, they implemented a new mechanism for dynamically over-sampling/under-sampling during training from the dataset to overcome the issue of the under represented classes. Their approach was tested on two datasets: NSL-KDD and AWID. The approach was compared with several other approaches that implemented supervised and reinforcement learning algorithms. They achieved an accuracy of 80%.

A novel approach using deep reinforcement learning to detect attacks in a network without requiring to solve an environment by directly using batches from two datasets (NSL-KDD and AWID) separately was suggested by Lopez-Martin et al. [12]. This approach proposes a new reward method for the model in the training process, whether the detection was correctly performed or not. They used four different approaches to implement their proposal, which are Policy Gradient, Double Deep Q-Network (DDQN), Actor-Critic, and Deep Q-Network (DQN). They claimed that the top performance was achieved by the Double Deep Q-Network approach and that they were able to decrease the overall computational time required compared to traditional machine learning approaches.

Suwannalai and Polprasert [13] proposed a multi agent deep reinforcement learning model to detect intrusions. They used the NSL-KDD dataset to train and test the proposed model. They created a multi-class model to evaluate their model capabilities on detecting each attack type in the dataset. The accuracy of the model was 80% and the F1 score was 79%. The model showed very low accuracy in detecting minority attack types.

An intrusion detection system that utilizes a reinforcement learning approach based on Deep Q-learning was implemented by Hsu and Matsuoka [14]. The system uses two modes: learning mode and detection mode. They tested their model on a real captured traffic from their campus network, NSL-KDD dataset, and the UNSW-NB15 dataset. Moreover, the model achieved an accuracy of 97.95%, 91.4%, and 91.8% respectively. Additionally, they compared the performance of their approach with three other traditional machine learning approaches and two published research papers.

Ma and Shi [15] suggested building a network intrusion detection system based on a deep reinforcement algorithm known as Deep Q-learning. They tested their approach using the NSL-KDD benchmark dataset. Moreover, a suggestion was made to increase the number of less represented classes (R2L and U2R) and reduce the number of over represented classes (NORMAL, DoS, and PROBE) in the dataset by creating synthetic data using various over and under sampling techniques like Synthetic Minority Over-Sampling Technique (SMOTE), Random Over Sampling (ROS), NearMiss1, and NearMiss2. The SMOTE was the highest performing technique when combined with a reinforcement learning framework with an accuracy of 82%.

The most relevant work from the literature are [12,14,15] since they used a reinforcement learning approach and tested their approach on the same dataset used in this research. Therefore, these research papers' results will be used later for performance comparison purposes.

3   Deep Q-Learning

Reinforcement learning is an evolutionary machine learning approach that simulates the learning process in living organisms [16]. Learning is a process in which living organisms increase their knowledge in the scope of different tasks by accumulating knowledge that enhances their capability on how to perform certain tasks better as their knowledge increases. Similarly, machine learning algorithms can emulate living organisms' learning process through their exposure to data, which in this scenario simulates the accumulative knowledge living organisms acquire over time. In reinforcement learning, the algorithm does not need to have prior domain knowledge and can learn over time by trial and error [17]. The algorithm in that scenario represents the brain in living organisms. The main purpose in reinforcement learning is trying to develop that brain, which in this case is called an agent, and this is where the algorithm resides. The agent's duty is to sum up the reward R_t over time t and get as many positive rewards as possible. This process simulates doing correct actions in certain tasks given the prior knowledge accumulated in the brain of a living organism. The agent interacts with an observation acquired from the environment (see Fig. 1), which is called a state S_t. In living organisms, such state represents the interaction with the real world whereby experience or knowledge is gained. Building a well-structured reinforcement learning environment is paramount for optimizing the agent's learning process and performance in a specific domain. For the reinforcement learning agent to take any action, it follows a strategy called an epsilon greedy strategy. We start by setting an exploration rate which is initially $\epsilon = 1$ and will decay by a certain rate at the beginning of each training loop, or in this context called episode. Then, a random number between "0" and "1" will be produced as a threshold. If that threshold is greater than ϵ, the agent will start exploitation to select its next action [18].

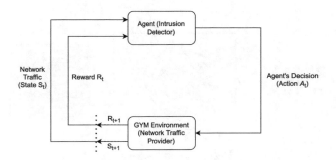

Fig. 1. Reinforcement learning workflow

The environment includes the dataset of captured network traffic features. It also evaluates the agent's actions (right or wrong). The observation represents the network traffic, and the agent represents the detection mechanism that decides whether the network traffic is an intrusion or not. The agent chooses its best action trajectory using a sequential decision-making process called Markov Decision Process (MDP). The sequential process starts by the agent observing a state from the environment. Depending on the state, the agent selects an action to perform. Afterwards, the agent gets a reward from the environment, and then another state is initiated. This entire process can be optimized in order for the agent to get the maximum accumulative reward, not just the immediate reward. Simply put, we are trying to map state-action pairs to rewards as represented in Eq. 1.

$$f(S_t, A_t) = R_{t+1} \tag{1}$$

The quality of the agent action in each step of the process is calculated by using the Bellman equation (Eq. 2). Q-Function is an indication of the quality of an action taken by the agent given a certain state. Therefore, it is referred to as the action-value function. The output of the function for any state-action is known as the Q-value which is the left hand side of the equation. The Bellman equation right hand side consists of two parts: the immediate reward R_{t+1} and the future Q-Values. A discount actor γ is added to make the process finite.

The $\max q_*\left(s', a'\right)$ is the maximum discounted future reward we can get from some states onwards if we are going to follow this particular behavior. In our approach, we used neural networks to work as a function approximator to obtain the optimal Q-values [19]. This approach is known as Deep Q-learning. We use neural networks to map states to action-Q-value pairs as shown in Fig. 2.

$$q_*\left(s, a\right) = E\left[R_{t+1} + \gamma \max_{a'} q_*\left(s', a'\right)\right] \tag{2}$$

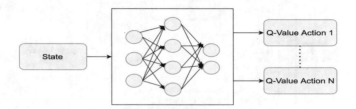

Fig. 2. Deep Q-learning

4   Methodology

This research proposes a novel approach for intrusion detection in network systems, which utilizes a deep reinforcement learning algorithm and network traffic present in the NSL-KDD intrusion detection dataset. The deep reinforcement learning approach used for learning is Deep Q-Learning. This approach includes building an RL agent using a reinforcement learning algorithm known as Deep Q-Network. It is used as part of the agent's structure to detect intrusions. Moreover, the approach includes creating and setting up an environment to guide the agent's learning process. The environment created uses the network traffic in the NSL-KDD dataset and was inspired by an environment proposed by Hari Koduvely [10]. The environment was custom-built to suit the developed RL approach.

4.1   Deep Reinforcement Learning Agent

A reinforcement learning agent is the brain where the algorithm resides. The agent's purpose is to select the appropriate action given a certain state to maximize the overall reward. This indicates that it is taking the best action available. Deep Q-learning is the chosen approach used in this research. Deep Q-learning uses two separate neural networks: the first one is called the main network or the policy network, and the second one is called the target network. The reason for using two separate neural networks is that when we feed a state to the neural network in order to obtain the appropriate action-Q-value pairs, the weights of the network are updated. With that constant change, this mapping process will be impossible because we are chasing a dynamic target that contradicts the whole purpose of mapping states to action-Q-value pairs. Instead, we use a second neural network with initially the same weights as those of the main network. The target network is used to obtain the target Q-values as shown in Fig. 3, and then, the weights will be updated to match the main network ones every certain number of iterations to ensure that the learning process is functioning properly [20].

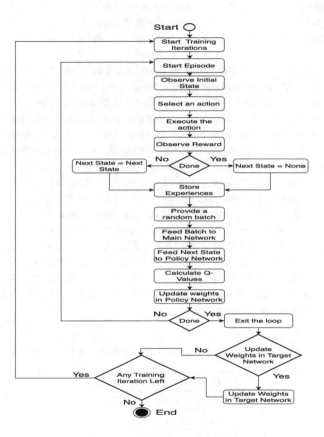

Fig. 3. Activity Chart of RL Agent

4.2   NSL-KDD Dataset

The NSL-KDD dataset, which is used by the Gym environment to provide observations to the agent, was obtained from the website of the Canadian Institute of Cyber Security, University of New Brunswick [21]. This dataset possesses the criteria to serve as a benchmark for today's modern intrusion detection systems [8]. Its training set contains 125,973 records of which 67,343 represent normal network traffic and 58,630 represent intrusions. Its testing set contains 22,544 records of which 9,711 represent normal network traffic and 12,833 represent intrusions [22]. The dataset contains 4 main attack categories (DoS, Probe, U2R, and R2L) and 39 different subcategories. Moreover, each record of the dataset represents 41 features of network traffic and a label.

4.3   Gym Environment

The environment is a very important element of the reinforcement learning process. Building an expressive environment is a necessity for the agent to be properly trained and perform the assigned task efficiently. In order to implement this environment concept, a specifically developed toolkit called OpenAI Gym toolkit is used.

The Gym environment structure is based on four different functions that are all in the same class: initialization, step, reset, and render. The environment will be referred to as an object called *env*. The initialization function is composed of several parameters. The two most relevant parameters are the action space, which is the available actions for the agent to take, and the observation space, which includes network traffic's features that will be observed by the agent. The step function will observe the action taken by the agent and return four different parameters: *next state, reward, done,* and *info*. These parameters represent the next state that the agent will observe, the current reward, a Boolean value indicating whether this episode is over or not, and some additional diagnostic information. The reset function provides a new random initial state. The render function outputs a graphical representation of the current situation in the environment. In this context, however, it is irrelevant because we do not have a graphical representation for the Gym environment. Moreover, all the possible actions, state labels, and rewards are provided in Table 1 to deepen the understanding of the previous concepts.

Table 1. State, Action and Reward as represented in the environment

State label	Action	Reward
Normal	0	1
Normal	1	−1
Intrusion	0	−1
Intrusion	1	1

First, the reset function is used to obtain the initial observation from the environment, which is then passed to the agent. Second, the agent will choose an action from the available ones. Third, the action chosen by the agent is passed to the environment step function. The step function will return the next observation, the reward for the agent's past action, a Boolean value indicating whether the training episode is over or not, and some diagnostic information.

This training process continues until it is terminated. Table 1 also explains the reward system given by the environment in correspondence with the action taken by the agent. When the agent observes a state provided by the environment, it responds by choosing an action from the available actions. The agent can either choose "0" or "1" which indicates whether this network traffic is a normal or an intrusion, respectively. Given this action, the environment judges the agent behavior by matching it with the label associated with each network traffic in the dataset. Thereupon, the environment assigns either a positive reward "1" or a negative reward "–1" in response to the correct or incorrect action of the agent, respectively.

5   Results

In this section, the evaluation approach, metric, and the results of the experiments will be discussed. The experiments were conducted using different number of training iterations, various neural network structures including different numbers of hidden layers and neurons, and multiple batch sizes. Additionally, we used 50% of the data in some experiments and 100% in others to prove that reinforcement learning can handle more attack representations and function properly with the increase in the number of attack types represented in the dataset, leading to the conclusion that RL can generalize. The reason for the variation proposed in the experiments is that obtaining an improved model is not an exact formula but rather based on a trial-and-error approach. At the end of the section, a general comparison with other reinforcement learning approaches implementing the NSL-KDD dataset from the literature is discussed in order to demonstrate performance comparison.

5.1   Evaluation Metrics

The most appropriate evaluation methods used for classification prediction models are confusion matrix, F1 score, accuracy, precision, and recall [23].

The confusion matrix [23] can be used for classification problems to show the actual classification (label) and the predicted one. Since all of the approaches in this research use binary classification, confusion matrix is considered the optimal option for evaluation. Although using confusion matrix is not suitable for reinforcement learning model evaluation, we can utilize the confusion matrix for our reinforcement learning approach since the data is labeled. Metrics were derived from the confusion matrix, which represent accuracy, precision, recall (sensitivity), and F1 score of the evaluated model in terms of True Positive (TP), False Negative (FN), False Positive (FP), and True Negative (TN) predictions.

5.2   Reinforcement Learning Model Evaluation

Reinforcement learning model evaluation differs from other machine learning approaches because of its unique structure. In order to evaluate a reinforcement learning model, we have two approaches: the performance of the policy estimated by the model and the learning curve that indicates the improvement of the agent's actions over time. The learning curve approach was followed in this paper. It can be identified as the accumulative reward as a function of the number of episodes. In other words, we are plotting the accumulative reward acquired by the RL agent over time and showing if it is increasing or decreasing. If the accumulative reward is increasing, it indicates that the RL agent's actions are improving during the learning process, which is the desired output.

5.3   Reinforcement Learning Experimental Results

In Fig. 4 below, the learning curve of the RL agent is plotted. It is showing a great improvement in the agent's performance over time, which validates the model. This means that the accumulative reward that the agent is receiving is increasing. This indicates that the agent's ability to choose the most appropriate action by maximizing its gain is improving over time.

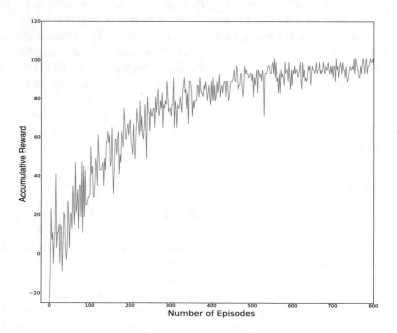

Fig. 4. RL agent learning curve

The results of the Deep Q reinforcement learning model are shown below. Although many experiments were conducted, only the top-performing ones are expressed. The other experiments were attempts to reach the optimal hyper-parameters. An experiment with only 50% of the dataset is performed to show the generalization property of RL when compared to the full dataset experiment.

Another experiment is shown using different hyper-parameters to prove that the RL model's hyper-parameters and neural network structure can affect its performance drastically.

We started by using 50% of the data with 400 training iterations, 100 steps per episode, and a batch size of 64. We used two hidden layers of size 50*10. The accuracy reached 86.80%. In the second conducted RL experiment, we used 100% of the data with 800 training iterations, 100 steps per episode, and a batch size of 64. We used two hidden layers of size 50*10 and the accuracy was 93.12%.

An aggregation of all the conducted experiments' results can be shown in Table 2 to give an overview.

Table 2. Aggregated results for the reported experiments

Experiment	F1 (%)	Recall (%)	Precision (%)	Accuracy (%)
RL - Experiment 1	84.45	86.55	82.45	86.8
RL - Experiment 2	96.36	95.9	96.83	93.12

5.4   Performance Comparison with Relevant Work

Table 3 is an aggregation of the top-performing experimental results of the proposed RL agent alongside state-of-the-art results collected from related researches in the literature. The related researches implemented RL approaches such as Deep Q-Network (DQN) [12,14], and Double Deep Q-Network (DDQN) [15] alongside the same NSL-KDD dataset. The table also expresses the value added by our approach (OGE-DQN) in enhancing and improving the reinforcement learning agent's performance in intrusion detection as shown.

Table 3. Aggregated Results Comparison with Relevant Work

Approach	F1 (%)	Recall (%)	Precision (%)	Accuracy (%)
OGE-DQN (Developed in this research)	96.36	95.9	96.83	93.12
DQN (Manuel Lopez-Martin et al. [12])	89.35	89.37	89.33	87.87
DQN (Hsu and Matsuoka [14])	–	90.2	92.8	91.4
DDQN (Ma and Shi [15])	82.43	82.09	84.11	82.09

6   Conclusion

Due to the massive shift to network-based technologies, cloud computing-based services are becoming the ultimate replacement and solution for handling and providing the infrastructure of computer systems, storage space, and computational power needed by agencies, companies, organizations, institutions, and governments. This transition introduced major concerns regarding the medium or network systems used to deliver these services and resources. Securing the network systems poses a challenge to customers and users of cloud computing services and other network-based technologies. The new attack types and continuous change in attack patterns introduced frequently pose a threat that makes it very difficult for traditional cyber-security and intrusion detection systems to keep up with those developments in attacks and the increasing scale of network systems.

In this research, we provided an answer to the intrusion problem in modern network systems by using deep reinforcement learning. We developed a novel approach for intrusion detection, and this approach had two key parts. The first part was the agent, which was built using Deep Q-learning algorithm and a deep neural network. The second part was the custom-built Gym environment that guided the learning process and provided network traffic from the dataset to the reinforcement learning agent. The dataset used to fuel the learning process was the NSL-KDD dataset, as it is used as a benchmark for most modern intrusion detection systems. The Gym environment presented is the first attempt to create a standardized Gym environment for intrusion detection using OpenAI Gym. Additionally, the custom-built Gym environment, which is our main contribution, contributed to the model's superiority by having expressive and inclusive features of intrusions. The superiority of our approach was validated and confirmed by comparing its results with other relevant work from the literature. The proposed Deep Q-learning solution achieved the highest performance with an accuracy of more than 93%, and it appears to be the most efficient solution among those compared. This high accuracy was achieved as a result of using an expressive Gym environment and a well-tuned model. Several experiments were carried out using several hyper-parameters and configurations by trial and error to obtain the optimal results. The experiments performed were reported and classified according to their configurations.

References

1. Cloud infrastructure market share. https://www.statista.com/chart/7994/cloud-market-share/ (2021)
2. Cloud infrastructure market share. https://www.statista.com/chart/18819/worldwide-market-share-of-leading-cloud-infrastructure-service-providers/ (2021)
3. Armin, J., Thompson, B., Ariu, D., Giacinto, G., Roli, F., Kijewski, P.: 2020 cyber-crime economic costs: no measure no solution. In: 2015 10th International Conference on Availability, Reliability and Security, pp. 701–710. IEEE (2015)
4. The hidden costs of cybercrime. https://www.csis.org/analysis/hidden-costs-cybercrime (2021)
5. Scarfone, K.A., Mell, P.M.: Sp 800-94, guide to intrusion detection and prevention systems (IDPS). Tech. Rep., Gaithersburg, MD, USA (2007)
6. Roesch, M., et al.: Snort: Lightweight intrusion detection for networks. In: Lisa, vol. 99, pp. 229–238 (1999)

7. Brockman, G., et al.: OpenAI Gym. arXiv preprint arXiv:1606.01540 (2016)
8. Sharafaldin, I., Gharib, A., Lashkari, A.H., Ghorbani, A.A.: Towards a reliable intrusion detection benchmark dataset. Softw. Netw. **2018**(1), 177–200 (2018)
9. Liang, C., Shanmugam, B., Azam, S., Jonkman, M., De Boer, F., Narayansamy, G.: Intrusion detection system for internet of things based on a machine learning approach. In: 2019 International Conference on Vision Towards Emerging Trends in Communication and Networking (ViTECoN), pp. 1–6. IEEE (2019)
10. Koduvely, H.: Anomaly detection through reinforcement learning (2018). https://doi.org/10.13140/RG.2.2.33673.29283
11. Caminero, G., Lopez-Martin, M., Carro, B.: Adversarial environment reinforcement learning algorithm for intrusion detection. Comput. Netw. **159**, 96–109 (2019)
12. Lopez-Martin, M., Carro, B., Sanchez-Esguevillas, A.: Application of deep reinforcement learning to intrusion detection for supervised problems. Expert Syst. Appl. **141**, 112963 (2020)
13. Suwannalai, E., Polprasert, C.: Network intrusion detection systems using adversarial reinforcement learning with deep Q-network. In: 2020 18th International Conference on ICT and Knowledge Engineering (ICT&KE), pp. 1–7. IEEE (2020)
14. Hsu, Y.F., Matsuoka, M.: A deep reinforcement learning approach for anomaly network intrusion detection system. In: 2020 IEEE 9th International Conference on Cloud Networking (CloudNet), pp. 1–6 (2020). https://doi.org/10.1109/CloudNet51028.2020.9335796
15. Ma, X., Shi, W.: AESMOTE: adversarial reinforcement learning with SMOTE for anomaly detection. IEEE Trans. Netw. Sci. Eng. 8, 943–956 (2020)
16. Hougen, D.F., Shah, S.N.H.: The evolution of reinforcement learning. In: 2019 IEEE Symposium Series on Computational Intelligence (SSCI), pp. 1457–1464. IEEE (2019)
17. Qiang, W., Zhongli, Z.: Reinforcement learning model, algorithms and its application. In: 2011 International Conference on Mechatronic Science, Electric Engineering and Computer (MEC), pp. 1143–1146. IEEE (2011)
18. 5 things you need to know about reinforcement learning. https://www.kdnuggets.com/2018/03/5-things-reinforcement-learning.html (2021)
19. Yildiz, B.: Reinforcement learning using fully connected, attention, and transformer models in knapsack problem solving. Concurrency and Computation: Practice and Experience (2021). https://doi.org/10.1002/cpe.6509
20. A hands-on introduction to deep Q-learning using openAI GYM in python. https://www.analyticsvidhya.com/blog/2019/04/introduction-deep-q-learning-python/ (2021)
21. NSL-KDD dataset. https://www.unb.ca/cic/datasets/nsl.html (2021)
22. Tavallaee, M., Bagheri, E., Lu, W., Ghorbani, A.A.: A detailed analysis of the KDD cup 99 data set. In: 2009 IEEE Symposium on Computational Intelligence for Security and Defense Applications, pp. 1–6. IEEE (2009)
23. Vihinen, M.: How to evaluate performance of prediction methods? measures and their interpretation in variation effect analysis. In: BMC genomics, vol. 13, pp. 1–10. BioMed Central (2012)

Comparison of Small-Sized Deep Neural Network Models for Hyperspectral Image Classification

Ekrem Tarık Karan[1]([✉]) [iD], Zümray Dokur[2] [iD], and Tamer Ölmez[2] [iD]

[1] School of Graduate, Satellite Communication and Remote Sensing, Istanbul Technical University, 34467 Maslak, Istanbul, Turkey
karane19@itu.edu.tr

[2] Department of Electronics and Communication Engineering, Istanbul Technical University, 34467 Maslak, Istanbul, Turkey

Abstract. Many different methods related to deep learning are used in a broad scope in areas such as identification, clustering, and classification of 1D, 2D, and 3D data. In contrast to traditional images, deep learning studies are also performed on hyperspectral images, which are structurally different by the hundreds of bands it contains. Although hyperspectral images are data in 3D format, they can be analyzed separately in 1D, 2D and 3D formats, which is advantageous for deep learning research. Although the presence of so many different types of information provides diversity for studies, on the other hand, it significantly affects the values and working times of the results. In the scope of this study, the classification of hyperspectral images using convolutional neural networks, a subtitle of deep learning, was carried out. The classification results obtained by convolutional neural networks with different characteristics and changing parameters were examined. In this way, the effect of other methods on the classification results was also discussed. Within the scope of the whole study, all methods were tested on Indian Pines, Salinas and Pavia University datasets, frequently preferred in the literature. The results obtained with the methods we used were compared with the results of other studies in the literature. It has been confirmed that the network models with the 3D convolution layer mentioned in the literature and our studies have better performance results. Furthermore, although our proposed models contain fewer parameters, unlike the models in the literature, they almost approach or exceed the other models' results.

Keywords: Hyperspectral image classification · Convolutional neural networks · Principal component analysis · Incremental principal component analysis

1 Introduction

Thanks to the hyperspectral image concept that emerged with the developments in remote sensing systems and sensors, it has become possible to obtain information about the remaining surface components other than the visible components. Hyperspectral imaging measures the energies of each object at different wavelengths and reveals the spatial

F. P. García Márquez et al. (Eds.): ICCIDA 2022, LNNS 643, pp. 244–256, 2023.
https://doi.org/10.1007/978-3-031-27099-4_19

and spectral properties of the objects. Objects are also distinguished with the aid of the different wavelength information contained in the pixels corresponding to each object. Hyperspectral images are three-dimensional images usually consisting of narrow bands of more than one hundred and about 5–10 nm wide. It usually consists of and is represented as a cube of images, with the spatial information of the object on the x and y axes and the spectral information on the z-axis. The hyperspectral image's formation and the different pixel wavelengths are visualized in Fig. 1 [1].

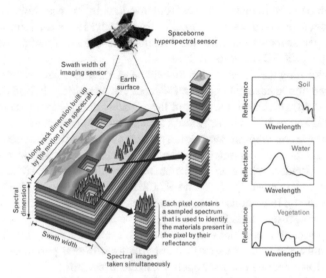

Fig. 1. Visualization of the hyperspectral image [1].

Classification studies on hyperspectral images were initially carried out using basic machine learning methods such as the Support Vector Machine and Random Forest. Then, more different techniques were introduced, beginning with the use of deep learning methods in the field of hyperspectral images. Convolutional neural network (CNN) models are often preferred. The primary purpose of models containing a 1D convolution layer is to classify hyperspectral images using spectral characteristics. When the studies in the literature are examined, the classification results obtained by models with a 1D convolution layer are compared with the traditional machine learning methods or with the classification results obtained by different CNN models. For example, Tao and Xinjie compared the performance results of a 1D convolution layer with the results of a model containing a Multilayer Perceptron (MLP) and a 2D convolution layer in their studies. Their study achieved classification results with an overall accuracy of 82.86%, 88.61%, and 99.52%, respectively, using MLP, 1D-CNN and 2D-CNN [2].

Network models containing a 2D convolution layer are preferred for using spatial information. It is thought that the pixels around each pixel in the input data in 2D format, that is, the adjacent pixels, have similar characteristics. For this reason, it is desirable that the filter also includes a different number of pixels around each pixel in the process. The window parameter, which can be adjusted in various sizes, is a situation that diversifies the studies in this field. For example, Tun *et al.* have prepared the input by choosing a 3 × 3 dimension in the deep learning model with a 2D convolution layer. Thus, they achieved a 99.49% success result in the classification on the Pavia University dataset [3].

Hyperspectral image data, three-dimensional data, is expressed as a hyperspectral image cube consisting of 1D spectral and 2D spatial information. This three-dimensional image cube, which has spatial characteristics on the X and Y axes and spectral features on the Z (or λ) axis, is also expressed as a voxel, which means volume element or volume pixel abbreviation. Using network models containing a 3D convolution layer, it has become possible to obtain many features from raw data. In this way, higher classification results have been received. For example, in their study, He *et al.* achieved a 95.45% classification performance result on the Indian Pines dataset, together with the CNN model containing the 3D convolution layer [4].

While performing the classification on hyperspectral images, methods other than the ones mentioned above have also been introduced. For example, a hybrid CNN model is often preferred, in which both the 2D and the 3D convolution layers are used together. However, the 2D convolution layer alone is not sufficient to extract features. In addition, while using the 3D convolution layer alone, the computational time increases. Therefore, using both a 2D and a 3D convolution layer together is proposed to eliminate such disadvantages. Ghaderizadeh *et al.* used models with only 2D and 3D convolution layers in their study and reached 96.76% and 97.68% classification results on the Salinas dataset, respectively. In response to these, they achieved 99.07% success on the Salinas dataset with their presented method [5].

This study aimed to conduct comprehensive research by trying all these methods. During this study, the scope was expanded by using different parameters while using other convolutional neural network models. First, the dimensions of hyperspectral images were reduced using the Principal Component Analysis method. Then, by adjusting different window sizes, the effect of window size was also measured. Within the scope of these studies, the datasets of Indian Pines, Salinas and Pavia University, often preferred in the literature, were selected.

2   Methods

2.1   Features in Hyperspectral Images and Dimension Reduction

Features in a hyperspectral image are readily available to the researchers in 3D format; in other words, the hyperspectral image consists of 1D and 2D data as a hypercube in 3D format. The 1D data of the hyperspectral image contains the spectral data. This spectral data consists of electromagnetic waves reflected from the surfaces in the area studied. 2D information, on the other hand, includes spatial data located on the x and y axes of the hypercube. Spatial data differs from spectral data in that there must be a spatial reference in return when generating. Therefore, 3D data, combining 1D and 2D data, is more valuable and functional because it contains spectral and spatial data.

The high level of information and features in hyperspectral images bring disadvantages, such as overlapping during the training phase of the deep learning model in classification studies. For this reason, the dimensions of hyperspectral images should be reduced. In some studies, feature selection is performed by removing unnecessary or unrelated data contained in the raw data. It is aimed to reduce the dimension of the raw data by storing a certain number of bands from hundreds of bands in hyperspectral images and removing the remaining bands from the image. For this reason, feature selection methods in the hyperspectral image field are also referred to as band selection methods. In some other studies, dimension reduction is accomplished by using a transformation method such as the Principal Components Analysis (PCA), which is a frequently preferred method in the literature in this field. Apart from the PCA method, there are unsupervised dimension reduction methods such as the folded-PCA, kernel-PCA, Kernel Entropy Component Analysis; and supervised dimension reduction methods such as Fisher's linear discriminant analysis and nonparametric weighted feature extraction [6, 7].

The PCA is an unsupervised feature extraction method to determine the correlation between bands containing spectral wavelengths found in hyperspectral images. When the PCA method is used, classification accuracies are significantly increased. However, the noise in the extracted features with low variance is not considered. For this reason, features tend to be corrupted by noise. In cases where the data to be applied PCA is significant and large to fit in the memory, the application of PCA after dividing the data is expressed as incremental PCA (IPCA). In this way, IPCA reveals more efficient memory usage [8].

2.2 Convolutional Neural Network

In deep learning, CNNs, proposed by LeCun in 1989, are preferred in computer vision and image processing applications such as target detection, image enhancement, *etc.* [9]. The CNN consists of three parts except for the output and input parts. The three parts are the convolution layer, the pooling layer, and the fully connected layer. The convolution layer, the first of these parts, is where the calculations occur. The parameters in the convolution layer are learnable filters or kernels. In this layer, different features are extracted from the input feature map. Then, the pooling function is performed in the pooling layer, as the name suggests. The pooling function is performed with many methods, such as maximum pooling, average pooling, and stochastic pooling. Thanks to the pooling layer, the features are maintained at the highest level while the parameters are reduced, and the training process is accelerated. Finally, in the fully connected layer, all neurons are fully connected to the neurons in the previous layer. This way, the computed values are obtained as a vector of features in the output layer. CNNs achieve high success rates because hidden layers do not entirely depend on the previous layers and do multiple computations between convolution and pooling layers. While performing these calculations, some mathematical models are preferred. These models are expressed as activation functions. Although many activation functions have been proposed in the literature, ReLU, sigmoid, or tanh activation functions are usually preferred. Figure 2 shows the visualization of the processes in the CNN architectures from an input pattern to the obtained output value [5, 10, 11].

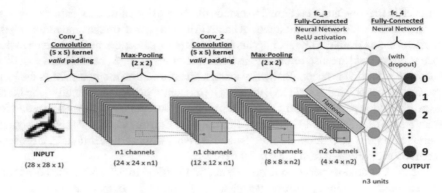

Fig. 2. The structure of the CNN [12].

3 Simulation Results

3.1 Datasets

One of the most common problems with hyperspectral images is the lack of publicly available datasets with ground truth data. For this reason, the most common datasets with ground truth information are considered in this study. These datasets were collected and made available to researchers by M. Graña *et al.* at Basque University. Within the scope of the experiments in this study, three of the most frequently used datasets in the literature were preferred [13].

First, the dataset referred to as Indian Pines (IP) was chosen. This data set, collected by the AVIRIS sensor in northwestern Indiana, consists of 224 spectral bands measuring 145×145 pixels in the wavelength range of 0.4–2.5×10^{-6} m. The IP dataset's ground truth, primarily agricultural and forested, includes 16 classes. The colourized image and ground truth of the IP dataset are shown in Fig. 3a.

The Salinas (SA) dataset was preferred as the second hyperspectral image dataset. The Salinas dataset was collected with the AVIRIS sensor, as in the Indian Pines dataset. This dataset, acquired over the Salinas Valley in California (USA), consists of 217×512-dimensional 3.7 m-pixel high spatial resolution images with 224 bands. The ground truth includes 16 classes of vegetables, vacant land, and vineyards. The colourized image and ground truth of the SA dataset are as in Fig. 3b.

Finally, the Pavia University (PU) dataset was preferred as the third hyperspectral image dataset. This dataset, collected by the ROSIS sensor on the Pavia region in northern Italy, consists of images of 610×610 pixels with 103 bands. Colourized image and ground truth of the PU dataset, which consists of 9 classes, are as in Fig. 3c.

40% of all these data sets are reserved for the test sets. The remaining 42% of the data was used in the training sets, and 18% was used in the verification set. All experiments were carried out using the Colab application, where Google provides an online working environment for researchers working on artificial intelligence.

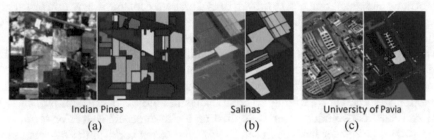

| Indian Pines | Salinas | University of Pavia |
| (a) | (b) | (c) |

Fig. 3. False colour representation of hyperspectral datasets and ground truths of (a) Indian Pines, (b) Salinas, and (c) Pavia University [14]

3.2 Evaluation Metrics

Within the scope of this study, overall accuracy and average accuracy values were computed to evaluate the classification performances of the hyperspectral images. Overall accuracy expresses the result obtained by dividing the total number of correctly classified items by the whole test items. Average accuracy is used to represent the average of each accuracy per class.

Equations 1 and 2 have general expressions for calculating the overall accuracy and the average accuracy, respectively.

$$\text{Overall Accuracy (OA)} = (TP + TN)/(TP + TN + FP + FN) \tag{1}$$

$$\begin{aligned}&\text{Average Accuracy (AA)}\\ &= (\text{sum of accuracy for each class predicted})/(\text{number of classes}) \end{aligned} \tag{2}$$

In the above equations, true positive (TP) expresses the number of truly positive results and the number of positively predicted results. The true negative (TN) value represents the number of truly negative values and negative predicted results. The number of results predicted as positive even though they are negative is expressed by the false positive (FP) value. Finally, the false negative (FN) value refers to the measurement of truly positive values as negative. [15, 16].

3.3 Classification Results

Classification studies were carried out within four different experimental headings. These four experiments are expressed as 1D-CNN, 2D-CNN, 3D-CNN and Hybrid-CNN models. In addition, different window sizes were preferred as a changeable parameter while performing these experiments. However, to make a better comparison within this study's scope, only the 11×11 window size results are presented. Besides, PCA and IPCA methods were preferred for the dimension reduction of the feature vectors. The size was reduced by storing 20 components in the Indian Pines dataset and 15 in the Salinas and Pavia University datasets. However, this situation differs for the model containing only the 1D convolution layer. In the model containing the 1D convolution layer, size reduction was applied by storing 20 components in all datasets. In addition, since the 1-dimensional convolution layer is applied on a one-dimensional vector element, no window size parameter has been used in this type of network. In the models used in all experiments, Adam's method, with a learning rate of 0.001, was preferred as the optimizer during compilation. When the studies in the literature were examined, and as a result of the studies carried out, the batch and epoch values were selected as 256 and 50, respectively.

The proposed 1D-CNN model has only four 1D convolution layers as convolution layers. Then, one flattening layer, three dense layers and two dropout layers were selected. The classification results with our models are available in Table 1. The success of the model we used is lower for the Indian Pines dataset but higher for the Salinas and Pavia University datasets.

Table 1. Comparison of the 1D-CNN model's classification results

Feature selections for datasets	Training time (sec.)	Testing time (sec.)	Overall accuracy (%)	Average accuracy (%)
PCA for Indian Pines	9.74	1.00	79.40	78.88
IPCA for Indian Pines	6.33	0.28	79.90	79.28
PCA for Salinas	34.96	1.82	95.52	97.94
IPCA for Salinas	30.38	0.71	95.53	97.91
PCA for Pavia Univ.	28.26	1.57	94.66	92.67
IPCA for Pavia Univ.	22.80	0.70	94.60	92.55

The structure of the model used to classify using the 2D convolution layer is similar to the network model containing the 1D convolution layer. Of course, the difference here is that the convolution layer is 2D. In this way, classification studies are carried out using spatial properties. The results of the proposed network model containing the 2D convolution layer are shown in Table 2.

Table 2. Comparison of the 2D-CNN model's classification results

Feature selections for datasets	Training time (sec.)	Testing time (sec.)	Overall accuracy (%)	Average accuracy (%)
PCA for Indian Pines	47.13	0.74	99.07	98.57
IPCA for Indian Pines	82.76	0.55	71.78	51.36
PCA for Salinas	322.86	2.58	99.60	99.79
IPCA for Salinas	262.99	2.34	98.39	99.07
PCA for Pavia Univ.	262.80	2.84	99.70	99.70
IPCA for Pavia Univ.	226.32	2.80	99.55	99.39

Network models with 3D convolution layers are used to use spectral features in addition to spatial features. Four convolution layers were used in the proposed network model containing a 3D convolution layer as in the other two models; the succeeding layers were also preferred as in the other network models. Accordingly, the results of this model are available in Table 3. The classification outputs produced by network models containing the 3D convolution layer that achieve the highest performance using the PCA size reduction method are shown in Figs. 4 and 5 for each dataset.

Table 3. Comparison of the 3D-CNN model's classification results

Feature selections for datasets	Training time (sec.)	Testing time (sec.)	Overall accuracy (%)	Average accuracy (%)
PCA for Indian Pines	862.95	5.25	99.44	99.66
IPCA for Indian Pines	836.59	5.30	98.22	97.38
PCA for Salinas	2546.01	25.40	99.99	99.99
IPCA for Salinas	2182.80	20.64	97.91	97.41
PCA for Pavia Univ.	1883.79	12.84	99.98	99.93
IPCA for Pavia Univ.	1736.92	23.11	99.93	99.91

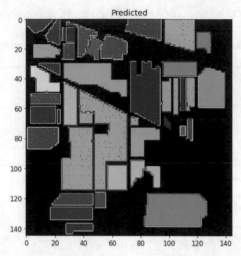

Fig. 4. Classification result using the 3D-CNN model and PCA on the Indian Pines dataset.

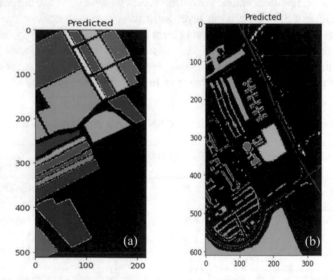

Fig. 5. Classification results using the 3D-CNN model and PCA on the (a) Salinas and (b) Pavia University datasets.

In addition to the 3D convolution layer, the 2D convolution layer is also used in the proposed hybrid network model to develop the network model containing only the 3D convolution layer. In this way, it is aimed to reduce the computational load of the 3D convolution layer. For this purpose, after the 3D convolution layers, the outputs were transferred to the 2D convolution layer using the reshaping layer. Then, in the continuation of the model, flattening, dropping, and fully connected layers were used as in the other layers. The results obtained within this study's scope are in Table 4.

Table 4. Comparison of the Hybrid model's classification results

Feature selections for datasets	Training time (sec.)	Testing time (sec.)	Overall accuracy (%)	Average accuracy (%)
PCA for Indian Pines	982.88	8.48	98.17	98.91
IPCA for Indian Pines	922.82	10.66	93.10	70.20
PCA for Salinas	2422.85	20.70	99.98	99.97
IPCA for Salinas	2076.54	11.36	88.90	84.60
PCA for Pavia Univ.	2123.10	20.74	99.99	99.98
IPCA for Pavia Univ.	1822.88	12.53	99.98	99.82

3.4 Comparison of the Proposed Network Models

The performance results obtained on all datasets, together with the total number of network parameters, are presented in Table 5. The table here shows the results obtained with the PCA method. When the obtained results were compared, it is observed that higher success rates were achieved in studies conducted using the PCA dimension reduction method. However, the IPCA method, which offers better memory usage, has revealed lower performance than the PCA method in terms of training and test times. Furthermore, as mentioned in the previous section, higher performance results were obtained with models containing 3D convolution layers. However, as shown in Table 5, the models with the highest parameter numbers are models with a 3D convolution layer.

For this reason, the computational load is increasing. Depending on the application, models with a 2D convolution layer or hybrid models might also be preferred because they show good performance results. Apart from this, the results of some studies in the literature are also shared in Table 5.

Table 5. Comparison of the proposed CNN models with some studies in the literature

	Methods	Number of param.	Training time (sec.)	Testing time (sec.)	Overall accuracy (%)	Average accuracy (%)
IP	1D CNN	240,032	9.74	1.00	79.40	78.88
	2D CNN	**208,424**	**82.76**	**0.55**	71.78	51.36
	3D CNN	995,456	862.95	5.25	**99.44**	**99.66**
	Hybrid CNN	350,336	922.82	10.66	93.10	70.20

(continued)

Table 5. (*continued*)

	Methods	Number of param.	Training time (sec.)	Testing time (sec.)	Overall accuracy (%)	Average accuracy (%)
	1D CNN in [17]	>250,000	–	–	83.4	–
	2D CNN in [18]	>400,000	–	–	89.48	86.14
	3D CNN in [19]	>994,166	–	–	97.75	94.54
	Hybrid CNN in [20]	5,122,176	846.0	4.8	99.75	99.63
SA	1D CNN	240,032	34.96	1.82	95.52	97.94
	2D CNN	**208,424**	**322.86**	**2.58**	99.60	99.79
	3D CNN	995,456	2546.01	25.40	**99.99**	**99.99**
	Hybrid CNN	350,336	2422.85	20.70	99.98	99.97
	1D CNN in [17]	>250,000	–	–	91.8	–
	2D CNN in [18]	>400,000	–	–	97.38	98.84
	3D CNN in [19]	>994,166	–	–	98.06	98.80
	Hybrid CNN in [20]	4,845,696	1530.0	9.0	100.0	100.0
PU	1D CNN	239,129	28.26	1.57	94.66	92.67
	2D CNN	**207,161**	**262.80**	**2.84**	99.70	99.70
	3D CNN	257,273	1883.79	12.84	99.98	99.93
	Hybrid CNN	257,273	2123.10	20.74	**99.99**	**99.98**
	2D CNN in [18]	>400,000	–	–	97.86	96.55
	3D CNN in [19]	>994,166	–	–	98.40	97.89
	Hybrid CNN in [20]	4,844,793	1218.0	6.6	99.98	99.67

When the results in Table 5 and the results presented in the previous tables are examined, it has been revealed that the PCA size reduction method is more advantageous than the IPCA method. Since not all information is conveyed in detail in the studies in the literature, a complete comparison could not be made in every field. However, when we examined within the scope of classification performance results, the models we used outweighed the studies in the literature. In addition, the number of parameters used in the hybrid CNN model is less, contrary to the studies in the literature. Therefore, high classification performance results were obtained by using fewer parameters.

It is necessary to express the differences between this study and other studies in the literature. For example, by using fewer parameters, the performance results in the hybrid model used by Roy were approached [20]. In this way, the parameter load is reduced. In addition, a smaller window size was used in contrast to studies in the literature using a window size of 11 × 11. Thanks to this situation, the processing load was reduced, and at the same time, it did not affect the performance results too much. In summary, it has been shown that with fewer data, similar or, in some cases, better performances can be achieved compared with the results in the literature.

4 Conclusions

Within the scope of this study, the effects of network models with different size convolution layers were investigated on the classification performance of the Indian Pines, Salinas and Pavia University datasets. In addition, the effect of different size reduction methods on the performance results was observed using two size reduction methods, PCA and IPCA. Apart from this, the scope of the study was limited in this area by keeping the window size constant.

When all the results were examined, it was concluded that the network model containing the 3D convolution layer achieved more successful classification outcomes. However, suppose a better result is desired in terms of computation time; in that case, it is necessary to use the hybrid network model. In addition, the most critical result and novelty obtained in this study are that it is possible to get results close to the results in the literature with fewer parameters by using the low-dimensional neural network model. With the models used, classification results were obtained at equivalent levels with the classification results in the studies in the literature.

It should also be noted that not every dataset will achieve the same degree of success. Therefore, it is necessary to adjust the network model and its parameters according to the intended work. In future studies, the scope of the study can be expanded by using different window sizes or other size reduction methods. In addition, the free availability of different datasets will ensure that the work continues in this field.

Acknowledgements. The environment in which deep learning applications are run seriously affects the results of the study. For this reason, we would like to thank the National High-Performance Computing Application and Research Center (UHeM) at Istanbul Technical University for providing its quality hardware.

References

1. Shaw, G.A., Burke, H.K.: Spectral imaging for remote sensing. Linc. Lab. J. **14**(1), 3–28 (2003)
2. Jiang, T., Wang, X.J.: Hyperspectral images classification based on fusion features derived from 1D and 2D convolutional neural network. Int. Arch. Photogramm. Remote Sens. Spat. Inf. Sci. **42**, 335–341 (2020)
3. Tun, N.L., et al.: Hyperspectral remote sensing images classification using fully convolutional neural network. In: 2021 IEEE Conference of Russian Young Researchers in Electrical and Electronic Engineering (ElConRus). IEEE (2021)

4. He, M., Li, B., Chen, H.: Multi-scale 3D deep convolutional neural network for hyperspectral image classification. In: 2017 IEEE International Conference on Image Processing (ICIP). IEEE (2017)
5. Ghaderizadeh, S., et al.: Hyperspectral image classification using a hybrid 3D-2D convolutional neural networks. IEEE J. Sel. Top. Appl. Earth Observ. Remote Sens. **14**, 7570–7588 (2021)
6. Liao, W.: Feature extraction and classification for hyperspectral remote sensing images. Dissertation, Ghent University (2012)
7. Uddin, M.P., Mamun, M.A., Hossain, M.A.: Feature extraction for hyperspectral image classification. In: 2017 IEEE Region 10 Humanitarian Technology Conference (R10-HTC). IEEE (2017)
8. Rasti, B., et al.: Feature extraction for hyperspectral imagery: the evolution from shallow to deep: overview and toolbox. IEEE Geosci. Remote Sens. Mag. **8**(4), 60–88 (2020)
9. LeCun, Y., et al.: Gradient-based learning applied to document recognition. Proc. IEEE **86**(11), 2278–2324 (1998)
10. Arora, D., Garg, M., Gupta, M.: Diving deep in deep convolutional neural network. In: 2020 2nd International Conference on Advances in Computing, Communication Control and Networking (ICACCCN). IEEE (2020)
11. Yaxue, Q.: Convolutional neural networks for literature retrieval. In: 2020 International Conference on Computer Vision, Image and Deep Learning (CVIDL). IEEE (2020)
12. Summer School Series: Lecture 5 by Rahul Sukthankar | Archana Swaminathan (2020). https://archana1998.github.io/post/rahul-sukthankar/. Accessed 01 Mar 2022
13. Hyperspectral Remote Sensing Scenes - Grupo de Inteligencia Computacional (GIC). http://www.ehu.eus/ccwintco/index.php/Hyperspectral_Remote_Sensing_Scenes. Accessed 01 Sept 2021
14. Evaluation Metrics for Binary Classification (And When to Use Them) - neptune.ai (2019). https://neptune.ai/blog/evaluation-metrics-binary-classification. Accessed 02 Mar 2022
15. Lee, H., Kwon, H.: Going deeper with contextual CNN for hyperspectral image classification. IEEE Trans. Image Process. **26**(10), 4843–4855 (2017)
16. Evaluation Metrics Machine Learning (2019). https://www.analyticsvidhya.com/blog/2019/08/11-important-model-evaluation-error-metrics/. Accessed 02 Mar 2022
17. Hsieh, T.-H., Kiang, J.-F.: Comparison of CNN algorithms on hyperspectral image classification in agricultural lands. Sensors **20**(6), 1734 (2020)
18. Makantasis, K., et al.: Deep supervised learning for hyperspectral data classification through convolutional neural networks. In: 2015 IEEE International Geoscience and Remote Sensing Symposium (IGARSS). IEEE (2015)
19. Ahmad, M., et al.: A fast and compact 3-D CNN for hyperspectral image classification. IEEE Geosci. Remote Sens. Lett. **19**, 1–5 (2020)
20. Roy, S.K., et al.: HybridSN: exploring 3-D–2-D CNN feature hierarchy for hyperspectral image classification. IEEE Geosci. Remote Sens. Lett. **17**(2), 277–281 (2019)

Time-Sensitive Embedding for Understanding Customer Navigational Behavior in Mobile Banking

Hakan Hakvar[1]([✉]), Cansu Cavuldak[1], Oğulcan Söyler[1], Yıldız Karadayı[1],
and Mehmet S. Aktaş[2]

[1] R&D Center, Fibabanka, Istanbul, Turkey
{hakan.hakvar,cansu.cavuldak,ogulcan.soyler,
yildiz.karadayi}@fibabanka.com.tr
[2] Yildiz Technical University, Istanbul, Turkey
aktas@yildiz.edu.tr

Abstract. The availability of digital products, mobile, and the internet has become very widespread in the financial world, which provides a lot of information regarding customer financial habits. Purchasing habits of customers and their tendency to buy a specific product can be determined by the footprints left by customers in digital banking applications. However, tendency based on visit frequency and time-dependency that might provide valuable data for representing customer behavior are generally missed out by generic embedding approaches. We commit this problem by suggesting a customer embedding framework to leverage the time-sensitive digital footprint of mobile banking customers to gain benefit in prediction scenarios.

The proposed framework utilizes only the digital footprints of customers generated on a mobile banking application. It helps us to represent customers who don't have much contextual history in the banking environment and accurately predict their investment tendency. We test customer embedding vectors generated by the proposed framework using real-world digital footprints of mobile banking customers in a prediction scenario in which the intention of customers towards a financial product is predicted. The proposed customer embedding framework has shown better performance over the plain usage of state-of-the-art embedding approaches Word2Vec and DeepWalk.

Keywords: Customer embedding · Customer behavior · Graph base embedding · Aging based embedding · Representation learning

1 Introduction

Digital banking applications offer customers much ease of use and possibilities. In this way, customers can easily and quickly carry out their transactions without visiting a branch or wasting time. Customers also provide data to companies while navigating and transacting in digital banking applications. It is necessary to make sense of these

F. P. García Márquez et al. (Eds.): ICCIDA 2022, LNNS 643, pp. 257–270, 2023.
https://doi.org/10.1007/978-3-031-27099-4_20

footprints left by the customers and convert them into structured data sets and analyze these datasets [1].

By using customers' footprints that they leave on digital banking applications, we start to know their behaviors and tendency toward different banking products; so we can predict what will be their next move and use this valuable knowledge for the benefit of customers and the bank too [1].

To better prediction of customer activities, based on the time-based dynamics of customer behavior in digital banking applications, compact and dynamic representations of customers are needed [1].

It is possible to make sense of customers' navigation in digital banking applications with embedding techniques [1]. Embedding techniques help to digitize the transitions of customers between pages and transform each mobile customer's behavior into numerical representations that can be used to accurately predict their future intention [1]. Word embedding techniques have been long applied within the field of natural language processing and text processing [2, 3]. It has also been observed that comparative analyzes of embedding methods in digital footprint studies are scarce and the use of embedding methods in a time-based dataset is insufficient. The scarcity of techniques that would help to represent customers' navigational footprint by taking the time and frequency dimensions into account is the main driver of our study.

In this study, we propose a framework based on customer interaction data with temporal dynamics for a prediction task in the banking ecosystem. We have designed our framework in that it leverages the state-of-the-art embedding algorithms like DeepWalk [4] and Word2Vec [5] and extended the generated embedding representations by applying aging factor and overall tendency of customers based on visit frequency. In the proposed framework, a new customer-based representation generation methodology is designed by applying novel data processing techniques based on the unique requirements of the banking ecosystem. Changing financial needs of customers and investment intent, which are highly time-dependent, are accurately modeled via the proposed methodology.

The remainder of this article is organized as follows: Sect. 2 includes the review of embedding techniques used. In Sect. 3, we talked about the problems we have addressed in this research. Section 4 gives the implementation of our framework and information about the data used. The full methodology and detailed information about the different approaches, which was used in our research are given in Sect. 5. In Sect. 6 comparison of results for different vector sizes and techniques is summarized. Results and which method is more successful are given in Sect. 7.

2   Literature Review

Word embedding is a presentation of the real value of both semantic and syntactic meanings of words in a huge corpus [2]. It's a powerful tool of natural language processing (NLP) techniques [2]. These techniques provide vector representation of any word in the corpus. They show a similarity in words, speech labelling, clustering and synthetic parsing within natural language processing [6]. Thanks to word embedding, many NLP-based analyses and predictions are used in supervised, semi-supervised and unsupervised machine learning algorithms by converting from words to vectors [2].

Due to its use for a different purpose, embedding techniques has diversified over time and divided into sub-branches like Word2Vec [5], Glove [7] in word embedding, and DeepWalk [4], Graph2vec [8], Node2Vec [9], Struct2vec [10] in graph embedding. Word2Vec technique was developed by Tomas Mikalov and his team. It used two shallow models are used to learn a vector-space presentation of words by mapping these occurrences and similar meanings to their neighbors in the vector space [11]. Word2Vec model has two approaches. One of them is Continuous Bag of Words (CBoW), and the other one is Skip-gram algorithm [11]. CBoW predicts a specific word via analyzing neighbor words. Every neighbor word converted one-hot encoded vectors as input of a neural network. The output of the network is the predicted word. Conversely, Skip-gram predicts neighbor words around the word add to input [12]. In Skip-gram, a specific word converted one-hot encoded vector as an input of a neural network. The output of the network are predicted neighbor words [11, 12] (Fig. 1).

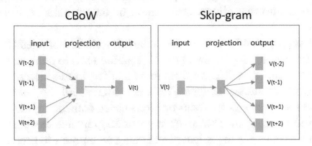

Fig. 1. CBOW and SkipGram

As opposed to word embedding, graph embedding uses graphs that consist of combined nodes, and represents nodes as vectors in low-dimensional space. Nodes in a graph can have similar attributes. Graph-based embedding modeling is used in plenty applications such as node classification, community detection, etc. [13]. Graph embedding approaches are obtained by computationally intensive eigenvalue decomposition techniques [14] (Fig. 2).

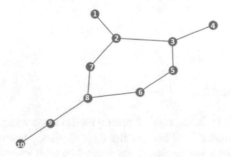

Fig. 2. Example of a node

Graph embedding approaches are more powerful techniques, and they succeed to solve community detections and social network problems. They ensure a powerful tool to create vectorized features for graph attributes, such as nodes, edges and subgraphs by protecting internal graph properties [15].

DeepWalk is the most recent and successful graph embedding technique [14]. Deep-Walk maintains higher order proximity by maximizing the likelihood of monitoring the last node and subsequent nodes in each random walk [16]. Briefly, DeepWalk is applied in two stages. The first stage is random walk provides that embedded representations of nodes are optimized by wandering near neighbors randomly [17].

And the second stage is Word2Vec Skip-gram applied to every created walk to vectorize. So, DeepWalk is applied to the successful word embedding to set the nodes, so that occurance frequencies of pairs in short random walks are protected [14]. DeepWalk uses Hierarchical Softmax which the Huffman binary tree is used as an alternative representation of the vocabulary. The gradient descent will be faster because of the Huffman tree structure that allows degradation of output units necessarily evaluated [18].

There exist surveys that overview the embedding methodologies [19]. We observe pattern-based embedding approaches [20, 21] that are used to analyze the navigational behavior of the customers. Different from these studies, in this study, we mainly focus on time-sensitive embedding strategies to better understand the behavior of the customers. We also see some studies focus on word-based embedding methodologies [22–25] to understand the customer behavior to generate graphical customer interface testing scripts. In this study, we investigate embedding techniques to predict the following behavior of the customers on digital banking applications (Fig. 3).

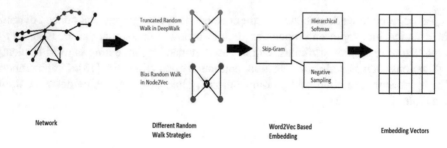

Fig. 3. DeepWalk structure

3 Research Problem

In this work, our aim is building an embedding methodology to better represent digital banking customers to understand and predict their tendency towards different financial products. For this purpose, we propose a framework, which best leverages the customer digital footprints to use for various prediction tasks in the banking ecosystem. By using only the digital footprints of customers, we are able to accurately predict the intent of customers who have only limited transaction history and contextual data.

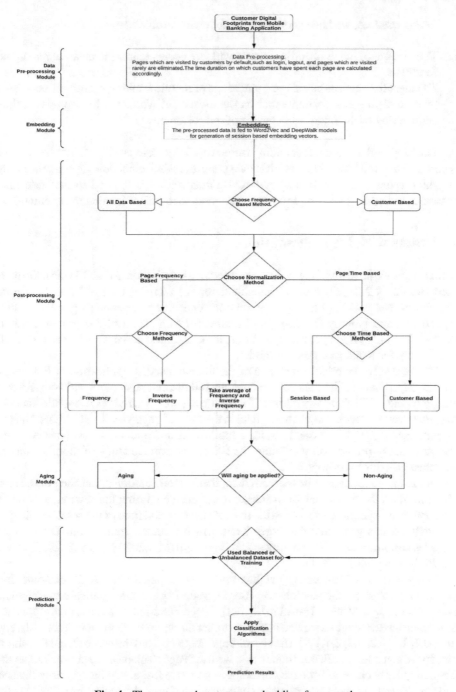

Fig. 4. The proposed customer embedding framework.

In this research, we have addressed the following problems:

1. The design of an embedding framework to better represent customer's digital footprints,
2. To extent the session-based representations of mobile banking customers generated by state-of-the-art algorithms such as DeepWalk and Word2Vec to leverage the time dependency of the financial behavior of the customers.

The proposed customer embedding framework is indicated in Fig. 4. The framework is composed of the following modules: data pre-processing, embedding, post-processing in which aging and visit frequency-based normalization is applied to generate final customer embedding vectors, and prediction module to utilize learned representations.

4   Framework Implementation

In this study, we have implemented the framework using Python-3 (3.6.13) [26]. We have used Gensim (4.2.0) [27] library implementations of Word2Vec CBoW and Skip-gram algorithms, NetworkX (2.5.1) [28] library for converting sessions to graphs, Pandas (1.1.5) [29] and Numpy (1.19.5) [30] libraries for data processing operations, scikit-learn (0.24.2) [31] library for Isolation Forest model and metrics, and Xgboost (1.5.2]) [32] library for Xgboost Classifier model.

We have used the clickstream data of the mobile banking application of Fibabanka for our experiments. The data set contains 36.160 distinct customers and consists of 5 attributes and 2.971.999 rows of data. To capture the online activities of mobile banking customers, we collect the clickstream data in the form of page views and actions, such as foreign exchange (FX) transactions, loan application, and opening a new bank account. We particularly focused on predicting the FX transaction tendency of mobile banking customers for our experiments.

In the mobile banking click-stream data, there is information about the unique customer number, the pages that the customer has clicked on during the session, how much time spent on a specific page during the session, the date and time of the session, whether the customer has generated the target event, which is an FX transaction. The data set includes customers with 50 or more sessions between 01.04.2022 and 15.06.2022, which arc 76 days in total (Table 1).

"Page Sequence" feature is a list that represents the pages in order the customer has navigated throughout the session, e.g. [Login, Main Page, Loan Application, Calculate Loan, Main Page, Money Transfer, Logout]. "Page Durations" is a list that represents time spent on each page navigated by the customer throughout the session in seconds, e.g. [0.5, 0.7, 1.1, 1.2, 0.5, 2.4, 0,2] The size of Page Sequence and the size of Page Durations must have equal length. In our framework, we use "Page Sequence" data to calculate the frequency for each page for each customer to generate frequency-based normalization and use "Page Durations" to calculate customer-based time duration normalizations for embedding vectors.

In the data preprocessing module in the framework shown in Fig. 4, pages with maximum frequency (such as login, and logout) and pages with minimum frequency

Table 1. Browsing data on mobile banking application pages of bank customers.

Feature	Explanation
Customer number	Unique customer number
Page sequence	Pages the customer browsed in the mobile app during the session
Page durations	How long the customer has been on a specific page (in seconds)
Session start date	The date and time the customer created the session via logging in to the mobile app
Foreign exchange event flag	Whether the customer executes the FX transaction during the session 1 = True, 0 = False

(pages visited very rarely by customers) are excluded from the clickstream dataset. After the elimination process, the time spent on each page at each session is calculated accordingly. Then, if there are any null values for page durations, we calculated the average page duration for sessions and set these average values in place of null ones.

In the embedding module, we have applied Word2Vec CBoW, which represents word embedding methodology, and DeepWalk, which represents graph embedding methodology to generate session-based representations. To effectively encode customer behavior and intent, we have extended generated embedding vectors by applying novel frequency and time-based normalization techniques in the following post-processing module.

During the experiments, our target task is to predict whether a customer executes a foreign exchange buy or sell transaction based on previous mobile sessions. In the dataset, we used "Foreign Exchange Event Flag" as the target label. We have experimented with three different embedding vector sizes: 20, 50, and 100. These vector sizes were defined empirically, by trial and error. Since our training dataset was relatively small with only about 36K distinct entities, we have chosen smaller vector sizes as opposed to vector sizes used in word embedding in NLP (Natural Language Processing) researches.

5 Methodology

Our main extension to the embedding vectors generated by Word2Vec (CBoW) and DeepWalk is to incorporate mobile banking customers' tendency on specific products by using two new dimensions: Time spent by customers on pages of a mobile banking application and visit frequency of pages by each customer. For this purpose, in the post-processing module of the proposed framework, we have normalized embedding vectors using two different approaches: frequency-based and time-based.

The frequency-based approach contains frequency, inverse frequency, and average frequency that takes the average of both techniques. In the frequency-based approach, a frequency weight is calculated according to the weights of the pages entered by each customer. Thanks to this technique, more importance is given to the most visited pages. In the inverse frequency technique, less importance is given to the most visited pages by taking the inverse of the frequency of the most visited pages by each customer. In the average frequency approach, we took the average of both techniques mentioned above.

There are different time-based embedding methods in the literature [33, 34]. However, the time-based weighting approach we used in our study is fed from an approach based on the time spent in sessions unlike the literature. In the time-based approach, two different time-based weighting techniques, customer and session-based, are considered. However, we used a customer-based time normalization approach in this study. It is a time-based normalization technique based on the total time spent by each customer for the duration of 76 days on relevant pages. After weight vectors are generated for each normalization techniques, session-based embedding vectors are normalized by multiplying them by weight vectors (Table 2).

Table 2. Example of a session.

Page sequence id	Page sequence
1	Foreign exchange, Transactions, loans, Products and loans, Foreign exchange

We obtained session vectors for two different embedding techniques. Our goal in the next step is, to sum up, the session vector data of each customer and convert them to customer-based embedding vectors to represent each customer with only one vector. The size of embedding vectors that represents the customer best for the target problem can be determined empirically.

At the aging module in the framework shown in Fig. 4, we have proceeded with two approaches while reducing session embedding vectors to customer-embedding vectors: Aging and non-aging. In the aging approach, session embedding vectors are aged and aggregated by applying the following formula:

$$Cust\_Based\_Seq\_Vec = a * Mean\big(X_{(t-n)}\big) + (1-a) * Mean\big(X_{(n)}\big) \qquad (1)$$

In this formula a denotes the aging coefficient. As it increases, recent customer sessions get more important than older sessions. Hence, the aging coefficient as a determines how important recent sessions are. n denotes the number of recent session vectors of a customer. t denotes the number of all sessions for each customer. $Mean(X_{(n)})$ indicates the average of recent n vectors. $Mean(X_{(t-n)})$ indicates average of older session vectors.

In the non-aging approach, customer embedding vectors are generated by taking the arithmetic average of session embedding vectors for each customer.

$$Cust\_Based\_Seq\_Vec = Mean\big(X_{(t)}\big) \qquad (2)$$

$Mean(X_{(t)})$ indicates average of all sessions for each customer.

After applying aging or non-aging aggregated techniques, we obtained customer-based vectors that each customer represented by a $(1 \times n)$ embedding vector, in which n is a framework parameter and represents the dimension of the embedding vector.

For our experiments, we use real-world digital footprints of mobile banking customers which is internal to Fibabanka. In the prediction module, we have utilized customer embedding vectors in predicting customer tendency toward a specific financial product scenario. We fit two machine learning algorithms on generated customer embedding vectors to predict customers' intent. One of them is Isolation Forest, which is an unsupervised machine learning technique used for anomaly detection. Isolation Forest, a proficient method to determine anomalies, assumes that the instances which move away from the data center are anomalies. It creates binary trees and gathers them by random sampling for a dataset [35]. The main target of Isolation Forest is to determine unusual samples which are different from the rest of the data. This method was chosen because our dataset is quite unbalanced (foreign exchange traders are in the minority in this dataset), and the Isolation Forest does a very good job of finding outliers. As the rate of those who create foreign exchange events is around 15% compared to those who do not, we have chosen to treat customers who perform FX transactions as outliers.

One of the reasons causing bias in the evaluation of the results is data imbalance for classification especially. In the case of one of the classes has a higher percentage than the percentage of another class data imbalance occurs [36].

The other algorithm applied to the prediction module is Xgboost. It is a supervised machine learning algorithm. It boosts operations to have accurate models. Boosting points the ensemble method of building many models in order, with each new model attempts for fixing the shortcomings in the former model. A decision tree is, addition of the each model to the ensemble, in tree boosting [37] (Fig. 5).

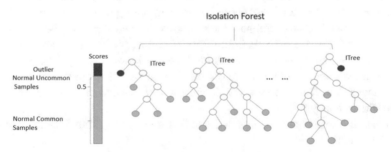

Fig. 5. Isolation forest structure

The Random under-sampling technique has been applied to create a balanced training dataset in which each sample has an equal probability of being chosen. Random sampling creates minority class size datasets from selection of samples randomly from the majority class. In our use case, minority samples are customers who execute foreign exchange transactions (Fig. 6).

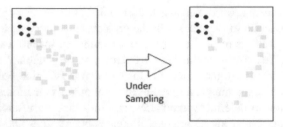

Fig. 6. Random under sampling

6 Experiments

For the experiments, we have used Word2Vec (CBoW) and DeepWalk algorithms as base models. These base model embeddings are fed directly into the prediction module without applying the proposed post-processing for the validation of the proposed framework. A comparison of prediction results under different vector sizes and normalization options, in which embedding results of base models in their plain form and the proposed frameworks' are utilized, are given below in Table 3, Table 4, and Table 5.

All models are trained on balanced and unbalanced datasets for different vector sizes. The proposed framework's prediction outputs are created by involving both aging and non-aging methodologies combined with frequency-based tendency analysis. After creating balanced and unbalanced datasets, training and test datasets follow the 80–20 rule. Our target task is to predict customers who execute foreign exchange transactions.

Just looking at accuracy is not enough when comparing model performance metrics. We have used the F1 score to compare the results as we tried to predict the minority class. Since the accuracy score predicts both minority and majority classes and the minority class is more difficult to predict, the results here can be misleading. The F1 score that is the harmonic mean of precision and recall is more sensitive to one of the two data having a low value. It is a more accurate indicator of prediction quality as the positive cases are the minority [36].

You can find experiment results in Table 3, Table 4, and Table 5 for the vector size of 100, 50 and 20.

Table 3. Experiment results for vector size of 100.

Vector Size 100		Word2Vec				DeepWalk			
		Unbalanced		Balanced		Unbalanced		Balanced	
Train(%80)- Test(%20)		Isolation Forest	Xgboost	Isolation Forest	Xgboost	Isolation Forest	Xgboost	Isolation Forest	Xgboost
		F1	F1	F1	F1	F1	F1	F1	F1
Aging	Frequence	10,35%	0,36%	22,67%	**24,24%**	12,56%	0,70%	22,07%	23,81%
	Inverse Frequence	12,82%	1,10%	**23,44%**	22,17%	14,19%	**1,41%**	22,53%	22,49%
	Average Frequence	10,37%	2,35%	22,60%	22,96%	13,53%	1,22%	22,22%	24,08%
	Customer Time	10,12%	0,55%	23,13%	23,19%	13,65%	1,06%	21,94%	**24,50%**
Non-Aging	Frequence	10,28%	0,37%	23,01%	21,70%	13,06%	0,88%	22,64%	23,91%
	Inverse Frequence	10,45%	**0,92%**	24,02%	23,02%	14,23%	0,71%	22,64%	**24,50%**
	Average Frequence	10,45%	0,73%	22,40%	22,30%	13,78%	0,88%	22,43%	23,82%
	Customer Time	10,17%	0,37%	23,46%	20,83%	13,65%	1,40%	22,35%	24,48%
Base Models		**14,51%**	0,73%	22,53%	23,17%	**14,67%**	0,81%	**23,57%**	24,20%

Table 4. Experiment results for vector size of 50.

Vector Size 50		Word2Vec				DeepWalk			
		Unbalanced		Balanced		Unbalanced		Balanced	
Train(%80)- Test(%20)		Isolation Forest	Xgboost	Isolation Forest	Xgboost	Isolation Forest	Xgboost	Isolation Forest	Xgboost
		F1	F1	F1	F1	F1	F1	F1	F1
Aging	Frequence	11,55%	0,55%	22,26%	22,73%	12,32%	0,53%	22,16%	24,11%
	Inverse Frequence	11,53%	0,73%	23,17%	22,49%	15,43%	0,18%	22,72%	23,51%
	Average Frequence	11,94%	0,55%	22,57%	22,96%	13,53%	0,36%	22,18%	24,65%
	Customer Time	11,43%	0,19%	22,73%	22,61%	12,95%	0,71%	22,16%	24,42%
Non-Aging	Frequence	12,1%	0,19%	22,84%	22,52%	12,63%	0,07%	22,25%	24,72%
	Inverse Frequence	12,1%	0,92%	24,01%	22,85%	13,70%	0,07%	23,20%	23,60%
	Average Frequence	10,9%	0,74%	23,39%	23,81%	13,09%	0,35%	22,17%	24,42%
	Customer Time	12,1%	0,91%	23,53%	21,80%	13,36%	0,71%	21,69%	24,42%
Base Models		14,04%	0,73%	22,54%	22,79%	14,67%	0,64%	23,67%	24,44%

Table 5. Experiment results for vector size of 20.

Vector Size 20		Word2Vec				DeepWalk			
		Unbalanced		Balanced		Unbalanced		Balanced	
Train(%80)- Test(%20)		Isolation Forest	Xgboost	Isolation Forest	Xgboost	Isolation Forest	Xgboost	Isolation Forest	Xgboost
		F1	F1	F1	F1	F1	F1	F1	F1
Aging	Frequence	11,33%	0,19%	22,58%	23,11%	13,51%	0,18%	22,49%	25,22%
	Inverse Frequence	11,82%	0,37%	23,68%	23,38%	15,44%	0,36%	23,49%	23,66%
	Average Frequence	11,10%	0,37%	22,69%	22,68%	12,77%	0,36%	21,88%	23,94%
	Customer Time	10,66%	1,11%	23,06%	21,34%	13,62%	0,36%	21,96%	24,72%
Non-Aging	Frequence	12,58%	0,19%	11,74%	21,18%	13,78%	0,35%	23,24%	24,05%
	Inverse Frequence	11,88%	0,56%	23,92%	22,53%	14,66%	0,18%	23,28%	23,63%
	Average Frequence	12,00%	0,37%	22,92%	22,80%	14,11%	0,89%	22,91%	24,75%
	Customer Time	12,00%	0,00%	22,84%	23,51%	13,33%	0,36%	21,76%	23,90%
Base Models		14,73%	0,19%	22,45%	21,92%	14,77%	0,22%	24,11%	24,55%

For Word2Vec Embedding;

- When the vector size was chosen as 100, the best performance in balanced data was for the aging, frequency-weighted Xgboost model. Likewise, in the unbalanced dataset, the Isolation Forest model, which generated the base Word2Vec embedding approach provided the best performance.
- When the vector size was selected as 50, the best performance in balanced data was for the non-aging, average frequency-weighted Xgboost model. Likewise, in the unbalanced dataset, the Isolation Forest model, which generated the base Word2Vec embedding approach provided the best performance.
- When the vector size was selected as 20, the best performance in balanced data was for the aging, inverse frequency-weighted Isolation Forest model. Likewise, in the unbalanced dataset, the Isolation Forest model that generated the base Word2Vec embedding approach provided the best performance.

For DeepWalk Embedding;

- When the vector size was chosen as 100, the best performance in balanced data was for the aging, time-weighted Xgboost model. Likewise, in the unbalanced dataset, the Isolation Forest model, which generated the base DeepWalk embedding approach provided the best performance.

- When the vector size was selected as 50, the best performance in balanced data was for the non-aging, frequency-weighted Xgboost model. Likewise, in the unbalanced dataset, the aging, inverse frequency-weighted Isolation Forest model provided the best performance.
- When the vector size was selected as 20, the best performance in balanced data was for the non-aging, average frequency-weighted Xgboost model. Likewise, in the unbalanced dataset, the aging, inverse frequency-weighted Isolation Forest model provided the best performance.

Experiment results show that our proposed framework has been more successful in the balanced dataset than base embedding methods. In the unbalanced dataset, the Word2Vec base embedding method was more successful. However, our proposed framework has been more successful than base embedding methods when we use DeepWalk algorithm in the core embedding module. We can also deduce that. The aging approach gave better results than the base embedding approaches in the Word2Vec embedding method in the balanced dataset. Overall, results show that the customer embedding methods generated by the proposed framework outperform the core embedding methods in lower dimensions. As hypothesized, our framework is able to capture better time-dependency and repetitiveness of customer behaviors using footprints in mobile banking.

7 Discussion and Future Work

In this study, we have proposed a customer embedding framework to represent banking customers better using the temporal and repetitive dynamics of navigational footprints generated in the mobile banking application. We have shown that recent embedding techniques may not be as effective in customer representation learning where context is time-dependent and contains rich information regarding customer preferences, as in the case of digital financial footprints in mobile banking.

We have designed our framework in that it leverages the state-of-the-art embedding algorithms like DeepWalk and Word2Vec and extends the generated embedding representations by applying the aging factor and the overall tendency of customers based on visit frequency.

In our intent prediction experiments, we have used the generated embedding vectors in a prediction use case to predict customers' tendency toward a financial product. According to the results, the proposed framework has successfully captured customers' intent compared to the based models in almost all test scenarios.

As an extension of this study, we plan to enhance the framework so that it can learn from multi-model data to accurately predict the intention of the customer towards a diverse set of products. Also, we want to extend the framework with time-series prediction capabilities using RNN-based algorithms to predict when the customer will use a specific product: in the next session, within the following ten sessions, etc.

Acknowledgment. This study was supported by Fibabanka Research and Development Center. The authors thank Fibabanka, which is a corporate banking company in Turkey, for making this study possible.

References

1. Chitsazan, N., Sharpe, S., Katariya, D., Cheng, Q., et al.: Dynamic customer embeddings for financial service applications, p. 1 (2021). https://doi.org/10.1145/3442381.3450020
2. Wang, B., et al.: Evaluating word embedding models: methods and experimental results, p. 1 (2019). https://doi.org/10.1017/ATSIP.2019.12
3. Levy, O., Goldberg, Y.: Neural word embedding as implicit matrix factorization. In: NIPS, pp. 2177–2185 (2014)
4. Perozzi, B., et al.: DeepWalk: online learning of social representations, pp. 701–710 (2014). https://doi.org/10.1145/2623330.2623732
5. Mikolov, T., et al.: Efficient estimation of word representations in vector space, pp. 4–5 (2013). https://doi.org/10.4236/apm.2016.66030
6. Ghannay, S., Favre, B., Esteve, Y., Camelin, N.: Word embeddings evaluation and combination, Slovenia, pp. 300–305. ELRA (2016). https://doi.org/10.1109/EUSIPCO.2015.736 2668
7. Penington, J., et al.: GloVe: global vectors for word representation, USA, pp. 1532–1543 (2014). https://doi.org/10.3115/v1/D14-1162
8. Narayanan, A., et al.: Graph2vec: learning distributed representations of graphs, Singapore, pp. 3–4 (2017). https://doi.org/10.48550/arXiv.1707.05005
9. Grover, A., Leskovec, J.: Node2vec: scalable feature learning for networks. In: SIGKDD, p. 8 (2016). https://doi.org/10.1145/2939672.2939754
10. Ribeiro, L., et al.: Struc2vec: learning node representations from structural identity. In: SIGKDD, pp. 385–394 (2017). https://doi.org/10.1145/3097983.3098061
11. Wu, L., et al.: Word mover's embedding: from word2vec to document embedding, Belgium, pp. 4524–4534 (2008). https://doi.org/10.18653/v1/D18-1482
12. Jang, B., et al.: Word2Vec convolutional neural networks for classification of news articles and tweets, p. 4 (2019). https://doi.org/10.1371/journal.pone.0220976
13. Zuckerman, M., et al.: Using graphs for word embedding with enhanced semantic relations, Hong Kong, pp. 32–41 (2019). https://doi.org/10.18653/v1/D19-5305
14. Rossi, R.A., et al.: Deep inductive graph representation learning, pp. 438–452. IEEE (2020). https://doi.org/10.1109/TKDE.2018.2878247
15. Makarrov, I., et al.: Survey on graph embeddings and their applications to machine learning problems on graphs, p. 39 (2020). https://doi.org/10.7717/peerj-cs.357
16. Rizi, F.S., et al.: Properties of vector embeddings in social networks, Germany, pp. 109–124 (2017). https://doi.org/10.3390/a10040109
17. Li, J., Zhu, J., Zhang, B.: Discriminative deep random walk for network classification, p. 1006. In: ACL (2016). https://doi.org/10.18653/v1/P16-1095
18. Hou, M.., et al.: Network embedding: taxonomies, frameworks and applications, pp. 4–5 (2020). https://doi.org/10.1016/j.cosrev.2020.100296
19. Tasgetiren, N., Aktas, M.S.: Mining web user behavior: a systematic mapping study. In: The 22nd International Conference on Computational Science and Its Applications (ICCSA 2022), Malaga, Spain (2022)
20. Olmezogullari, E., et al.: Pattern2Vec: representation of clickstream data sequences for learning user navigational behavior. CCPE Special Issue (2021). https://doi.org/10.1002/cpe.6546
21. Olmezogullari, E., et al.: Representation of click-stream DataSequences for learning user navigational behavior by using embeddings. In: IEEE Big Data, User Understanding from BigData Workshop (2020). https://doi.org/10.1109/bigdata50022.2020.9378437
22. Oz, M., et al.: On the use of generative deep learning approaches for generating hidden test scripts. IJSEKE J. **31**(10), 1447–1468 (2021). https://doi.org/10.1142/s0218194021500480

23. Oguz, R.F., et al.: On the use of deep learning approaches for extracting information from large scale graph data: case study on automated UI testing. In: Euro-Par 2021, Lizbon, Portugal (2021)

24. Uygun, Y., et al.: On the large-scale graph data processing for user interface testing in big data science projects. In: 6th International Workshop to Improve Big Data Science Project Team Processes (2020). https://doi.org/10.1109/bigdata50022.2020.9378153

25. Erdem, I., et al.: Test script generation based on hidden Markov models learning from user browsing behaviors. In: IEEE Big Data 2021, Florida, USA (2021). https://doi.org/10.1002/cpe.6546

26. Python official website. www.python.org

27. Python Gensim library official website. pypi.org

28. Python Networkx library official website. networkx.org

29. Python Pandas library official website. pandas.pydata.org

30. Python Numpy library official website. numpy.org

31. Python Scikit-Learn library official website. scikit-learn.org

32. Python Xgboost library official website. xgboost.readthedocs.io

33. Nguyen, G.H., Lee, J.B., et al.: Continuous-time dynamic network embeddings. In: WWW 2018, pp. 969–976 (2018) https://doi.org/10.1145/3184558.3191526

34. Xiang, Y., et al.: Time-sensitive clinical concept embeddings learned from large electronic health records. In: CHIP, China (2018). https://doi.org/10.1186/s12911-019-0766-3

35. Sadaf, K., Sultana, J.: Intrusion detection based on autoencoder and isolation forest in fog computing, pp. 167059–167068 (2020). https://doi.org/10.1109/ACCESS.2020.3022855

36. Kulkarni, A., et al.: Foundations of data imbalance and solutions for a data democracy, p. 4 (2021). https://doi.org/10.48550/arXiv.2108.00071

37. Mitchell, R., et al.: Accelerating the Xgboost algorithm using GPU computing, p. 2 (2017). https://doi.org/10.7717/peerj-cs.127

Do Lunar Cycles Affect Bitcoin Prices?

Ugurcan Erdogan$^{(\boxtimes)}$ ⓘ, Alperen Berk Isildar ⓘ, Tugba Gurgen Erdogan ⓘ,
and Fuat Akal ⓘ

Computer Engineering Department, Hacettepe University, Ankara, Turkey
ugurcanerdogan3306@gmail.com, alperenberkisildar@gmail.com,
tugba@cs.hacettepe.edu.tr, akal@hacettepe.edu.tr

Abstract. People may want to benefit from other areas besides scientific and financial-based methods when making investment decisions. For example, it is thought that the lunar cycles guide the investors in some way and this is reflected in the stock market prices. In this paper, we examined the most dominant coin on the cryptocurrency exchange, Bitcoin, and observed the effects of the lunar cycles on its price. We compared Bitcoin opening and closing prices at the beginning and end of the lunar cycles by using McNemar's test. As a result of our analyses, we concluded that the lunar cycles have no statistically significant impact on Bitcoin price changes.

Keywords: Cryptocurrencies · Lunar cycles · Data science

1 Introduction

Most people can often consult investment and stock market experts when making investment decisions. Frequently, the details of the position to be taken in the stock market are determined by concluding with fundamental and technical analysis, trends, patterns, and other interpretation tools. Sometimes, instead of benefiting from the contributions of financial instruments, they may also be interested in techniques or advice from other fields. A few of them are astrological events and related recommendations [1]. The positions of the planets in the universe and their relations with other planets, the movements of the moon around the earth, and the traits that the signs reveal in people at certain periods or enable them to show themselves dominantly are examples of these events or recommendations.

William Delbert Gann was a businessman who developed various technical analysis methods using geometry, astronomy, astrology, and ancient mathematics about 100 years ago and engaged in financial activities in this direction. He suggested various stock market analysis techniques based on these methods [2,3]. There was no crypto exchange yet in those years. Today, crypto traders try to catch certain trends and patterns by applying Gann's analyses to various coins.

Many people believe that the lunar cycles cause changes in the biological and mental structures of people and thus affect their investment decisions. Due to

F. P. García Márquez et al. (Eds.): ICCIDA 2022, LNNS 643, pp. 271–280, 2023.
https://doi.org/10.1007/978-3-031-27099-4_21

this astrological-based belief, there is an increase in the markets from the day after the full moon to the new moon. On the contrary, from the next day of the new moon until the full moon, the stock market is in decline. It is also mentioned that half moons can undertake the task of correcting or completing.

In this study, we investigated whether lunar cycles have any effect on investors' behavior leading to Bitcoin price changes. Bitcoin is the most dominant coin in the cryptocurrency markets. It has become an alternative instrument for investors. Due to its liquidity, volume, and relatively less volatile nature as compared to other coins, we selected Bitcoin to use in our study. Then, we obtained all Bitcoin prices for the periods between January 2020 and December 2021. Afterward, the two-week periods and dates consisting of the transition periods from the full moon to the new moon and from the new moon to the full moon were tabulated. To explain more clearly, the prepared dataset consists of three columns and 48 rows. Columns have the following dates and information, respectively; the day after the start of the cycle, the first day of the next cycle, and the cycle name so we can distinguish which cycle these dates relate to. In total, 48 rows contain information on two types of cycle transitions, each of which consists of 24 samples. Since the start and end date of these periods are known, Bitcoin price at the end of the period was subtracted from the start date price and it was determined that the price would increase or decrease in two weeks. The dataset for all these dates regarding the month was accessed via [4], a calendar application on the internet.

Finally, by labeling these periods in the determined direction, (similar to the predicted labels produced by machine learning models), we applied McNemar's test [5]. At the last step, it was observed whether they had a distribution that supported the said belief.

The rest of the paper is organized as follows: Sect. 2 presents background information on Bitcoin, lunar cycles, and McNemar's test and the related work. Section 3 provides an overview of the followed methodology during this data science project. Section 4 gives a detailed explanation of the analyses and the results. The last section discusses the findings of the analyses on Bitcoin data providing the overall conclusion and recommendations for future work.

2   Related Work

The crypto exchange has recently become widespread with people's access to the internet and has become a market that they refer to as an idea against the changing economies of countries after the pandemic. Investors who use cryptocurrencies as investment tools have adapted the tools used in stock exchanges, commodity exchanges, or various exchanges for their use to profit in easy and low-risk ways. Although these tools are numerous, there are indicators and analysis forms that contain serious mathematical calculations. In addition to all these, Gann's technical analysis methods, which managed to keep mathematical tools in it, basically deal with angles and trend lines drawn on the tables. Using these concepts, he provided investors with ideas about the direction of the stock market through time, price, and repetitive patterns. It has shown that the timings of the lunar cycles and the expected patterns to form between these cycles can offer unrealistic ideas about the future prices of Bitcoin.

The way the patterns are interpreted in the stock market and the types of forecasting tools are increasing day by day. Research on this is still limited in the cryptocurrency exchange. There is a lot of research on the interpretation of historically older stock markets. Especially in studies such as [6], which analyzes natural and unnatural events based on calendars, it has been tried to establish links between these events and prices. This study, which is based on the behavioral and psychological effects of the lunar cycles on humans and makes concrete claims in this direction, has concluded by examining the relationship between stock returns and lunar cycles.

Another work [7], examines the impact of lunar cycles on cryptocurrencies rather than stock returns. As a method, research was conducted on the "windows", based on the full moon instead of the 14-day period of this work. As a result, it was stated that there is no correlation between cryptocurrencies and lunar cycles. They argue that there is no consistent evidence for the impact of some famous calendar effects other than lunar cycles on cryptocurrencies as well. Also, Hamurcu [8] shows that, particularly in some weekly period samples, certain days of the week have significant effects on Bitcoin prices. In addition, in terms of volatility, the months of March and September also have special effects on this value according to this work. Lopez-Martin [9], on the other hand, examines the effect of the holy month of Ramadan on the return and conditional volatility values of some cryptocurrencies. While it has been shown that in this work, the month of Ramadan has no effect on the returns and conditional volatility values of some coins, it has been shown that there may be an effect on some values of other coins. Finally, Kumar [10] investigates the turn of the month effect on Bitcoin, Ethereum, and Litecoin, which are the most dominant coins in the crypto market.

Our contribution to literature and the main difference from similar works is the specific focus we put on determining the effects of the moon's movements on Bitcoin prices.

3 Methodology

In this work, a simple variation of the CRISP-DM model [11] is adopted as the methodology. Figure 1 depicts the overview of the implemented workflow.

3.1 Business Understanding

In the business understanding step, we studied the cryptocurrencies and lunar cycles to gain insights into the concepts that we will work with.

Firstly, we researched the cryptocurrency markets to find an appropriate coin to use in our study. We decided to use Bitcoin as it is the most dominant, prevalent, and robust one among all coins. Then, we searched for an API (application

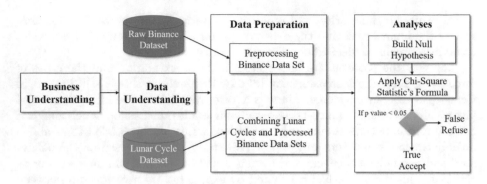

Fig. 1. The overview of CRISP-DM methodology

programming interface) to get the Bitcoin price data for a specific date range. We used the Binance API [12] because of its ease of use and Python support. API key on Binance was created and implemented in the source code. Afterward, we pulled Bitcoin prices from January 2020 to December 2021 via the Binance API. We refer to this dataset as the Bitcoin dataset.

Secondly, we investigated lunar cycles. We tried to understand how we can describe and use them. We found out that lunar cycles are identified by start and end dates, durations of periods, and cycle names. The names of the cycles were also very important to know how these cycles are expected to affect Bitcoin. Looking at the applications of Gann's technical analysis and the posts of social media influencers, there is an increase in the markets from the day after the full moon to the new moon. On the contrary, from the next day of the new moon until the full moon, the stock market is in decline. It is also mentioned that half moons can undertake the task of correcting or completing. With this situation, it has also been determined how the data will be separated. We created a lunar cycle dataset ourselves since we could not find a publicly available one on Internet. To achieve this, we used a public calendar application [4] where the details of the past lunar cycles are available.

3.2   Data Understanding

In the data understanding step, we explored the datasets we obtained for further understanding.

The lunar cycles dataset consists of 48 rows and three columns as listed in Table 1. Each row represents a lunar cycle period, which contains the day after the start of a cycle, the last day of the cycle, and the cycle's name. Each cycle lasts two weeks. The entire data set contains data for two years. The first five entries are shown in Table 2. For example, the starting phase name for the first row is Full Moon. The date of the next day after the start of this cycle is January 11th, 2020, and the date of the last day of this cycle is January 25th, 2020.

The Bitcoin dataset consists of 709 rows and 12 columns. The processed Binance dataset's column names and types are listed in Table 3. We used six of

Table 1. Lunar cycles dataset column names and types.

Column name	Data type
Phase's next date	Datetime64[ns]
Next phase's first date	Datetime64[ns]
Starting phase name	str

Table 2. Lunar cycles dataset's first five rows.

Cycle's next date	Next cycle's first date	Starting cycle name
11.01.2020	25.01.2020	Full moon
26.01.2020	09.02.2020	New moon
10.02.2020	23.02.2020	Full moon
24.02.2020	09.03.2020	New moon
10.03.2020	24.03.2020	Full moon

them which are `Open Time`, `Open`, `High`, `Low`, `Close` and `Close Time` columns. Each row represents general price information about Bitcoin for a day which contains which day it is, named `Open Time`, Bitcoin's first price when that day is started, named `Open`, the highest price of that day, the lowest price of that day, the last price when that day is over, named `Close`, and the closing date which is the same as open date. The first five entries are shown in Table 4.

For example, the open date of the first row is 11.01.2020, Bitcoin's first price is $8198.86 and the last price is $8020.01 of that day. The highest price is $8286.34 and the lowest price is $8003.16 during that day. The close time is the same as open date which is 11.01.2020.

Table 3. Processed Binance dataset's column names and types.

Column name	Data type
Open time	Datetime64[ns]
Open	Float64
High	Foat64
Low	Float64
Close	Float64
Volume	Float64
Close time	Datetime64[ns]
Quote asset volume	Float64
Number of trades	Int64
TB base volume	Float64
TB quote volume	Float64
Ignore	Object

Table 4. Bitcoin dataset's first five rows.

Open time	Open	High	Low	Close	Close time
11.01.2020	8198.86	8286.34	8003.16	8020.01	11.01.2020
12.01.2020	8020.01	8197.00	7960.00	8184.98	12.01.2020
13.01.2020	8184.97	8196.00	8055.89	8110.34	13.01.2020
14.01.2020	8110.34	8880.00	8105.54	8810.01	14.01.2020
15.01.2020	8814.64	8916.48	8564.00	8821.41	15.01.2020

3.3   Data Preparation

In data preparation step, we performed the necessary conversions and combined the lunar cycles and Bitcoin datasets properly to construct our final dataset to be used in the statistical testing.

In the Binance dataset, the values of the `Open Time` and `Close Time` columns were of type int64. We converted them in to DateTime objects. Also, `Open`, `High`, `Low` and `Close` were non-numeric columns. We converted them to floats. In the lunar cycles dataset, we did not require any further processing as we prepared that dataset ourselves. Also note that, there was no missing values in both datasets.

Afterward, we combined both datasets and do the necessary transformations. To achieve this, we created a new data frame with two columns by using the lunar cycles begin and end dates. We named those columns `Open Date` and `Close Date`, respectively. We, then, looked at the Bitcoin price changes for those dates at each row. We added two more columns to the data frame named `Status` and `Lunar Cycle`. If the Bitcoin price on Close Date was higher than the price at Open Date, we set the value of Status to "1" for that row. If that price was lower than the price at Open Date, we set the value to "0". We set the values in the Lunar Cycle column to "1" or "0" according to the expected price change value for each date range, using the information from the Lunar Cycle table.

At the end of the data preprocessing step, the final dataset seen in Table 5, consisted of 48 rows and four columns.

Table 5. Final dataset.

Open date	Close date	Status	Lunar cycle
2020-01-11	2020-01-25	1	1
2020-02-10	2020-02-23	0	1
2020-03-10	2020-03-24	0	1
2020-04-09	2020-04-23	1	1
2020-05-08	2020-05-22	0	1
...	...	...	...
...	...	...	...

4 Analyses

The McNemar's test is a non-parametric test for paired nominal data. This test is used when it is desired to find proportional variation for paired data. This test is also known as McNemar's Chi-Square test as the test statistic has a chi-square distribution. The chi-squared distribution can also be explained as follows. In probability theory and statistics, the chi-squared distribution with k-degrees of freedom is the distribution of a sum of the squares of k independent standard normal random variables. The degree of freedom, often abbreviated as "DF" or "DOF", refers to the number of independent variables or parameters of a system. In this study, two independent variables are being worked on. While one of them is the phases of the moon, the other is the knowledge of the increase and decrease in the value of Bitcoin. Therefore, we used McNemar's test for our analyses.

For McNemar's test to be carried out, two hypotheses must be put forward. These are the null hypothesis and alternative hypothesis. The opposite of the expected result is given for the null hypothesis in value, and the opposite of this null hypothesis is tried to be proven. The alternative hypothesis is the hypothesis that will be accepted when the null hypothesis is rejected. As the expected result of this study is that "moon phases have no effect on cryptocurrency investors", the null hypothesis is "moon phases have an effect on cryptocurrency investors".

McNemar's test returns a p-value result, and that p-value means that in null-hypothesis significance testing, the p-value is the probability of obtaining test results at least as extreme as the result observed, under the assumption that the null hypothesis is correct. A very small p-value means that such an extreme observed outcome would be very unlikely under the null hypothesis. In order for the null hypothesis to be accepted, the resulting p-value must be below 0.05.

The most understandable way to apply McNemar's test is to create a cross table with the data before the test is run. The cross table created from the data in this project is given in Table 6.

Table 6. An example of a cross table.

X Y	No	Yes	All
No	a	b	a+b
Yes	c	d	c+d
All	a+c	b+d	a+b+c+d

Cells b and c are used to calculate the test statistics; these cells are called "discordant." The formula of McNemar's test is explained in Eq. 1.

$$X^2 = \frac{(b-c)^2}{b+c} \tag{1}$$

The cross table created from the data in this project is given in Table 7."0" in the lunar phases section of the cross table means that the moon is in the new moon phase in that interval, and "1" means that the moon is at the full moon in that interval. "0" in the Status section means Bitcoin's value has fallen within that range, and "1" means Bitcoin's value has risen within that range. The expectation here is that Bitcoin will fall when the moon is in a new moon phase, and Bitcoin will rise when the moon is in a full moon phase.

Table 7. Lunar cycles and Bitcoin risen status cross table.

Lunar phase	0	1	All
Status			
0	14	10	24
1	10	14	24
All	24	24	48

When the moon phases and the exchange of Bitcoin are put together on a cross table, it is clearly seen that there is no correlation between the moon phases and the behavior of cryptocurrency investors, but for proof, it is necessary to refer to the above mentioned McNemar's test.

When McNemar's test is applied to the given statistics, the resulting formula is described in Eq. 2.

$$X^2 = \frac{(10 - 10)^2}{10 + 10} \tag{2}$$

The chi-square statistic's value obtained was found to be "0" as a result of this equation. And according to this statistic value, p-value will be found 1.0. This result gives a certain result that null hypothesis should be rejected because $p - value > 0.05$ means that null hypothesis should be rejected. In this situation, the hypothesis that put forward which is "Moon phases have no effect on cryptocurrency investors" should be correct according to McNemar's test.

5   Conclusion

In this paper, we created a dataset containing lunar cycle information in tabular form. Then, by merging this dataset with Bitcoin data fetched from Binance API, we created two data frames to see if the cycles of the moon do not have a considerable impact on the behavior of cryptocurrency investors by using McNemar's test.

Two different distributions can be used to perform the test. Since the data used in this study are independent, the chi-square distribution was used. For McNemar's test to be applied, a null hypothesis had to be put forward, and

"Moon phases have an effect on cryptocurrency investors." null hypothesis was put forward.

Continuity correction can also be applied when applying McNemar's test. While reaching the results, both the result of the application of the continuity test and the result of the non-applied version was obtained. The p-value obtained as a result of the tests is approximately "0.823" if continuity correction is applied and "1.0" if continuity correction is not applied. In both cases, whether we apply the continuity correction or not, the p-value of the test is greater than 0.05, which implies that we need to reject the null hypothesis.

Considering our findings, we statistically showed that cycles of the moon have no significant effect on the cryptocurrency investors.

We would like to point out that the results of this study may have different interpretations due to many variables. We can explain some of these variables as follows. We do not have any related work to compare our results directly with. However, we could further investigate the effect of lunar cycles on Bitcoin prices by integrating lunar phase information into a machine learning model to get a fair comparison. Then, we could explain whether the lunar cycles we use have a complementary or correcting effect on the increase or decrease patterns instead of a direct effect on the price. We could also reach different results by choosing other tests instead of the statistical test we use. It may be possible to turn some columns (for example, volume) that we did not use in the Bitcoin dataset into meaningful information and include them in the analysis by investigating the effect of price increase or decrease. We could select Bitcoin data for a broader range and perform the analyses for a bigger dataset. Finally, we could use other coins to repeat our experiments to see if we could generate similar results.

References

1. Mitchell, V.W.: Using astrology in market segmentation. Manag. Decis. **33**(1), 48–57 (1995)
2. Hyerczyk, J.A.: Pattern, Price and Time: Using Gann Theory in Technical Analysis. John Wiley & Sons, Hoboken (2009)
3. MacLean, G.: Fibonacci and Gann Applications in Financial Markets: Practical Applications of Natural and Synthetic Ratios in Technical Analysis. John Wiley & Sons, Hoboken (2005)
4. Moon Phases 2022 - Lunar Calendar. https://www.timeanddate.com/moon/phases/ (2020-2021). Accessed 01 Apr 2022
5. Lachenbruch, P.A.: Mcnemar Test. Statistics Reference Online, Wiley StatsRef (2014)
6. Yuan, K., Zheng, L., Zhu, Q.: Are investors moonstruck? Lunar phases and stock returns. J. Emp. Financ. **13**(1), 1–23 (2006), https://www.sciencedirect.com/science/article/pii/S0927539805000691
7. Qadan, M., Aharon, D.Y., Eichel, R.: Seasonal and calendar effects and the price efficiency of cryptocurrencies. Financ. Res. Lett. **46**, 102354 (2022). https://www.sciencedirect.com/science/article/pii/S1544612321003597
8. Hamurcu, C.: Examining the existence of day-of-week and month-of-year anomalies in bitcoin. Kırklareli Üniversitesi İktisadi ve İdari Bilimler Fakültesi Dergisi 11(1), 162–183 (2022)

9. Lopez-Martin, C.: Ramadan effect in the cryptocurrency markets. Rev. Behav. Financ. **14** (2022)
10. Kumar, S.: Turn-of-the-month effect in cryptocurrencies. Manag. Financ. ahead-of-print(ahead-of-print) (2022)
11. Chapman, P., Clinton, J., Kerber, R., Khabaza, T., Reinartz, T., Shearer, C., Wirth, R.: The crisp-DM user guide. In: 4th CRISP-DM SIG Workshop in Brussels in March. vol. 1999 (1999)
12. Binance API. https://www.binance.com/en/binance-api (2020–2021). Accessed 01 Apr 2021

Stock Price Prediction in Response to US Dollar Exchange Rate Using Machine Learning Techniques

Muhammad Atif Saeed[(✉)] and Akhtar Jamil

National University of Computer and Emerging Sciences, Islamabad, Pakistan
atifsaeedarts@gmail.com, akhtar.jamil@nu.edu.pk

Abstract. Stock market is a place where shares of public limited companies are traded. The stock exchange allows the end-user to buy and sell the shares and other security/financial instruments. Previously, different mathematical techniques are used to predict the movements in the stock markets or shares based on different market factor i.e. demographic factors, country's financial position, political factors. The purpose of the research is to find the relationship or impact of the US Dollar fluctuation on the stock price movement. For the analysis purpose use two different dataset, first, movement in the USD as independent data member and fluctuation in stock price as a dependent data member. For this research, deploy different machine learning analytical tools to analyze the prices and predict the near future price or fluctuation of the prices based on different factors of the market.

Keywords: Stock price · US Dollar exchange rate · Stock price prediction · Machine learning · Artificial Neural Network

1 Introduction

The stock market is one of the key trading platforms for every country, where different public sector companies list their shares. Shares are basically small commodities which are also called paper commodities and used to raise money from the general public [3]. Initially, the public sector companies will list their shares at a price called IPO or Initial Public Offering [4]. This is the actual or base price of the stocks which actually public sector company raised or collected from the general public by selling of the shares. After this, these stocks are the property of the sole person who buy these shares and they can sell these shares at any price on the stock exchange i.e. Pakistan Stock Exchange (PSX) [2].

Many factors involved in the stock price movement i.e. company internal news and events [6], sentiments of the investors, national or international interest rates, political movements and events, natural climates, demographics changes, changes in monetary policies and movement in the US Dollar exchange rate [6]. The main focus behind the research is to correlate the stock price and exchange rate and also

© The Author(s), under exclusive license to Springer Nature Switzerland AG 2023
F. P. García Márquez et al. (Eds.): ICCIDA 2022, LNNS 643, pp. 281–290, 2023.
https://doi.org/10.1007/978-3-031-27099-4_22

trying to figure out the impact of USD exchange rates on the stock price movement. For this purpose two different type of datasets are used i.e. stock market data as well as USD exchange rate historical data. If you want to predict the Pakistani stock market, you have to analyze or account for the above mentioned factors to predict stock prices using different machine learning techniques. Pakistani Stock market is one of the highly unpredictable markets because Pakistan is one of under developed country and many unseen factors are affecting the stock prices [6], but in this research main focus point is to correlate the impact of USD exchange rate on the stock prices as well as the historical data of the stock market to predict the stock price for short term as well long term.

In the last decade, Machine Learning (ML) is one of the powerful analytical and prediction tools [6] which is used to predict the stock price movement. Researchers are widely used different prediction or classification machine learning techniques for the analysis of financial data to predict stock prices and this will help the investors to make better decisions about the investment, where to invest, and how much they have to invest. This decision will help them to take better financial advantage or profit from their investment decision.

2 Related Work

Stock market prediction is a challenging topic because of time series data [16] and researchers stated that the stock market is not predicable because of many unpredictable factors involved in the price fluctuation i.e. social, political, and behavioral factors involved in stock price prediction [11]. Two major factors are involved in the prediction of the price one is the fundamental analysis which is based on companies performance, market position [4], expenses, and other market factors which decided the price movement [10] and the other one is technical market analysis which is based on previous stock prices, values and different past events which affect the stock price movements [10].

In the past, stocks were predicted by financial experts, but now data scientists also started to predict and design different prediction models for the prediction of future stock prices [10] using different machine learning, deep learning, and neural network techniques to develop the prediction models, to check their accuracy and enhance the performance of prediction models [7].

Financial market is the vast field and there is a lot of research was done in this field, these are the some of the major machine learning, deep learning and neural network techniques which are used by the research for the stock price prediction and analysis.

Linear Regression [13] [12] is one of the most used machine learning techniques which is used to predict the linear and independent values [13] and also used for minimizing the error with the help of gradient descent loss function [12]. While the uses Linear Regression you must have to carefully choose the attribute/features from your dataset because if you will not choose the correct data for the input the result will be useless for you [12].

LSTM is one of the mostly used technique in the literature, which is one of the most advanced and latest RNN versions [13] which is mostly used for classification, time-series analysis [10], and many other prediction related problems. The benefits behind the LSTM is there are two different memories, long-term memory is used to learn the weights of the different attributes, short-term memory refers to the internal cells [11], which stores the time-series data at different time intervals [10]. The LSTM model predicts better results and correlation of the non-linear time series in the stock prediction process [14].

Another mostly used machine learning technique in literature is SVM, which is one of the supervised learning techniques which is used for classification and regressions purposes [10]. SVM is used to find the optimal separating hyper-plane between the two different classes [5]. As per researchers, financial market have different nature of data which non-linear, so to predict a stock price or market trend, set a linear line and with the help of quadratic formula or equation, try to set the data on a linear line [11] with the help of linear optimality [11].

In [5], [10] authors also used some traditional classifiers for the prediction of the stock pricing i.e. Naïve Bayes Classifiers & KNN because these two classifiers are computationally faster. Naïve Bayes Classifiers are predicted the class membership based on the probabilities [5]. This is also a kind of supervised ma chine learning technique and each class is independently evaluated and separated from the set of distributing classes [10]. In KNN the data are divided into two different vectors [5] i.e. training dataset and testing dataset [10] and classify the selected data to a specific class on the basis of Euclidean distance [5]. The concept of ANN is used by most researchers for the modeling or processing of nonlinear data [9]. There are two different types of ANN techniques i.e. single layer technique or multi-layer technique (input layer output layer) and all the nodes are connected with each other and there are also two hidden layers [10]. Nowadays, ANN is widely used in neural network techniques in the field of financial prediction and financial planning [11]. In the ANN model, there are multiple layers, the first layer received the raw data as input and generates the output, that output data will be the input to the next layer, the next layer received this data as input and calculated some numbers and transfer it to the next layer, and so on [11]. This will help to predict better prices with a high level of accuracy and fewer errors [11]

The literature also indicated that Deep Learning techniques are widely used to predicts the stock price movement. While using deep learning to predict the stock pricing mostly four major stages are taken i.e. prepare the training data, selecting the appropriate network architecture, after the selection [3], the network will be trained, training of a network is an iterative process until get the desired results and the last stage is training improvement methods [11].

In literature, some of the researchers also used ensemble techniques [15], they suggest if the researcher will use a hybrid model or a combination of different machine learning, neural network techniques that will impact more and the result will be more accurate. [3,7] some of the researchers used machine learning techniques with sentiment analysis [7] because sentiments are one of the key factors in price movements. Some others used SVM with ensemble adaptive neuro-fuzzy inference system (ENANIS) [9,15,16] and another one which is combined with the feature weighed with SVM and KNN [8] and researchers suggest that price predicting with ensembling techniques are better results with higher accuracy compared to a single model or without neural networks [7].

3 Data Set

The data used in this study are two different types of data one is the stock opening, closing prices, and changes in the prices for the Pakistani Stock Exchanges. For the experimental purpose in this research used one of the major Pakistani fertilizer company i.e. Fauji Fertilizer's stock prices data from January 2011 to December 2021 Table 1. The second Type of data is the Dollar exchange rate on daily basis from January 2011 to December 2021 with open, closing, and changes in the rate Table 2.

Research data has been collected from [1,2] and applied different data preprocessing techniques to make the data suitable for our study and identify the suitable attributes for the prediction and analysis of the stock prices.

Table 1. Pakistan stock price data [2]

Open	High	Low	Close	Change
31,722.16	31,800.90	31,597.31	31,626.19	−21.38
31,874.78	31,958.58	31,612.55	31,647.57	−203.61
31,748.75	31,904.30	31,749.43	31,851.48	91.36
32,049.85	32,104.67	31,745.72	31,759.82	−288.86
32,166.21	32,390.77	32,044.01	32,048.68	−93.15

Table 2. Dollar exchange rate [1]

Price	Open	High	Low	Change
158.8	159	159.25	158.755	−0.03
158.75	158.75	159.215	158.59	0.06
158.65	160.05	160.075	158.49	−0.28
159.1	160.075	160.14	158.92	−0.03
159.15	159.3	160.24	158.85	0.14

4 Methodology

To predict the stock prices or Stock Exchange movement on timely and accurately manner is still a complex problem because there are many factors involved to predict the stock price i.e. many political factors, economic factors. But with the help of different machine learning techniques, we can better relate the past data with our current scenarios and predict the future stock prices.

4.1   Artificial Neural Network (ANN)

ANN is widely used by researchers to predict time series data or to model non-linear data. Firstly, this model was proposed by McCulloch and Pitts (1943). Neural Networks either Artificial Neural Network which is based on a single perceptron [6] or Recurrent Neural Network which is the base of LSTM or Deep Neural Network have been widely used for prediction or analysis purposes i.e. stock price predictions, bankruptcy, or different financial decision-making problems. In the neural network-based models, there are two different networks i.e. feedforward [14] networks that send the data or values from input layers to output layers with initiated weights and later on the backpropagation [14] of the loss to update the weights balance and minimize the errors [6] or loss and predict the stock price. ANN has the ability to tackle the complex nonlinearity within the data and can also handle noisy or missing data.

4.2   Support Vector Machine/Regression (SVM/SVR)

SVM/SVR is another powerful tool that is widely used to predict stock price movement or prediction. Before the popularity of convolution and deep learning models [6], SVM is widely adopted by researchers to predict or classify time series data. SVM/SVR also learns a hyperplane that maps the training sample with maximum distance space and distinguishes it with high dimensional space [14]. SVM models are suitable when you have multiple inputs and you need to accurately predict the stock prices or movement. This model also has the ability to tackle the overfitting problem.

4.3   Random Forest (RF)

Random Forest are decision tree based algorithms widely used for classification and regression problems. This model are firstly introduced by Leo Breiman in 2001 and later on widely used by researchers for the classification and prediction of problems. The working of the RFs is based on two different features i.e. bagging which is a bootstrapping and feature composition technique used to combine different weak learners to produce one strong learner [6] and the other feature is the random selection on each stage or node. RFs work well with continuous and discrete data but this model needs more computations and other resources because this model creates multiple trees that need a lot of resources. This model will also take more time to train than decision trees because in RFs there are multiple trees [14].

4.4   Decision Tree (DT)

The decision tree is one of the simplest model to classify the stock and researchers widely used this technique to predict the stock market or other classification or regression problems [8]. This model works by splitting the data into different subsets based on conditions or target variables or output parameters.

4.5   Logistic Regression (Logit)

Logistic regressions is one of the widely used linear models, in this model a logistic function is applied to predict the stock price movement either 0 or 1. This is able to tackle the complex nonlinearity within the data but sometimes this is also sensitive to outliers [8]. Logit is suitable if you want to classify the stock either their prices are moving up or down.

4.6   Linear Regression (LR)

Linear Regression [12] [11] is one of the most used machine learning techniques which is used to predict the linear and independent values [12] and is also used for minimizing the error with the help of gradient descent loss function [11]. While the uses Linear Regression you must carefully choose the attribute/features from your dataset because if you will not choose the correct data for the input the result will be useless for you [11].

5   Model Evaluation

In the last section, we applied multiple machine learning models and techniques to predict the stock price or future movement in the stock market, so there are different metrics that are used to evaluate the result of the model. There are two different types of metrics used.

5.1   Regression Metrics

Machine learning techniques that are used to predict the stock price or future continuous values. To measure such types of models with Mean Absolute Error (MAE), Root Mean Absolute Error (RMAE), Mean Squared Error (MSE), and Root Mean Squared Error (RMSE) [8].

Mean Absolute Error (MAE): Normally for regression nature problems or for continuous values mean absolute error is used. In this metric, the error prediction is the sum of the expected variable and the actual results, divided by the number of instances or points.

$$\sum_{i=1}^{D} |x_i - y_i| \tag{1}$$

Mean Squared Error (MSE): MSE is the square average error that is normally used to calculate the loss for minimum square regression [14]. Basically, MSE is the sum of the difference between the expected results and actual results divided by the number of instances or points [8].

$$\sum_{i=1}^{D} (x_i - y_i)^2 \tag{2}$$

Root Mean Square Error (RMSE): RMSE is used to calculate the difference between the two results i.e. [8] Expected results and actual results, so the values from the RMSE are very close to training and assessment datasets.

$$\sqrt{\frac{1}{n}\Sigma_{i=1}^{n}\left(\frac{d_i - f_i}{\sigma_i}\right)^2} \tag{3}$$

6 Experimental Results

In this research, trained five different Machine Learning model i.e. Linear Regression (LR), Logistic Regression (LGR), Decision Tree (DT), Random Forest (RF), Support Vector Machine (SVM) and Artificial Neural Network (ANN).

We have dataset of about 4000+ entries of records from last ten years with eight different features i.e. Date, Price, Open, High, Low, Volume, Change, USD Rate, but after checking the correlation among the features there are only four features that are highly correlated and pass the minimum threshold of 0.5, so we only used these four features to train and test the above mentioned models.

For the implementation purpose, divided the dataset into two different parts i.e. training and testing. So we have used 80% of the data for training of the models and the remaining 20% of data are used for testing purposes.

For implementation purposes, there are many different options available i.e. python, R, Matlab, and Java [14]. But in the previous decades' Python is one of the prominent language for the implementation of machine learning models. So We also used Python to the implement above six machine learning models to predict and classify the stock prices and movements [8]. Python has a lot of packages to implement these models i.e. Keras, TensorFlow, PyTorch, and scikit-learn. But in our paper, we are applying basic machine learning techniques, so we have used the scikit-learn package for these all experimental results. But for the deep learning models, TensorFlow and PyTorch are the best packages to implement.

For the implementation, the hardware is used core i7 intel CPU but for the deep learning bases models we need more computation resources, so we need GPU for the implementation and processing of multiple inputs. We have used three different metrics to test the performance of different models i.e. Mean Absolute Error (MAE) (1), Mean Squared Error (MSE) (2), Root Mean Squared Error (RMSE) (3) (Fig. 1).

After applying the above mentioned models, Random Forest is performing better on our data. The test scores for ANN is 94, for RF 99.95, SVM 78.53, Decision Tree 100, Linear Regression 78 and for Logistic Regression 87 (percent) so Decision Tree and Random Forest are performing better on the data (Table 3).

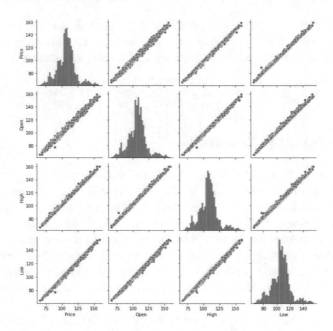

Fig. 1. Correlation plot using pair plot

Table 3. Results comparison table

Models	MAE	MSE	RMSE	R2	Test	Train
ANN	0.69	0.94	0.97	0.99	0.94	0.70
RF	0.50	0.55	0.74	0.99	0.995	0.996
SVM	0.40	0.80	0.89	0.79	0.785	0.768
DT	0.389	0.778	0.882	0.219	1.00	0.80
Linear	–	–	–	–	0.78	–
Logistic	–	–	–	–	0.87	–

7   Conclusion

The purpose of this study is to predict stock prices and check the impact of dollar prices on the stock movement so we have selected the share prices of one of the most powerful company from the Pakistan Stock Exchange and implemented six different machine learning models i.e. ANN, RF, SVM, DT, Linear & Logistic Regression. We have also tested the model with two different versions i.e. using continuous values as well as numeric values and we see models are working well on numeric values. So with different metrics we identified that RF and DT are the model which are performing well on the data (Fig. 2).

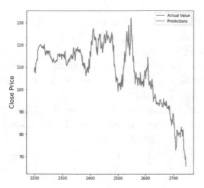 

Fig. 2. Predicted & Actual Price Plot using Random Forest & Predicted & Actual Price Plot using Decision Tree

8   Future Work

As we discussed above, many factors are involved in the stock price movement, and in this research, we have only accounted for the US Dollar exchange rates. So this factor enhances the accuracy of the stock price prediction. However, still there are many factors that are involved in stock price movement i.e. political factors, economic indicators, and market sentiments. So in our future work also account for these other factors to predict stock price movement.

References

1. Kaggle. http://www.kaggle.com/. Accessed 12 Feb 2022
2. Pakistan stock exchange limited. https://www.psx.com.pk/. Accessed 16 Feb 2022
3. Agrawal, M., Shukla, P.K., Nair, R., Nayyar, A., Masud, M.: Stock prediction based on technical indicators using deep learning model. Comput. Mater. Continua **70**(1), 287–304 (2022)
4. Akbar, M., Iqbal, F., Noor, F.: Bayesian analysis of dynamic linkages among gold price, stock prices, exchange rate and interest rate in Pakistan. Resour. Policy **62**, 154–164 (2019)
5. Ghazanfar, M.A., Alahmari, S.A., Aldhafiri, Y.F., Mustaqeem, A., Maqsood, M., Azam, M.A.: Using machine learning classifiers to predict stock exchange index. Int. J. Mach. Learn. Comput. **7**(2), 24–29 (2017)
6. Jiang, W.: Applications of deep learning in stock market prediction: recent progress. Expert Syst. Appl. **184**, 115537 (2021)
7. Jing, N., Wu, Z., Wang, H.: A hybrid model integrating deep learning with investor sentiment analysis for stock price prediction. Expert Syst. Appl. **178**, 115019 (2021)
8. Kumar, D., Sarangi, P.K., Verma, R.: A systematic review of stock market prediction using machine learning and statistical techniques. Materials Today: Proceedings (2021)
9. Lv, B., Jiang, Y.: Prediction of short-term stock price trend based on multiview rbf neural network. Computational Intelligence and Neuroscience **2021** (2021)

10. Nabipour, M., Nayyeri, P., Jabani, H., Shahab, S., Mosavi, A.: Predicting stock market trends using machine learning and deep learning algorithms via continuous and binary data; a comparative analysis. IEEE Access **8**, 150199–150212 (2020)
11. Nikou, M., Mansourfar, G., Bagherzadeh, J.: Stock price prediction using deep learning algorithm and its comparison with machine learning algorithms. Intell. Syst. Account. Financ. Manage. **26**(4), 164–174 (2019)
12. Pahwa, K., Agarwal, N.: Stock market analysis using supervised machine learning. In: 2019 International Conference on Machine Learning, Big Data, Cloud and Parallel Computing (COMITCon), pp. 197–200. IEEE (2019)
13. Parmar, I., et al.: Stock market prediction using machine learning. In: 2018 First International Conference on Secure Cyber Computing and Communication (ICSCCC), pp. 574–576. IEEE (2018)
14. Thakkar, A., Chaudhari, K.: A comprehensive survey on deep neural networks for stock market: the need, challenges, and future directions. Expert Syst. Appl. **177**, 114800 (2021)
15. Worasucheep, C.: Ensemble classifier for stock trading recommendation. Appl. Artif. Intell. **36**(1), 2001178 (2022)
16. Zhang, J., Li, L., Chen, W.: Predicting stock price using two-stage machine learning techniques. Comput. Econ. **57**(4), 1237–1261 (2021)

A Service of RSU Communication in Internet of Vehicles (IoV) in Urban Environment

Raneen I. Al-Essa$^{(\boxtimes)}$ and Ghaida A. Al-Suhail

Department of Computer Engineering, University of Basrah, Basrah, Iraq
engpg.raneen.imad@uobasrah.edu.iq

Abstract. Nowadays, Internet of Vehicles (IoV) is an emerging new technology that allows smart automobiles to communicate while traveling at high speeds via cloud platform and wireless access technologies. It arose from the concept of creating a vehicular ad hoc network (VANET) integrated with Internet of Things (IoT) to achieve specific applications, such as Intelligent Transport System (ITS) and smart cities. Therefore, IoV introduces a smart version of Vehicular Communication System (VCS) that vehicles and Road Side Units (RSUs) are essentially used for communication on highways and in urban cities. This study investigates the communication performance of the static Road Side Unit (RSU) which acts as an important component of IoV/VCS system in urban environment. A DSRC/WAVE stack mechanism is considered for beaconing and channel switching between CCH and SCH of control and service channels. Here, RSU also manages the Wave Short Messages (WSMs) and periodic beaconing, which are used to send WAVE service announcements. The simulation has been verified via implementing one RSU in urban city to evaluate the quality metrics of its communication mechanisms with vehicles nodes in terms of packet deliver rate (PDR), throughput, busy time and Signal-to-Noise Plus Interference Ratio (SNIR) lost packets. The results are obtained using OMNeT++ joined with Veins framework to make intercommunication with SUMO traffic road simulator.

Keywords: IoV · ITS · VANET · VCS · RSU · DSRC/WAVE · OMNeT · SUMO · Veins

1 Introduction

Recently, the new technological revolution has an impact on all industries and the desire for smarter, more connected devices is dramatically growing. Over this, the Internet of Things (IoTs) has widely deployed around us in numerous industrial applications. Among these, the Internet of Vehicles (IoV) is classified as a subset of Internet of Things (IoT). It is a new emerging data communication standard for smart vehicles and fixed or moving equipment such as road side units (RSUs), infrastructure mobile networks, pedestrians or sensors via cloud platform and wireless access technologies such as IEEE 802.11p, Wi-Fi, 4G/LTE/5G [1, 2]. IoV is a more advanced version of the Vehicular Ad hoc Network (VANET) that is primarily intended to provide safe

F. P. García Márquez et al. (Eds.): ICCIDA 2022, LNNS 643, pp. 291–303, 2023.
https://doi.org/10.1007/978-3-031-27099-4_23

driving and efficiency through real-time communication by connecting every vehicle to a mobile node or wireless router, resulting in a large network. In other words, each vehicle is viewed as a smart object with its own computing, storage, and networking capability based on hybrid of the VANET and IoT paradigms to improve road safety and vehicle security with providing services on demands of car drivers and passengers [3, 4]. More specifically, IoV and traffic safety implementations are used to significantly reduce accidents and fatalities on the streets, primarily by providing warning systems such as intersection collision, lane change assistance, emergency vehicle, wrong-way driver, crash risk, dangerous location, and signal contravention [5]. However, IoV or VCS might well disrupt network performance, for example, by modifying or inserting false information into the network, potentially resulting in life-threatening accidents to deal with any threat. Thus IoV/VANETs should meet three basic requirements: authentication, integrity, and conditional privacy [6, 7].

On the other hand, one of important components in IoV infrastructure are RSUs which are designed to play critical role in providing measurements, continuous communication range, permanent connectivity, processing, and cloud computing. For example, traffic checking RSUs, which are equipped with cameras to measure the vehicles speed on roads, are one of the most common RSUs nowadays. As well as, Toll collection RSUs, smart bus stations, adverts, and motion sensing RSUs are also other types of RSU. Basically, RSU is fixed device equipped with at least a network device using 75 MHz Dedicated Short Range Communication (DSRC)/ IEEE 802.1p with spectrum at 5.9 GHz. It serves a vital device in collecting and analyzing traffic data received from smart vehicles. However, RSUs communicate in two ways: they either use DSRC to connect with static or mobile RSUs and vehicles on the road, or they serve as gateways that are either directly or indirectly connected to the Internet. In general, the RSU deployment techniques are categorized into two types: (*i*) Static RSUs which cover short limited transmission ranges but with highly infrastructure cost; and (*ii*) Dynamical (mobile) RSU based on using some vehicles equipped to act as RSU, parked cars, bus lines and UAVs as RSUs [8].

Accordingly, this paper emphasizes the influence of the static RSU and vehicles communication on the quality of service performance in IoV/VCS system in urban environment. The DSRC/WAVE stack is CCH beaconing (continuous access) mode and channel switching (alternating access) mode between CCH and SCH of control and service channels.

The rest of paper is organized as follows: Sect. 2 focuses on some of recent related works. Section 3 presents briefly the IoV classifications. In Sect. 4, architecture and challenges are also described; and Sect. 5 focuses on DSRC/WAVE protocol mechanism. In Sect. 6, the proposed methodology is defined and its results analysis is depicted in Sect. 7. Finally, Sect. 8 summarizes the outcomes and future directions.

2   Related Work

A notable number of studies in comparing and evaluating RSUs performance in IoV/VCS have been conducted. Table 1 summarizes some recent studies in this field that evaluated the performance of vehicular networks using various simulators and investigated the effect of various network parameters. Among these studies, recent works are presented as follows. In [9], the communication reliability and quality of the Wi-Fi and DSRC was examined a realistic mobility simulation environment for the city of Casablanca with two different scenarios for 46 and 97 nodes and one static RSU. The performance was evaluated taking three metrics: throughput, packet delivery ratio and end-to-end delay using NS2.35 simulator. In Ref. [10], two scenarios were modeled one with a road accident and one without, the average number of received Base Safety Message and Wave Short Message packets from each node (vehicle) was measured over a given

Table 1. Existing related work

Reference	Year	Communication/simulator	Performance metrics	Type
Ref. [9]	2018	• Wi-Fi and DSRC • One RSU • NS2.35 Simulator	• End-to-end delay • Packet Delivery Ratio • Throughput	Analytic
Ref. [10]	2020	• WAVE • OMNET++	• Packet Error Ratio • Number of received packets	Analytic
Ref. [11]	2020	• Hybrid Vehicular Network (HVN) • Ad hoc Networks enabled by DSRC and infrastructure-based Cellular V2X (C-V2X) • Matlab	• Number of VHOs performed • Latency • Throughput • Packet Delivery Ratio (PDR)	Analytic
Ref. [12]	2020	• RSU-Fog • Matlab with NS3	• Bidders' utility • Average communication latency	Analytic
Ref. [13]	2021	• Warning message reception probability sent by accident vehicle, vehicle mobility and time delay • Matlab and NS2	• (0–160) number of RS • Relationship between RSU number and vehicles velocity, warning message probability	Analytic
Ref. [14]	2022	• A comparison of HNR, PTCCR and OPBR • NS2 Simulator	• E2E delay • PDR • Drop ratio	Analytic & Survey
Ref. [15]	2022	NaN	NaN	Survey

simulation time. Based on the simulation results, it is determined which of the used antennas has the best performance and which is most suitable and effective for use on the VANET network in an urban area, whereas OMNET++ simulator was used in this study.

On the other hand, Ref. [11] proposed a Hybrid Vehicular Network (HVN) architecture and protocol stack that manages to combine ad hoc networks enabled by DSRC technology and infrastructure-based (C-V2X) technologies. Ref. [12] proposed a Vickrey-Clarke-Groves method for controlling computing resources in a fog-enabled ITS, and used a distributed blockchain system for transaction verification among RSUs.

In 2021, Ref. [13] presents a practical model for computing the optimal RSU number in smart IoV based on a single and clusters of vehicles. The model incorporated three highway situations, and they also derived closed form expressions for various situations, which were validated by extensive simulations. For further details, IoV architectures, including layer types, layer functions, application areas, and protocols are also found in [14]. In addition, Ref [15] provided in-depth a survey of RSU deployment in IoV network including two types of RSU deployment: static and dynamic, which are based on vehicles mobility. The comparison study confirms that factors such as road shape, specificity of road segments such as accident-prone ones, network access methods, traffic characteristics, and vehicle allocation over time and space all have a significant impact on the performance of different RSU placement solutions.

In this regard, this study investigates the performance of static RSU and vehicles in IoV/VCS when an accident urban scenario is proposed via the DSRC/WAVE using channel switching and periodic beaconing.

3 IoV Classifications

The IoV includes many types of vehicular communications as shown in Fig. 1:

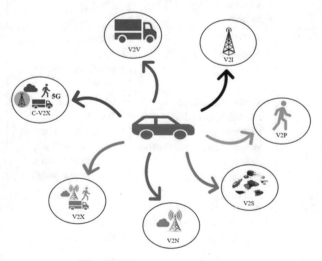

Fig. 1. IoV communication types

- **Vehicle-to-Vehicle (V2V):** Vehicles can communicate with other vehicles directly without the use of communication infrastructure; all vehicles collaborate and forward information on each other's behalf.
- **Vehicle-to-Infrastructure (V2I):** Vehicle communication pattern that enables vehicles to exchange data with hardware, software, and firmware components that support systems such as overhead RFID readers and cameras, traffic lights, signage, lane markers, streetlamps, and parking meters [16].
- **Vehicle-to-Personal devices (V2P):** Mobile vehicles could communicate directly with Pedestrians and bicyclists handled devices.
- **Vehicle-to-Sensors (V2S):** It refers to Automobile with sensor side communication.
- **Vehicle-to-Network (V2N):** Cellular communication between vehicles and a cellular infrastructure to aid in vehicular traffic functions a distinct use case to C-V2X, given that the latter only supports direct communication [17].
- **Vehicle-to-Everything (V2X):** Technology makes every mobile vehicle on the road smarter and safer by allowing them to connect with the traffic system, which includes other vehicles and infrastructure [18].
- **Cellular-Vehicle-to-Everything (C-V2X):** It differs from other types of vehicular networking technologies in that it uses the cellular network to support data exchange. The term "cellular" refers to how smart vehicles frequently send and receive data over 4G or 5G networks.

4   IoV Architecture and Challenges

The IoV architecture is composed of three layers, as shown in Fig. 2 [19].

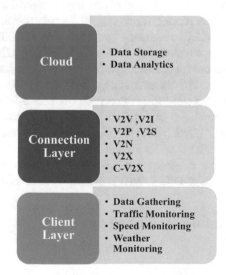

Fig. 2. IoV layers

On the other hand, IoV challenges are described in the following items:

Security: IoV security is a serious issue when an intrusion attack constitutes the IoV system. It can endanger safety since hackers can get control over vehicles, resulting in traffic accidents [20].

Reliability: A reliable connection is essential for the Internet of Things with autonomous vehicles, and network bottlenecks, DoS attacks, and communication malfunctions can all significantly disrupt the infrastructure's operation. Manufacturers must ensure that all nodes can transmit and receive data despite of vehicle speed or location.

Data Storage: Many smart vehicles currently communicate with one another and attempt to connect to the cloud on a regular basis. As a result, depending on a centralized cloud server to manage traffic and massive data storage across billions of nodes is inefficient for the IoV [21]. As a result, blockchain is used for solving centralized organization problems in the IoV, because it enables distributed real-time collection, handling, and storage of vehicle information [22].

Real Time Data Processing: Due to the large volume of data flowing between connected vehicles, insufficient storage or system delays can impede cloud computing and harm the system. In situations where only parallel data processing is insufficient, a good integration of parallel and sequential data processing is required. IoV necessitates parallel data processing and big data analytics. [23].

5 DSRC/WAVE Protocol Mechanism

Vehicular communication systems consist of one-hop or multi-hop communication that is carried out using various routing protocols. VCSs that operate on dedicated short-range communications (DSRC) frequency bands are critical enablers for the emerging intelligent transportation systems market (ITS). The IEEE 802.11P and IEEE 1609 specifications are the most mature set of DSRC/WAVE network standards which are intended to provide a comprehensive ITS communication stack. IEEE 802.11p governs dedicated short range communications (DSRC), which is an optimized version of IEEE 802.11a with a dedicated frequency band of 5.9 GHz and modified timing values to meet IoV requirements. The WAVE/ IEEE 1609.4 stack employs a split phase approach for multi-channel operation, in which all nodes switch to a well-known CCH and SCH at well-known interval times to exchange service announcements [24] which is called channel switching. The architecture of DSRC/WAVE communication is depicted in Fig. 3.

6 Proposed Simulation Methodology

6.1 Simulation Tools

The simulation of City Scenario run by using OpenStreetMap, GatcomSUMO and analyzed by OMNeT++ 5.5.1 simulator, Veins 5.0 frameworks with SUMO 1.6.0 as shown in Fig. 4 and Fig. 5.

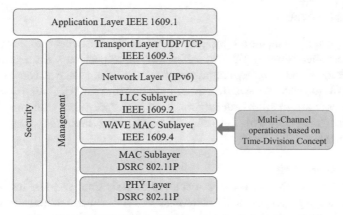

Fig. 3. DSRC/WAVE protocol communication architecture

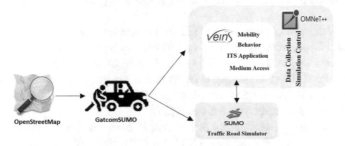

Fig. 4. Simulation configuration

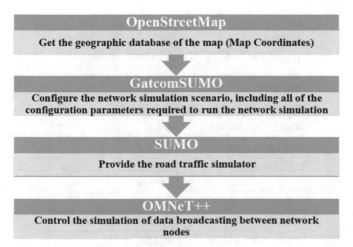

Fig. 5. Simulation architecture

6.2 Network Model

In this work, to implement the proposed network model the network setup is defined using simulation parameters as described in Tables 2 and 3, respectively. The network is established by taking Basrah city map coordinates from OpenStreetMap and downloading it in GatcomSUMO to generate the required configuration files for SUMO and Omnet simulation as follows (config.xml, *.luanchd.xml, *.net.xml, *poly.xml, *rou.xml, *.sumo.cfg, omnet.ini and *.ned) [25, 26]. Thus Fig. 6 illustrates the Basrah city map used for the proposed scenario.

Experimental Scenario

We have considered that an accident occurred at 100s time and last for 50s. During the accident the vehicles and RSU communicates for accident announcement to send information to other vehicles to stop or taking another path. In this scenario three communication configurations default, channel switching and beaconing are applied for three different numbers of vehicles 30, 60 and 90, respectively, as follows:

- **Default:** Nodes are sending only WSM for an accident announcement.
- **Beaconing** (*Continuous Access*)**:** It is performed by IEEE 802.1p for only CCH where each node resides (i.e.,100ms) 100% of the time on CCH for beaconing. In this mode, nodes broadcast Short Messages (WSM) based on Basic Safety Messages (BSM), for safety advertisement. BSM includes basic information for the vehicle's status like speed, position and predicated path, etc. [27].
- **Channel Switching** (*Alternating Access*)**:** It is performed by IEEE 1609.4 when a node switches every 50ms between CCH and SCH. Here, WSM are broadcasted with Wave service advertisement (WSA) over the concept of time division between CCH and SCH [28].

Table 2. OMNeT simulation parameters

Parameter	Value
OMNeT++	V 5.5.1
Veins	V 5.0
Simulation time	600 s
MAC layer type	IEEE 802.11p & IEEE 1609.4
Data rate	6 Mbps
Beacon generation rate	10 Hz
Transmission power	20 mW
Accident time	Start:100 s, Duration: 50 s

Table 3. Traffic parameters

Parameter	Value
SUMO	V 1.6.0
GatcomSUMO	V 1.0.4
Simulation area	$(1000 \times 2000)\ \text{m}^2$
Vehicle length	5 m
Vehicle width	2 m
Maximum speed	50 Kmph (city roads speed)
Acceleration/deceleration	9 Kmph/16 Kmph
# RSU	1
Network size	30, 60 and 90 nodes

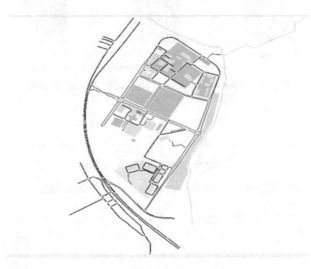

Fig. 6. Basrah city map

7   Analysis of Simulation Results

The performance of the proposed network has been obtained for PDR (Packet Delivery Ratio), throughput, the total busy time and SNIR lost packets (signal to noise plus interface ratio lost packets) for (30, 60 and 90) nodes by taking three modes default, channel switching and beaconing. The performance metrics are defined as follows:

- **PDR**: The packet delivery ratio is the ratio of successfully received packets by the destination to total packets sent by the source.
- **Throughput:** It is described as the total number of packets successfully received by the destination in a given period of time.

- **Busy Time:** The total time that the MAC treated the channel as busy.
- **SNIR Lost Packets:** The number of lost packets because of bit errors caused by noise interference at destination nodes [29].

The packet delivery ratio in beaconing is higher than the other two modes in all node density values, as shown in Fig. 7. The default configuration is similar to channel switching in 30 nodes density and beaconing in 90 nodes density, but it is significantly greater than channel switching and lower than beaconing in 60 nodes density.

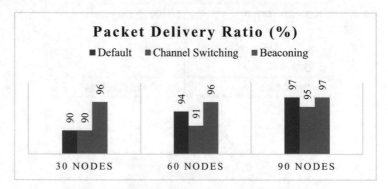

Fig. 7. Packet delivery ratio (%) vs number of vehicles

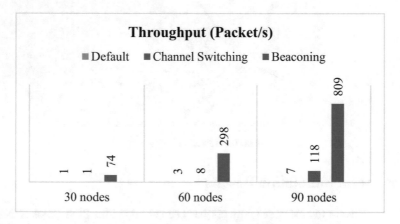

Fig. 8. Throughput (Pkt/s) vs number of vehicles

Furthermore, as illustrated in Fig. 8, the throughput value increases as the density of vehicle nodes increases due to more packets being sent. Besides that, beaconing communication has a higher throughput than default and channel switching. In addition, as observed in Figs. 9 and 10, the busy time and SNIR lost packets value increased as node density increased, and they registered higher values in beaconing than default and channel switching due to signal interference. In other words, it means that once the number of vehicles increases the number of interfered message increases.

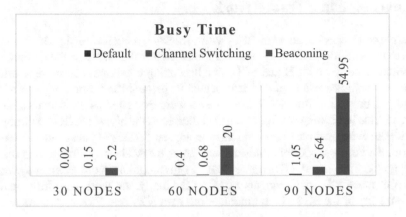

Fig. 9. Busy time (s) vs number of vehicles

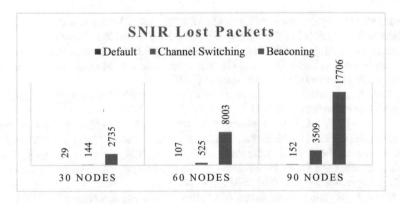

Fig. 10. SNIR lost packets vs number of vehicles

8   Conclusion and Future Work

In this paper, the performance of RSU with vehicles has been investigated in an urban scenario by using the Basra city map via the DSRC/WAVE stack as the default, channel switching between CCH and SCH, and beaconing configurations in three different numbers of vehicles when an accident occurs to prevent the traffic congestion. PDR, throughput, busy time, and SNIR lost packets were evaluated as performance metrics. As a result, the beaconing configuration is dedicated to the highest PDR and throughput, but it takes the longest busy time and has the highest SNIR lost packets in the communication. For Future work, this study can extend on a few ideas. (1) Integrating machine learning algorithm-based VCS framework. (2) Investigating more extensive scenarios for two or more RSU deployments in IoV. (3) Finally, the work may investigate the service quality of the 5G/LTE communications over V2V and V2R environments.

References

1. Kezia, M., Anusuya, K.V.: mobility models for Internet of Vehicles: a survey. Wirel. Pers. Commun. **125**, 1857–1881 (2022). https://doi.org/10.1007/s11277-022-09637-7
2. Aramice, G.A., Miry, A.H. Salman, T.M.: Internet of Vehicles: architectures, protocols and communication technologies. In: 2022 2nd International Conference on Computing and Machine Intelligence (ICMI 2022), pp. 1–5 (2022)
3. Aljabry, I.A., Al-Suhail, G.A.: A simulation of AODV and GPSR routing protocols in VANET based on multimetrices. IJEEE **17**, 66–72 (2021)
4. Hussain, S.M., Yusof, K.M., Hussain, S.A., Khan, A.B.: An efficient interface selection scheme (DSRC/LTE) of vehicles for data dissemination enabling V2V communication to support Internet of Vehicles (IoV). In: Reddy, V.S., Prasad, V.K., Wang, J., Reddy, K.T.V. (eds.) Soft Computing and Signal Processing. AISC, vol. 1340, pp. 573–581. Springer, Singapore (2022). https://doi.org/10.1007/978-981-16-1249-7_54
5. Abbas, M.T., Muhammad, A., Song, W.-C.: Road-aware estimation model for path duration in Internet of Vehicles (IoV). Wirel. Pers. Commun. **109**(2), 715–738 (2019). https://doi.org/10.1007/s11277-019-06587-5
6. Alouache, L., Nguyen, N., Aliouat, M., Chelouah, R.: Survey on IoV routing protocols: Security and network architecture. Int. J. Commun. Syst. **32**, e3849 (2019)
7. Mansour, M.B., Salama, C., Mohamed, H.K., Hammad, S.: VANET security and privacy - an overview. Int. J. Netw. Secur. Appl. (IJNSA). **10**(2), 13–34 (2018)
8. Iqbal, R.: Challenges in designing ethical rules for infrastructures in Internet of Vehicles. Int. J. Adv. Comput. Sci. Appl. (IJACSA) **9**(3), 11–15 (2018). https://thesai.org/Downloads/Volume9No3/Paper_3-Challenges_of_Ethical_Rules_for_IoV.pdf
9. Fitah, A., Badri, A., Moughit, M., Sahel, A.: Performance of DSRC and WIFI for intelligent transport systems in VANET. Procedia Comput. Sci. **127**, 360–368 (2018)
10. Zherka, A.: VANET network performance analysis in the case of a road accident scenario, vol. 17 (2020)
11. Mir, Z.H., Toutouh, J., Filali, F., Ko, Y.-B.: Enabling DSRC and C-V2X integrated hybrid vehicular networks: architecture and protocol. IEEE Access **8**, 180909–180927 (2020)

12. Lee, Y., Jeong, S., Masood, A., Park, L., Dao, N., Cho, S.: Trustful resource management for service allocation in fog-enabled intelligent transportation systems. IEEE Access **8**, 147313–147322 (2020)
13. Zhang, H.: A practical method of optimal RSUs Deployment for smart Internet of Vehicle (2021)
14. Seth, I., et al.: A taxonomy and analysis on Internet of Vehicles: architectures, protocols, and challenges. Wirel. Commun. Mob. Comput. **2022** (2022). Article ID 9232784, 26 pages
15. Guerna, A., Bitam, S., Calafate, C.T.: Roadside unit deployment in Internet of Vehicles systems: a survey. Sensors **22**(9), 3190 (2022)
16. Kanthavel, D., Sangeetha, S., Keerthana, K.P.: An empirical study of vehicle to infrastructure communications - an intense learning of smart infrastructure for safety and mobility. Int. J. Intell. Netw. (IJIN) **2**, 77–82 (2021)
17. Tahir, M.N., Leviäkangas, P., Katz, M.: Connected vehicles: V2V and V2I road weather and traffic communication using cellular technologies. Sensors **22**(3), 1142 (2022)
18. Klingler, F. Dressler, F., Cao, J., Sommer, C.: MCB – A multi-channel beaconing protocol. vol. 36, pp. 258–269, ISSN 1570–8705 (2016)
19. Lone, F.R., Verma, H.K., Sharma, K.P.: Evolution of VANETS to IoV: applications and challenges. Tehnički glasnik **15**(1), 143–149 (2021)
20. Sharma, N., Chauhan, N., Chand, N.: Security challenges in Internet of Vehicles (IoV) environment. In: 2018 1st International Conference on Secure Cyber Computing and Communication (ICSCCC), December 2018, pp. 203–207. IEEE (2018)
21. Fadhil, J.A., Sarhan, Q.I.: Internet of Vehicles (IoV): a survey of challenges and solutions. In: 2020 21st International Arab Conference on Information Technology (ACIT), pp. 1–10 (2020)
22. Chen, W., Wu, J., Lin, H., Chen, W., Zheng, Z.: A secure and efficient blockchain-based data trading approach for Internet of Vehicles. IEEE Trans. Veh. Technol. **68**, 9110–9121 (2019)
23. Tuyisenge, L., Ayaida, M., Tohmé, S., Afilal. L.: Network architectures in Internet of Vehicles (IoV): review, protocols analysis, challenges and issues. In: Proceedings of the 5th International Conference, IOV 2018, Paris, France, 20–22 November 2018 (2018)
24. Eenennaam, M.V., Venis, A., Karagiannis, G.: Impact of IEEE 1609.4 channel switching on the IEEE 802.11p beaconing performance. In: 2012 IFIP Wireless Days, pp. 1–8 (2012)
25. Aljabry, I.A, Al-Suhail, G.A, Jabbar W.A.: A fuzzy gpsr route selection based on link quality and neighbor node in VANET. In: 2021 International Conference on Intelligent Technology, System and Service for Internet of Everything (ITSS-IoE) (2021)
26. Al-Essa, R.I., Al-Suhail, G.A.: Mobility and transmission power of AODV routing protocol in MANET. In: 2022 2nd International Conference on Computing and Machine Intelligence (ICMI), pp. 1–5 (2022)
27. Meyo, Z., Ii, N., Nlong, J.M., Ndoundam, R.: An Adaptation of DSRC protocol for V2V communications in developing countries: end-to-end delay evaluation. Int. J. Comput. (IJC) **39**, 1–13 (2020)
28. Song, C.: Performance analysis of the IEEE 802.11p multichannel MAC protocol in vehicular ad hoc networks. Sensors **17**(12), 2890 (2017)
29. Gama, O., et al.: Evaluation of push and pull communication models on a VANET with virtual traffic lights. Information **11**(11), 510 (2020)

A Comparative Study on Subdural Brain Hemorrhage Segmentation

Tuğrul Hakan Gençtürk[1]([✉]) [iD], İsmail Kaya[2] [iD], and Fidan Kaya Gülağız[1] [iD]

[1] Kocaeli University, 41001 Kocaeli, Turkey
hakan.gencturk@kocaeli.edu.tr
[2] Niğde Ömer Halis Demir University, 51240 Niğde, Turkey

Abstract. Brain hemorrhages are one of the most dangerous disease groups. If not detected early, it can lead to death or severe disability. The most common method used to detect bleeding is the evaluation of computed tomography (CT) images belonging to the bleeding area by specialist physicians. Considering the difficulty of access to neurosurgery specialists and the lack of expertise of other doctors in emergency intervention on the subject, there is a need for decision support mechanisms to assist physicians in the diagnosis and treatment process. Artificial intelligence-based systems to be used for this purpose can accelerate the diagnosis and treatment process while reducing the burden on physicians. In this study, the suitability of Mask Region-Based Convolutional Neural Network (Mask R-CNN), Cascade Region-Based Convolutional Neural Network (Cascade R-CNN), Mask Scoring Region-Based Convolutional Neural Network (MS R-CNN), Hybrid Task Cascade (HTC), You Only Look At Coefficients (YOLACT), Instances as Queries (QueryInst), and Sample Consistency Network (SCNet) methods, investigated for the problem of detection and segmentation of subdural brain hemorrhages. The performance of the methods was determined over the images in the CQ500 dataset. This is one of the few studies that perform segmentation of subdural cerebral hemorrhages using CT images from an open dataset. The results were evaluated according to Intersection Over Union (IoU) and Mean Average Precision (mAP) metrics. Experimental results showed that two methods could detect and segment subdural hemorrhages more accurately than the others.

Keywords: Deep learning · CNN · Instance segmentation · Subdural hemorrhage

1 Introduction

The phrase "silent pandemic" is used to characterize how the medical community perceives traumatic brain injury (TBI) [1] and subdural hematoma (SDH) can be classified as a subtype of TBI.

SDH is the accumulation of blood between the arachnoid membrane around the brain as well as the dura mater underneath the skull. There are three main types of SDH. These are acute, chronic, and subacute. Acute hematomas are hyperdense on CT scans compared to normal brain tissue, and they often develop after a severe head injury. Subacute hematoma is a hemorrhage with isodense as brain tissue that occurs after one week when the hemorrhage resorbs. Chronic hematoma is a slowly progressing hemorrhage that occurs in elderly patients, usually because of a rupture of the cerebral veins. It appears more hypodense than that of a normal brain on CT images because of the late diagnosis. Regardless of the type of bleeding, a buildup of blood inside the subdural space causes brain compression. This condition can lead to neurological problems, loss of consciousness, and death. Consequently, early detection of SDHs is essential for reducing mortality and enhancing treatment.

However, especially small SDHs are challenging to detect as the SDH sites are directly contiguous to the bone [2]. Today, general practitioners and emergency department specialists must interpret whole-body CT scans. This situation causes inadequacies, especially in places with no specialists with the widespread use of tomography. Knowing whole-body tomography is out of the education of emergency specialists and practitioners. For this reason, specialized radiologists are now being assisted by telemedicine. This can be done with a certain margin of error and delay due to the workload of radiologists who do not have clinical knowledge. For this reason, if a doctor with a good command of pathology is not encountered in the emergency room, it causes delays and errors in diagnosis and treatment.

Artificial intelligence-based intelligent systems are also present in medicine with the development of technology to support physicians in the diagnosis and treatment phase. In terms of TBI, it is possible to detect and segment the bleeding area using artificial intelligence systems. Thus, the total volume of hematoma can be measured quickly, and the data obtained with traditional methods can be quantified. An automated diagnostic system can accelerate identifying cases that emergency physicians need to focus on. Each of these conditions will accelerate the early diagnosis process and increase the chances of treatment. Artificial intelligence-based systems are expected to detect bleeding and present it regionally to the physician. This problem is called image segmentation and can be performed on CT and magnetic resonance (MR) images. The process of image segmentation refers to dividing images into several discrete areas based on their features. Recent studies in the medical field have shown that segmentation can successfully detect typical anatomical structures, tumors, and ischemic injuries [3–6]. Still, segmenting regions with low contrast tissue along their borders is challenging.

This study evaluated the performances of current object segmentation models in detecting and segmenting brain hemorrhages from CT images. For this purpose, Mask R-CNN [7], Cascade R-CNN [8], MS R-CNN [9], HTC [10], YOLACT [11], QueryInst [12], and SCNet [13] methods were applied. The following list describes the study's contributions.

- An open dataset containing subdural brain hemorrhages has been marked appropriately for segmentation.
- Application of current segmentation methods to the detection and segmentation problems of subdural hemorrhages provided.
- The shortcomings of current methods in detecting brain hemorrhages are listed, and the most appropriate method is determined.
- Pathology determination was made with fewer errors than current telemedicine methods.

The subsequent part of the study is structured as follows. After first section recent studies on hematoma segmentation and detection presented. The third section describes the methods used in the study. In the fourth part, the data set and assessment criteria are described, and the outcomes are compared. The article is completed with the conclusion part.

2 Related Works

As described previously, a manual review of CT scans leads to deficiencies in diagnosis and treatment. To prevent this situation, many CT-based hematoma segmentation algorithms have been proposed for intracranial hematoma detection.

Yao et al. [14] customized a convolutional network architecture, U-Net, for hematoma segmentation. It has been found that the final segmentation performance can be improved by removing the up/down sampling layers in the network architecture used. As a result of the experimental studies, the Dice score for hematoma segmentation was 0.62.

Tu et al. [15] developed a clustering algorithm using Fuzzy C-means and Active Contours Without Edges (ACWE) techniques. In the study preprocessing phase, the skull's bony regions were deleted from the CT scan. In the hematoma segmentation stage, clustering was performed first. They applied the ACWE method using the result of the clustering process as the initial contour. The Dice Score Coefficient (DSC) value obtained with the proposed approach is over 90% compared to manual results.

Liu et al. [16] proposed a segmentation model for bleeding images. They used U-Net architecture. The dataset contains 82 anonymous cases with five types of hemorrhages, including intraventricular, intraparenchymal, subarachnoid, epidural, and subdural. In the preprocessing stage, all data were converted into normalized gray-scale data. The noise was also added to the randomly resized data. The MIOU of the model was 0.8871, and the dice score was 0.9362.

Yao et al. [17] studied on a method to identify and segment hematoma regions by using CT images. In the study, all photos were segmented to be expressed in super pixels. The features obtained from the super pixels are fed to a Support Vector Machine classifier for hematoma region segmentation. Used dataset contains of 35 head CT scans which are gathered for TBI. As a result of the proposed method, a dice score value of 0.60 was obtained.

Farzaneh et al. [18] proposed a method for SDH segmentation and radiographic severity assessment by analyzing head CT scans. The study used U-Net model to detect features based on the data. An average dice score coefficient of 75.35% was obtained as an overall segmentation score, and a dice score coefficient of 79.97% was obtained for moderate and severe SDHs.

Yao et al. [19] introduced a novel convolutional neural network for segmenting hematoma pictures extracted from CT scans of the head obtained 24 h after damage. To do this, 120 CT images of various patients from the Experimental Clinical Treatment (PROTECT III) dataset were chosen at random. A radiologist manually labeled the boundary of the brain hematoma. A dice score coefficient value of 0.697 was obtained with the proposed hematoma segmentation network.

When the studies were analyzed, it was found that there is a limited number of datasets that can be used openly in this field. Therefore, most of the studies has been used problem-specific datasets. At the same time, marking the boundaries of the hematoma area is complex and time-consuming. For this reason, it was found that open datasets containing CT hematoma images only collect information about the presence and type of hematoma and do not include the boundaries of the area that has hematoma. This has led to a limited number of studies in this field.

Within the scope of the study, 500 images containing subdural hemorrhage in the CQ500 dataset were used. First, these images were marked by an expert neurosurgeon. Then, a comparison of different methods was made on the drawn images. Considering the lack of marked datasets, this study will fill an essential gap in this field.

3 Methods

This paper compares the performance of eight different architectures of seven other current segmentation methods in detecting subdural brain hemorrhages. In the following subsection of the article, descriptions of these methods and their main differences are given.

3.1 Mask R-CNN

Mask R-CNN [7] is an object segmentation method built on the Faster R-CNN [20] algorithm. Two steps comprise the Fast R-CNN. The first stage that proposes potential object bounding boxes is called Region Proposal Network (RPN). [21] The second phase is known as Region of Interest Pooling (RoIPooling). During this phase, features are taken from each candidate box and bounding box regression is carried out. Like Faster R-CNN, Mask R-CNN employs a two-step process. In the second step, a binary mask is simultaneously generated for each Region of Interest (ROI) to estimate the class and box bounders.

The network of Mask R-CNN used in this study consists of three modules. The first module performs feature extraction and ROI generation, consisting of ResNet50 + Feature Pyramid Network (FPN) + RPN. The second module fixes the sizes of the ROIs obtained from the first module and prevents misalignments. This module is called Region of Interest Align (RoiAlign). The third module can be referred to as mask acquisition. The fixed-size ROIs received from the second module are sent to the region segmentation network in this module. In this study, the Mask R-CNN method was run in two different architectures using Resnet-50 and Resnet-101 in the backbone network. Figure 1 shows a visual representation of the Mask R-CNN architecture used in this study.

3.2 Cascade R-CNN

Cascade Mask R-CNN [8] uses a cascade learning method that gradually improves the object position and the segmentation accuracy.

The first stage, as in Mask R-CNN, receives ROI features and performs classification and regression. Each predictor's output is considered input for the next predictor in the

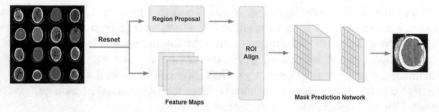

Fig. 1. General architecture of mask R-CNN

Cascade R-CNN architecture. Moreover, each regressor is augmented with the localization distribution predicted by the preceding regressor rather than the starting distribution itself. In the last stage of cascade networks, segmentation is conducted with regression and classification. The general architecture of the Cascade R-CNN is given in Fig. 2.

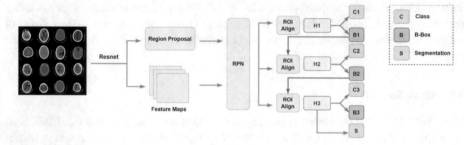

Fig. 2. General architecture of cascade R-CNN

In the network design described in this research, ResNet-50 serves as the network's backbone. From the CT image it gets as input, the ResNet-50 backbone builds feature maps. The generated characteristics are sent to the RPN, which forecasts candidate locations with potential. In the initial phase of Cascade R-CNN, the network output obtains RPN module recommendations and FPN module feature maps, then classifies.

3.3 Mask Scoring R-CNN

MS R-CNN [9] is an improved version of Mask R-CNN [7]. To increase the accuracy, the MS R-CNN adds a MaskIoU head at the network's output. This allows the generated masks to be scored. The mask scoring process consists of two stages. First, a mask is generated for the correct sample, followed by the development of MaskIoU for the backdrop [9].

MS R-CNN has four primary stages, the first 3 which are equivalent to Mask R-CNN. First, features are extracted using a backbone architecture built with ResNet-101 + FPN. In the second stage, candidate ROIs are proposed as the output of RPN. In the third step, features are extracted from each candidate ROI using RoIAalign, generating a segmentation mask output. In the fourth stage, MaskIoU, a score is generated between the predicted and actual mask. The overall architecture of the MS R-CNN is given in Fig. 3.

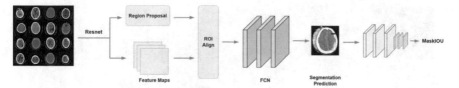

Fig. 3. General architecture of MS R-CNN

3.4 Hybrid Task Cascade

HTC [10] is object detection and segmentation architecture [12] that aims to determine the connection between the tasks of segmentation, bounding box detection, and mask estimation. In contrast to conventional segmentation techniques, this approach improves the information flow across mask branches by applying mask features out from preceding layer to the present stage. It aims to discover more contextual information by combining an additional semantic segmentation branch with mask and bounding box branches [10]. Figure 4 shows the general architecture of the HTC.

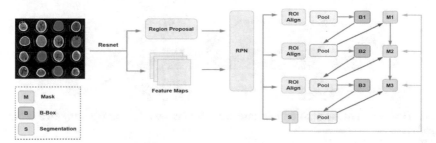

Fig. 4. General architecture of HTC

3.5 YOLACT

YOLACT aims to add a mask offshoot into the existing single-stage object detection model, as in the Mask R-CNN, Faster R-CNN relationship [11]. For this purpose, the segmentation process is divided into two parallel tasks. The first task can be summarized as generating a set of prototype masks, and the second task estimates the mask coefficients per sample. It uses the (Fully Convolutional Networks) FCN architecture to generate prototype masks and calculates a set of linear combination coefficients per sample. It then creates sample masks by linearly combining prototypes with mask coefficients [11]. The overall architecture of the YOLOACT is given in Fig. 5.

3.6 QueryInst

QueryInst [12] is a query-based segmentation method. It uses six dynamic mask heads and one query-based object detection output that can run in parallel. It can be built on

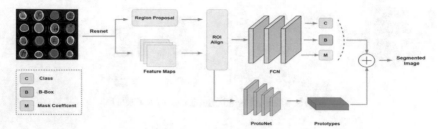

Fig. 5. General architecture of YOLACT

any query-based object detection network. Sparse Region-Based Convolutional Neural Network (Sparse R-CNN) [22] having six query stages is used by default.

In addition to object detection, QueryInst uses a query-based mask prediction head. With this dynamic design, information flows to the mask header in the queries directed by the parallel mask layer, and in this way, communication is strengthened for sample segmentation [12]. Figure 6 shows two of the six query stages in the overall architecture of the QueryInst method.

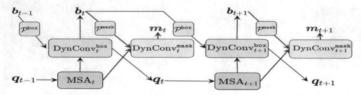

Fig. 6. Overview of QueryInst architecture with dynamic mask head (showing two of the six stages). [12]

3.7 SCNet

SCNet is a sample segmentation method that ensures that the intersection regions obtained during training are close to the distribution at inference time. SCNet is a cascaded object segmentation method [13]. Moreover, it tries to improve the shortcomings of currently used cascaded methods. Cascade Mask R-CNN [8], one of the most basic methods, employs more than one object detector for object segmentation. This way, it improves segmentation and object detection compared to non-gradual methods. HTC [10], proposed as an improved version of Cascade Mask R-CNN, offers an architecture that allows information to flow between the layers responsible for creating the segmentation mask.

Cascade Mask R-CNN uses all box outputs for mask predictions during training. At inference time, however, only the production of the last box stage is used for mask estimates. SCNet, on the other hand, shows that performing the training and inference phases consistently improves performance even further. In addition, many methods for detection and segmentation base their estimation on features extracted from a small region by a pooling layer. These layers allow the detector to focus on the relevant part of the image. However, objects are, in some cases, visually ambiguous when standing alone. To overcome this limitation, SCNet also examines the connection of each object with its global context before the final estimation [13]. Figure 7 shows the general architecture of the SCNet method.

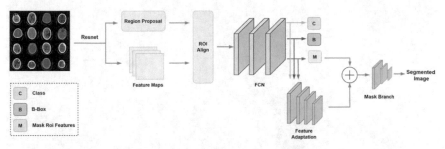

Fig. 7. General architecture of SCNet

4   Experimental Study

4.1   CQ500 Dataset

An openly available dataset, CQ500 [23], was used to train and evaluate the methods used in this paper. The CQ500 dataset contains CT images of findings such as bleeding, fracture, midline shift, and mass effect. The manual evaluation of three experts labels the images. The dataset contains 491 scanned CT sets and approximately 200,000 slices suitable for segmenting TBIs.

The article uses 500 CT images from 14 patients labeled by experts with SDH. 191 of these scans contained hemorrhage. The dataset does not contain images labeled as suitable for segmentation. Therefore, the scans having bleeding were first marked by an expert doctor. Figure 8 shows some of the marked scan images as an example. The data was then partitioned as 80% training and 20% testing. As a result of the partitioning, 152 scan images were used for training, and 39 images were used for testing the model. A server with Intel Xeon 2.2 GHz processor, 28 GB Ram, and 16 GB NVIDIA Tesla P100 GPU was used. The methods were implemented using Python language. Figure 8 shows the dataset's marking process on the sample scan images.

4.2   Performance Metrics

The IoU value is used as a criterion to assess the approaches' precision. The IoU gauges the extent to which two borders overlap. The method computes the overlap between the expected and actual boundaries (the actual object boundary). Traditionally, cutting-edge datasets utilize an IoU threshold equal to or higher than 0.5 to identify the prediction as either a true positive or a false positive. This research evaluated the accuracy of detection and segmentation using an IoU value of 0.5.

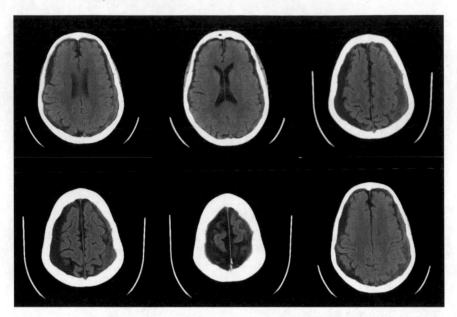

Fig. 8. Sample scans annotated from the CQ500 data set

The mAP measure was used to determine the accuracy. This statistic is the product of the accuracy and sensitivity measurements of the bounding boxes that have been detected. Calculating each class's average precision (AP) (SDH and background) independently and then averaging across classes yields the mAP result. A detection is only regarded as a genuine positive if the IoU exceeds a specified threshold. This research computes the mAP value for the bounding boxes determined for object identification and the segmentation masks. In the scope of the study, following formulas were used to calculate the accuracy.

$$IOU = \frac{A \cap B}{A \cup B} \tag{1}$$

IOU is the intercept ratio of predicted and expected segmentation outputs. A and B in the formula are the outputs of the model and the ground truth. For an image segmentation to be considered true or false, the IOU intersection ratio must be higher than a specified threshold value. After all images are evaluated according to the threshold value can the precision-recall be computed.

Precision refers to the total number of accurate instances generated by the model and is determined as follows:

$$P = \frac{\text{True positive}}{\text{True positive } + \text{ False positive}} \tag{2}$$

The number of total positive occurrences the model may generate is measured by a recall, which is calculated as follows:

$$R = \frac{\text{True positive}}{\text{True positive } + \text{ False negative}} \tag{3}$$

The area under the PR curve is used to determine the average accuracy. The Formula (4) shows the calculation of AP:

$$AP = \sum_{n=1}^{N} [R(n) - R(n-1)] \cdot maxP(n) \tag{4}$$

N is the total number of PR points determined. mAP is computed using the following equation:

$$mAP = \frac{1}{n} \sum_{n=1}^{n} AP_i \tag{5}$$

AP_i is the average precision of class i, and n is the number of classes.

4.3 Results and Discussion

Architecture details of the used methods are given in Table 1. In the table, the LR abbreviation represents the Learning Rate value for the methods. For (Stochastic Gradient Descent) SGD, an LR value between (0.02 and 0.002) gives better results than the traditional values between (00.1 and 0.0001) [24].

For this reason, the LR value was taken as 0.02 except for QueryInst, which uses the Adam optimizer, and YOLACT, which is stated in the paper to be trained with an LR of 0.001. Table 2 lists the accuracy values obtained for eight different architectures. A visual comparison of the accuracy values is also shown in Fig. 9. The images used in this study are challenging in terms of segmentation because the bleeding regions to

Table 1. Architecture details of the methods.

Method	Backbone	Neck	Head	Opt	LR
Mask R-CNN 50	Resnet50	FPN-RPN	FCBBoxHead FCNMaskHead	SGD	0.02
Mask R-CNN 101	Resnet101	FPN-RPN	FCBBoxHead FCNMaskHead	SGD	0.02
Cascade R-CNN	Resnet50	FPN-RPN	3*FCBBoxHead FCNMaskHead	SGD	0.02
MS R-CNN	Resnet101	FPN-RPN	FCBBoxHead FCNMaskHead	SGD	0.02
HTC	Resnet50	FPN-RPN	3*(FCBBoxHead HTCMaskHead)	SGD	0.02
YOLACT	Resnet50	FPN	YOLACTHead YOLACTProtonet	SGD	0.001
QueryInst	Resnet50	FPN-RPN	SparseRoIHead 6*(DIIHead-DynamicMaskHead)	Adam	0.0001
SCNet	Resnet50	FPN-RPN	3*SCNetBBoxHead SCNetMaskHead GlobalContextHead	SGD	0.02

Table 2. Accuracy values of the methods

Method	Detection mAP 50	Segmentation mAP 50
Mask R-CNN 50	0.8540	0.7960
Mask R-CNN 101	0.7930	0.7840
Cascade R-CNN	0.8470	0.8270
MS R-CNN	0.8670	**0.8780**
HTC	0.8520	0.8250
YOLACT	0.8490	0.8590
QueryInst	0.7770	0.8090
SCNet	**0.8710**	0.8080

be segmented have similar texture characteristics to the background brain tissue and are very small in size and irregular shape.

Nevertheless, each selected method has more than 78% accuracy in terms of detection and segmentation value. While SCNet was the most successful method of detecting the hemorrhagic region, MS R-CNN was the most successful method in segmentation. The highest success was obtained with the MS R-CNN method when evaluated in terms of both criteria. Figure 9 shows the success of the MS R-CNN method more clearly.

Table 2 and Fig. 9 list the results from the oldest to the most current methods. However, when we examine the results, it is seen that the QueryInst method, considered among the most up-to-date methods, has the lowest detection accuracy, and the segmentation success is lower than the other techniques. The most obvious reason for this situation is that the QueryInst architecture follows a query-based learning approach, unlike other methods. While query-based methods provide communication between objects, they sparse the connections within the architecture. While this method gives better results for some problems, it is inadequate for the data set used in this study and the situation we address. The fact that the number of queries used by the QueryInst architecture is much lower than the number of proposals utilized by Cascade Mask R-CNN and HTC may be a contributing factor to this conclusion. Because in this case, the training examples may be insufficient and the inferences to be obtained from the training data may produce results on a narrower scale.

Table 3 shows sample segmentation results for MS R-CNN and YOLACT, two methods with high detection and segmentation success. Table 4 shows sample segmentation results for Mask R-CNN 101 and QueryInst, two methods with low detection and segmentation success. The column given as GT in the tables refers to the Ground Truth image marked by the expert. When the results in Table 3 are analyzed, it is seen that MS R-CNN and YOLACT methods can detect the bleeding area with high accuracy. They only have deficiencies in detecting the bleeding's boundary points (start and end line). When the results in Table 4, especially those of the Mask R-CNN 101 method, are analyzed, it is seen that the method cannot clearly distinguish between brain tissue and hemorrhage compared to other methods. However, all methods were accurate enough to

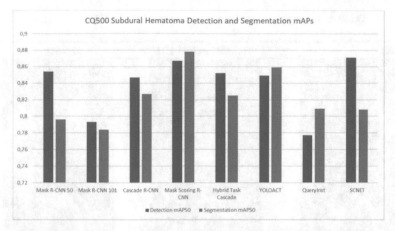

Fig. 9. Graphical representation of the values obtained from the methods

Table 3. Example results for methods with high segmentation success

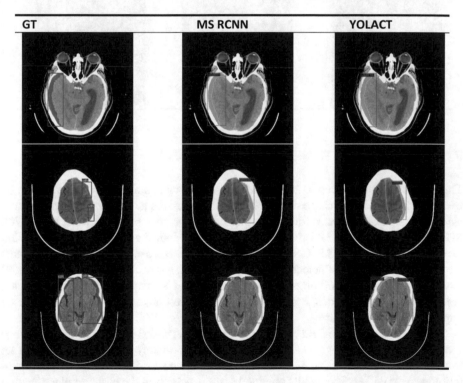

detect the presence of a hemorrhage. The results show that segmentation methods will provide great convenience to physicians regarding early detection and diagnosis.

Table 4. Example results for methods with low segmentation success

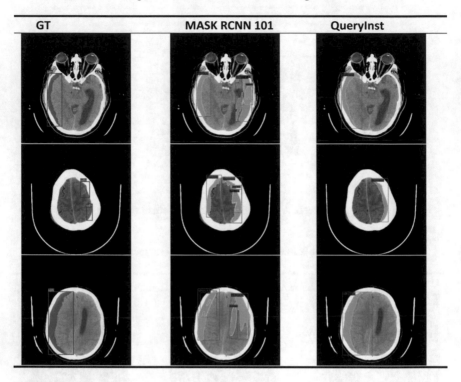

5   Conclusion

This article compares deep learning-based instance segmentation methods that can be used for subdural hemorrhage detection on CT images. As a result of the experimental study, the highest accuracy for hemorrhage detection was obtained from the 0.8780 SCNet method with a mAP value of 0.8710, and the highest accuracy for segmentation was obtained from the MS R-CNN method with a mAP value of 0.8780. When the results are considered in terms of detection and segmentation, it is seen that the most successful method is MS R-CNN. Although the QueryInst method is an up-to-date method, it failed compared to other methods, especially in the detection phase, due to the query-based approach in its architecture.

With this study, in the field of cerebral hemorrhage, which has limited studies due to the difficulties in the image labeling phase, both appropriately marking an open data set to segmentation problem and the determination of the most suitable method for segmentation were provided. The results obtained in the article form the basis for classification and segmentation studies on brain hemorrhages. This study will be a good starting point for new works in this field by comparing the results of current instance segmentation models for brain hemorrhage segmentation. In future works, an alternative approach specific to the problem and the dataset used can be developed to improve the accuracy values obtained from experimental studies. While analyzing the brain scans, metadata obtained from the DICOM images can be used to improve the accuracy. In future studies, the effect of this data on the result can be examined. Likewise, the effect of the window used in DICOM to RGB conversion can be studied.

References

1. Traumatic brain injury in the United States : emergency department visits, hospitalizations, and deaths, 2002–2006. https://stacks.cdc.gov/view/cdc/5571. Accessed 03 June 2022
2. Mikrogianakis, A., Valani, R., Cheng, A.: The Hospital for Sick Children Manual of Pediatric Trauma, 1stm edn. Lippincott Williams & Wilkins, USA (2007)
3. Zhang, Y., Liu, S., Li, C., Wang, J.: Application of deep learning method on ischemic stroke lesion segmentation. J. Shanghai Jiaotong Univ. Sci. **27**(1), 99–111 (2022)
4. Joskowicz, L., Cohen, D., Caplan, N., Sosna, J.: Automatic segmentation variability estimation with segmentation priors. Med. Image Anal. **50**, 54–64 (2018)
5. Zeineldin, R.A., Karar, M.E., Coburger, J., Wirtz, C.R., Burgert, O.: DeepSeg: deep neural network framework for automatic brain tumor segmentation using magnetic resonance FLAIR images. Int. J. Comput. Assist. Radiol. Surg. **15**(6), 909–920 (2020)
6. Isensee, F., Jäger, P.F., Full, P.M., Vollmuth, P., Maier-Hein, K.H.: nnU-Net for brain tumor segmentation. In: International MICCAI Brainlesion Workshop, pp. 118–132. Springer, Cham (2020). https://doi.org/10.1007/978-3-030-72087-2_11
7. He, K., Gkioxari, G., Dollár, P., Girshick, R.: Mask R-CNN. In: IEEE International Conference on Computer Vision, pp. 2980–2988. IEEE (2017). https://doi.org/10.1109/ICCV.2017.322
8. Cai, Z., Vasconcelos, N.: Cascade R-CNN: high quality object detection and instance segmentation. IEEE Trans. Pattern Anal. Mach. Intell. **43**(5), 1483–1498 (2019)
9. Huang, Z., Huang, L., Gong, Y., Huang, C., Wang, X.: Mask scoring R-CNN. In: IEEE/CVF Conference on Computer Vision and Pattern Recognition (CVPR), pp. 6402–6411. IEEE, Long Beach (2019). https://doi.org/10.1109/CVPR.2019.00657
10. Chen, K., et al.: Hybrid task cascade for instance segmentation. In: IEEE/CVF Conference on Computer Vision and Pattern Recognition, pp. 4974–4983. IEEE, Long Beach (2019). https://doi.org/10.1109/CVPR.2019.00511
11. Bolya, D., Zhou, C., Xiao, F., Lee, Y. J.: Yolact: real-time instance segmentation. In: IEEE/CVF International Conference on Computer Vision, pp. 9157–9166. IEEE, Seoul (2019)
12. Fang, Y., et al.: Instances as queries. In: Proceedings of the IEEE/CVF International Conference on Computer Vision, pp. 6910–6919. IEEE, Montreal (2021)
13. Vu, T., Kang, H., Yoo, C. D.: Scnet: training inference sample consistency for instance segmentation. In: AAAI Conference on Artificial Intelligence, pp. 2701–2709. AAAI, USA (2021)
14. Yao, H., Williamson, C., Soroushmehr, R., Gryak, J., Najarian, K.: Hematoma segmentation using dilated convolutional neural network. In: 40th Annual International Conference of the IEEE Engineering in Medicine and Biology Society (EMBC), pp. 5902–5905. IEEE, USA (2018). https://doi.org/10.1109/EMBC.2018.8513648
15. Tu, W., Kong, L., Karunamuni, R., Butcher, K., Zheng, L., McCourt, R.: Nonlocal spatial clustering in automated brain hematoma and edema segmentation. Appl. Stoch. Models Bus. Ind. **35**(2), 321–329 (2019). https://doi.org/10.1002/asmb.2431
16. Liu, H., et al.: Brain hematoma segmentation based on deep learning and data analysis. In: ITM Web of Conferences, pp.1–8. EDP Sciences, China (2021)
17. Yao, H., et al.: Brain hematoma segmentation using active learning and an active contour model. In: International Work-Conference on Bioinformatics and Biomedical Engineering, pp. 385–396. Springer, Cham (2019)

18. Farzaneh, N., et al.: Automated segmentation and severity analysis of subdural hematoma for patients with traumatic brain injuries. Diagnostics **10**(10), 773 (2020). https://doi.org/10.3390/diagnostics10100773
19. Yao, H., Williamson, C., Gryak, J., Najarian, K.: Automated hematoma segmentation and outcome prediction for patients with traumatic brain injury. Artif. Intell. Med. **107**, 101910 (2020). https://doi.org/10.1016/j.artmed.2020.101910
20. Ren, S., He, K., Girshick, R., Sun, J.: Faster r-cnn: towards real-time object detection with region proposal networks. In: Advances in Neural Information Processing Systems 28, pp. 91–99. MIT Press, Canada (2015)
21. Girshick, R.: Fast r-cnn. In: Proceedings of the IEEE international conference on computer vision, pp. 1440–1448. IEEE, Santiago, Chile (2015)
22. Sun, P., et al.: Sparse r-cnn: end-to-end object detection with learnable proposals. In: Proceedings of the IEEE/CVF Conference on Computer Vision and Pattern Recognition, pp. 14454–14463. IEEE, USA (2021)
23. Development and validation of deep learning algorithms for detection of critical fndings in head CT scans. https://arxiv.org/abs/1803.05854. Accessed 23 June 2022
24. Swa object detection. https://arxiv.org/abs/2012.12645. Accessed 23 June 2022

Deep Learning Based Spectrum Sensing Method for Cognitive Radio System

Ahmed T. Hussein[1] , Didem Kivanc[1](✉), Hikmat Abdullah[2] ,
and Muntaser S. Falih[3]

[1] Department of Advanced Electronics and Communication Technology, Istanbul Okan
University, Istanbul 34959, Turkey
`didem.kivanc@okan.edu.tr`
[2] Department of Information and Communication Engineering, Al-Nahrain University,
Baghdad 10072, Iraq
`Iraqhikmat.abdullah@coie-nahrain.edu.iq`
[3] Department of System Engineering, Al-Nahrain University, Baghdad 10072, Iraq

Abstract. Cognitive radio (CR) network is the promised paradigm to resolve the
spectrum shortage and to enable the cooperation in heterogeneous wireless networks in 5G and beyond. CR mainly relays on Spectrum Sensing (SS) strategy by
which the vacant spectrum portion is identified. Therefore, the sensing mechanism
should be accurate as much as possible, as long as the subsequent cognition steps
are mainly depended on it. In this paper, an efficient and blind SS algorithm called
Deep Learning Based Spectrum Sensing (DBSS) is proposed. This algorithm utilizes the deep learning approach in SS by using Convolutional Neural Network
(CNN) as a detector instead of energy thresholding. In this algorithm, the computed
energies of the received samples are used as dataset to feed the optimized CNN
model in both training and testing phases. The proposed algorithm is simulated
by MATLAB, the simulation scenarios divided into: CNN optimization (training)
and SS. The last scenario shows the detection ability of the proposed algorithm
for PU under noisy environment. The simulation results show that the proposed
algorithm reached high detection probability (Pd) with low sensing errors at low
SNR. In addition, high recognition ability to identifying Primary User (PU) signal
form noise only signal is achieved as well. Finally, the proposed algorithm is validated with respect to real spectrum data that supported by SDR in an experimental
signal transmission and reception scenario.

Keywords: Cognitive radio · Spectrum sensing · Deep learning · Probability of
detection

1 Introduction

With the rapid advancement of modern wireless communication technologies in both
military and civilian applications, the radio frequencies spectrum have become highly
scarce [1]. This dramatic change led to the development of a new technology namely
the Cognitive Radio (CR) system [2]. Cognitive Radio figured out as a trustworthy

© The Author(s), under exclusive license to Springer Nature Switzerland AG 2023
F. P. García Márquez et al. (Eds.): ICCIDA 2022, LNNS 643, pp. 319–331, 2023.
https://doi.org/10.1007/978-3-031-27099-4_25

and valuable solution for this issue through utilizing the unused spectrum portion in an efficient manner [3]. CR based on the scenario of two players, the first player is called the Primary User (PU) who is the official owner of the licensed spectrum, and sometimes PU called the licensed user. While the second player is called a Secondary User (SU) who used the licensed spectrum in an opportunistic manner without harmful effects on the PU [4]. The fundamental idea behind CR is spectrum reuse, which permits unlicensed users to use a portion of the licensed spectrum band as long as they observe the restriction to not interfere with the licensed users' Quality of Service (QoS) [5]. According to [6], Fig. 1 depicts the standard cognitive cycle of a CR.

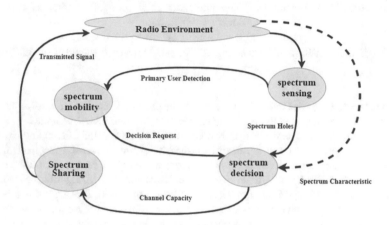

Fig. 1. The cognitive cycle.

The immediate identification of opportunistic white spaces is the concatenate of spectrum sensing. Therefore, the sequence of sensing is used to estimate PU activity statistics such as the duration of idle/busy periods, their minimum duration, mean, higher-order moments, and distribution. Following the idle/busy intervals, the PU activity statistics and occupancy patterns are used to perform spectrum sensing, which is the common technique for determining the condition of the spectrum occupation of the PUs [7]. Consequently, when the PUs are absent, then the SUs can pass through these channels. On the contrary, if the PU is already present, then SU working on sharing the frequency bands with the PU via limiting their communication power. There are two types of Spectrum Sensing (SS) schemes: parametric and non-parametric [8]. The SS that requirement from the PU prior activity information then this SS scheme is called parametric sensing. Non-parametric approaches ability to work without known any prior information about the PU. Resulting in, the non-parametric SS being preferred over parametric SS [9]. While there are various traditional nonparametric SS methods such as matched filter, cyclostationary, and energy detection, these methods are widely utilized since the computational is simple.

The most notable, on the other hand, drawbacks of the traditional techniques are taking a long time period, needing prior information about PU spectrum and noise uncertainty [10]. Therefore, to overcome the problems of traditional techniques, industry and academia have gained commonly interest in technologies of treating the spectrum sensing techniques that for adopting the best of cognitive radio network based on the Machine Learning (ML) and Deep Learning (DL) [11, 12] respectively.

In this paper, we exploited transfer learning from computer vision to improve the robustness of identifying various types of signals while solving SS problems using a CNN as a classifier of two categories. Since the CNN has a powerful tool, which is the extraction of features automagically and their model could optimize easier. In our context, we use generated signals that represent the PU signals to train CNN model in order to predict the input data is signal or noise.

The rest sections of this paper are structured as follows: Sect. 2 describes related works about SS, detection based on Machine Learning (ML) and Deep Learning (DL) in CR. Section 3 presents the discussion of how to formulate a SS problem as a binary hypothesis testing and modeled Dl algorithms as a classifier of two-category. Section 4 illustrates the simulation results and performance parameters used, as well as, the experimental results of real spectrum data that supported by SDR. Finally, the conclusion section of discussed the limitations and future research scope.

2   Related Work

There are many works proposed to enhance SS in CR system based on machine learning and deep learning. Some of these efforts are:

In [13], the authors proposed a SS algorithm based on transfer learning to enhance dataset resolution that in turn increasing the detection opportunity. Although, this method improved the sensing quality, but still suffering from high computation complexity resultant from the data pre-processing and transforming. In [14], the authors presented the first investment of deep learning approach in SS. In this method, the role of CNN is applied at Fusion Center (FC) for the received sensing data from the individual nodes. This method enhances the decision making at FC, but ignored the improvement of detection quality at each sensing node. In [15], SS method based on hybrid deep learning model (CNN-LSTM) is proposed. The detection accuracy is well improved by this method and hence the spectrum utilization is also optimized. The main shortage related to this method, the system testing scenario with real time spectrum data is not applied. In [16], proposed same sensing same model as in [15], this method enhanced the detection as compared with Conventional Energy Detection (CED) method and real time spectrum signals are not validated as well. In spite of this improvement, the detection performance in [16] at low SNR values need extra enhancement to recognizing PU signal from noise only signals. In [17] SS was implemented based on real signals by through the ARDUINO UNO card that generated real signals such as ASK and FSK modulation types. Then uses a deep learning CNN model to detect the transmitted signals in MATLAB software. Results indicated that the proposed technique outperformed ED, ANN, and SVM in terms of classification accuracy and training speed. The drawback in this paper is the use of real signals to train the network CNNs, whereas it is preferable to use the data generated from the simulator to train CNNs to show that they detect real signals well.

In this paper, a blind SS algorithm based on efficient CNN model is proposed. In this algorithm, CNN model is trained and tested offline by dataset includes energies of PU and noisy signals. Additionally, the detection performance is also tested by real time signals generated and captured by SDR transmitter and receiver devices in realistic wireless channel.

3 Problem Formulation

In detection theory, SS is the way by which the spectrum can be identified either busy or idle. There are two hypotheses; H_1 and H_0 as shown in Eq. (1) that represents PU is present and absent respectively [18].

$$Y(n) = \begin{cases} w(n); & H_0 \\ s(n) + w(n); & H_1 \end{cases} \tag{1}$$

where Y(n) is the received sample, w(n) is the Additive White Gaussian Noise AWGN sample collected from the transmission channel, s(n) is the PU signal sample. SS in CR system can be formulated as binary classification problem to identifying the testes statistics that is defined in the following equation:

$$T = \frac{1}{N_s} \sum_{n=1}^{N_s} |y(n)|^2 \tag{2}$$

where Ns is the total number of received samples. The distribution of T is chi-squared with 2Ns degrees of freedom [19]. In CED, the SS process is achieved by comparing T with predefined threshold value. Furthermore, the spectrum status is defined based on thresholding result, if it greater than or equals threshold this is PU, otherwise it is a spectrum hole.

Instead of using CED for many reasons, the most one is noise uncertainty problem. Deep leaning can be appalled to solve this problem and countermeasure noise fluctuation effect. So, SS problem can be formulated as a binary classification problem of H_1 and H_0 classes using CNN with ANN classifier and Y(n) as input dataset as shown in Fig. 2.

Fig. 2. Deep learning block diagram.

The received signals Y(n) are squared and labeled according to its type. The labeling operator is Zn, this value is assigned to one when PU signal is present and to zero when PU is absent as in the following equation.

$$Zn = \begin{cases} 0, & for\ H_0 \\ 1, & for\ H_1 \end{cases} \tag{3}$$

The training dataset consists of vectors each one includes the signal samples and its type according to the assigned label as (Y (n), Zn). The vector length is limited to n samples that must be inserted during training process at each row of dataset.

3.1 The Proposed Algorithm

In this paper, an efficient SS algorithm based on CNN Called Deep learning Based Spectrum Sensing (DBSS) is proposed. This algorithm transfers the detection mechanism of CR sensing engine as a binary classification problem using deep learning classifier. The measured energy of the received samples is used as train and test data in the classification process. Figure 3 shows the general workflow of this algorithm.

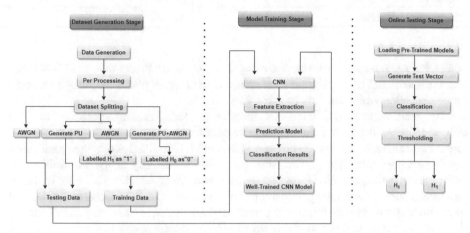

Fig. 3. Block diagram of the proposed model.

3.2 Dataset Generation

The first stage of this algorithm as shown in Fig. 3 presents the dataset preparation. Dataset consists of measured energies (squared samples) of both the PU signal and only noise. PU signal is supposed to be QPSK added to AWGN channel, while noisy signal is presented by random noisy signal generated by AWGN model. After that, the whole dataset is split into train and test dataset. The train dataset contains the labeled part of origin dataset when the PU signal samples are labeled with "1" and noise samples are labeled with "0". The test dataset contains groups PU and noisy signal samples without labeling part. Table 1 shows dataset parameters.

3.3 The CNN Model

The detection mechanism in this work is achieved by using deep neural network with CNN model as feature extractor and back propagation neural network as a binary classifier. Due to the high recognition accuracy of CNN in many applications such as computer vision allow us to utilizing this facility in PUs signal detection and recognized it for noisy signal (Spectrum Holes). As shown in Fig. 4, the proposed CNN model has 22 layers, including an input layer, six convolution layers, five layers of batch normalization, three layers of Rectified Linear Unit (Relu), three of max pooling layers, a fully

Table 1. Dataset parameters.

Parameter	Value
PU	QPSK
Signal vector dimension	1*20
No. of signal and noise	21892
No. of class "1" signals	11892
No. of class "0" signals	10000

connected layer, a softmax layer, and a classification layer. In addition, the CNN model is one dimensional and configured with training parameters obtained through extensive cross-validation which are lists in Table 2.

The main issue in DL based classification is model training approach. The well trained model will increase the opportunity of achieving high detection accuracy with low false positive. However, the proposed CNN model is trained and refined until reaching the required detection performance (detection probability at specific SNR). This optimization cycle generated based on try and error under detector limitation like computation capability and energy consumption. Stochastic Gradient Descent (SGD) with momentum is used as error minimizing method and data matrix was zero-padded. Finally, the successful trained CNN model is saved to be used later in online sensing scenario.

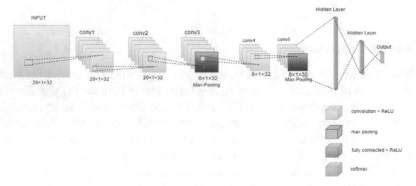

Fig. 4. CNN architecture.

4   The Simulation Results

In this section, the proposed algorithm is simulated in MATLAB (version R2021a) to illustrate the performance of the proposed CNN model. The simulation process is divided into; CNN model training and online signal detection scenario which is performed by Monte Carol method.

Table 2. Hyper-parameters of the proposed CNN networks.

Parameter	Value
Filters per Conv layer	32
Filter size	5
Neurons per FC layer	128 & Sample length & 2
Optimizer	Sgdm
Learning rate	0.001
Mini-patch size	256
activation function	ReLU and Softmax
Dropout ratio	0.1

Figure 5 shows the training performance of CNN model, it that evident to see the training is fast. For example, it reached full performance (100%) with free error in short time (6 min with 10 epoch during 85 iterations). Therefore, the trained network is ready to recognizing the channel if busy or empty. The next figure reveals the confusion matrix in testing performance. The testing scenario with little training time has shown efficient classification of the performance for the proposed CNN model as given in the confusion matrix in Fig. 6. Whereas, the classification of the signal obtained with accuracy reaches 100% and the classification of the noise is 100%. As far as the performance for the proposed CNN then concluded that the enteral system model is efficacious with high accuracy which is 100%. After completed training stage, the trained network now is ready to recognizing the channel if busy or empty.

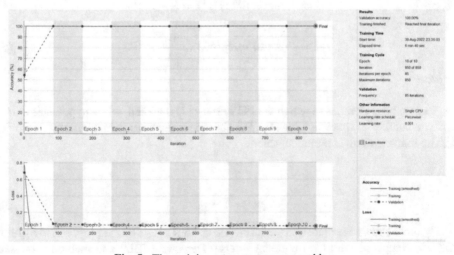

Fig. 5. The training process accuracy and loss.

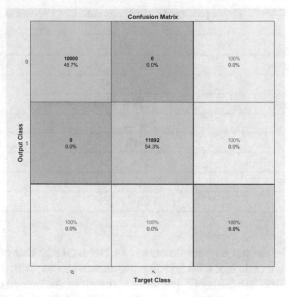

Fig. 6. Confusion matrix of the proposed PU detection system.

The detection performance is performed by detection probability (P_d) while the detection accuracy is assessment by probability of false alarm (P_{fa}). The desired spectrum sensing algorithm is that can reach high detection performance with minimum at low SNR regime with minimum detection error and miss detection probability (P_{md}). Following equations reveals the detection performance formulas respectively:

$$Pd = P(decisionH1/H1) = \frac{K_p}{K} \times 100 \tag{4}$$

$$P_{fa} = P(decisionH_1/H_0) = \frac{K_n}{K} \times 100 \tag{5}$$

where:

K_p: The number of times in which the PU presence is declared, while hypothesis is H_1;
K_n: The number of times in which the PU absence is declared, while hypothesis is H_0;
K: The number of all captured signals.

$$pm = 1 - pd \tag{6}$$

Figure 7 shows the detection performance at SNR range of the proposed algorithm (DBSS) compared with Conventional Energy Detector (CED) and method proposed in [12]. From this it can be seen DBSS reaches high detection probability than other compared method, for instance at (SNR = −18 dB) the P_d in DBSS method is enhanced by ratio of 70% and 50% over than in CED and method in [12], respectively.

Figure 8 shows the performance curve of P_{md} vs SNR of the proposed method as compared with CED and the proposed method in [12]. From this figure, the superior performance of the proposed method is clear and it achieved low miss detection error especially at low SNR levels. For example, at (SNR = -18 dB) the P_{md} in the proposed method reached free detection error earlier than the other compared method in this figure.

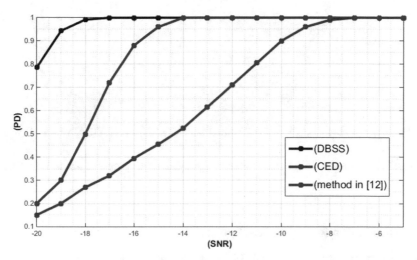

Fig. 7. Pd versus SNR.

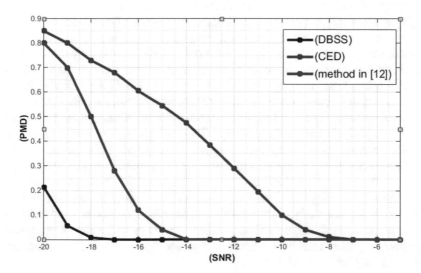

Fig. 8. P_{md} vs SNR.

Another important comparison is the noise uncertainty measurement. This evaluation shows the resistance of a sensing algorithm against noise fluctuation. This scenario performed by inserting the proposed sensing algorithm with noise only signal. This evaluation mimics the case when the channel is empty and the detector may decide PU is presence that in turn causes extra spectrum underutilization. Figure 9 shows P_d vs number of iterations for the noisy signal of the proposed algorithm and CED. From this figure, it can be noted, the proposed algorithm superior performance and achieved high resists against noise fluctuation than CED.

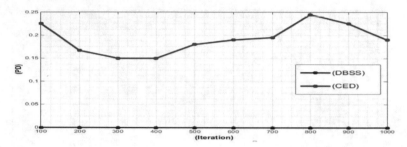

Fig. 9. P_d vs number of iterations for the noisy signal of the proposed method and CED.

4.1 System Validating Using Real Spectrum

In this section, the validation of DBSS based on real time spectrum data is presented. The robust sensing algorithm that can archives the same detection performance for both simulation and real transmission scenarios. In this scenario, an experimental and realistic communication suit is carried out based on Software Defined Radio (SDR). The scenario consists of transmission and receiving sides are separated by distance of 2m. The transmitter part performed by a PC supported with GUN radio software for PU signal generation using flowgraph blocks and HachRF one SDR to transmit the signal. On the same way, the receiver side represented by a PC with MATLAB software for signal processing (capturing and detecting) and RTL SDR to receive the transmitted signal. Figure 10 reveals the layout and the experimental visions of the proposed scenario.

Regarding SS, the captured signal from RTL is injected to the well trained model of DBSS algorithm as a testing vector to decide wither the PU is absent or present, Table 3 shows the design parameters that were set. The evaluation approach is achieved by taking many snapshots of real signals, detecting them by DBSS and then calculating the detection probability based the loop cycle with respect to the number of received signals farms. Consequently, in case of PU is present the P_d was 0.987, while in case of PU is absent it reached zero detection. All these results show the detection accuracy of the proposed algorithm and how it achieved superior signal recognition comparing with other sensing algorithm like CED.

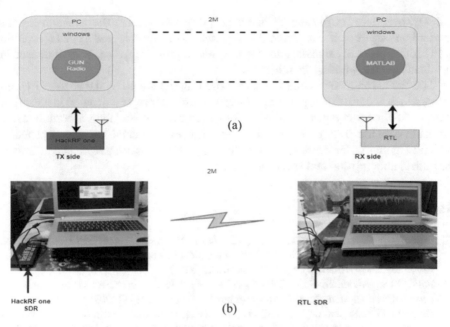

Fig. 10. (a): sensation's layout, (b): the experimental work

Table 3. Parameters of experimental scenario.

Parameter	Value
PU	QPSK
No. of frames	512
No. of samples per frame	20
SNR	(−30 to −20) dB
Distance	2 m
Channel	Realistic (Faing + AWGN)
Up converter carrier	100 MHz

5 Conclusion

In this paper, a blind spectrum sensing algorithm based on deep learning has been presented. This algorithm formulates the spectrum sensing problem as a binary classification problem and investigates the CNN robustness in this issue. However, the PU and noisy signals used as training and testing dataset during CNN model synthesizing. This model has been optimized with the required detection characteristics and then applied to the detection scenarios. The simulation results showed that; the proposed algorithm achieved and outperformed the other compared sensing method especially at low SNR region. For example, the detection probability in the proposed algorithm enhanced by ratio of 70%

and 50% over than in CED and method in [12] respectively, also, the detection error has been minimized as well. Moreover, the proposed algorithm achieved a good recognition to PU form noise only signals, thus, the false alarm probability is heavily diminished in turn leads to high white space utilization.

The limitation of this work is the proposed model identifies the state of PU only without their modulation type. Future investigation will propose a model include modulation classification types. Finally, the proposed system validation according to real spectrum data has been presented. The presented experimental scenario depicted the detection ability of the proposed in reaching high detection accuracy and performance for both simulating and real spectrum signals.

References

1. Joykutty, A.M., Baranidharan, B.: Cognitive Radio Networks: Recent advances in spectrum sensing techniques and security. In: 2020 International Conference on Smart Electronics and Communication (ICOSEC), pp. 878–884. IEEE (2020)
2. Falih, M.S., Abdullah, H.N.: DWT based energy detection spectrum sensing method for cognitive radio system. Iraqi J. Inf. Commun. Technol. 3(3), 1–11 (2020)
3. Bharathy, G.T., Rajendran, V., Tamilselvi, T., Meena, M.: A study and simulation of spectrum sensing schemes for cognitive radio networks. In: 2020 7th International Conference on Smart Structures and Systems (ICSSS), pp. 1–11. IEEE (2020)
4. Wang, W.: Spectrum sensing for cognitive radio. In: 2009 Third International Symposium on Intelligent Information Technology Application Workshops, pp. 410–412. IEEE (2009)
5. Basile, F., et al.: Interleukin 31 is involved in intrahepatic cholestasis of pregnancy. J. Matern. Fetal Neonatal Med. 30(9), 1124–1127 (2017)
6. Mor, T., Arora, D., Mehta, T.: Simulation of probability of false alarm and probability of detection using cyclo stationary detection technique in coginitive radio. Int. J. Electron. Commun. Technol. 6 (2015)
7. Gao, J., Yi, X., Zhong, C., Chen, X., Zhang, Z.: Deep learning for spectrum sensing. IEEE Wirel. Commun. Lett. 8(6), 1727–1730 (2019)
8. Soni, B., Patel, D.K., López-Benítez, M.: Long short-term memory based spectrum sensing scheme for cognitive radio using primary activity statistics. IEEE Access 8, 97437–97451 (2020)
9. Liu, C., Wang, J., Liu, X., Liang, Y.-C.: Deep CM-CNN for spectrum sensing in cognitive radio. IEEE J. Sel. Areas Commun. 37(10), 2306–2321 (2019)
10. Goyal, S.B., Bedi, P., Kumar, J., Varadarajan, V.: Deep learning application for sensing available spectrum for cognitive radio: an ECRNN approach. Peer-to-Peer Networking Appl. 14(5), 3235–3249 (2021)
11. Shah, H.A., Koo, I.: Reliable machine learning based spectrum sensing in cognitive radio networks. Wireless Communications and Mobile Computing 2018 (2018)
12. Zheng, S., Chen, S., Qi, P., Zhou, H., Yang, X.: Spectrum sensing based on deep learning classification for cognitive radios. China Communications 17(2), 138–148 (2020)
13. Peng, Q., Gilman, A., Vasconcelos, N., Cosman, P.C., Milstein, L.B.: Robust deep sensing through transfer learning in cognitive radio. IEEE Wirel. Commun. Lett. 9(1), 38–41 (2019)
14. Lee, W., Kim, M., Cho, D.-H.: Deep cooperative sensing: cooperative spectrum sensing based on convolutional neural networks. IEEE Trans. Veh. Technol. 68(3), 3005–3009 (2019)
15. Xie, J., Fang, J., Liu, C., Li, X.: Deep learning-based spectrum sensing in cognitive radio: a CNN-LSTM approach. IEEE Commun. Lett. 24(10), 2196–2200 (2020)

16. Yang, K., Huang, Z., Wang, X., Li, X.: A blind spectrum sensing method based on deep learning. Sensors **19**(10), 2270 (2019)
17. Liu, C., Wang, J., Liu, X., Liang, Y.-C.: Deep CM-CNN for spectrum sensing in cognitive radio. IEEE J. Sel. Areas Commun. **37**(10)
18. Saber, M., Chehri, A., El Rharras, A., Saadane, R., Wahbi, M.: A cognitive radio spectrum sensing implementation based on deep learning and real signals. In: The Proceedings of the Third International Conference on Smart City Applications, pp. 930–941. Springer, Cham (2020)
19. Solanki, S., Dehalwar, V., Choudhary, J.: Deep learning for spectrum sensing in cognitive radio. Symmetry **13**(1), 147 (2021)
20. Falih, M.S., Abdullah, H.N.: A combined spectrum sensing method based DCT for cognitive radio system. Int. J. Electr. Comput. Eng. (2088–8708) **10**(2), 1935 (2020)

Colour Image Encryption Based on HACS and Knuth-Durstenfeld Algorithm

Renjith V. Ravi[1]([✉])[ID], S. B. Goyal[2][ID], Chawki Djeddi[3,4][ID],
and Vladimir Kustov[5][ID]

[1] Department of Electronics and Communication Engineering,
M.E.A Engineering College, Kerala, India
renjithravi@meaec.edu.in
[2] Faculty of Information Technology, City University, Petaling Jaya 46100, Malaysia
sb.goyal@city.edu.my
[3] Department of Mathematics and Computer Science, Larbi Tebessi University,
Tebessa, Algeria
c.djeddi@univ-tebessa.dz
[4] LITIS Lab, Rouen University, Rouen, France
[5] Saint Petersburg Railway Transport University of Emperor Alexander I,
Saint Petersburg, Russia

Abstract. This research proposes an algorithm for an image encryption strategy based on a HACS and a scrambling algorithm to solve the issues of limited key space, insecure encryption frameworks, and easy-to-break present image encryption techniques utilising chaotic systems and bit wise XoR encryption. To begin, the image is encrypted using the sequence generated using HACS. The image is disordered using the shuffling method, then the bit wise XoR sequence technique is utilised to dilute the image's pixel value. Experiments demonstrate that the scheme's key space exceeds 2327 and that it is very sensitive to small changes in keys. The encrypted images' histogram is equally distributed. Almost every neighbouring pixel has a correlation coefficient of zero. The UACI and NPCR values are both close to ideal, and the entropy values of the encrypted images are all close to eight. Experiments show that the encryption method described in this paper can survive extensive cryptanalysis, such as statistical and differential types. The research's results show that the system and algorithm have made encryption more effective and that the suggested method is useful and doable for image encryption.

Keywords: Hidden attractor chaotic system · Knuth-dustenfed algorithm · Image cryptography · Knuth-dustenfed shuffling · Image encryption

1 Introduction

The growth of digital advancements and widescale usage Internet has greatly facilitated people's work and lifestyles. Digital media has significantly improved

F. P. García Márquez et al. (Eds.): ICCIDA 2022, LNNS 643, pp. 332–343, 2023.
https://doi.org/10.1007/978-3-031-27099-4_26

people's lives as carriers of works such as photos, movies, music, and books due to easier access, convenience in copying, rapid distribution, and a few other advantages. However, as can be shown, particular harmful activities targeted at intercepting valuable information by leveraging network flexibility and options explored have severely harmed the ambitions of communication parties. As a result, developing secure information transfer technology is critical. Digital images that can be seen are a required component of digital media data. As a result, image encryption has become a popular method for securely transmitting digital images over the Internet. However, the computation of data encryption techniques like DES and AES takes a long time. Because of several unique aspects of images that are insufficient or unsuitable for real-time image transmission. As a result, a wide range of image data encryption using chaotic mechanisms, DNA sequences, 'cellular automata', magic cube, and other factors have been developed. Fridrich [3] presented the first general architecture for permutation (Shuffling) and diffusion-based chaos theory-based image encryption in 1998. The permutation approach is used to separate the connections between close pixels in a given image by altering the orientation of the pixels while keeping the histogram intact. The diffusion shifts the histogram by changing the pixel values.

Image encryption methods frequently employ diffusion and permutation. The diffusion procedure expands the image component data directly to encompass the entire range of text. The shuffling or permutation process, on the other hand, by shifting the pixel location, may disrupt the strong association of nearby digital image pixels, which is required for the electronic encryption process. Several academics have worked extensively on permutation algorithms and proposed numerous practical permutation techniques, including the Arnold transforms, Baker transforms, and E-curve transforms. These traditional approaches have aided image encryption research significantly. However, further investigation has revealed several flaws. For instance, Arnold's transformation and Baker's transformation appear periodic. Finally, the approaches presented in have flaws like weak randomization. The Knuth-Durstenfeld method, on the other hand, is highly random. In, [4], [5] and [8] employs the Knuth-Durstenfeld permutation algorithm for cryptographic operations on the images. In contrast, the Knuth-Durstenfeld technique is used only for the generation of keys in [4], not to permute the images. Furthermore, it is a universal image encryption technique rather than a chaotic-based method. Our study uses the Knuth-Durstenfeld method to permute pixels. A hidden attractor chaotic system (HACS) [5,8] generates keys. Comparing to specific other permutation techniques on chaos theory-based image encryption, this Knuth-Durstenfeld permutation method in our algorithm has more excellent nature in randomness.

2   Materials and Methods

Concerns about information security are relevant to almost every facet of human activity and daily living in today's contemporary civilization. The problem of information being leaked to unauthorised parties is becoming a far more serious issue [10]. The creation of procedures that efficiently preserve information security has emerged as a prominent focus of study in recent years. Conventional algorithms for encrypting data, such as RSA and DES, have been utilised extensively in the process of encrypting textual information; however, due to developments in the electronics industry and the field of computer engineering, the skills of human computers have advanced at an accelerated rate. As a direct consequence of this, conventional techniques for encrypting data run the danger of being broken. On the other hand, despite the vast improvements in the computational power of computers, people are still in need of faster encryption rates. The amount of image data is much greater than that of text data, and there is a strong association between neighbouring pixels. As a result, there is a need for the development of cryptosystems that are both more effective and safe.

Image encryption may be accomplished using either the scrambling encryption technique or the pixel grayscale value encryption method. The process of applying algorithms to modify the location of pixels is referred to as scrambling encryption. The connection between pixels may be reduced due to the use of encryption, but the values of the pixels themselves cannot be altered. This paper proposes an algorithm using scrambling based on Knuth-Dustenfed Sect. (2.2) scrambling and XoR diffusion using 4D HACS mentioned in Sect. 2.1.

2.1 Chaotic System

This work employs the four-dimensional hyperchaotic system with implicit attractor and the generalised chaotic system derived from the Lorenz equation (non-diffusion type). The system features chaos with double scroll, dynamics (periodic), and quasiperiodic behaviors. The research uses a four-dimensional HACS created by extending the generalised non-diffusion Lorenz equation. There are no equilibria in the system, although it may show chaos with double scroll, dynamics (periodic), and quasiperiodic. Furthermore, coexisting hidden attractors, such as hyperchaotic and periodic, may be seen for specific parameter values. The Eq. 1 is a description of the novel 4-dimensional HACS [2,6,9]:

$$\begin{cases} \dot{p} = x(y-p) \\ \hat{q} = -pr - zq + fs \\ r = -y + pq \\ s = -eq \end{cases} \tag{1}$$

For which z, y, x, e, and f are just the chaotic system's real parameters, and $k \times m \neq 0$. So the system has no equilibrium conditions when $b \neq 0$.

According to hyperchaos theory, at least two Lyapunov exponents must be positive for a 4-dimensional hyperchaos system. The system will be in various states depending on the starting circumstances and settings. Table 1 provides specific instances.

Different paths emerge from a bit of change in the system's starting state. The experiment uses parameter z as an example. Table 2 shows the effect of the

Table 1. The impact of the system's starting value and parameters [5]

Begining value	Parameters	State of the system
(00.20, 0.10, 0.750, −2.0)	x = 10.0, y = 25.0, z = −2.50, e = f = 1	Double-scroll hyperchaos
(00.20, 0.80, 0.750, −2.0)	x = 10.0, y = 25.0, z = −4.660, e = f = 1	Chaos
(00.20, 0.80, 0.750, −2.0)	x = 10.0, y = 25.0, z = 2.0, e = f = 1	Periodic orbits
(00.20, 0.10, 0.750, −2.0)	x = 10.0, y = 25.0, z = −4.660, e = f = 1	Hyper-chaos

variable z on the performance of the algorithm. With the changing of parameters, chaotic systems display varied dynamic features.

Table 2. The effect of changing system parameters [5]

Value in the range of z	System state
(−07.450, −04.960) (−04.940, −4.680) (−04.660, −4.120) (−00.460, 0.240)	Chaos
(−04.960, −04.940) (−04.680, −4.660) (−04.120, −0.460) [01.840, 1.880]	Hyper-chaos
(−00.240, 0.1540)	Quasi-periodic orbits
[00.1540, 1.840) [01.880, 02.840]	Periodic
[02.840, 8.540]	Quasi-periodic
(08.540, 9.0)	Chaos

Hidden Attractor Chaotic System (HACS). A 4D hyperchaotic system should have a minimum of two Lyapunov exponents with positive value, according to hyperchaotic theory. The Lyapunov exponents used in this system are $P_1 = 0.91160$, $P_2 = 0.02350$, $P_3 = 0$, and $P_4 = 0.8.42203.00$ when the parameters are having values assigned as $x = 10.0$, $y = 25.0$, $z = 2.50$, $f = 1.0$, and $e = 1$ and the starting conditions are assigned as $(0.20, 0.10, 0.750, 2.0)$ and a hyperchaotic attractor with two scrolls exists in the system.

2.2 Knuth-Durstenfeld (K-D) Algorithm for Pixel Permutation

By associating with integers in the existing array and reducing the unnecessary $K(n)$ space, Knuth and Durstenfeld enhanced the Fisher-Yates scramble method [8] [7]. The method's primary method is related to the Fisher-Yates permutation technique. First, we choose a number at random from the collected data and place it at the ending position of the array. As a result, the array's actual number has already been processed. Second, Knuth and Durstenfeld enhanced the Fisher-Yates shuffling method by interacting with integers in the actual array, thereby reducing the unnecessary $K(n)$ space. The fundamental concept behind the method is nearly identical to the Fisher-Yates permutation algorithm. First, we select a number at random from the data sources and place it at the ending position of the array. As a result, the last number in the array has been analysed.

The following are the specifics:

1. Make an array called *Arr* with the size of an array as n, to hold the values.
2. Make an random integer x between 0 and $n - 1$.
3. The value of variable *Arr* subscripted as x should be output.
4. Substitute the element at the end for the element subscripted with x.
5. Create a random integer between 0 and $n - 2$, as in step -2.
6. The value of array named *Arr* is assigned as x will be output.
7. Substitute the second-to-last element for the element named with x. Repeat steps 1–3 until all n components have been handled.

$K(n)$ is the time complexity, while $K(1)$ is the space complexity. This is a disordered in-place algorithm. The Fisher-Yates shuffle algorithm's space complexity has been reduced from $K(n)$ to $K(1)$. In addition, the Fisher-Yates shuffle algorithm's runtime complexity has been improved from $K(n \times n)$ to $K(n)$.

3   Encryption Process

Figure 1 depicts the encryption technique presented in this article in action. The algorithm's particular encryption method is as follows, assume the size of the plaintext image I as $M_1 \times M_2$.:

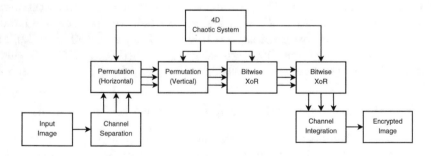

Fig. 1. Proposed encryption approach

1. The double-scroll hyperchaos of the HACS is employed in this work. The value of hash of the plaintext, the parameters and beginning values of the chaotic map (given in the second row of Table 1) forms the algorithm's key.

2. To eliminate the system's transitory impact, the chaotic system iterates 1000 times using the key from step 1. The produced chaotic sequence is separated into six groups to improve the encryption system's sensitivity: $C_1(a,b)$, $C_2(a,c)$, $C_3(a,h)$, $C_4(b,c)$, $C_5(b,h)$, $C_6(c,h)$.
3. There are two variables defined: hash and index. First, the value of hash of the plaintext image is calculated using the SHA-256 algorithm. Then, the hash value in hexadecimal nature is translated to a decimal number and appended to acquire the hash value. The Formula shows the specific approach in Eq. 2.

$$index = \mod(hash, 6) + 1 \tag{2}$$

The rest of the hash divided by 6 is shown by $Mod(hash, 6)$. $C_j(1)$ and $C_j(2)$ are represented by S_1 and S_2 Eq. 3. When index $= n$ then $C_j = C_n, j = n$

$$\begin{cases} j = 1, S_1 = A, S_2 = B \\ j = 2, S_1 = A, S_2 = C \\ j = 3, S_1 = A, S_2 = D \\ j = 4, S_1 = B, S_2 = C \\ j = 5, S_1 = B, S_2 = D \\ j = 6, S_1 = C, S_2 = D \end{cases} \tag{3}$$

Here S_1 and S_2 are treated as stated above in Formula in Eq. 4 to accomplish the scrambling effect, and the outcomes processed are assigned as Column vector and Row vector, accordingly:

$$Vector(j) = \mod \left(\text{floor} \left((S_n(j) + 100) \times 10^{10} \right), M_1 \times M_2 - j + 1 \right) + 1, (n = 1, 2) \tag{4}$$

4. The algorithm for shuffling(Permutation) utilises the chaotic sequence S_1 produced in step 3 to jumble the plaintext image. P Row is a one-dimensional vector created from the original image matrix P. Formula in Eq. 5 depicts the scrambling process.

$$P\_ \text{Row} (S_1(j)) = P\_ \text{Row}(M_1 \times M_2 - j + 1) \tag{5}$$

The resulting vector with P number of Rows is transposed and extended for getting a one-dimensional vector with P number of Columns.
5. S and P pair scrambles through the chaotic sequence according to the formula Eq. 5. The sequence processed with P Column is then re-converted into an $M_1 \times M_2$ size matrix P. Finally, the variable temp is determined using the Formula in Eq. 6.

$$temp = \mod \left(\sum_{j=1}^{M_1 \times M_2} P_1, 256 \right) \tag{6}$$

6. The starting values and parameters of chaotic map will be fixed to the values mentioned in step 1. Then, the chaotic system's parameters and starting values are iterated $1000 + M_1 \times M_2$ times, eliminating the chaotic system's transitory impact. Finally, their values are kept in the chaotic starting valued sequences generated from the chaotic system.

7. Each of the elements works to generate 4-vectors S_a, S_b, S_c, and S_h using the formula in Eq. 7 pair of four number of chaotic sequences.

$$\begin{cases} S_x(j) = \mod\left(A_1(j)'10^{10}, 8\right) + 1 \\ S_y(j) = \mod\left(B_1(j)'10^{10}, 8\right) + 1 \\ S_z(j) = \mod\left(C_1(j)'10^{10}, 8\right) + 1 \\ S_h(j) = \mod\left(D_1(j)'10^{10}, 256\right) \end{cases} \quad (7)$$

Here the variable $'j'$ represents the element in j^{th} position of four chaotic sequences, $j \in [1, M_1 \times M_2]$. The matrix P_1 is converted into a one-dimensional vector $E(j)$. The jth element of four chaotic sequences, $j \in [1, M_1 \times M_2]$, is represented as $'j.'$ P_1 is transformed into a one-dimensional vector $E.(j)$.

4   Results and Discussion

This section shows the results of the proposed encryption technique for a typical test images of size 256×256. Figure 2 depicts the results of the experiments. Figure 2d, Fig. 2e, Fig. 2f depicts the ciphertext as an apparently random image with no discernible text content. Both of these images are unrelated each other. The plaintext and encrypted ciphertext images have a structural similarity of less than 1. The effectiveness of the algorithm's decryption and encryption capabilities is shown by the results.

4.1   Statistical Attack Analysis

Analysis of Histogram. To withstand statistical assaults, the ciphertext image should have a uniform histogram, since the image histogram indicates the distribution of pixel intensity levels. Figure 3 shows the histograms of the original and ciphertext images. The grey level is represented in the abscissa of the histogram, and the frequency of occurrence of those grey levels is represented in the ordinate.

The results of the studies show that the ciphertext image pixels have a uniform distribution. This means that the post-encryption frequency value for each pixel is very close. The statistical rule of the ciphertext image will be unavailable to the attacker.

Correlation Analysis. The nearby pixels in plaintext image has a significant association in all three directions. The image produced by the encryption technique resists statistical assaults only when the coefficient of correlation between neighbouring pixels in the ciphertext image is very less enough. Adjacent pixels

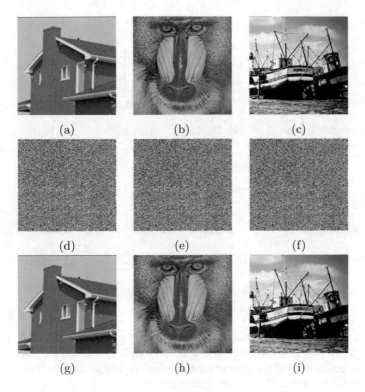

Fig. 2. Results of proposed encryption and decryption algorithms. (a), (b), (c) Plain images, (d), (e), (f) Corresponding Encrypted images, (g), (h), (i) Corresponding Decrypted images.

from each direction in the plaintext image are chosen at random. Correlation coefficients are determined in the encrypted image. The relationship between neighbouring pixels in the original and encrypted images is investigated. Formula in Eq. 8 Eq. 9, Eq. 10 and Eq. 11 are the calculation formula for the correlation coefficient r_{ay}:

$$r_{ab} = \frac{\text{cov}(a, b)}{\sqrt{D(a)D(b)}} \tag{8}$$

$$\text{cov}(a'b) = \frac{1}{T} \sum_{j=1}^{T} (x_j - E(a))(b_j - E(b)) \tag{9}$$

$$E(a) = \frac{1}{T} \sum_{j=1}^{T} x_j \tag{10}$$

$$D(a) = \frac{1}{T} \sum_{j=1}^{T} (a_j - E(a))^2 \tag{11}$$

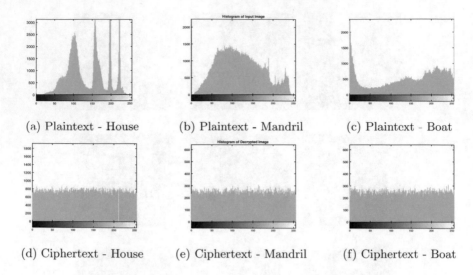

(a) Plaintext - House   (b) Plaintext - Mandril   (c) Plaintext - Boat

(d) Ciphertext - House   (e) Ciphertext - Mandril   (f) Ciphertext - Boat

Fig. 3. Histogram analysis

where T is the total number of pixels, a and b are the grey values of neighbouring pixels, $E(a)$ is the pixel's average value, the variance is $D(a)$, the correlation function is $cov(a, b)$, and the correlation coefficient is r_{ab}; the greater the absolute value, the connection will be more stronger.

Table 3. Correlation coefficients in three directions

Images	Plaintext			Ciphertext		
	H	V	D	H	V	D
House	0.9741	0.9637	0.9531	−0.0081	0.0023	−0.0168
Mandril	0.9636	0.9482	0.9036	0.0143	0.0124	−0.0038
Boat	0.9818	0.9567	0.9518	−0.0261	−0.0251	0.0045

The coefficient of correlation between the plaintext and ciphertext images are shown in Table 3 in all the three directions. In the original image, the correlation coefficients of neighbouring pixels are near one. The coefficient of correlation of encrypted images are near 0, suggesting that pixels in opposite orientations in the original image have a substantial connection. After the encryption procedure is run, the correlation of neighbouring pixels is still removed.

The test image *Lena* and its encrypted versions are used as case studies to more thoroughly examine the correlation between neighbouring pixels after and before the encryption process. The correlation of their neighbouring pixels is

Table 4. Comparison of CC of *Boat* Image with other works in the literature

Reference	Ciphertext		
	H	V	D
Proposed	−0.0261	−0.0251	0.0045
Belazi et al. [1]	−0.0100	−0.0124	−0.0185
Jin et al. [5]	−0.0130	0.0111	−0.0182

shown horizontally, vertically, and diagonally along the comparison with other works in the literature are shown in Table 4. The value of data at the randomly positioned point is represented by the abscissa, and the value of random points in adjacent position is represented by the ordinate.

The plaintext image's neighbouring pixel points are continuously dispersed. The values of adjacent pixel points of the ciphertext image, on the other hand, are randomly distributed. They are dispersed across two-dimensional (2D) space. Here, the ciphertext image hides the data properties of the original image by removing the correlation of neighbouring pixels.

Information Entropy: The Information entropy or shannon's entropy is the most essential measure of unpredictability or randomness. Here, the information source is specified as m, and we may compute information entropy using the Formula in Eq. 12.

$$E(p) = \sum_{i=0}^{T-1} p\,(p_i) \log \frac{1}{p\,(p_i)} \tag{12}$$

where T stands for the total number of symbols, $p_i \in p; p(p_i)$ stands for the probability of symbols, and its theoretical value $E(p) = 8$ is calculated using Eq. 12. Attackers will have a harder time decoding cypher images if the number is near to 8. The information entropy is shown in Table 5. The entropies in Table 5 are close to 8, suggesting that the suggested method has a good information entropy quality.

Table 6 shows that the encrypted image's Shannon entropy is more than 7.99, which is near the theoretical value of 8 and higher than the literature. As a result, the encrypted image produced by the encryption technique suggested in this research has sufficient unpredictability and security to withstand statistical assaults.

Differential Attack Analysis. In this section, we have made a comparison of the image encryption approach used in this work towards others described in the literature [1], [5], [8] and calculate the average UACI and NPCR values, as shown in Table 7 and Table 8. We must create the encryption mechanism highly

Table 5. Entropy values obtained

Test image	Plaintext image	Ciphertext image
House	7.0686	7.9966
Mandril	7.6764	7.9986
Boat	7.1923	7.9978

Table 6. Comparison of entropy values obtained for *Boat* image

Reference	Entropy
Propose Work	7.9978
Belazi et al. [1]	7.9980
Jin et al. [5]	7.9986

sensitive to the plaintext image to make it resistant to differential attack. Slight modifications in the original image may have a big impact on the encrypted image. The anti-differential cryptanalysis capacity of an encryption method is evaluated using a variety of metrics, the most common of which are UACI and NPCR.

Table 7. NPCR and UACI Values obtained

Input image	NPCR	UACI
House	99.67%	33.71%
Mandril	99.8127%	33.6453%
Boat	99.67%	33.81%

Table 8. Comparison of NPCR and UACI Values obtained for *Boat* image

Reference	UACI	NPCR
Proposed Work	33.81%	99.67%
Belazi et al. [1]	33.5367%	99.6102%
Jin et al. [5]	33.79%	99.58%

5 Conclusion

We developed a novel algorithm for secure image transmission based on a HACS, the Knuth-Durstenfeld permutation, and bit wise XoR function in this study. Here, the HACS is utilized to generate the chaotic sequences for permutation and diffusion in image encryption, overcoming the disadvantages of using self-excited attractor chaotic maps in the past. The HACS's chaotic sequence shows randomness, indicating that it is appropriate encryption. Because of Knuth-Durstenfeld technique provides a high level of randomness, it is better suited for disrupting the association between nearby pixels in the image in this study. The bit wise XoR method is used for diffusion procedures. Extensive attacks, statistical attacks, differential attacks, as well as experimental simulations and comparisons, have all confirmed the proposed approach's security. This scheme has a large keyspace and nonvulnerable keys. As a result, it is resistant to repeated attacks. This scheme's histogram is consistent. The coefficient of correlation is close to zero, and the value of entropy is nearly 8. As a result, the method is impervious to statistical attacks. Because both UACI and NPCR values are approaching their optimal values, the proposed technique can withstand differential assaults. The preceding data demonstrate that the proposed communication strategy is effective and feasible. However, there are some areas that could be investigated and improved.

References

1. Belazi, A., Abd El-Latif, A.A., Belghith, S.: A novel image encryption scheme based on substitution-permutation network and chaos. Signal Process. **128**, 155–170 (2016)
2. Danca, M.F.: Hidden chaotic attractors in fractional-order systems. Nonlinear Dyn. **89**(1), 577–586 (2017). https://doi.org/10.1007/s11071-017-3472-7
3. Fridrich, J.: Image encryption based on chaotic maps. In: 1997 IEEE International Conference on Systems, Man, and Cybernetics. Computational cybernetics and Simulation, vol. 2, pp. 1105–1110. IEEE (1997). https://doi.org/10.1109/ICSMC.1997.638097
4. Güvenoğlu, E., Tüysüz, M.A.A.: An improvement for knutt/durstenfeld algorithm based image encryption. In: 2015 23nd Signal Processing and Communications Applications Conference (SIU), pp. 1761–1764. IEEE (2015). https://doi.org/10.1109/siu.2015.7130194
5. Jin, X., Duan, X., Jin, H., Ma, Y.: A novel hybrid secure image encryption based on the shuffle algorithm and the hidden attractor chaos system. Entropy **22**(6), 640 (2020). https://doi.org/10.3390/e22060640
6. Pham, V.T., Volos, C., Jafari, S., Kapitaniak, T.: Coexistence of hidden chaotic attractors in a novel no-equilibrium system. Nonlinear Dyn. **87**(3), 2001–2010 (2017). https://doi.org/10.1007/s11071-016-3170-x
7. Ravi, R.V., Goyal, S.B., Djeddi, C.: Colour image encryption based on fisher-yates algorithm and chaotic maps. In: Djeddi, C., Siddiqi, I., Jamil, A., Ali Hameed, A., Kucuk, İ (eds.) MedPRAI 2021. CCIS, vol. 1543, pp. 63–76. Springer, Cham (2022). https://doi.org/10.1007/978-3-031-04112-9_5
8. Wang, S., Wang, C., Xu, C.: An image encryption algorithm based on a hidden attractor chaos system and the knuth-durstenfeld algorithm. Opt. Lasers Eng. **128**, 105995 (2020). https://doi.org/10.1016/j.optlaseng.2019.105995
9. Wei, Z., Wang, R., Liu, A.: A new finding of the existence of hidden hyperchaotic attractors with no equilibria. Math. Comput. Simul. **100**, 13–23 (2014). https://doi.org/10.1016/j.matcom.2014.01.001
10. Xihua, Z., Goyal, S., Tesfayohanis, M., Verma, C.: Blockchain-based privacy-preserving approach using svml for encrypted smart city data in the era of ir 4.0. J. Nanomaterials **2022** (2022)

Turkish Sign Language Recognition by Using Wearable MYO Armband

Muhammet Düzenli$^{(\boxtimes)}$ [ID], Kerem Salur [ID], Kübra Erat [ID], and Pınar Onay Durdu [ID]

Kocaeli University, İzmit, Kocaeli, Turkey

Abstract. In this study, it is aimed to automatically recognize Turkish Sign Language based on EMG signals gathered by a wearable device namely MYO armband. EMG, acceleration, and gyroscope data were recorded and then they were subjected to preprocessing and feature extraction. The data set formed as a result of these processes was classified by using Support Vector Machine, k-Nearest Neighbor, and Random Forest algorithms. In addition, the effect of EMG and inertial measurement unit (IMU) sensors on overall performance were evaluated. Additionally, the effects of the window size and sliding interval variables used in the sliding window method were also evaluated. As a result, the highest performance was obtained by using the RF algorithm with a window size of 100 and a scroll size of 20, with 96% when EMG and IMU sensors used together. It can be concluded that the MYO armband can be used successfully in recognition of Turkish Sign Language.

Keywords: Turkish sign language · MYO armband · Machine learning · Wearable devices

1 Introduction

Sign language is a nonverbal communication method that is widely used in communication with individuals with hearing and speech impairments [1]. This method helps these individuals to communicate with others with visual expression using hand, facial movements, and body language. Although the development of sign languages occurs as a result of a natural process as in the development of any language and they are used universally, the number of disabled people or healthy individuals who know sign language is quite a minority compared to the world population [2]. In addition, the limited number of sign language interpreters further restricts the communication of individuals with hearing and speech impairments in daily life [3]. There are various sign languages that are based on the alphabets of different countries all around the world. Hearing impaired individuals in Turkey and the Turkish Republic of Northern Cyprus use Turkish Sign Language (TSL). According to the Disabled and the Elderly Statistics Bulletin published by the Ministry of Family and Social Services [4], there are 1.343.000 citizens in Turkey who state that they have difficulty in hearing or speech.

The importance and use of gesture-based user interfaces are increasing each day based on the developments of virtual reality and wearable devices. Thus, these wearable

© The Author(s), under exclusive license to Springer Nature Switzerland AG 2023
F. P. García Márquez et al. (Eds.): ICCIDA 2022, LNNS 643, pp. 344–357, 2023.
https://doi.org/10.1007/978-3-031-27099-4_27

technologies are used in many fields. The recognition of gestures can be done by vision sensors [5, 6] or by body-movement sensors such as electromyography (EMG) [7, 8]. The MYO armband, developed by Thalmic Labs, is one of the commercial gesture control devices that detect biosignals from body movements [1, 7–9]. It has 8 EMG sensors and nine-axis Internal Measurement Unit (IMU) including a gyroscope, accelerometer, and magnetometer [10]. It can measure the EMG signals directly by its sensors and provides an easy-to-use way that does not have the limitations of vision-based methods [5], such as being sensitive to occlusion, light, or skin color.

The main contribution of this work is the recognition of some words selected from the TSL Dictionary published by the Ministry of National Education [11] were done by implementing machine learning methods. The selected words were defined as some emergency words (allergy, vaccine, doctor, flu, patient, nurse, vomiting, appointment, x-ray, and serum) that can enable disabled individuals to communicate especially in health institutions. EMG, accelerometer, and gyro sensors' data were gathered by the MYO armband, and then they were subjected to preprocessing and feature extraction. The data was classified with different machine learning algorithms and the accuracy of the classification results was compared. The effects of EMG and IMU sensors and different features combination on overall performance were also evaluated. Additionally, the effects of the window size and sliding interval values, which were determined to have a direct effect on the classification results, used in the sliding window method were also evaluated.

The rest of this paper is organized as follows. Section 2 summarizes the previous research of sign language classification by using the Myo armband. Section 3 describes the experimental framework such as dataset, pre-processing, feature extraction and selection, and also classification algorithms used in this paper. Section 4 reports the analysis results of different methods and discussion. Finally, Sect. 5 presents the conclusion and proposes future work.

2   Related Works

There are some previous studies on the recognition of various sign languages using EMG signals obtained by MYO armband. For instance, [12] conducted a study that aimed to recognize the American Sign Language alphabet based on the collected data from ten subjects by using the MYO armband. The signals gathered were classified with Support Vector Machine (SVM) and Ensemble Learning Classifier after preprocessing, filtering, and feature extraction stages. In the study, the ensemble learning classifier was reported to have the highest performance with a rate of 79.35%.

In another study [13] which was conducted in Sri Lankan (Sinhala) Sign Language, both EMG and IMU sensors' data of the MYO armband were used. 12 gestures in four categories were selected and classified. These categories were determined as static, only hand movements, only finger movements, and both hand and finger movements. The selected movements were trained and tested both subject dependent and independent. A supervised machine learning technique was used for the gesture recognition phase by implementing an artificial neural network (ANN) and it was determined that the accuracy of the subject independent data was above 90%.

346   M. Düzenli et al.

There are also some studies implemented in TSL. [9] conducted a study that aimed to recognize the TSL equivalents of the numbers in the range of 0–9 with EMG signal data and to compare the machine learning algorithms. In the scope of the study, the sliding window method was applied as preprocessing of the data collected by the MYO armband, then feature extraction was performed and finally k-nearest neighbors (KNN), SVM, and ANN algorithms were used for classification. As a result of the study, it was seen that SVM has yielded the highest success rate of 86%.

In a more recent study [14] focusing on TSL, the performance of time-domain and time-frequency domain features of EMG and the IMU data gathered by using MYO armband were evaluated. The study was conducted with ten subjects with 36 static TSL gestures. Random forest (RF) classifier with different input sets (time-domain, time-frequency domain, and IMU) was trained. Using all the input sets, the highest accuracy achieved was 54.2%. In addition, various gender-dependent models were trained, and it was reported that the overall accuracy of the classifiers did not perform well for the female subjects which might be due to their weaker muscular strength.

3 Method

Within the scope of the study, the Turkish Sign Language recognition with Myo armband was carried out. The stages are implemented in the study are shown on Fig. 1.

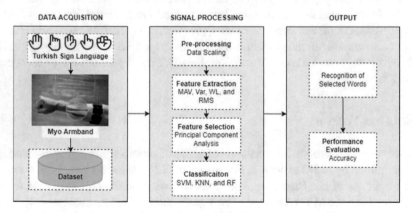

Fig. 1. Flowchart of this study

3.1 Data Acquisition

The MYO armband developed by Thalmic Labs was used to collect EMG data in the study. The MYO armband is a wearable device with 8 EMG sensors to measure hand and arm muscle movements and 9 IMU sensors that measure acceleration, position, and motion intensities [10]. EMG sensors sample at a frequency of 200 Hz, and IMU sensors at a frequency of 50 Hz. Figure 2 shows the MYO armband.

Fig. 2. MYO armband

EMG and IMU sensors were used together in the study to recognize the TSL equivalents of some words. Therefore, the data to be classified must be collected in the most
accurate and consistent way. To achieve this, data collection was performed with a single
person in a fixed position holding the armband at the same height, and the same position
at the beginning of the movement. Since a single MYO armband was used in the study,
it was paid attention that the selected words required the gesture of one arm. Six of the
words were selected to be performed with static hand gestures while two of them required
performing two consecutive movements. The gestures with label presented in the first
two rows of Fig. 3 were all static gestures while the gestures in the third row required
performing two different movements consecutively. The words were selected as emergency words that impaired individuals might require in health institutions. These words
were defined as vaccine, doctor, vomit, allergy, serum, nurse, patient, flu, appointment,
and x-ray.

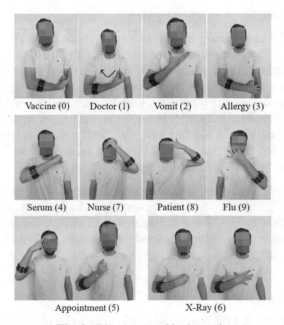

Fig. 3. TSL signs used in the study

When the previous studies were examined, it was observed that the data were collected in various ways. For instance, some studies [15–17] collect data from a movement repeatedly over a period, while some studies [9, 13] collect a certain number of movements independent of time. Likewise, some studies [9, 17] were conducted with a single participant, while some studies [18, 19] were conducted with more than one participant.

In this study, data were recorded by counting the movements independent of the time as in [9], and [13]. Data were collected with 200 Hz sample rate from the EMG sensors and 50 Hz sample rate from the IMU sensors. Each movement belonging to the word was repeated 5 times and data was collected during this time. The length of the dataset for each movement is approximately 1000 rows. Figure 4 presents the signal changes that occurred on the EMG and IMU sensors during the sensor reading of the "Doctor" sign.

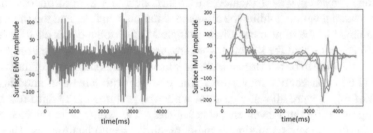

Fig. 4. EMG and IMU sensors data of "doctor" sign

3.2 Pre-processing

Since the EMG signals recording at 200 Hz sample rate on the MYO armband are very sensitive, instantaneous deviations may occur in the data during the recording of sign language movements [20]. Since the EMG and IMU sensors are used together in the study, the raw data collected from the device must go through certain processes to be cleaned from noisy data before processing. Standardized scaling was implemented. The number of features has been reduced to preserve data integrity with the principal component analysis method. Finally, the data were classified by machine learning algorithms. The data were scaled using the "StandartScaler" in the scikit-learn library which is a widely used machine learning library for Python.

Sliding Window Method. Sliding window method is applied to determine the segment where the muscle contracted due to the gestures [21]. In this method, the algorithm is run according to the variables based on the determined window size and sliding interval. Figure 5 illustrates how the sliding window method works on EMG data on one of the TSL words "Doctor" used in this study.

3.3 Feature Extraction

Feature extraction enables the size of the existing large-size raw data set to be reduced without losing the information and features in it [22]. Various formulas are used to

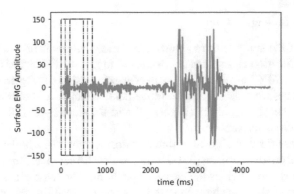

Fig. 5. Illustration of Sliding Window Method

determine the characteristics of the values in the data set to be different to reduce the data size with the least possible data loss. In this way, the time complexity of machine learning models is reduced [23].

The features extracted in this study with descriptions and mathematical formulas are given in Table 1. These features are mean absolute value (MAV), variance (VAR), waveform length (WL), and root mean square (RMS).

Table 1. Mathematical formulas of the features

Feature	Definition	Formula
MAV	It is the expression of the contraction level of muscles during movements in a positive plane	$\frac{1}{N}\sum_{i=1}^{N}|a(i)|$
VAR	It is the square root of the standard deviation, which is a measure of how far the values diverge from the mean or expected value. It determines how much the data is spread out	$\frac{1}{N-1}\sum_{i=1}^{N}|x_i - \mu|^2$
WL	It refers to the complexity measure of EMG and IMU signals. It is calculated as the sum in the absolute value of the differences between two successive samples	$\sum_{i=1}^{N}|x_i - x_{i-1}|$
RMS	It is calculated as the square root of the average of squared value of the signal It is used to measure the magnitude of changing signals	$\sqrt{\frac{1}{N}\sum_{i=1}^{N}a(i)^2}$

3.4 Feature Selection

Feature selection is the reduction of data size by eliminating features that have little effect on the features extracted from the data set, that are irrelevant to the data set, or that repeat themselves. Principal component analysis (PCA) method [24] was used to project the data dimensions into a lower-level space. 72 features were reduced to 18 features with 96% data protection in this study.

3.5 Classification Algorithms

The main aim of the study is to recognize determined words from the TSL dictionary. Therefore, classification algorithms are needed to match the EMG and IMU sensor data produced from the MYO armband. Since, the dataset consists of labeled data, supervised learning algorithms were decided to be implemented. Support Vector Machine (SVM) [9, 10, 17], K-Nearest Neighbors (KNN) [9] and Random Forest (RF) [14, 16] were chosen as in previous researches.

SVM is a classification algorithm that transforms data into a higher dimension and analyzes this data on a hyperplane [25]. The algorithm extracts a hyperplane to categorize the existing data set and divides the classes equally into the hyperplane.

The KNN algorithm is an algorithm used in both classification and regression studies, based on the method of comparing data of unknown classes with other data in the training set and performing a distance measurement. According to the calculated distance, it is aimed to classify the next data by analyzing their distance or closeness to the existing data [26]. The Manhattan Distance method was used as the distance function in the study.

RF is a supervised machine learning method that generates multiple decision trees using randomly selected training data and many subsets of their variables and allows each decision tree to make an individual prediction [27].

4   Results and Discussion

4.1 Analysis of Sliding Window Method

It has been observed in previous studies [20, 28, 29] that different window sizes affect the recording and processing of EMG signals. In this context, various window sizes and sliding interval values were tested based on the values defined in the previous studies examined. Table 2 illustrates that there is a proportional relationship between the selected window size and the sliding interval as the smaller the sliding interval value in the window, the higher the classification performance. However, when the window size increases and the sliding interval is reduced the time complexity increases, too. Therefore, in order to provide optimum accuracy and lower time complexity, it was decided to use the window size as 100 and the sliding interval as 20.

4.2 Analysis of Feature Extraction Methods

In the previous studies examined [19, 30] it was reported that there is a direct effect of feature extraction methods on classification performance used together or separately. Therefore, within the scope of the study, it was also aimed to evaluate the effect of different features on the performance of classification algorithms. Table 3 represents the effect of different features on classification performance.

4.3 Analysis of Used Sensors

TSL signs chosen in the study required the contraction of the hand and arm muscles that will activate the EMG sensors and the change of the arm movements that will activate the

Table 2. Effect of window size and sliding interval on classification performance

Window size	Sliding interval	SVM	RF	KNN	Mean
400	300	%83	%83	%74	%80,00
400	200	%88	%85	%75	%82,67
400	100	%93	%90	%86	%89,67
400	80	%93	%91	%88	%90,67
400	50	%96	%95	%91	%94,00
400	40	%97	%96	%92	%95,00
400	20	%98	%99	%97	%98,00
200	150	%79	%78	%70	%75,67
200	100	%86	%84	%78	%82,67
200	50	%92	%91	%87	%90,00
200	40	%93	%93	%87	%91,00
200	20	%96	%97	%93	%95,33
100	50	%82	%87	%82	%83,67
100	**20**	**%90**	**%96**	**%91**	**%92,33**
50	40	%72	%83	%75	%76,67
50	20	%78	%91	%85	%84,67
20	20	%71	%84	%78	%77,67

Table 3. The effect of different features on classification algorithms

Sensor	Features	SVM	RF	KNN
EMG	'MAV', 'VAR', 'WL', 'RMS'	%65	%83	%80
IMU	'MAV', 'VAR', 'WL', 'RMS'	%82	%96	%94
EMG+IMU	**'MAV', 'VAR', 'WL', 'RMS'**	**%95**	**%98**	**%97**
EMG	'MAV', 'VAR', 'WL'	%64	%83	%77
IMU	'MAV', 'VAR', 'WL'	%82	%97	%94
EMG+IMU	'MAV', 'VAR', 'WL'	%92	%98	%95
EMG	'MAV', 'VAR'	%64	%83	%80
IMU	'MAV', 'VAR'	%81	%96	%95
EMG+IMU	'MAV', 'VAR'	%93	%98	%96

IMU sensor. In this context, the performance evaluation of the separate and combined use of sensors for classification was made.

As a result of the performance analysis, it has been determined that the distinguishing factor of the sign language movements were the positional and accelerational movements of the arm. For this reason, when the IMU sensor was used alone (85.33%), more successful results were gathered than when the EMG sensor was used alone (69.33%). Therefore, when two sensors are active at the same time provided more accurate data since both positional changes and force changes on the muscle occur at the same time. As can be seen in Table 4, it has been observed that the use of the two sensors together provides an average of approximately 7% accuracy performance increase compared to the use of the IMU sensor alone, and approximately 23% accuracy performance increase compared to the use of the EMG sensor alone. It has been determined that an average of 92.33% accuracy was achieved with the use of IMU and EMG sensors together.

Table 4. Effect of used sensors on performance

Sensor	SVM	RF	KNN	Mean
EMG	%58	%77	%73	%69,33
IMU	%72	%94	%90	%85,33
EMG+IMU	**%90**	**%96**	**%91**	**%92,33**

4.4 Analysis of Classification Algorithms

Three different classification algorithms were used in the study. In the experiments conducted to compare data classification algorithms, EMG and IMU sensors were used together, k-fold cross validation was used as k = 10 and, while applying the sliding window method, the window size was determined as 100 and the sliding interval as 20. The dataset is divided into 25% testing and 75% training as in [16]. Table 5 presents the recall, precision, and f-score values for each class and the classification algorithm. Labels 0–9 represent the selected words in Fig. 3, while label 10 indicates the rest position in the beginning. According to that table, label 3 and 10 have the highest recall, precision, and f-score ratio for all classifiers. Label 4 and 9 have the lowest precision and f-score ratio with SVM. It has been observed that recall, precision, and f-score values for other labels have close ratios for all three classifiers.

Support Vector Machine (SVM). In the SVM classification used in the study, a linear function was used as the kernel. Figure 6 shows the confusion matrix of the SVM. For instance, when the table is read for the movement with the label "7", 49 of the 57 samples were predicted right, 4 of them were predicted as the movement with label "8", 3 of them were predicted as the movement with the label "4", and one of them was predicted as the movement labeled "1". The classification performance of the algorithm for the movement with label "7" was calculated as 86%, as it was correctly predicted 49 out of 57 samples.

Table 5. Performance of each hand gesture class

Hand gesture classes		0	1	2	3	4	5	6	7	8	9	10
Recall	SVM	0.83	0.85	0.88	0.99	0.86	0.88	0.78	0.80	0.88	0.90	0.91
	KNN	0.92	0.96	0.91	0.99	0.91	0.86	0.86	0.89	0.83	0.99	0.99
	RF	0.99	0.98	0.97	0.99	0.98	0.88	0.88	0.93	0.92	0.97	0.99
Precision	SVM	0.86	0.85	0.99	0.99	0.60	0.96	0.93	0.85	0.90	0.64	0.99
	KNN	0.93	0.88	0.97	0.98	0.89	0.90	0.98	0.89	0.91	0.77	0.99
	RF	0.98	0.96	0.97	0.99	0.90	0.94	0.99	0.96	0.95	0.82	0.99
F-score	SVM	0.85	0.85	0.93	0.99	0.71	0.92	0.85	0.83	0.89	0.75	0.95
	KNN	0.93	0.92	0.94	0.99	0.90	0.88	0.92	0.89	0.87	0.87	0.99
	RF	0.99	0.97	0.97	0.99	0.94	0.91	0.93	0.94	0.93	0.89	0.99

K-Nearest Neighbors (KNN). In the KNN algorithm used in the study, the neighborhood value was used as 5 and the metric value was "Manhattan". The confusion matrix of the KNN classification algorithm for the movement labeled "5" is given in Fig. 7 and it was calculated as 91%, as it was correctly predicted 50 out of 55 samples.

Random Forest (RF). In the RF algorithm, which is another classification model used in the study, a forest with 100 trees and 2 fragmentation rates was created as a parameter. Figure 8 shows the confusion matrix resulting from the classification for the movement labeled "4" as an example. The classification performance of the algorithm for the movement with label "4" was calculated as 90%, as it was correctly predicted 58 out of 64 samples.

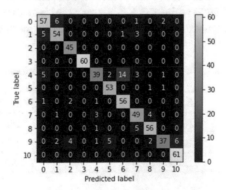

Fig. 6. SVM confusion matrix

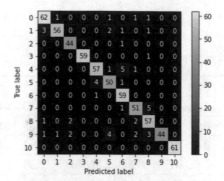

Fig. 7. KNN confusion matrix

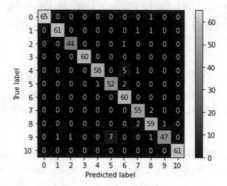

Fig. 8. RF confusion matrix

The comparative accuracy performance of the classification algorithms used is given in Table 6. As seen in Table 6, it was observed that the SVM algorithm achieved a maximum accuracy of 90% while the KNN algorithm had 91% accuracy and the RF algorithm has the highest accuracy rate of 96% under the same conditions and variables.

Table 6. Classification accuracy of algorithms

Classifier	Sensor	Accuracy
SVM	EMG+IMU	%90
KNN	EMG+IMU	%91
RF	**EMG+IMU**	**%96**

5 Conclusion

Within the scope of the study, recognition of some words selected from the TSL was conducted using machine learning methods on EMG signals gathered by the MYO armband. Both EMG and IMU sensors of the MYO armband were used to recognize ten emergency words selected in the scope of the study.

The effects of sensors used in classification on accuracy performance were analyzed separately and together. It has been determined that the classification performance was low on contrary to the previous study [9] which was implemented with a different dataset that included only the EMG sensors' data. Similarly, in another study [12] which uses only EMG signals, while the SVM classifier reaches 60.85% accuracy, it was observed that the accuracy was 58% when only EMG signals were used in our dataset. This was because the chosen words included the movement of the hand and arm at certain angles rather than a fixed muscle movement. Therefore, the IMU sensor played a more important role than the EMG sensor. In the previous study [16] using EMG and IMU signals together, 10% performance improvement was observed compared to the study using only EMG signals. Similarly, in our study, it has been determined that the use of the two sensors together provided an approximately 7% increase in performance compared to the use of only the IMU sensor, and 23% increase in performance compared to the use of only the EMG sensor.

Additionally, it was observed that there was a direct effect of the sliding window method used in the study on the algorithm performance rates. Therefore, the detailed analysis was carried out by using different parameters and as a result, it was decided to use the window size as 100 and the sliding interval as 20 in this study. Three different classification algorithms which were SVM, KNN, and RF were used in this study. And it was revealed that the maximum accuracy was achieved by the RF classifier as 96% when using EMG and IMU sensors' data together. It was noted that a 78% success rate was achieved in the previous study [16] using the RF classifier by using the signals from the EMG and IMU sensors' data together.

Based on the results of the study it can be concluded that the MYO armband can be used in recognition of sign language. However, since only one MYO armband was used in the study, the words that can be made with one hand were selected for the study. Therefore, in future studies, the number of defined words can be increased by using two armbands simultaneously. In addition, an application interface can be developed to display the movements recognized as a result of classification algorithms.

References

1. Vachirapipop, M., Soymat, S., Tiraronnakul, W., Hnoohom, N.: Sign Translation with Myo Armbands. In 2017 21st International Computer Science and Engineering Conference (ICSEC), pp. 1–5. IEEE (2017)
2. Amor, A.B.H., Ghoul, O., Jemni, M.: Toward sign language handshapes recognition using Myo armband. In: 2017 6th International Conference on Information and Communication Technology and Accessibility (ICTA), pp. 1–6. IEEE. (2017)

3. Vachirapipop, M., Soymat, S., Tiraronnakul, W., Hnoohom, N.: An integration of Myo Arm-bands and an android-based mobile application for communication with hearing-impaired per-sons. In: 2017 13th International Conference on Signal-Image Technology & Internet-Based Systems (SITIS), pp. 413–417. IEEE (2017)

4. T.C Aile ve Sosyal Hizmetler Bakanlığı, 2021, "Engelli ve Yaşlı Hizmetler Genel Müdürlüğü Engelli ve Yaşlı Bülteni Nisan-2021". https://www.aile.gov.tr/media/78170/eyhgm_istati stik_bulteni_nisan_2021.pdf. Accessed 29 June 2022

5. Khan, R.Z., Ibraheem, N.A., Meghanathan, N.: Comparative study of hand gesture recogni-tion system. In: Proceedings of International Conference of Advanced Computer Science & Information Technology in Computer Science & Information Technology (CS & IT), vol. 2, no. 3, pp. 203–213 (2012)

6. Pisharady, P.K., Saerbeck, M.: Recent methods and databases in vision-based hand gesture recognition: a review. Comput. Vis. Image Underst. **141**, 152–165 (2015)

7. Wibawa, A.D., Sumpeno, S.: Gesture recognition for Indonesian Sign Language Systems (ISLS) using multimodal sensor leap motion and myo armband controllers based-on naïve bayes classifier. In 2017 International Conference on Soft Computing, Intelligent System and Information Technology (ICSIIT), pp. 1–6. IEEE (2017)

8. Kim, H.J., Lee, Y.S., Kim, D.: Arm motion estimation algorithm using MYO armband. In: 2017 First IEEE International Conference on Robotic Computing (IRC), pp. 376–381. IEEE (2017)

9. Kaya, E., Kumbasar, T.: Hand gesture recognition systems with the wearable myo armband. In: 2018 6th International Conference on Control Engineering & Information Technology (CEIT), pp. 1–6. IEEE (2018)

10. Benalcázar, M.E., Jaramillo, A.G., Zea, A., Páez, A., Andaluz, V.H.: Hand gesture recognition using machine learning and the Myo armband. In: 2017 25th European Signal Processing Conference (EUSIPCO), pp. 1040–1044. IEEE (2017)

11. T.C Milli Eğitim Bakanlığı, Özel Eğitim ve Rehberlik Hizmetleri Genel Müdürlüğü, 2015, "Türk İşaret Dili Sözlüğü". https://orgm.meb.gov.tr/alt_sayfalar/duyurular/1.pdf. Accessed 30 June 2022

12. Savur, C., Sahin, F.: American Sign Language Recognition system by using surface EMG signal. In: 2016 IEEE International Conference on Systems, Man, and Cybernetics (SMC), pp. 002872–002877. IEEE (2016)

13. Madushanka, A.L.P., Senevirathne, R.G.D.C., Wijesekara, L.M.H., Arunatilake, S.M.K.D., Sandaruwan, K.D.: Framework for Sinhala Sign Language recognition and translation using a wearable armband. In: 2016 Sixteenth International Conference on Advances in ICT for Emerging Regions (ICTer), pp. 49–57. IEEE (2016)

14. Seddiqi, M., Kivrak, H., Kose, H.: Recognition of Turkish Sign Language (TID) using sEMG sensor. In: 2020 Innovations in Intelligent Systems and Applications Conference (ASYU), pp. 1–6. IEEE (2020)

15. Bhatti, S.A.: Finger movement classification via machine learning using EMG Armband for 3D Printed Robotic Hand (Doctoral dissertation, University of Minnesota) (2019)

16. Galea, L.C., Smeaton, A.F.: Recognising Irish sign language using electromyography. In: 2019 International Conference on Content-Based Multimedia Indexing (CBMI), pp. 1–4. IEEE (2019)

17. Tepe, C., Demir, M.C.: Real-Time Classification of EMG Myo Armband Data Using Support Vector Machine. IRBM (2022)

18. Merzoug, B., Ouslim, M., Mostefai, L., Benouis, M.: Evaluation of dimensionality reduction using PCA on EMG-based signal pattern classification. Eng. Proc. **14**(1), 23 (2022)

19. Javaid, H.A., Rashid, N., Tiwana, M.I., Anwar, M.W.: Comparative analysis of emg signal features in time-domain and frequency-domain using myo gesture control. In: Proceedings of the 2018 4th International Conference on Mechatronics and Robotics Engineering, pp. 157–162 (2018)

20. Zhang, Z., Su, Z., Yang, G.: Real-time chinese sign language recognition based on artificial neural networks. In: 2019 IEEE International Conference on Robotics and Biomimetics (ROBIO), pp. 1413–1417. IEEE (2019)

21. Bakırcıoğlu, K., Özkurt, N.: Classification of EMG signals using convolution neural network. Int. J. Appl. Math. Electron. Comput. **8**(4), 115–119 (2020)

22. Bharti, K.K., Singh, P.K.: Hybrid dimension reduction by integrating feature selection with feature extraction method for text clustering. Expert Syst. Appl. **42**(6), 3105–3114 (2015)

23. Fatih, O.N.A.Y., Ahmet, M.E.R.T.: Değişken Kuvvetli EMG Sinyallerinin Çok Değişkenli Görgül Kip Ayrışımı ile Analizi ve Sınıflandırılması. Int. J. Adv. Eng. Pure Sci. **32**(3), 229–238 (2020)

24. Too, J., Abdullah, A.R., Mohd Saad, N., Tee, W.: EMG feature selection and classification using a Pbest-guide binary particle swarm optimization. Computation **7**(1), 12 (2019)

25. Pisner, D.A., Schnyer, D.M.: Support vector machine. In: Machine learning, pp. 101–121. Academic Press (2020)

26. Jiang, L., Cai, Z., Wang, D., Jiang, S.: Survey of improving k-nearest-neighbor for classification. In: Fourth International Conference on Fuzzy Systems and Knowledge Discovery (FSKD 2007), vol. 1, pp. 679–683. IEEE (2007)

27. Belgiu, M., Drăguţ, L.: Random forest in remote sensing: a review of applications and future directions. ISPRS J. Photogramm. Remote. Sens. **114**, 24–31 (2016)

28. Caliskan, A., Badem, H., Çil, Z.A.: Determination of window size and sliding interval for EMG signals by using genetic algorithm. In: 2019 Medical Technologies Congress (TIPTEKNO), pp. 1–4. IEEE (2019)

29. Smith, L.H., Hargrove, L.J., Lock, B.A., Kuiken, T.A.: Determining the optimal window length for pattern recognition-based myoelectric control: balancing the competing effects of classification error and controller delay. IEEE Trans. Neural Syst. Rehabil. Eng. **19**(2), 186–192 (2010)

30. Nawaz, R., Cheah, K.H., Nisar, H., Yap, V.V.: Comparison of different feature extraction methods for EEG-based emotion recognition. Biocybernetics Biomed. Eng. **40**(3), 910–926 (2020)

Effects of Different Parameters on Sleep Quality

Ece Kunduracioglu ⓘ, Ece Korkmaz(✉) ⓘ, Tugba Gurgen Erdogan ⓘ,
and Fuat Akal ⓘ

Computer Engineering Department, Hacettepe University, Ankara, Turkey
ecekorkmaz55@gmail.com, tugba@cs.hacettepe.edu.tr, akal@hacettepe.edu.tr

Abstract. Sleep is not linear and contains different stages. The quality of sleep and time spent in each sleep stage directly affect a person's daily life. Sleep quality can be defined as how well an individual rested during the time he or she slept. In order to measure the sleep quality, sleep monitoring is used and the main focus of most of the works about sleep monitoring in the literature is the effects of environmental factors on sleep. Some of the previous works consider the effects of physical exercise or addiction on sleep quality. The works that are focusing on the effects of diseases on sleep quality usually examine a single disease. Unlike previous works in the literature, the main goal of our work is to examine different factors - such as multiple medications and different diseases - altogether and compare these factors to each other. In this paper, we analyzed patient data to reach our goal. A variation of the Cross Industry Standard Process for Data Mining (CRISP-DM) methodology is used to conduct analyses. K-means clustering algorithm is used for clustering the data. Random forest classifier is used in order to determine feature importance. Our analyses of patients showed that diseases they have, medications they use, and their age and gender have the most impact on sleep quality. Diseases that change breathing patterns and depression affect the sleep quality mostly. Different types of the antidepressants such as SSRI and trazodone have an important impact. Also, epilepsy is one of the factors that affect sleep quality.

Keywords: Sleep quality · Sleep stages · CRISP-DM · Data science

1 Introduction

Sleep is not linear, it contains different stages: rapid eye movement (REM) and non-rapid eye movement (NREM) sleep. NREM is divided into three stages which are N1, N2, and N3 [1]. The quality of sleep and time spent in each sleep stage directly affect a person's daily life. Sleep quality can be defined as how well an individual rested during the time he or she slept. Sleep quality has four attributes: sleep efficiency, sleep latency, sleep duration, and wake after sleep

---

E. Kunduracioglu and E. Korkmaz—Contributed equally to this work as first authors.

© The Author(s), under exclusive license to Springer Nature Switzerland AG 2023
F. P. García Márquez et al. (Eds.): ICCIDA 2022, LNNS 643, pp. 358–369, 2023.
https://doi.org/10.1007/978-3-031-27099-4_28

onset. There are other factors that may affect the sleep quality. Some of these factors are categorized as physiological, psychological, environmental factors, and family/social commitments. Physiological factors include age of the individual, and time spent in each sleep stage. Some diseases that are included in this work such as depression can be considered as one of the psychological factors. Poor sleep quality may have negative consequences on the individual's daily life such as fatigue [2]. Recently, there are a lot of research about this topic done by both physicians and data scientists [3].

Due to the current significance of this topic, we focus on sleep monitoring. Our main goal is to analyze the effects of different parameters on sleep quality. We examine the effects of different diseases, problems, medications, and gender on sleep quality. In the current research, we worked with ISRUC-SLEEP Dataset [4] which contains the records of sleep stages of patients. The dataset also contains various information about the diseases that the patients have and medications that are used.

In this research, we adapted the Cross Industry Standard Process for Data Mining (CRISP-DM) methodology [5] to investigate the effects of the different parameters in sleep quality to conduct a data science project. The methodology includes five steps which are business understanding, data understanding, data preparation, modeling, and evaluation. Firstly, we get domain understanding by working together with domain experts, they observed our results during the project and make some suggestions to us. Also, two experiments are conducted in order to detect important features that affect sleep quality. In the first experiment, the main focus is the sleep stages of the patients. In the second experiment, a clustering model for classifying patients is used. After the data is classified, the important features are extracted by developing a random forest classifier model [6]. In both experiments, we observed similar results for the important features. These features are some diseases a patient has, some medications he or she uses, and the patient's age and gender.

The remaining of this article is organized as follows: Sect. 2 summarizes the background on the sleep quality and the related works that have investigated the effects of parameters on sleep quality. Section 3 overviews the followed methodology and its steps including inputs, details, and outcomes of each step. Section 4 presents the results of the current research in relation to the formulated goals. Section 5 includes a summary of our findings, concludes the article, and plans for future works.

2   Related Work and Background

Related literature shows that sleep stages can be used to determine the quality of sleep [7]. There are many factors that affect the sleep quality, including gender and age, medication a patient takes such as selective serotonin reuptake inhibitors (SSRI), and diseases a patient has such as sleep apnea and depression. Some of these factors are used to measure sleep quality in literature, and we give some examples of these factors in the following.

2.1 Stages of Sleep

Sleep is not linear, sleep period of an individual is composed of different stages. A person goes through these different stages multiple times a night [1].

NREM can be considered in three different stages: N1, N2, N3. N1 is the first stage of the sleep cycle and is a transition period between wakefulness and sleep [8]. During N2, the body temperature decreases, the muscles relax, eye movement stops, and the breathing slows down along with the heart rate. In N3, deep and slow brain waves known as delta waves begin to emerge. During this stage of sleep which is also known as delta sleep, the sleeping individual is not easily disturbed by any change in the environment.

Unlike NREM, brain activity increases during **REM**. The brain activity gets similar to the brain activity of an awake person. However, the body of the sleeping individual is paralyzed except for the eyes and the vital organs. Even though dreams can occur in any sleep stage, they mostly occur and more intense during the REM stage [9]. REM sleep is believed to be essential to cognitive functions like memory, learning, and creativity. The human brain processes emotions during REM sleep. Dreaming and memory consolidation occur during this stage of sleep as well, but it is important to note that they also occur during the NREM stages. Memory consolidation includes processing new learnings and motor skills and maintaining the memories made during the day. It can be noted that healthy adults usually spend 20–25% of their sleep duration in the REM stage [7].

2.2 Factors that May Affect Sleep

There are several factors that may affect sleep quality. These are gender, age, sleep apnea (SAOS), snoring (roncopatia), bipolar disorder (D. Afectiva), SSRI, benzo-diazepines (BZD), trazadone, depression, cheyne-stokes syndrome, and epilepsy.

There are many studies that try to explain the effects of gender on the sleep quality. It is known that gender also affects diseases a person may have more tendency to catch different diseases. These diseases or life stages affect sleep quality as well. For example, women who entered menopause may have trouble sleeping [9].

According to a study done in Spain, poor sleep quality is very common among adults, particularly among females. This paper also notes that there is a direct relationship between age and a decrease in sleep quality, and this relationship also appears to be more consistent in females. It is also shown that the amount of sleep males and females get is not very different, but the sex affects the sleep quality of the subjects [10]. Zhang and Wing [11] showed that females are 1.41 times more likely to suffer from insomnia than males at all ages but especially at advanced ages.

Sleep apnea is one of the sleep disorders that affect the breathing of the patient. The patient's breathing may pause or become shallow during the sleep period, and this can occur more often than normal [12]. It can be related to obesity, having a large neck, age, genetics, smoking, drinking alcohol, and sleeping on your back [12]. As stated before, it causes a patient to have trouble breathing while sleeping, hence it affects the sleep quality.

Obstructed air movement may occur during breathing while the patient is sleeping. The obstruction of air movement causes vibrations of respiratory structures, and these vibrations generate a sound. The obstruction and the sound it causes are known as snoring. Snoring can be a sign of sleep apnea. Its causes are similar to sleep apnea's causes. Even if it usually does not cause a patient to stop breathing entirely, it lowers sleep quality, because patients wake themselves up without realizing it.

Bipolar disorder (manic depression) is a mental health condition that affects the individual's moods. His or her moods can change from an extreme mood to another [12]. Bipolar disorder frequently includes sleep disturbance [13]. Hence, the patients with bipolar disorder are more likely the have poorer sleep quality than healthy people.

Selective serotonin reuptake inhibitors (SSRI) are mainly prescribed to treat depression, especially in severe cases. Since SSRIs have fewer side effects compared to other antidepressants, they are one of the first-choice medicines to prescribe in order to treat depression. It's thought that SSRIs work by increasing serotonin levels in the brain. It's thought to have a good influence on mood, emotion, and sleep [12].

Benzodiazepines (BZD) are prescribed to treat conditions such as anxiety, insomnia, and seizures [12]. Although they are given to patients to treat insomnia, they may lead to adverse effects in adults with insomnia as well [14].

Trazodone is an antidepressant that is prescribed to treat depression, anxiety, and insomnia. Trazodone was commonly prescribed medicine for insomnia in the early 2000s, but it is important to note that studies that are done to determine its effect on sleep quality have been in depressed individuals.

Sleep disturbances are an integral aspect of depression [15]. People with depression may find it difficult to fall asleep and stay asleep during the night. They can also have excessive daytime sleepiness or even sleep too much. Hence, their sleep quality may decrease. Depression and sleep quality are dependent on each other: Poor sleep quality may cause depression and depression may cause poor sleep quality.

It is also important to note that "emotional memories" are processed during the REM stage of sleep [16]. Depression can cause "negative emotional memories" to be processed and make patients wake up more frequently or have trouble sleeping.

Cheyne-stokes respiration is a rare abnormal breathing pattern that mostly occurs during sleeping, but it can also occur while the patient is awake. The pattern involves a period of fast, shallow breathing followed by slow, heavier breathing and moments without any breath at all, similar to sleep apnea [17]. It may be directly related to sleep apnea and snoring. It causes patients to wake up frequently during the night and decreases the quality of sleep.

Epilepsy is the condition of recurrent, unprovoked seizures. Epilepsy has numerous causes, each reflecting underlying brain dysfunction [18]. Anti-epileptic drugs (AEDs) can cause some side effects concerning the sleep quality of the patient. The effect of AEDs can change from tiredness to insomnia or disrupted sleep, depending on the patient and the dose. Some AEDs may increase the length of deep sleep. Hence, the sleep quality of the patient may increase [19].

3 Methodology

In our research an adapted version of CRISP-DM methodology is used. Implementation of this adaptation is done with Python, using Jupyter Notebook. The process splits the research into five stages, which are business understanding, data understanding, data preparation, modeling, and evaluation. Our adaptation of CRISP-DM methodology and the steps that we followed are briefly shown in Fig. 1.

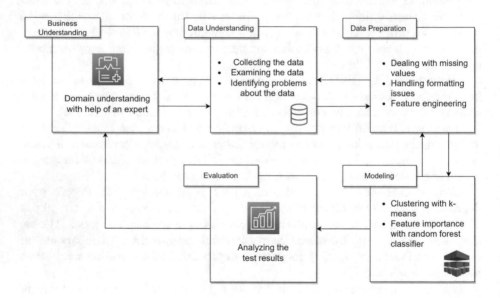

Fig. 1. CRISP-DM methodology diagram

3.1 Business Understanding

The main goal of this step can be stated as understanding the problem and finding business requirements for the problem. The risks are determined, and the initial plans of the research are created in this step. The perspective of a domain expert about the problem can be beneficial [5].

The aim of this study is to do data science research with sleep monitoring data for finding the most important factors that affect sleep quality.

3.2   Data Understanding

In this step, the goal can be summarized as collecting the data, examining it, and identifying the problems about the data. This step is about understanding the data that will be worked on [5].

In this work, we chose to use the ISRUC-SLEEP public dataset [4]. Because, the ISRUC-SLEEP Dataset contains considerably more various features, like diagnosis and medications, than other publicly open datasets. In addition, before choosing to use the ISRUC-SLEEP dataset for our work, we consulted with a physician who also examined the dataset and approved it for our use. The dataset is a polysomnographic dataset. Polysomnography can be explained as a sleep study and it is used for diagnosing sleep disorders [20]. The dataset contains information in both Portuguese and English. In the dataset, patients are referred to as subjects. The dataset contains both healthy subjects and subjects that have sleep disorders and/or are prescribed with some sleep medications [4].

The ISRUC-SLEEP dataset contains three different subgroups. The first subgroup contains 100 adult subjects who have been diagnosed with sleep problems, some of the subjects have other diagnoses, and some of them are using different medications. The second subgroup is gathered from eight subjects with sleep-related diseases. This subgroup's features are slightly different. Two different recordings on different dates for each subject are gathered for this subgroup. Due to these differences from other subgroups, we did not use the second subgroup in our experiments. The third subgroup contains 10 healthy subjects. Subjects of the third subgroup do not have any diseases related to sleep, and it is stated that this subgroup can be used as a control group. In our experiments, we chose to use the first and the third subgroups. For both subgroups, there is one data acquisition session per subject, and two observations which are created by two experts. Features of the first subgroup and the third subgroup are compatible, and as mentioned before, the third subgroup can be used as a control group [4].

The diseases and medications of five patients are shown in Table 1 and the sleep stages of five patients are listed in Table 2. The first sample shows a male patient who is 64 years old. He is diagnosed with sleep apnea (SAOS). He also has depression, and he uses SSRI and trazadone which are shown in the medication column. Any EEG alterations are not observed in this recording. The patient's sleep cycles are observed by two experts. The first expert observed the REM percentage of this patient as 13.41%. The second expert observed the REM percentage of the patient as 13.98%.

The fourth sample shows a male patient who is 27 years old. The patient is diagnosed with upper airway resistance syndrome (SRVAS: síndrome da resistência das vias aéreas superiores) and epilepsy. There is no information on whether he uses any medication or not. His EEG alterations were noted as paroxistic activity left frontoparietal. The first expert observed awaken stage (W) percentage of this patient as 2.91%. The second expert observed the REM percentage of the patient as 2.70%.

Table 1. Five samples of the dataset with demographics, diagnosis, medications and other information related to sleep quality of the subjects

Subject	Age	Sex	Diagnosis	Other problems	Epoches	Date of recording	Medication	EEG alterations
1	64	M	SAOS	Depression	880	18/5/2009	SSRI; tradzona	
2	52	M	SAOS	Restless leg syndrome	964	2009-08-06		
3	38	M	REM Sleep Behaviour Disorder	PLMS	943	21/5/2009	Risperidona; Tegretol	
4	27	M	SRVAS	Epilepsy	963	18/6/2009		Paroxistic activity left fronto-parietal
5	58	F	SAOS	Insomnia	875	25/5/2009	BZD	

Table 2. Five samples of the dataset with the percentage of time spent in each sleep stage (N1, N2, N3, REM) and awaken stage (W) are shown. The percentages with .1 show the second expert's observations.

Subject	W%	N1%	N2%	N3%	REM%	W%.1	N1%.1	N2%.1	N3%.1	REM%.1
1	30.00	8.30	22.05	26.25	13.41	30.91	4.77	24.09	26.25	13.98
2	25.41	11.93	35.79	16.29	10.58	21.99	11.93	44.40	11.10	10.58
3	14.00	17.50	26.09	18.35	24.07	12.73	4.45	39.24	15.69	27.89
4	2.91	6.75	44.24	22.22	23.88	2.70	3.43	43.61	25.65	24.61
5	33.83	12.34	30.29	18.74	4.80	35.66	4.69	35.66	18.86	5.14

3.3   Data Preparation

This step is about preparing the data for building a machine learning model. It is almost in the nature of data science projects that the dataset to be used is ideal, and complete for model building tasks. The data preparing steps are essential and the most time-consuming part of data-science research on healthcare data which is highly complex, multi-disciplinary, and ad hoc [21,22]. Therefore, before using the data we should correct it by filling in the missing values or doing spelling checks, and parsing augmented data. In addition, some feature engineering can be applied to the data. For this purpose, we used various data preparation methods.

First, it can be noted that most of the algorithms cannot handle null values, hence we do not want any null values in our dataset. However, in the ISRUC-SLEEP Dataset, if the patient does not have other problems or does not take any medication, the doctors left that feature as a null value. Therefore, we filled the null values with 'no problem', or 'not use' phrases. In addition, the patients in subgroup 3 do not have any diagnosis, thus we fill the null values in the 'Diagnosis' column as 'none'. Additionally, the missing values in the age column are filled with K-Nearest Neighbors Imputer of Sklearn library [23]. With this method, missing values are filled with the mean value of the k-nearest neighbors.

Second, it can be stated that the attributes of the subjects are not collected in the same format. If the patient has more than one problem or uses more than one medication, the doctors add the multiple features within one column by separating them with a semicolon ';' or comma ','. Also, the data contains different spellings of the same value. Therefore, we apply some string operations to create the set with distinct values. We used one-hot encoding to change categorical features to binary values.

At last, we did some feature engineering on our data. We observed that two different experts divide patients' sleep into similar stages. In order to make our data clear and more understandable, we decided to use the average value of the observations of the experts.

It can also be noted that some features are not used in experiments. The unnecessary features such as the date of the recording are extracted from the data. In addition, since it overshadowed the other attributes the age attribute is not used in the experiments.

3.4   Modeling

In this step, the most suitable model is selected. For the first experiment, the data is split into five different classes according to the REM amount. In addition, we also want to cluster our data. K-means is a well-known and most used clustering algorithm [24]. It was suitable for our dataset, as the dataset is not hierarchical and it is noisy. Therefore, we clustered our data with k-means. Different k values are tried and their inertia values are calculated to determine which K value is more suitable for the dataset. From the inertia values, we chose five clusters for the dataset. K is chosen according to the point at which inertia starts decreasing linearly. The graph can be seen in the Fig. 2.

We conducted two different experiments. In the experiments, the subjects are classified into five groups. In the first experiment, the subjects are divided into classes by directly using the REM value, and in the second experiment, the subjects are divided into classes by using a clustering model. Since the data is classified, a random forest classifier is used for finding important features. A random forest creates decision trees on randomly selected data samples. It gets predictions from each tree and performs voting in order to select the best solution. Random forests are considered a highly accurate and robust method [25]. We used the random forest classifier from the Scikit Learn library with default values [26]. The random forest classifier of scikit learn library has a feature importance attribute that uses Impurity-based feature importances. By using the library we found the importance values of all features.

3.5   Evaluation

In this step, the main goal is to evaluate the test results in both the business domain and model performance [5]. Since we did two experiments, we achieved two different outcomes. We evaluated the results individually, and we compared them with each other.

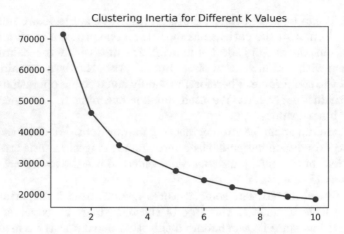

Fig. 2. The K-means clustering inertia graph

4   Analyses and Evaluation

The feature importance values of all features for clustered data and the data
with REM groups are examined. For finding the most important features we
chose to use a threshold. After different experiments, we decided to use 0.03
as the threshold value. In Fig. 3 the features that have bigger importance than
the threshold can be seen. Orange bars represent the most important features of
clustered data. Blue bars represent the most important features for REM groups.

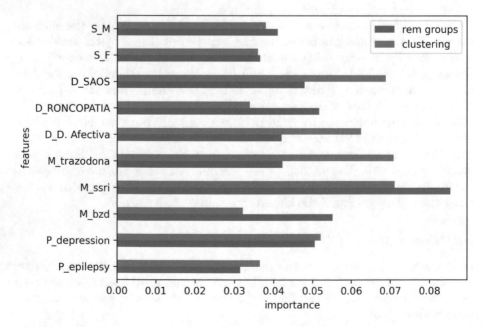

Fig. 3. Most important features of REM groups and clustering

From the figure, it can be observed that there are 11 features, but two of these features (S_M, S_F) represent gender. Therefore, we can assume that this figure shows the ten most important features, except age. Even though the importance values of the features are different in the two experiments, the most important features are the same for both experiments, such as SSRI and depression.

5 Conclusion

This research shows that some diseases and medications affect the sleep quality of a patient more than other factors. We observed that some medications affecting sleep quality are antidepressants, such as trazodone and SSRI. It is also observed that depression affects the sleep quality of a patient, which is compatible with the literature. Additionally, diseases that change the breathing pattern of a patient affect sleep quality. We can say that our analyses are supported by both observer and theory triangulation methods.

Healthcare data is more complex and limited compared to other domains. It can be noted that patients have legal rights to preserve their personal data, and they may choose to not share any information about their health records. This is one of the reasons why the number of samples in other domains outnumbers the healthcare data. In addition, more domain knowledge is required when working with healthcare data.

The dataset we used in our work contains data in both Portuguese and English. Due to this language variety, different abbreviations were used for the same diseases. We were obliged to do some additional research in order to learn about these abbreviations and understand the data correctly.

It is also important to note that healthcare data usually depends on human observations. Different human agents may make different observations on the same patient. As an example, in our dataset, two experts made slightly different observations on the same patients. Since these observations are subjective, results for similar samples might be slightly different.

In the future, the experiments can be extended with different models and algorithms. In addition, different datasets can be tried with our approach and the results can be compared with our current most important features. Furthermore, our work can be adapted to software for medical usage.

The dataset contains many patients with sleep apnea as a primary disease. Literature and our work show that sleep apnea may affect the patient's sleep quality. In the future, research about how sleep apnea affects the sleep quality of a patient can be done. Patients can be grouped considering if a patient has only sleep apnea or he or she has other diseases. In addition, different sleep stages can be used for grouping patients.

Acknowledgements. The authors thank O. Pinar Tayfun, M.D., general practitioner for her valuable suggestions.

References

1. Foulkes, W.D.: Dream reports from different stages of sleep. Psychol. Sci. Public Interest **65**(1), 14 (1962)
2. Nelson, K.L., Davis, J.E., Corbett, C.F.: Sleep quality: an evolutionary concept analysis. In: Nursing Forum, vol. 57, pp. 144–151. Wiley Online Library (2022)
3. Mendonca, F., Mostafa, S.S., Morgado-Dias, F., Ravelo-Garcia, A.G., Penzel, T.: A review of approaches for sleep quality analysis. Ieee Access **7**, 24527–24546 (2019)
4. Khalighi, S., Sousa, T., Santos, J.M., Nunes, U.: Isruc-sleep: a comprehensive public dataset for sleep researchers. Comput. Methods Programs Biomed. **124**, 180–192 (2016)
5. Wirth, R., Hipp, J.: Crisp-dm: towards a standard process model for data mining. In: Proceedings of the 4th International Conference on the Practical Applications of Knowledge Discovery and Data Mining, vol. 1, pp. 29–39. Manchester (2000)
6. Breiman, L.: Random forests. Mach. learn. **45**(1), 5–32 (2001)
7. Pacheco, D.: Natural Patterns of Sleep. https://healthysleep.med.harvard.edu/healthy/science/what/sleep-patterns-rem-nrem. Accessed 30 July 2022
8. The four stages of sleep. https://www.verywellhealth.com/the-four-stages-of-sleep-2795920. Accessed 29 July 2022
9. Patel, A.K., Reddy, V., Araujo, J.F.: Physiology, sleep stages. In: StatPearls [Internet]. StatPearls Publishing (2021)
10. Madrid-Valero, J.J., Martínez-Selva, J.M., Couto, B.R.d., Sánchez-Romera, J.F., Ordoñana, J.R.: Age and gender effects on the prevalence of poor sleep quality in the adult population. Gaceta sanitaria **31**, 18–22 (2017)
11. Zhang, B., Wing, Y.K.: Sex differences in insomnia: a meta-analysis. Sleep **29**(1), 85–93 (2006)
12. https://www.nhs.uk/conditions . Accessed 29 June 2022
13. Kaplan, K.A., Talbot, L.S., Gruber, J., Harvey, A.G.: Evaluating sleep in bipolar disorder: comparison between actigraphy, polysomnography, and sleep diary. Bipolar Disord. **14**(8), 870–879 (2012)
14. Holbrook, A.M., Crowther, R., Lotter, A., Cheng, C., King, D.: Meta-analysis of benzodiazepine use in the treatment of insomnia. CMAJ **162**(2), 225–233 (2000)
15. Thase, M.E.: Depression and sleep: pathophysiology and treatment. Dialogues in clinical neuroscience (2022)
16. Walker, M.P., van Der Helm, E.: Overnight therapy? the role of sleep in emotional brain processing. Psychol. Bull. **135**(5), 731 (2009)
17. Pacheco, D.: What is Cheyne-Stokes breathing? https://www.sleepfoundation.org/sleep-apnea/cheyne-stokes-respirations. Accessed 29 June 2022
18. Stafstrom, C.E., Carmant, L.: Seizures and epilepsy: an overview for neuroscientists. Cold Spring Harb. Perspect. Med. **5**(6), a022426 (2015)
19. Society, E.: Sleep and epilepsy. https://epilepsysociety.org.uk/about-epilepsy/epileptic-seizures/seizure-triggers/sleep-epilepsy. Accessed 29 July 2022
20. Clinic, M.: Polysomnography (sleep study). https://www.mayoclinic.org/tests-procedures/polysomnography/about/pac-20394877. Accessed 30 July 2022
21. Erdogan, T.G., Tarhan, A.K.: Multi-perspective process mining for emergency process. Health Informatics J. **28**(1), 14604582221077196 (2022)
22. Rebuge, Á., Ferreira, D.R.: Business process analysis in healthcare environments: a methodology based on process mining. Inf. Syst. **37**(2), 99–116 (2012)
23. Sklearn: sklearn.impute.KNNImputer. https://scikit-learn.org/stable/modules/generated/sklearn.impute.KNNImputer.html. Accessed 05 Aug 2022

24. Na, S., Xumin, L., Yong, G.: Research on k-means clustering algorithm: an improved k-means clustering algorithm. In: 2010 Third International Symposium on Intelligent Information Technology and Security Informatics, pp. 63–67. IEEE (2010)
25. Navlani, A.: Understanding Random Forests Classifiers in Python Tutorial. https://www.datacamp.com/tutorial/random-forests-classifier-python. Accessed 30 July 2022
26. Sklearn: sklearn.ensemble.RandomForestClassifier. https://scikit-learn.org/stable/modules/generated/sklearn.ensemble.RandomForestClassifier.html. Accessed 30 June 2022

Predictive Maintenance on Industrial Data Using Soft Voted Ensemble Classifiers

Ümit Dilbaz[1]([✉]) [iD] and Mustafa Özgür Cingiz[2] [iD]

[1] Intelligent Systems Engineering Department, Bursa Technical University, Bursa, Turkey
umitdlbz@gmail.com
[2] Computer Engineering Department, Bursa Technical University, Bursa, Turkey
mustafa.cingiz@btu.edu.tr

Abstract. The IoT sector leads improvements in the Industry 4.0 revolution. Failure-prone machinery puts operations and production costs at risk. A sudden failure results in high downtime expenses and a drop in output. Predictive maintenance is a crucial area to research in order to increase industrial productivity. The purpose of the paper is to present the proper method for implementing predictive maintenance using the soft voting undersampling approach. We also highlight the impact of ensemble learning on an imbalanced dataset that is crucial problem for modelling the failure-prone machinery dataset. This paper suggests a method for enhancing predictive learning by selecting six different machine learning algorithms, including decision tree, random forest, Gradient Boosting Machines (GBM), XGBoost, Light GBM, and CatBoost classifiers. The soft voted model is proposed to enhance the performance of machine learning classifiers, which have produced better results in terms of the Fowlkes-Mallows Index (FMI) and Cohen's Kappa score. The results of our study are close to the results of predictive maintenance studies in the literature.

Keywords: Predictive maintenance · Undersampling · Ensemble learning

1 Introduction

The internet of things (IoT) is a new topic related to the operation of various internet devices embedded with electrical components, detectors, and sensors. This word has been applied to the components of the home, industrial, consumer, commercial, economic, and infrastructural areas broadly. Industrial IoT is concerned with using technology in the manufacturing sector to maximize labour productivity. Unplanned stops in a production area have a negative impact on the industrial organization's production quality and costs as well as its ability to compete in the industry. This makes it crucial to constantly check on the condition of machinery and equipment and to take action before unanticipated downtime occurs.

Technology advancements have increased the variety and flexibility of statistical instruments, enabling us to measure a variety of machines and systems in metric units. It is now possible to anticipate the occurrence of the fault condition by the combination of the predictive techniques. Predictive maintenance is necessary to increase device functionality, boost output, and improve workflow. Industrial equipment collects a variety of characteristics to maintain its operation.

Many terminologies and categories of maintenance management techniques have been developed in the literature. The expert categorizes maintenance strategies as "Run to Failure," "Preventive Maintenance," or "Predictive Maintenance" [1]. Predictive maintenance, also known as "online monitoring," "condition-based maintenance," and "risk-based maintenance," has a long history of depiction [2]. Predictive maintenance allows machines to diagnose, decide, and act based on the system's knowledge and dataset training. Various techniques for improving predictive maintenance have been modelled in several research. Datasets commonly used in the methodologies are substantially skewed because the datasets are assumed to contain an equal number of examples as bias in datasets influences predictive maintenance. To deal with the imbalanced datasets, several sampling approaches, such as undersampling and oversampling, are proposed [3]. Various algorithms, such as SMOTE, ADASYN, SMOTE + Tomek, SMOTE + SVM, SMOTE + ENN and Borderline SMOTE, are used to overcome the problem of imbalance. To reduce the imbalance of the dataset for minority classes, undersampling approaches can be generally used in combination with oversampling methods. Various approaches are utilized in undersampling, one of the simplest being to delete the samples from the majority class without considering their relevance.

Ensemble modelling, which combines two separate models into one to obtain high accuracy and improved training outcomes, can also be used to improve results on datasets, it includes bagging, stacking and boosting [4]. The voting techniques of hard voting and soft voting are used in ensemble methods to make predictions. Soft voting concentrates on the cumulative probability of the most predictions, whereas hard voting entails selecting the prediction with the biggest number of votes. The conjunction of undersampling and voting approach is proposed in this article, the model implementation is divided into two steps which uses ensemble modelling by combining 6 different machine learning (ML) algorithms to obtain high performance through predictive maintenance.

Imbalanced data is a concern for predicting machine failure since it causes the model to be biased and overfitted. Undersampling is utilized to balance the dataset in this study, and it is supplemented with other approaches such as voting and ensembled procedures. Our research employed a synthetic yet imbalanced dataset to show a real-world phenomenon to reduce unexpected machine failure.

In order to prevent the data loss of the majority class, subdatasets are repeatedly created using the undersampling method in this study. By repeatedly merging each subdataset created from the majority class and the minority class, several multiple small balanced datasets are obtained. All of these balanced subdatasets are trained using a specific ML algorithm in the first step, then the results are voted on. And for the second step, the outcomes of six different ML models that were obtained in the same manner are once more voted for by an ensemble model in order to produce a final prediction and accurate results.

The contribution of our paper can be listed as:

- In our study, an imbalanced data set was chosen, which is a great challenge in terms of classification performance, but on the other hand it represents the nature of the manufacturing environment realistically.
- While balancing the dataset by downsampling method, most prominent disadvantage is losing the vast amount of data. In our approach this disadvantage of downsampling eliminated by using each instance of train data in multiple subdatasets.
- The ensemble learning approach was experimented in this study and their performances were investigated in terms of several performance metrics, such as recall value, FMI score, Cohen's Kapp score, AUC score and F1 score.

Furthermore, the article is divided into six sections. Predictive maintenance, its uses, fieldwork, and problem assessment are all described in Sect. 1 of this article. Section 2 discusses the article's contribution in relation to related work and different methods used to address the identified issue. The proposed methodology is declared in detail in Sect. 3. The case study's findings are presented in Sect. 4. In Sect. 5, the proposed model results are evaluated and discussed. Lastly, the ideas for the future work are presented in Sect. 6.

2   Related Works

Although there are many ways to implement predictive learning, anomalies in datasets have frequently made the results prone to error and difficulty in machine's fault prediction has been observed. To remove the imbalance in the datasets, the authors have suggested a number of different techniques. The work in the field of predictive maintenance is discussed in this section. Authors in [5] has suggested combining the Auto Encoders and Long Short-Term Memory (LSTM) to use the deep learning technique to address the issues in anomaly detection. The deep learning framework examines the anomalies and their patterns while correlating the progression with multiple datasets by analysing the obtained datasets from multiple sensors. The system will be able to identify more classes of anomalies from the input data with the aid of the unsupervised learning technique using auto encoders.

Authors in [6] presented the Explainable Artificial Intelligence (XAI) method and provided a synthetic yet realistic predictive maintenance dataset for evaluation because public access to predictive maintenance datasets is uncommon. The authors described, trained, and evaluated an explainable model and an explanatory interface, and their descriptive abilities were contrasted against the dataset with 10,000 points and 6 features. The predictive maintenance phenomenon in the real world is accurately portrayed by the data model. The decision tree has been used to represent the model performance, and the results show significant performance improvement. The NASA repository is the basis for the machine learning algorithm that the authors have suggested in [7]. To forecast the lifespan of the turbofan engine, algorithms are proposed. The time-series dataset is used and trained in parallel with ten machine learning algorithms, and the dataset includes both training and test datasets. The dataset already contained the Remaining Useful Lifetime (RUL) value. Each dataset is subjected to the application of all ten machine learning algorithms, including decision trees, random forest, and support vector machine (SVM) among others. The outcomes are compared by comparing the predicted value of RUL to the actual value of RUL. By analysing vibration signals using K-Nearest Neighbour (KNN) and SVM, the authors in [8] proposed a methodology for predicting the pattern of behaviour of bearings in wind turbines. To classify the fault pattern, bearing behaviour has been observed. Several vibration sensors were tracked to collect the data for the analysis. The Hilbert Transform is used for signal processing. Four ML algorithms were applied to the data evaluation process. The accuracy of the KNN was observed to be 87%, which was higher than that of the SVM, Euclidean distance, and K-means in the pattern of bearings. The proposed method was limited to fault prediction.

The authors in [9] suggested using artificial neural network (ANN) to predict potential faults in a wind turbine test rig that is intended to replicate the functionality and operation of an original wind turbine to detect the fault in a turbine. The high shaft and slow shaft speeds were adjusted according to the wind turbine's high rpm. Three test cases were run to verify the operational functionality, and through data acquisition, signals were monitored, and vibration analysis was performed to distinguish between the healthy and faulty conditions, ANN was used by applying gradient descent, and for evaluation mean squared error was used. The predictive performance accuracy was 92% using ANN. The Authors in [10] outlined the process for using the K-fold cross-validation method for generalization to train the dataset obtained from vibration signals by a neural network. The authors in [11] suggested using a random forest approach to predict industrial motor downtime. By using Azure Machine Learning Studio to train a random forest approach, 95% accurate results were achieved on a dataset of 50 thousand readings and 15 different machine features. The proposed predictive maintenance enables the adoption of real-time dynamical decision rules for maintenance management from the tested cutting machine. Preliminary results demonstrate the proper behaviour of the approach.

The authors in [12], proposed LSTM based predictive maintenance by using the hyperparameter optimization with genetic algorithms. In the LSTM model developed in this study, several feature vectors were obtained by using each feature by itself and combination of some features. The feature vectors are trained separately to find best model. The feature vector that consisted of kurtosis, skewness and crest factor gave the highest validation accuracy score. In the study [13], a hybrid deep learning model proposed. The authors compared the regular LSTM model and convolutional neural network-based LSTM model which resulted with the higher performance of hybrid model (CNN-LSTM). The authors in [14], studied on XGBoost, gradient boosting, AdaBoost, random forest, multilayer perceptron and support vector machine algorithms to develop a predictive maintenance system for the manufacturing environment. Principal component analysis (PCA) was performed to reduction of high feature space. The best performance was obtained with random forest model.

3 Proposed Methodology

The following are the subsections for this section. The dataset used in this study is described in Sect. 3.1. The proposed data balancing method is described in detail in Sect. 3.2. The proposed model is presented along with an explanation of the system's flowchart in Sect. 3.3.

3.1 Dataset Description

The dataset being used is artificial, but it accurately represents the real predictive maintenance data that industrial encounter. It has 10.000 row data points and is taken from [6]. Table 1 contains a summary of the dataset.

Table 1. Summary of dataset

Feature	Description
Type	Product quality variant (Low, Medium, High) – Categorical feature
Air Tempeature [K]	Ambient temperature
Process Temperature [K]	Operation temperature
Rotational Speed [rpm]	The speed at a power of 2860 W
Torque [Nm]	Torque of the machine
Tool Wear [min]	2, 3 and 5 min
Machine Failure	Machine Failure Status (Binary Label – 0 refers to machine non-failure, 1 refers to machine failure)

Table 2. Randomly chosen samples from datasets

Type	Air temperature	Process temperature	Rotational speed	Torques	Tool wear	Machine failure
M	298.1	308.6	1551	42.8	0	0
L	298.2	308.7	1408	46.3	3	0
L	298.1	308.5	1498	49.4	5	0
L	298.2	308.6	1433	39.5	7	0
H	296.9	307.8	1549	35.8	206	1
L	296.9	307.5	2721	9.3	18	1
H	296.6	307.7	1386	62.3	100	1
M	296.7	307.8	1258	69.0	105	1

Table 2 shows some sample which are chosen randomly from dataset. The "Type" is categorical feature and it processed with one hot encoding method in this study. The other features are numerical. The "Machine Failure" label is binary. The value "0" refers non-failure while the value "1" refers to failure situation with the machine.

The dataset is imbalanced, as is typical for industrial data that displays the status of machine failure. 9661 labels refer to "machine non-failure," while 339 labels refer to "machine failure." The performance of the model in this situation depends heavily on how the imbalanced data are handled. In a [15] the author used oversampling methods based on SMOTE to solve this issue.

3.2 Balancing the Dataset with Undersampling

To prevent the data leakage phenomenon, the dataset is divided into train and test portions of 80% and 20%, respectively, before any process is applied to the entire dataset. For the model's prediction, the test dataset with its 2000 datapoints is retained. 8000 data points make up the train dataset, which is still imbalanced. (7736 data points for machine non-failure, 264 data points for machine failure.)

The problem of imbalanced data was addressed in this study by repeatedly undersampling members of the majority class and combining them with members of the minority class. By employing this technique, numerous balanced subdatasets with an equal sample of machine failure and machine non-failure classes can be obtained.

3.3 Proposed Model

Several subdatasets and various machine learning algorithms are used to propose the 2-Stage Cascade Voting Model. The proposed model primarily uses hard voting and soft voting. In a hard vote, the voting classifier counts the number of instances of each class before allocating a test instance to the class that received the majority of votes from the classifiers. Soft voting uses a probability term to categorise the test instance that averages the probabilities for each class. Two stages can be used to examine the proposed model:

Stage 1. As shown in Fig. 1, the "k" pieces balanced subdatasets were created by merging instances from the majority class and minority class simultaneously. The "k" value which determines the number of subdatasets is obtained by dividing the number of instances in the majority class by the number of instances in the minority class in train datas. (k = 29 for this study, where was calculated by dividing 7736 in 264) The decision tree model has been used to independently determine predictions and prediction probabilities for each subdataset. To determine the final prediction probability for the first stage, the prediction probabilities of each decision tree model were added to the soft voting. Utilizing a different decision boundary, the final prediction's probability is assessed.

Stage 2. As shown in Fig. 2, six different machine learning algorithms, including Decision Tree, Random Forest, Gradient Boosting Machines, Extreme Gradient Boosting (XGBoost), Light GBM, and CatBoost, are repeated in parallel during the first step. Six machine learning models are soft-voted to create the final ensemble model.

376   Ü. Dilbaz and M. Ö. Cingiz

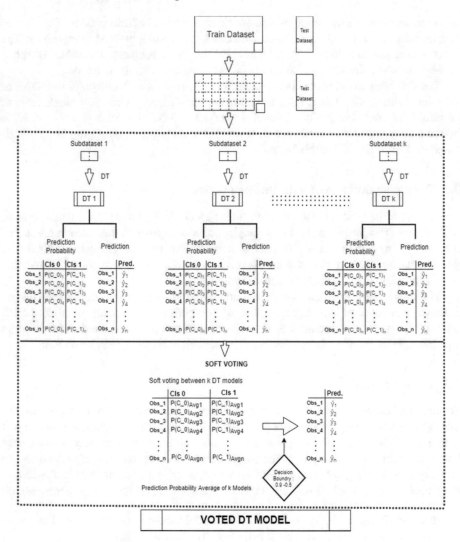

Fig. 1. Proposed model work flow (Stage-1)

For each step, the FMI, Cohen's Kappa Score, and recall values are taken into account when determining the decision boundary for the prediction optimization. The stage 2 can be examined in a flow chart as in Fig. 2

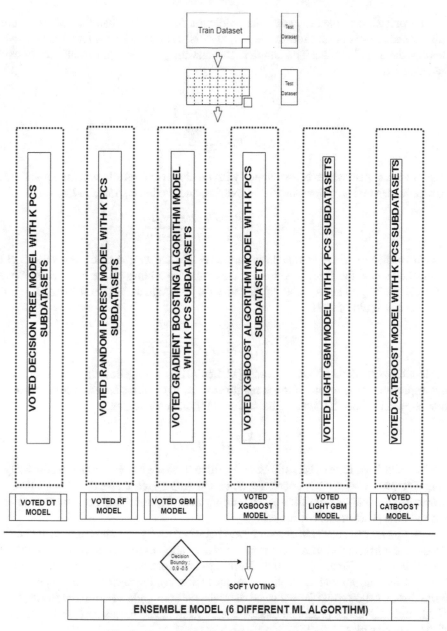

Fig. 2. Proposed Model Work Flow (Stage-2)

4 Results

The proposed model's performance is assessed using a variety of performance metrics such a, precision, recall, F1 score, FMI, Cohen Kappa and AUC score.

The precision is used to assess the prediction for classes when the data is severely imbalanced. Precision is used to identify results' relevance while recall is utilised to identify the actual results. The simplest formula for precision and recall are shown in Eqs. 1 and 2:

$$Precision = \frac{TP}{TP + FP} \tag{1}$$

$$Recall = \frac{TP}{TP + FN} \tag{2}$$

F1 score is a metric which is formulated with precision and recall combinations. It is the harmonic mean of recall and precision metrics, and it can be shown in Eq. 3:

$$F1 = 2x\frac{Precision \times Recall}{Precision + Recall} \tag{3}$$

A performance metric to evaluate the similarity of clusters obtained from various clustering methods is the FMI. It addresses the similarity among clusterings and it is characterized as the geometric mean of the pairwise precision and recall metrics. The Eq. 4 shows the formula of FMI

$$FMI = \sqrt{\frac{TP}{TP + FP}x\frac{TP}{TP + FN}} \tag{4}$$

The level of agreement is measured using Cohen's Kappa's and how the model performed in comparison to the extent of those random events in the class based on how frequently they happened. The Eq. 5 shows the formula for Cohen's Kappa.

$$k = \frac{P_o - P_e}{1 - P_e} \tag{5}$$

The True Positive Rate (TPR) and False Positive Rate are plotted on the Receiver Operating Characteristic (ROC) curve. AUC, or the entire area under the ROC, illustrates how well a model performs for a given random classification. These parameters are used to depict the performance of model.

An imbalanced dataset in this study was balanced using an undersampling technique. Six different ML algorithms were simultaneously applied to subdatasets to produce the ensemble model. First, 10 machine learning classifiers were tested by balancing the dataset with random undersampling in order to observe the performance improvement. Table 3 shows the results for balanced datasets, random undersampling with LR, KNN, SVM, and MLPC has produced unsatisfactory results, and in this case, more than 90% of the data was lost.

Due to the unsatisfactory results with the random undersampling method, the LR, KNN, SVM, and MLPC are not included in the proposed model scenario. The statistical values of algorithms like logistic regression in terms of FMI have produced values of 0.833 and 0.187 for Cohen's Kappa. Better outcomes in terms of FMI and Cohen's Kappa value are obtained by applying a 2-stage cascade soft voting model to an imbalanced dataset. Table 4 displays the outcomes for the suggested model as well as the outputs of each machine learning classifier that was voted on as well as the ensemble classifier.

Table 3. Classifier performances with random undersampling

Random Seed = 42 (for experiment reproducibility)	Balanced data (with random undersampling)								
Machine learning classifiers	Machine non-failure			Machine failure			FMI	Cohen's Kappa	AUC Score
	Precision	Recall	F1 Score	Precision	Recall	F1 Score			
Logistic Regression (LR)	0.99	0.83	0.90	0.14	0.78	0.23	0.833	0.187	0.901
K-Nearest Neighbour (KNN)	0.99	0.80	0.89	0.13	0.85	0.23	0.815	0.181	0.894
Support Vector Machine (SVM)	0.99	0.77	0.87	0.12	0.88	0.21	0.793	0.161	0.902
Multilayer Perceptron Classifier (MLPC)	0.99	0.67	0.80	0.09	0.87	0.16	0.731	0.099	0.926
Decision Tree (DT)	1.00	0.89	0.94	0.23	0.93	0.37	0.891	0.332	0.867
Random Forest (RF)	1.00	0.89	0.94	0.23	0.93	0.37	0.891	0.331	0.965
Gradient Boosting Machines (GBM)	1.00	0.89	0.94	0.23	0.93	0.37	0.893	0.338	0.960
XGBoost	1.00	0.89	0.94	0.24	0.94	0.38	0.894	0.344	0.961
LightGBM	1.00	0.90	0.94	0.24	0.93	0.38	0.895	0.343	0.961
CatBoost	1.00	0.90	0.95	0.25	0.93	0.39	0.899	0.351	0.964

The voted models perform better than directly random undersampling scenarios, as shown by FMI and Cohen's Kappa score. Additionally, the ensemble model, which includes all of the voted classifiers has the highest score with a 0,977 FMI score and a 0,704 Cohen's Kappa score.

In comparison to the suggested model, the values for conventional ML algorithms listed in Table 3 have demonstrated poor performance. In terms of FMI score, the values as displayed in Table 4 have outperformed the KNN, LR, and SVM. The ensemble model's score of 0.977 is higher than the score of the conventional algorithms as a whole.

Table 4. Proposed model performances using soft voting

Random Seed = 42 (for experiment reproducibility)	Voted multiple submodels (soft voting)								
Classification models	Machine non-failure			Machine failure			FMI	Cohen's Kappa	AUC Score
	Precision	Recall	F1 Score	Precision	Recall	F1 Score			
Voted Decision Tree (DT)	0.99	0.99	0.99	0.68	0.63	0.65	0.974	0.640	0.980
Voted Random Forest (RF)	0.99	0.98	0.98	0.56	0.72	0.63	0.967	0.615	0.972
Voted Gradient Boosting Machines (GBM)	0.99	0.99	0.99	0.69	0.69	0.69	0.976	0.681	0.980
Voted XGBoost	0.99	0.98	0.99	0.59	0.77	0.67	0.970	0.656	0.980
Voted LightGBM	0.99	0.98	0.99	0.63	0.81	0.71	0.974	0.697	0.982
Voted CatBoost	0.99	0.99	0.99	0.64	0.67	0.66	0.972	0.640	0.980
Ensemble Model (Voted Ensemble of 6 ML Model)	0.99	0.99	0.99	0.71	0.72	0.72	0.977	0.704	0.981

For the comparison of the classification with imbalanced dataset, by using random undersampling, classification of balanced dataset has been observed and classification with proposed model in this study, ROC curves and AUC scores are added to the article.

The curve for imbalanced data and the true positive rate in relation to the false positive value for traditional algorithms' scores is shown in Fig. 3.

Figure 4 shows the results for algorithms using balanced data by random undersampling, which resulted higher AUC scores with LR, KNN, SVM, MLPC and DT models but lower AUC scores with DT, RF, GBM, XGBoost, Light GBM and CatBoost models comparing to the values in Fig. 3. Although the LR, KNN, SVM and MLPC scores outperform comparing to the Fig. 3, they are still not better than the rest of the model in terms of FMI and Cohen's Kappa scores.

In Fig. 5, the ROC curve and AUC score performances of six voted machine learning and one ensemble model can be seen.

The proposed model appears to have made a satisfactory improvement in performance based on the ROC curve and AUC scores in Fig. 5. Although the ensemble model has the highest FMI and Cohen's Kappa scores, the AUC score difference between the

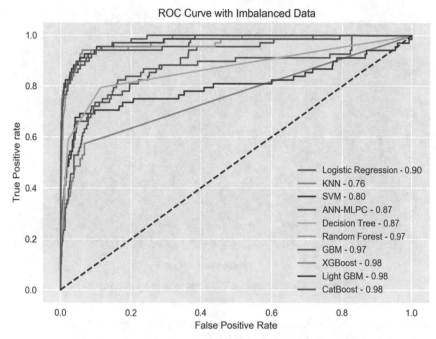

Fig. 3. ROC curve and AUC scores with imbalanced dataset

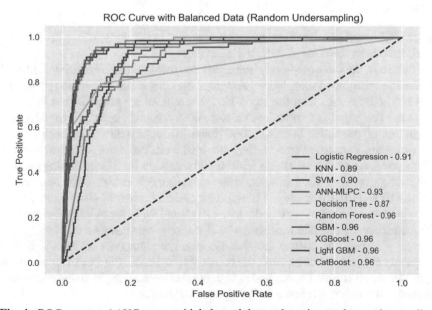

Fig. 4. ROC curve and AUC scores with balanced dataset by using random undersampling

382   Ü. Dilbaz and M. Ö. Cingiz

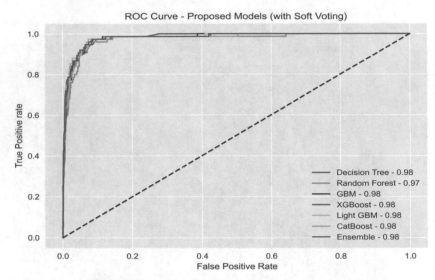

Fig. 5. ROC curve and AUC scores with proposed model

ensemble model and voted machine learning models is not particularly significant. Their AUC values fluctuated between 0.97 and 0.98.

5   Discussion

The SMOTE-based oversampling approach was used in the study, which used the same dataset, to address the issue of an imbalanced dataset.

In our study, the same case study was conducted using an undersampling method that combines the merging of the minority and small majority classes. This study's proposed model has better results than the study [15] that utilized the same dataset in terms of FMI score. The paper has formulated a soft voted ensemble approach to enhance the performance of predictive learning. Traditional ML approaches have been modelled many times but, in terms of results, they provided mediocre scores and data anomalies were observed. An engine oil dataset with 40,000 points has been trained using a cascade-correlation neural network approach, with the 89% result rate that has outperformed ANN and SVM [16].The study has been performed on the automotive industry by [17] ensemble approach incremental model using KNN and SVM for training and prediction and gave better performance with above 90% accuracy between year 2007 to 2008.

Since artificial intelligence can handle enormous amounts of data and makes machines intelligent, it has allowed the industry to flourish. The model's performance and the suggested soft voting method, which uses undersampling in an ensemble model improved this concept, as shown in the figures and statistics.

6   Conclusion and Future Work

When estimating predictive maintenance, it is very possible to run into a dataset that is imbalance, and this is one of the trickiest aspects of predicting machine failure. In this paper we have proposed a soft voted ensemble modelling to enhance the predictive maintenance in industry. The imbalanced dataset has been made balanced by 2 stage cascade voting model by dividing into subdatasets of majority and minority instances.

In this study, six voted machine learning algorithms including decision tree, random forest, GBM, XG Boost, Light GBM, and CatBoost were used in parallel for training and in order to prevent data loss subdatasets were created. All of the stage 1 and stage 2 prediction probabilities that were voted on were equally weighted. Soft voting was used to obtain the prediction value, and different parameters were tracked. The classification of the data was then carried out by utilising the FMI and Cohen's Kappa score, which displays the statistical performance of an algorithm in terms of values. The ROC curve with balanced and imbalanced dataset has been observed, demonstrating the astounding effectiveness of the suggested method.

In order to achieve better performance in the future, the dominance weight can be added while taking each of the voted machine learning classifiers' individual performances into consideration. With a natural dataset from an industrial site, the weighted model approach can be used.

References

1. Carvalho, T.P., Soares, F.A.A.M.N., Vita, R., da P. Francisco, R., Basto, J.P., Alcalá, S.G.S.: A systematic literature review of machine learning methods applied to predictive maintenance," Comput. Ind. Eng. **137**, 106024 (2019). https://doi.org/10.1016/j.cie.2019.106024
2. Coanda, P., Avram, M., Constantin, V.: A state of the art of predictive maintenance techniques. IOP Conf. Ser. Mater. Sci. Eng. **997**, 12039 (2020). https://doi.org/10.1088/1757-899X/997/1/012039
3. Mohammed, R., Rawashdeh, J., Abdullah, M.: Machine learning with oversampling and undersampling techniques: overview study and experimental results, pp. 243–248 (2020). https://doi.org/10.1109/ICICS49469.2020.239556
4. Rokach, L.: Ensemble Methods for Classifiers. In: Maimon, O., Rokach, L. (eds.) Data Mining and Knowledge Discovery Handbook, pp. 957–980. Springer, Boston (2005). https://doi.org/10.1007/0-387-25465-X_45
5. Kamat, P., Sugandhi, R.: Anomaly detection for predictive maintenance in industry 4.0-A survey. In: E3S Web Conf., vol. 170, pp. 1–8 (2020). https://doi.org/10.1051/e3sconf/202017002007
6. Matzka, S.: Explainable artificial intelligence for predictive maintenance applications. In: Proc. - 2020 3rd Int. Conf. Artif. Intell. Ind. AI4I 2020, pp. 69–74 (2020). https://doi.org/10.1109/AI4I49448.2020.00023
7. Mathew, V., Toby, T., Singh, V., Rao, B.M., Kumar, M.G.: Prediction of Remaining Useful Lifetime (RUL) of turbofan engine using machine learning. In: IEEE Int. Conf. Circuits Syst. ICCS 2017, vol. 2018-Janua, no. Iccs, pp. 306–311 (2018). https://doi.org/10.1109/ICCS1.2017.8326010
8. Durbhaka, G.K., Selvaraj, B.: Predictive maintenance for wind turbine diagnostics using vibration signal analysis based on collaborative recommendation approach. In: 2016 Int. Conf. Adv. Comput. Commun. Informatics, ICACCI 2016, pp. 1839–1842 (2016). https://doi.org/10.1109/ICACCI.2016.7732316
9. Biswal, S., Sabareesh, G.R.: Design and development of a wind turbine test rig for condition monitoring studies. In: 2015 Int. Conf. Ind. Instrum. Control. ICIC 2015, no. Icic, pp. 891–896 (2015). https://doi.org/10.1109/IIC.2015.7150869

10. Sampaio, G.S., de A. V. Filho, A.R., da Silva, L.S., da Silva, L.A.: Prediction of motor failure time using an artificial neural network. Sensors **19**(19), 5–7 (2019). https://doi.org/10.3390/s19194342

11. Paolanti, M., Romeo, L., Felicetti, A., Mancini, A., Frontoni, E., Loncarski, J.: Machine Learning approach for Predictive Maintenance in Industry 4.0. In: 2018 14th IEEE/ASME Int. Conf. Mechatron. Embed. Syst. Appl. MESA 2018 (2018). https://doi.org/10.1109/MESA.2018.8449150

12. Kim, D., Choi, J.: "Optimization of Design Parameters in LSTM Model for Predictive Maintenance", MPDI. Appl. Sci. **11**, 6450 (2021). https://doi.org/10.3390/app11146450

13. Nasser, A., Al-Khazraji, H.: A hybrid of convolutional neural network and long short-term memory network approach to predictive maintenance. Int. J. Electr. Comput. Eng. (IJECE) **12**(1), 721–730 (2022). ISSN: 2088-8708, https://doi.org/10.11591/ijece.v12i1.pp721-730

14. Ayvaz, S., Alpay, K.: Predictive maintenance system for production lines in manufacturing a machine learning approach using IoT data in real-time. Elsevier, January 2021, https://doi.org/10.1016/j.eswa.2021.114598

15. Sridhar, S., Sanagavarapu, S.: Handling data imbalance in predictive maintenance for machines using SMOTE-based oversampling. In: Proc. - 2021 IEEE 13th Int. Conf. Comput. Intell. Commun. Networks, CICN 2021, pp. 44–49 (2021). https://doi.org/10.1109/CICN51697.2021.9574668

16. Phillips, J., Cripps, E., Lau, J.W., Hodkiewicz, M.R.: Classifying machinery condition using oil samples and binary logistic regression. Mech. Syst. Signal Process. **60**, 316–325 (2015). https://doi.org/10.1016/j.ymssp.2014.12.020

17. Sankavaram, C., Kodali, A., Pattipati, K.R., Singh, S.: Incremental classifiers for data-driven fault diagnosis applied to automotive systems. IEEE Access **3**, 407–419 (2015). https://doi.org/10.1109/ACCESS.2015.2422833

Volume Forecasting in Supply Chain: A Mixed Study of Boosting and Prophet Algorithms

Furkan Oruc[✉] [ID], Ismet Yildirim [ID], and Gizem Cidal [ID]

DHL Supply Chain Turkey, 34758 Atasehir, Istanbul, Turkey
{furkan.oruc,ismet.yildirim,gizem.cidal}@dhl.com

Abstract. With the current technology development and the rapid rise of machine learning and forecasting algorithms, the supply chain evolves in a different era. In this paper, a forecasting approach is proposed, which has a strong capability of predicting the future volume size of products and customers. Therefore, in this paper, we propose a volume forecasting method that is compared with some well-known time series forecasting techniques from both statistical boosting methods using a specific dataset. These methods include LightGBM, XGBoost gradient boosting decision three algorithms, and Prophet algorithm. The experimental results indicate that RMSE, MAE, and R2 scores of products and customers trained by XGBoost, LightGBM, and Prophet, an interface that shows how the best model is chosen with statistical methods and how the three different algorithms forecast next day's or next week's volume efficiently.

Keywords: Supply chain · Machine learning · XGBoost · LightGBM · Demand forecasting · Time series

1 Introduction

With the new technologies and latest achievements in the supply chain area, forecasting is a necessary method to be able to follow business methods and future actions in logistics. The more industry and supply chain processes grow, the more quantitative data is collected in the aspect of data analysis and prediction. Gaur et al. [6] state that a supply chain is coordination of different managements and business methods. Those different managements bring about new technologies in data analytics and machine learning. A large amount of data collected and analyzed help us to manage business needs in logistic and supply chain to be able to understand customer behaviour in terms of logistic warehouses.

Supply Chain Management consists of market efforts to manage goods, services, and transportation from warehouses to customers. However, there may be seen some uncertainties during logistics and warehouse operations as weekly, monthly, or daily product orders and demand of customers in a wide range of time. A variety of statistical data management methods has been applied so far; nevertheless, with the new artificial

Supported by DHL Supply Chain Turkey.

intelligence and machine learning algorithms, we can perform prediction of the volume size and forecast the demand for further days or weeks.

In the paper, we would like to analyze the order dataset acquired by the Gebze facility of DHL Supply Chain Turkey and predict the volume size of customers and products using exploratory data analysis and time series forecasting methods. The statement of the problem is to predict and analyze the daily volume size of customers and products of the Gebze facility in a wide range for further applications. In this approach, we used and analyzed XGBoost, LightGBM, and Prophet algorithms to predict and forecast the volume size concerning analysis and comparison of time series algorithm methods. The paper is based on XGBoost, LightGBM, and Prophet algorithm outputs and time series methods used, We compared the models and analyze the metrics quantitatively. Finally, these algorithms output prediction and forecast results of the volume size of customers and products listed in the order dataset.

In the paper, first, we will examine and analyze what our dataset contains and how the previous work of our colleagues is applied. Later, we mention why we need these logistics and supply chain forecasting approaches. After that, we briefly mention pre-processing methods of our models, such as cleaning, normalization, and encoding, in the methodology section. Lastly, we indicate, analyze and compare our models with different metrics with a logic that compares the models according to RMSE and chooses the best model among them.

1.1 Dataset

The raw dataset has been data manipulated for 72 unique customers and 273 unique products of a FMCG Company. The complete dataset is divided into two parts: customer and product. Customer and product datasets include features such as weekly holidays, entry date and which quantile it belongs, which month and week it is ordered, and various event days of orders delivered by each customer and product.

2 Literature Review

2.1 Order and Demand Prediction

One of the approaches that are used to predict and analyze the orders is designed by Boone et al. [3]. They point out novel methods to perform big data forecasting in the supply chain. The authors try to manage a sales forecasting method using big data analytics based on privacy, security, and governance. They try to create an approach to forecasting as a qualitative method rather than a quantitative one. However, Aviv et al. [2] propose a pioneer time series forecasting framework for inventory management. They use linear state space forms and the Kalman filter as a prerequisite of the framework. It is based on product demand forecasting and it is supposed to use product demand characteristics. Lastly, they also make a good point to develop a cost estimation by using time series forecasting approach with Kalman estimation and correlation between each product and evaluation performances.

Another approach for volume forecasting is developed as a time series forecasting model with ARIMA as a supply chain model defined by Gilbert et al. [7]. It is one of the traditional approaches used in forecasting. However, it is a backbone for understanding how time series work in algorithms with a modern approach. They use a traditional mathematical approach of ARIMA as an auto-regressive moving average model presented by Box et al. [4].

$$z_t = \mu + \phi_1(z_t - 1 - \mu) + \phi_2(z_t - 2 - \mu) + \ldots + \phi_p(z_t - p - \mu) \qquad (1)$$

Equation 1 shows auto regressive ARIMA forecasting model with Z_t is the demand at time t, μ is the process average and α_t is the time series of an independent random variable.

Another approach to using demand forecasting is comparing two algorithms' performance on machine learning demand forecasting. Feizabadi et al. [5] develop two different machine learning forecasting algorithms using a simple neural network and ARIMAX. Both time series and their explanatory factors are developed in their model. In the end, they suggest that there is a performance difference between traditional ML methods and new demand forecasting methods, which two new methods are superior. Another approach focuses on forecasting in the supply chain: [12] researched a time series forecasting method with a more empirical approach, a multi-echelon distribution network. They also used a grouped time series forecasting structure with hierarchical forecasting. Ultimately, it is analyzed the relationship and probability of correlation between time series characteristics and their effectiveness. Results show that a bottom-up hierarchy is needed for better performance of these approaches.

2.2 Forecasting and Prediction with Boosting Algorithms

In addition to common approaches, [11] states a product-based prediction model comparing two boosting algorithms: XGBoost and LightGBM. The authors focus on predictive modeling with a product sales dataset. In this approach, algorithm parameters, number of estimators, and other dataset features are compared. They propose a result that there is no significant difference between these two approaches and gamma has the largest effect on the feature importance of the XGBoost algorithm. In the model evaluation and result analysis part, the authors show why the RMSE metric is the best value to compare boosting algorithms. On the other hand, learning rate has a wider impact on forecasting according to RMSE results of their volume as sales January to October 2015 as a time series forecast.

There is also a mixed approach using both classical machine learning methods and boosting algorithms: Zohdi et al. [16] presents a study of demand forecasting using a limited and intermittent demand prediction. The novel part of their approach is to compare and combine KNN, decision tree, boosting, and multi-layer perceptron algorithms. Their metrics are mean squared error, mean absolute error, coefficient of determination, and computational time. In the end, they conduct an analysis of variance and Kolmogorov–Smirnov technique as Yacshie et al. [15] mentioned.

2.3 Novel Methods and the Logic Behind Volume Forecasting

In addition to traditional and boosting technologies, there are novel and latest methods to use time series volume forecasting. Abbasimehr et al. [1] designed and developed an optimized method with LSTM (Long-Short Term Memory), utilized methods for this approach are KNN (K nearest neighbor), ANN, and RNN. The LSTM method mentioned in [1] takes several arguments such as input and output value, cell states, biases and weight matrices of input gate, recurrent weights, and output results since it is a prototype model of a neural network. The proposed methodology is firstly preparing the data with data processing libraries, generating a list of hyperparameters, splitting their dataset as a test, and train and training the model with LSTM on N layers. Their method compares the results by RMSE (Root Mean Squared Error) and chooses the best model as the most efficient one.

3   Logistic Operation Needs

Before getting into the details about volume forecasting methods and experimental methods, we need to mention the importance of this approach in terms of supply chain and business needs logistically. Demand forecasting is defined by Seyedan et al. [14] as focusing, planning, and managing market strategies. Therefore, we planned to develop a model that will take care of our warehouse demands. After holding various conversations with blue-collar workers from warehouse departments. We analyzed and decided that they need to know the future volume size of orders by each product or customer. It is also quite necessary to know how they will interpret and use our forecasting results as a feasible dashboard. Therefore, for dashboard visualization and responsiveness. We are also in the progress of designing a user interface which is called AiLand.

4   Methodology, Models and Results

Figure 1 describes the steps of our proposed method as a flowchart. First of all, we define the input data set, and to be able to use it for time series forecasting, we clean the data in case it might have noisy and missing values. The normalization part begins after getting the time series. After data preparation is completed, our dataset is split to train and test datasets. In the training part, this operation is repeated three times for three different models (XGBoost, LightGBM, and Prophet) with tuned hyperparameters. Lastly, trained models give us the RMSE scores of daily forecasting of each product and customers to decide for us to optimize the best model among the list of models.

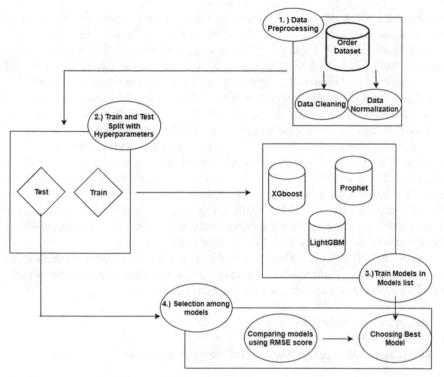

Fig. 1. The proposed methodology for volume forecasting using time series forecasting approach

4.1 Data Cleaning

Large datasets might contain noisy and missing values. Data cleaning and data normalization are two crucial steps of data cleaning and exploratory data analysis. Therefore, we applied to the entire order dataset EDA, then normalization.

4.2 Data Normalization

Time series and normalization during the normalization process, the following expression is used as: $s_n(t_i)$:

$$s_n(t_i) = \frac{s(t_i) - \min(s)}{\max(s) - \min(s)} \qquad (2)$$

Equation 2 shows the normalization formula used in the dataset where min(s) minimum and max(s) are the maximum value of time series s at a time t.

4.3 Modeling

One of the methods that machine learning algorithms utilize is gradient boosting. Gradient boosting is a predictive model approach allowing optimization [13]. In this approach, we are going to work with the time series regression method to forecast the volume sizes using two gradient boosting methods XGBoost and LightGBM and Prophet.

Our methodological approach model is trained as three different models on time series for daily forecasting of volume size of products and customers. The approach is also predictive and will be using big data analytics for further research purposes.

XGBoost As [8] state, XGBoost is a gradient boosting machine algorithm to predict the target variable using deep considerations of optimization and decision trees in machine learning. When we train the model, our XGBoost training model parameters are tuned and tuned hyperparameters and their value ranges for this algorithm are demonstrated in Table 1. Also, The optimization is carried out by minimizing the loss function [11]. Feature importance method in Table 2 and also applied in the order dataset and most three features with the highest impact on time series forecasting with XGBoost algorithms are IsWeekend, WeekOfYear8, and Quarter1 respectively. Those features also return a boolean value whether the specified entry date is a weekend or not, the 8th week of the year, and the first quarter of the specified year.

LightGBM LightGBM is also a gradient boosting algorithm that uses decision trees and was developed by Microsoft (see it in [10]). LightGBM uses a histogram algorithm [11]. The steps that we used in the LightGBM algorithm are:

1. Initializin n decision trees,
2. Training Regressor
3. Updating weights and getting the final results of regressor.

Table 1. XGBoost hyperparameters and value ranges

Hyperparameters	Range values
Eta	[0, 1]
Gamma	[0, 5]
Reg lambda	[0, 2]
Reg alpha	[0, 2]
Max depth	[1, 100]
Max child weight	[1, 10]

Table 2. XGBoost algorithm's feature importance

Features	F1 value importance
IsWeekend	5.29644528
WeekOfYear8	2.32381728
Quarter1	1.68349216

Initializing n decision steps and updating weights can be expressed as Liang et al. [11] state:

$$F_n(x) = a_0 F_0(x) + a_1 F_1(x) + \ldots + a_n F_n(x) \tag{3}$$

In the Eq. 3 where α is power constant of the regressor, and F function is the loop function that takes parameters in Table 2. In the LightGBM algorithm, we also applied hyperparameter tuning to our training model, tuned hyperparameters and their value range in the LightGBM training model are demonstrated in Table 3. Since LightGBM is a boosting tree method, there are different leaves and fractions, feature fractions are counted as 15 and the total number of leaves is changeable from 0 to 4. Therefore, maximum depth size is dependent on this and it should be tuned as n − 1 where n is the number of leaves.

Table 3. LightGBM hyperparameters and value ranges

Hyperparameters	Range values
Num leaves	[0, 4]
Feature fraction	[1, 15]
Bagging fraction	[1, 16]
Max depth	[0, 3]
Min data in leaf	[1, 100]

Prophet Unlike Gradient Boosting Decision Tree Algorithms (GBDT), Prophet is not a gradient boosting algorithm to be considered. The prophet algorithm is also a very common method to apply in time series forecasting for supply chain management projects. Guo et al. [9] used a novel algorithm which is called PROPHET-SVR; another aspect of the Prophet algorithm that integrates prophet and support vector regression algorithms. We just applied the prophet algorithm but similarly, both PROPHET-SVR and Prophet are used to forecast seasonality and seasonal fluctuations. In Table 4, tuned hyperparameters are shown as can be seen in our other training models. Seasonality hyperparameter means various cycle types of seasonal trends in a time series forecasting [9].

Table 4. Prophet hyperparameters and value ranges

Hyperparameters	Range values
Change point prior scale	[0.001, 0.5]
Seasonality prior scale	[0.01, 10]
Holidays prior scale	[0.01, 10]
Seasonality mode	[additive, multiplicative]

4.4 Utilized Metrics and Experimental Results

After training our three different models (XGBoost, LightGBM, and Prophet), We decided to make a performance comparison to be able to choose the best performance model among them for each customer and product of the order dataset mentioned in Sect. 1.1.

$$RMSE = \sqrt{\frac{1}{n}\sum_{i=1}^{n}(d_i - f_i)^2} \tag{4}$$

$$MAE = \left(\frac{1}{n}\right)\sum_{i=1}^{n}|y_i - x_i| \tag{5}$$

$$R^2 = 1 - \frac{\sum_{i=1}^{n}(t_i - y_i)^2}{\sum_{i=1}^{n}(y_i - z_i)^2} \tag{6}$$

Three different comparison metrics were applied to the list of models. One of them is Root Mean Squared Error (RMSE) in Eq. 3, second one is Mean Absolute Error (MAE) in Eq. 4, and the last one is R^2 Score in Eq. 5. These three metrics are applied to three different training models for each customer and product. We decided to choose the RMSE score for the best model logic application in the approach. The method compares each customer and each product according to the RMSE score and gives the best output among three training models in a daily database interface (Fig. 2).

Figure 3 shows that from an 80% train dataset and 20% test dataset as time series where LightGBM algorithm forecasts unlabeled 20% data (there is no data leaking) as it is shown with red.

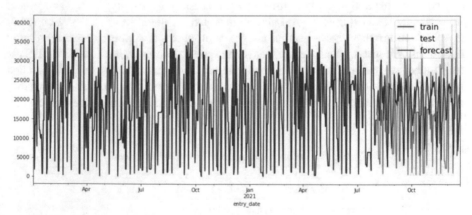

Fig. 2. An example time series forecasting with test and train lines of January 2021

customer	mae_xgb	mae_pro	mae_lgbm
0	TR04B010	747.011964	1101.515368	1414.818488
1	TR94M080D	185.259221	170.235182	322.344940
2	TR05F001	37.013904	10.634083	68.213749
3	TR003001	51.001722	153.027097	88.483180
4	TR09S001	78.243879	196.293611	134.277658
5	TR41D001	147.046921	238.044003	268.750861
6	TR04S001	315.852365	345.169867	553.444250
7	TR06S002	90.029985	128.381415	163.124173
8	TR05B002	105.833232	241.850496	186.460494
9	TR07S001	84.584975	123.954546	154.433230
10	TRE09B002	330.882809	774.205925	559.440722
11	TR06M000	145.685810	243.828382	248.401727
12	TR04E002	387.080578	1342.247371	709.406713
13	TR05L001	153.433377	163.214008	265.362591
14	TR01S001	143.959008	217.763997	255.512790
15	TR04B004	659.790164	1240.406325	1129.097220
16	TR06B001	136.579378	131.507268	256.890074
17	TR04R002	133.101003	74.405710	228.901851

customer	r2_xgb	r2_pro	r2_lgbm
0	TR04B010	0.162041	-0.195394	-0.654922
1	TR94M080D	0.022788	-0.007776	-0.118212
2	TR05F001	-0.403409	0.005751	-1.658310
3	TR003001	-0.093606	-2.314488	-0.334647
4	TR09S001	-0.113526	-1.172840	-0.420487
5	TR41D001	-0.367465	-0.388491	-1.237189
6	TR04S001	-0.085766	-0.046299	-0.409428
7	TR06S002	-0.123092	-0.157302	-0.447556
8	TR05B002	-0.036558	-1.373910	-0.255565
9	TR07S001	-1.041232	-0.362191	-3.438096
10	TRE09B002	0.103117	-0.779197	0.169419
11	TR06M000	0.024013	-0.338395	0.016547
12	TR04E002	-0.441331	-13.866692	-1.578836
13	TR05L001	-0.330556	-0.045002	-1.138001
14	TR01S001	-0.064524	-0.459098	-0.381865
15	TR04B004	-0.096030	-0.791653	-0.370733
16	TR06B001	-0.399831	-0.490134	-1.495170
17	TR04R002	0.031875	-0.038574	-0.045042

customer	rmse_xgb	rmse_pro	rmse_lgbm
0	TR34B010	1187.907064	1418.818060	2243.244900
1	TR94M080D	355.514648	361.031496	552.642449
2	TR30F001	49.603432	41.751039	85.402800
3	TR003001	161.726124	281.551507	250.042873
4	TR05S001	254.045040	354.874146	396.640887
5	TR41D001	385.141987	388.091592	630.802039
6	TR04S001	710.176895	697.150381	1120.210728
7	TR06S002	251.676004	255.480383	393.537230
8	TR05B002	318.370345	481.801257	496.580375
9	TR07S001	259.136008	210.911629	453.035543
10	TRE09B002	726.755889	1023.605447	1092.140409
11	TR06M000	290.827502	340.569435	437.035051
12	TR04E002	700.703340	2250.398570	1177.076951
13	TR05L001	346.127415	306.745360	565.563744
14	TR01S001	338.442019	392.560798	530.961338
15	TR04B004	1282.370595	1638.587514	1996.366201
16	TR05D001	251.031802	259.002285	422.729720
17	TR04R002	264.451497	273.904350	405.468418

Fig. 3. Score comparison of different models left to right R2 score, Mean Absolute Error (MAE), Root Mean Square Error (RMSE)

It is been computed three different results for three different training algorithms. Figure 3 shows each sample customer's R2, MAE, and RMSE scores with the comparison of XGBoost, Prophet, and LightGBM algorithms. For example, the R2 Score of Customer TR34B010 is approximately 0.10 on XGBOOST, −0.19 on Prophet, and −0.65 on LightGBM respectively. Since the closest score to 1 is the best for the R2 score, XGBoost's R2 score for that customer number is highlighted in green color. Hence, the best result for each customer is highlighted in green. The same strategy has also been applied for the product part of the dataset. We observed that XGBoost and Prophet's metric results are similar for all three metrics, unlike LightGBM. In this approach, Light-GBM gives us more inefficient and unfeasible results in terms of RMSE, MAE and R2 Score.

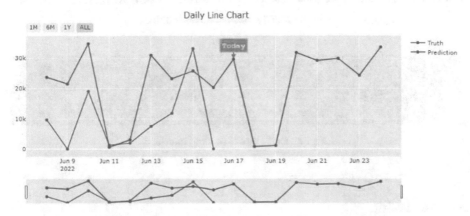

Fig. 4. An Example of Daily Forecast of XGBoost Algorithm

Figure 4 is a plot visualization of truth volume size values as a green line chart and XGBoost's daily predicted volume size as the gray chart from June 9, 2022, to June 17, 2022. In addition, we also measured these predictive approaches with other algorithms. After analyzing the benchmarks and metric results. It is decided to choose the RMSE score as the main performance metric for comparison in the daily and weekly predictive database interface.

The Logic of Choosing the Best Model

It has been developed and designed an approach that shows whether an order is completed on a specific date (daily or weekly) and chooses the best model name metric based on Root Mean Squared Error with its prediction volume size value. In Fig. 5, it is shown a version table that is listed model names and their last update and creation date. Figure 6 indicates the status of model prediction for a product and the chosen best model according to RMSE score in addition to the creation and update date for each product id number. There is also the same approach with a different time scale. While Fig. 6 works out the training model list weekly updated mode, Fig. 7, works out and prints the results daily.

version_num	completed	models	create_date	update_date
1	1	xgboost,prophet	2022-06-07 17:20:59	2022-06-29 00:05:27
2	1	xgboost,prophet	2022-06-12 15:36:21	2022-06-29 00:05:27
3	1	xgboost,prophet	2022-06-17 20:10:57	2022-06-20 17:23:52
4	0	xgboost,prophet	2022-06-27 12:28:12	NULL

Fig. 5. Model versions

product_id	version_num	status	model_saved	best_model_name	best_model_metric_rmse	create_date	update_date
61002995	2	Completed	1	prophet	908	2022-06-13 13:41:15	2022-06-13 14:00:09
61002995	3	Completed	1	prophet	921	2022-06-18 18:06:57	2022-06-18 18:25:50
61002995	4	Completed	1	prophet	906	2022-06-28 10:25:18	2022-06-28 10:44:20

Fig. 6. Weekly model training interface

product_id	date	version_num	prediction	truth	model_name	create_date	update_date
61000761	2022-06-17	2	141	384	prophet	2022-06-16 14:24:05	2022-06-18 08:01:03
61000761	2022-06-18	2	38	0	prophet	2022-06-16 14:24:05	2022-06-18 08:01:03
61000761	2022-06-19	2	27	0	prophet	2022-06-16 14:24:05	2022-06-19 08:01:03
61000761	2022-06-20	2	63	0	prophet	2022-06-16 14:24:05	2022-06-20 08:01:11

Fig. 7. Daily model training interface

5   Conclusion

In this study, we proposed an approach that firstly gets the dataset and does exploratory data analysis with data cleaning and normalization. Secondly, Train, test, and validation approach comes out, in this step, it is used 80% of the entire data is training, 10% is used for testing and 10% remaining is used for validation. To determine whether the performance of the proposed approach among three different algorithms is significantly better than the performances of the other methods, statistical tests are carried out. Among these statistical performance metrics, Root Mean Squared Error has been chosen after calculating Mean Absolute Error and R2 scores for each customer and product. Thirdly, three different algorithms (XGBoost, Prophet, and LightGBM) have been trained and hyperparameters tuned to choose the best model. Since our approach is for forecasting the future volume, we need a high-performance metric with a high-performance algorithm. In conclusion, among these three algorithms, we also claim that Prophet has the most green-lighted score of RMSE per product and customer compared to LightGBM and XGBoost.

6 Future Work

For further research, we are going to study neural network-based latest generation algorithms such as LSTM, GRU, and AdaBoost. In this approach, the order dataset was local. Additionally, our dataset will become globally wider and include much more customers, products, and different kinds of features from the Europe-Middle East and Africa region datasets.

References

1. Abbasimehr, H., Shabani, M., Yousefi, M.: An optimized model using lstm network for demand forecasting. Comput. Ind. Eng. **143**, 106435 (2020)
2. Aviv, Y.: A time-series framework for supply-chain inventory management. Oper. Res. **51**(2), 210–227 (2003)
3. Boone, T., Ganeshan, R., Jain, A., Sanders, N.R.: Forecasting sales in the supply chain: consumer analytics in the big data era. Int. J. Forecast. **35**(1), 170–180 (2019)
4. Box, G.E.P., Jenkins, G.M., Reinsel, G.C., Ljung, G.M.: Time Series Analysis: Forecasting and Control. Wiley, New York (2015)
5. Feizabadi, J.: Machine learning demand forecasting and supply chain performance. Int. J. Log. Res. Appl. **25**(2), 119–142 (2022)
6. Gaur, M., Goel, S., Jain, E.: Comparison between nearest neighbours and Bayesian network for demand forecasting in supply chain management. In: 2015 2nd International Conference on Computing for Sustainable Global Development (INDIACom), pp. 1433–1436. IEEE (2015)
7. Gilbert, K.: An arima supply chain model. Manage. Sci. **51**(2), 305–310 (2005)
8. Gumus, M., Kiran, M.S.: Crude oil price forecasting using xgboost. In: 2017 International Conference on Computer Science and Engineering (UBMK), pp. 1100–1103. IEEE (2017)
9. Guo, L., Fang, W., Zhao, Q., Wang, X.: The hybrid Prophet-SVR approach for forecasting product time series demand with seasonality. Comput. Industr. Eng. **161**, 107598 (2021)
10. Ke, G., et al.: LightGBM: a highly efficient gradient boosting decision tree. In: Advances in Neural İnformation Processing Systems, vol. 30 (2017)
11. Liang, Y., et al.: Product marketing prediction based on xgboost and lightgbm algorithm. In: Proceedings of the 2nd International Conference on Artificial Intelligence and Pattern Recognition, pp. 150–153 (2019)

12. Mircetic, D., Rostami-Tabar, B., Nikolicic, S., Maslaric, M.: Forecasting hierarchical time series in supply chains: an empirical investigation. Int. J. Prod. Res. **60**(8), 2514–2533 (2022)
13. Munoz, A.: Machine learning and optimization (2014). https://www.cims.nyu.edu/~munoz/files/ml_optimization.pdf. Accessed 2 Mar 2016. [WebCite Cache ID 6fiLfZvnG]
14. Seyedan, M., Mafakheri, F.: Predictive big data analytics for supply chain demand forecasting: methods, applications, and research opportunities. J. Big Data **7**(1), 1–22 (2020)
15. Yacshie, B.T.P.W.B., Prasetyo, Y., Arianto, A.C.: Walk back tuning and paper tuning: how do they improve archery accuracy? J. Sport Area **7**(1), 59–68 (2022)
16. Zohdi, M., Rafiee, M., Kayvanfar, V., Salamiraad, A.: Demand forecasting based machine learning algorithms on customer information: an applied approach. Int. J. Inf. Technol. **14**, 1937–1947 (2022)

Blockchain-Based System for Traceability of Counterfeit Medicine

Kritika Sharma$^{(\boxtimes)}$ (iD) and Ravreet Kaur$^{(\boxtimes)}$ (iD)

Department of Computer Science and Engineering, University Institute of
Engineering and Technology, Panjab University, Chandigarh, India
Sharmakritika804@gmail.com, ravreetkaur@pu.ac.in

Abstract. Healthcare supply chain management systems (HSCM) in the
world have recently undergone a tremendous change due to the unprece-
dented pandemic eruption in 2019. It was observed that non-automated
HSCM system is making it difficult to combat rising menace of substandard and falsified products. The presence of counterfeit medication is a
serious problem that needs to be totally eliminated because it prevents the
intended ailment from being treated. Blockchain has emerged as a promising technology to automate global supply chain management systems.
With the support of its decentralized and distributed structure, recent
technological developments will lead to improved medical transactions,
automated supply chain management (SCM) and more effective traceability. This article offers a thorough analysis of the automated HSCM system
in use, both with and without blockchain technology. This study identifies the drawbacks of existing solutions. As a result, this research suggests
a blockchain-based smart contract system that will enhance the supply
chain's visibility, traceability, and verification of medications. By altering
the block sizes, number of peers, and number of clients in an organization,
the presented system has been evaluated in terms of metrics, transaction
throughput, and transaction latency.

Keywords: Blockchain · Counterfeit medicine · Healthcare ·
Permissioned blockchain · Hyperledger fabric · Supply chain
management

1 Introduction

Need of the hour demands automated healthcare supply chain management system. Crucial issues, namely provider preference items, lack of automated SCM
systems and hidden costs, leads to failure of healthcare, which puts large number
of human lives at risk [1]. According to a WHO fact sheet, one out of every ten
medical items in low- and middle-income nations is subpar or fake [2]. Counterfeit medicine is particularly that medicine which has been deliberately misla-
belled. Counterfeit medicine can arise before, during and after packaging. This
paper will focus on eradicating the ways to generate counterfeit medicine after
packaging i.e. while transmitting and before delivery to consumer. At present,

F. P. García Márquez et al. (Eds.): ICCIDA 2022, LNNS 643, pp. 397–410, 2023.
https://doi.org/10.1007/978-3-031-27099-4_31

the tracking and screening to halt counterfeit medicine in non-automated SCM systems is cumbersome and unsuccessful, pertaining to WHO statistics as mentioned earlier [2].

Recently during the covid time it become very difficult to trace the PPE (Personal Protective Equipment) kits and oxygen cylinders, which leads to its urgent need in the society [3,4]. Counterfeit medicine is not only accountable for financial loss, but, it is extremely significant in terms of loss of human lives due to misinformation to its consumer. Many governments are seeing the promise of blockchain and are working to standardize its use. It came into existence on 31st October 2008 [5]. For the prevention of counterfeit medicine, blockchain technology is the best solution available [6]. As per research reports, blockchain investments have risen nearly six times in the year of 2021. Blockchain technology's transaction system is giving promising results to exclude middleman in supply chains. Blockchain based solutions for HSCM systems are able to perform information traceability, prevent tampering, decentralisation, data protection and integrity [7].

The remainder of this paper is structured as : Sect. 2, provides review of existing HSCM systems and their drawback. Section 3, explains the proposed methodology along with work flow of the proposed system using flowchart and diagrams. In Sect. 4, experiments performed and results obtained are presented. Finally, in last Sect. 5, conclusion and suggestions for future work are stated.

2   Literature Review

The issue in HSCM systems undertaken for study of this research work is the existence of counterfeit medicines. Blockchain can be used to combat counterfeiting by increasing traceability and visibility of the drug supply [8]. In order to verify the legality of medicine bottles, author Michael Paik [9] proposed non blockchain system called Smart-Track that includes a bar code or RFID (Radio Frequency Identification) code for traceability. M. Jansen-Vullers [10] and Y.S. Kang [11], also proposed a non blockchain approach using gozinto graph modelling and RFID respectively. In Table 1, the drawbacks of non blockchain based systems is presented with reference to blockchain based system characteristics.

Table 1. Non blockchain based systems

S. No.	Characteristics	Smart track [9]	Development of generic RFID traceability [10]	Managing traceability info in manufacturer [11]
1	Information traceability	✓	✓	✓
2	Preventing tampering	✗	✗	✗
3	Decentralisation	✗	✗	✗
4	Data protection	✗	✗	✗
5	Integrity	✗	✗	✗

Since traditional traceability systems are centralised and not decentralised, it is exceedingly difficult to prevent data tampering. As a result, these systems are not safe and robust. The identity mechanism of blockchain shares medical data while keeping the one's private data secret. Blockchains come in two forms: permissioned and permissionless. In permissionless blockchain, the identities of all participants are unknown [12,13]. In this paper permissioned blockchain where all the identities are known is used. Quorum [14], Ripple [15], Stellar [16], Corda [17] and Hyperledger Fabric [18] are some popular permissioned blockchains. This blockchain is faster, more energy efficient and fault tolerant [19]. For better understanding a comparison table (Table 2) is shown below.

Table 2. Permissioned and permissionless blockchain platforms.

S. No.	Property	Permissioned	Permissionless
1.	Access	Authorized members only	Open
2	Scalability	High	Limited
3	Performance	Faster	Slower
4	Transaction cost	No or very low	High

On the other-hand as per the existing research work, researchers are working on using blockchain-based solutions for traceability of drug in SCM systems. Y. Huang [8] offered a method for securing the privacy and validity of traceable data that uses bitcoin-based blockchain technology. Drugledger reconstructs the entire service architecture through service provider separation, ensuring service delivery while ensuring the validity and privacy of traceable data. Ahmad Musamih and Khaled Salah [20] proposed a method based on the Ethereum blockchain that makes use of smart contracts and off-chain storage that is decentralised for effective product traceability in the healthcare supply chain. Ilhaam A. Omar and Raja Jayaraman [22] presented a summary of the contracting process for the healthcare supply chain, which involves numerous parties including manufacturers, GPOs, distributors, and healthcare providers. Using Ethereum smart contracts and a decentralised storage system, it increases the effectiveness of the contracting procedure in the healthcare supply chain. The existing solutions require intensive secure mechanisms due to their ability to store data off-chain and public permissioned platform. A thorough research determined that blockchain is a developing technology that has the potential to significantly improve the healthcare industry [23]. Table 3 shows the related work done in the field of managing healthcare supply chain data from counterfeiting using different blockchain technologies.

Table 3. Related work using different blockchain technologies

Sr. No	Title, Author, Year	Blockchain	Mode of operation	Curren cy	Off-chain data storage	Programmable Module
1.	Drug Ledger: A practical blockchain system for drug traceability and regulation, Y.Huang, J.Wu and C.long. 2018[8]	Bitcoin	Permissionless	BTC	No	None
2.	Blockchain based approach for drug traceability in health care supply chain management, Ahmad Musamih, Khaled Salah 2019[20]	Ethereum	Permissionless	Ether	Yes	Smart contract
3.	Enhancing Blockchain Traceability with DAG-based Tokens, Hiroki Watanabe, Tatsuro Ishida 2019[21]	Ethereum	Permissionless	Ether	Yes	DAG (directed acyclic graph)
4.	Automating Procurement contracts in healthcare supply chain management using blockchain supply chain, Ilhaam A. Omar, Raja Jayaraman 2021[22]	Ethereum	Permissionless	Ether	Yes	Smart contract

3   Proposed Methodology

This section introduces a blockchain-based method for tracking down counterfeit medications. Hyperledger fabric 1.4.2 blockchain is used for management of supply chain.

3.1   System Architecture

In this section, a blockchain-based architecture of the system for detecting counterfeit drugs is suggested. Blockchain has two main types Permissionless blockchain and Permissioned blockchain [24], In permissioned blockchain, trusted parties can write or fetch any information. This blockchain has four modules i.e. Manufacturer, Distributor, Retailer, Consumer. After creating a new product, the manufacturer stores all of its data in the blockchain. It generates a hash key for each special medicine and stores it as a block in the blockchain. This block of information can be visible to the connected parties like distributors, retailers, and also to consumers. The information related to the distributor request or

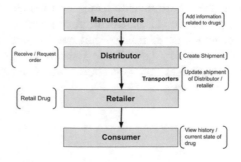

Fig. 1. Proposed System Architecture

receive is also added in the blockchain as the process of sending the drugs to the consumer starts. At the end, consumers can easily fetch information about the medicine-like where is it coming from?, when is it made? and when will the medicine be expired?, etc. It improves the visibility and security of the medicine. For implementing a permissioned blockchain many networks are available. E.g. Hyperledger Fabric [25], Ethereum [26], Bitcoin blockchain [5].

3.2 Proposed Algorithms

The proposed solution is composed of modules to register new company, add new drug and creating new shipment, as shown below in Algorithms 1, 2 and 3.

Algorithm 1. Pseudo-code to register a new company

```
BEGIN
if New Node == contract deployer then
    if No such company ID exists then
        registerCompany(ledger,companyCRN,companyName,location,organizationRole)
        companyIdKey=ledger.CreateCompositeKey(companyCRN,companyName)
        fetchCompanyDetails= ledger.getState(companyIdKey)
        if fetchCompanyDetails != null: then
            throw error: "Company already exist"
            if organizationRole in ["Manufacturer", "Distributor", "Retailer"] then
                if organizationRole == "Manufacturer" then
                    hierarchyKey = 1;
                else if organizationRole == "Distributor" then
                    hierarchyKey = 2;
                else
                    hierarchyKey = 3;
                    newCompanyObject = [companyID: companyIdKey,
                    name: companyName,
                    location: location,
                    organizationRole: organizationRole,
                    hierarchyKey: hierarchyKey];
                end if
            else if organizationRole == "Transporter" then
                newCompanyObject =
                companyID: companyIdKey,
                name: companyName,
                location: location,
                organizationRole: organizationRole,
            else
                throw error: "Please enter valid organization role"
            end if
            ledger.putState(companyIdKey, newCompanyObject)
    end if
END
```

Algorithm 2. Pseudo-code to add a new drug

BEGIN
if manufacturer == legitimate node **then**
    Enter drug details
    Create an unique medication ID using SHA256
    Data is kept.
    drug's distinctive ID is mapped to a saved address.
elseNo new drugs can be made
end if
New drug is created
END

Algorithm 3. Pseudo-code for creating PO

BEGIN
Request to create new PO created
if buyerOrgDetails.organizationRole === "Retailer" &&
    sellerOrgDetails.organizationRole === "Distributor"||
    buyerOrgDetails.organizationRole === "Distributor"&&
    sellerOrgDetails.organizationRole === "Manufacturer" **then**
    newPO = poID: poIDKey,
    drugName: drugName,
    quantity: quantity,
    buyer: buyerKey.value.key,
    seller: sellerKey.value.key,
    ledger.putState(poIDKey, newPO);
else
    throw error: "Please make sure that the transfer of drug takes place in
    a hierarchical manner and no organization in the middle is skipped. ",
end if

Shipment is created.
END

3.3   Technology Stack

Proposed solution is implemented using the following tools:

3.3.1   Docker

Docker creates multiple linux machines which act as peers on the blockchain. For improved performance we are using Windows Subsystem for Linux as our backend for the docker which provides kernel level access to a linux environment while running Windows as host OS. For the HyperVisor platform, Hypervisor 2 is used which allows efficient hardware resource sharing over the peers. Furthermore the hyperledger platform used is easily deployable on the docker using the docker images provided by the developer.

3.3.2   NodeJS

We are using NodeJS as our javascript runtime because of its core being implemented on V8 engine by google which has the leading performance. NodeJS allows us to test and develop our smart contracts while also giving us added functionality of livuv which provides us many features, notably access to the file system of our OS and ability to run npm packages. NodeJS also allows us to host our server as well on the javaScript.

3.3.3   ExpressJS

In our case expressJs is used to create a web app which provides an endpoint to interact, communicate with peers and ensure that all the peers connected are authorized. It acts as a face of the blockchain.

3.3.4   Hyperledger

Hyperledger provides the ability to create peers, distributors, manufactures etc. out of the box. It provides ability to make distributed ledgers based on blockchain using javaScript, while using GoLang as its backend, both of which are modern languages, js being dynamic and goLang being statically typed, which makes it flexible. Hyperledger fabric 1.4.2 blockchain is used for management of supply chain.

4   Results and Discussion

All of the experimental findings for our suggested algorithm are presented in this section. Version 1.4.0 of the blockchain framework and version 2.0 of Hyperledger Fabric are used for the experiments.

4.1   Measurement Parameters

4.1.1   Transaction Throughput

The block size and the block interim both affect how many transactions may be processed at once. To ensure that the most peers in the system have received the current block before the creation of the next block. The rate at which the blockchain system transmits valid trades within the allotted period is known as transaction throughput. Average throughput can be calculated as follows:

$$TransactionThroughput = \frac{total committed transaction}{total time}$$

4.1.2   Transaction Latency

The amount of time between an exchange being submitted and being confirmed by the system is referred to as a transaction's latency. There are two ways to look at transaction latency: from the perspective of the number of peers at which the exchange is perceived to have settled and from the perspective of the proportion of perceptions equal to or below which the estimation is substantial (Percentile). Typically, Average latency, which is calculated as follows:

$$AverageTransactionLatency = \frac{sum\,of\,transaction\,latency}{total\,committed\,transactions}$$

4.2   Result Analysis

4.2.1   Examining the Effect of Block Size

By modifying the block size (20, 30, and 50) while comparing different transaction send rates, we were able to assess the effect of block size on performance (range from 10 tps to 225 tps).

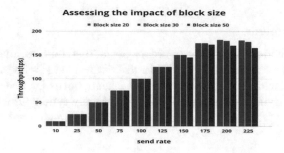

Fig. 2. Transaction throughput for block size (20, 30 and 50)

The experimental findings are represented in terms of average transaction throughput in Fig.2. When the block size was increased to 20, the number of transactions increased linearly as the transmit rate increased. The transaction throughput increased in the other two scenarios, though, and the transmit rate eventually reached roughly 175 tps. Whenever the transmit rate exceeded this, the growth in transaction throughput halted. The flow for each block size set was 181 tps and 178 tps, respectively, at a transmission frequency of 225 tps.

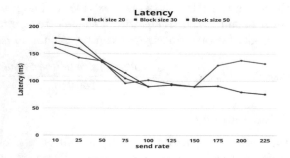

Fig. 3. Transaction latency for block size (20, 30 and 50)

Figure 3 displays the findings from the transactions lag experiment. The send rate increases and the transaction delay decreases. The transaction latency, for instance, dropped from 160 ms to 70 ms when the block size was modified to 30 tp block and the transmit rate increased from 25 to 100 tps. Smaller blocks are associated with improved application throughput and latency performance, according to the results of this experiment.

4.2.2   Examining the Influence of the Variety of Endorsers Peers

Here, performance is first being evaluated in relation to the quantity of peers who have endorsed it. In every run there will be a different variance each time, because of the OS memory allocation to different resources. For removing or minimizing the effect of this the same set of transactions is executed for 3 times and after that an average is taken as a result.

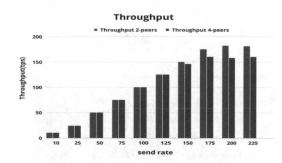

Fig. 4. Transaction throughput for 2 peers and 4 peers.

In Fig. 4, 1 organization 2 peers and 1 organization 4 peers are examined since it will aid in analysing the performance of system with various or maximum number of peers at various transaction send rates ranging from 10 to 225 tps. The transaction throughput grew linearly as the send rate was raised until it reached 150 tps. The transaction growth gradually slows down after 150 tps and stays constant at around 160 tps. So observation is that, when peers are increased 100% then performance will start decreasing by 30%. Suppose if 3 peers are evaluated, then peers may increase 50% and performance will decrease by 15%. So that's why here 4 peers used for evaluating performance with increased number of peers.

In Fig. 5, observed that the transaction latency will start increasing as the number of endorsed peers increases and maximum decrease will be 131 tps approximately. The quantity of recommended peers significantly affects performance. Therefore, increasing the number of peers may increase transaction latency but decrease transaction throughput. So for an efficient system latency must not be very high otherwise it will affect performance speed of a system.

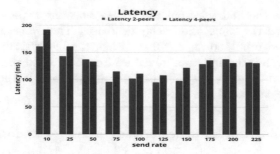

Fig. 5. Transaction latency for 2 peers and 4 peers.

4.2.3   Examining the Consequences of the Clientele

The effect of client volume on performance was examined in this experiment.

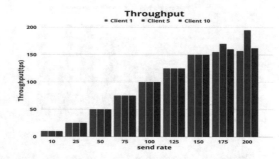

Fig. 6. Transaction throughput for clientele

The graph displays the throughput of transactions for various clients minimum 1 and max 10 clients, at various transaction send rates. The throughput increased proportionally as the transmit rate increased when there were 5 clients. After 150 tps, the transaction speed for the other two scenarios grow, on the other hand, considerably fell and nearly levelled off. The transaction throughput for each scenario at a transmit rate of 200 tps was 157 tps, 195 tps, and 162 tps. Figure 7 illustrates, as the number of customers increased, transaction delay increased as well. For instance, when there were 10 clients and the transmission rate went from 125 to 200, the latency rose from 130 to 550 ms. According to the findings of this experiment, performance is significantly impacted by the quantity of clients. Although more clients can boost throughput, too many customers can cause a large rise in latency due to an increase in network traffic.

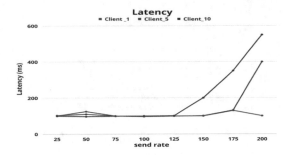

Fig. 7. Transaction delay for clientele

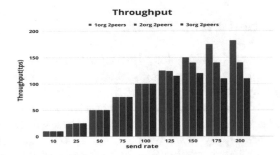

Fig. 8. Transaction throughput for no. of organizations (1 org 2 peers, 2 org 2 peers, 3 org 2 peers)

4.2.4 Examining the Effect of the Variety of Organizations

The effect of the number of organizations on performance was examined in this experiment. The average transaction throughput of various organizations (1, 2, and 3) is plotted in Fig. 8 over a range of (From 10 to 200) transactions per second. When there was only one organization, the transaction throughput rose proportional to the number of organizations. However, the growth in transaction throughput substantially slowed and almost stopped when the send rate exceeded 125 tps. The transaction throughput for each scenario at a transmit rate of 150 tps was 149.5 tps, 140 tps, and 120 tps, respectively.

According to Fig. 9, A network with more organizations creates significantly more transaction delay than a network with fewer organizations. For instance, the delay increased from 200 to 430 ms when there were three organizations and the transmission rate was doubled from 125 to 200 tps. This experiment demonstrates how a network might lose performance and see an increase in delay as the number of organizations grows. To examine the impact on endorsement policy, we built up a prototype network of three organizations, each with 2 endorser peers.

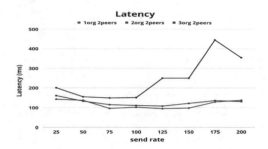

Fig. 9. Transaction latency for no. of organizations (1 org 2 peers, 2 org 2 peers, 3 org 2 peers)

5 Conclusion and Future Work

The adoption of blockchain technology in conjunction with other cutting-edge technologies is crucial to the drug supply chain and has the potential to completely change the way that healthcare is provided today. In this paper an hyperledger fabric-based blockchain system for counterfeit medicine in a healthcare supply chain system is proposed. The solution can successfully trace counterfeit medicine in healthcare supply chain management systems. It was observed that smaller blocks suggest better application throughput and latency performance. Proposed solution has been tested for different scenarios, namely, varying number of peers, number of clients and multiple organizations. Proposed solution went upto 4 peers to evaluate performance and it was identified that increasing the number of peers increased transaction latency but decreased transaction throughput. Further throughput of transactions for various clients minimum 1 and max 10 clients was observed. Finally, proposed solution was also tested by varying number of organizations and it was identified that if send rate is more than 125 tps, the growth in transaction throughput significantly slowed and nearly levelled off. In future, our proposed solution can be used as a benchmark for comparison with the solutions that are proposed in similar settings. Also the proposed system can be evaluated at real time by a pharmacompany and will work on improving the performance and scalability limitations of proposed solution.

References

1. Puri, S., Ranjan, J.: Study of logistics issues in the Indian pharmaceutical industry. Int. J. Logist. Econ. Global. **4**(3), 150–161 (2012)
2. WorldHealthOrganization. https://www.who.int/news-room/fact-sheets/detail/substandard-and-falsified-medical-products. Accessed 5 Aug 2022

3. Tesfaye W.F.: How do we combat bogus medicines in the age of the COVID-19 pandemic? Am. J. Trop Med. Hyg. **103**(4), 1360–1363 (2020). PMID: 32815510; PMCID: PMC7543841. https://doi.org/10.4269/ajtmh.20-0903

4. S. Bergman, F.: Permissioned blockchains and distributed databases: a performance study. Concurr. Comput. Pract. Ex. **32** (12) (2019)

5. Satoshi Nakamoto. Bitcoin: a peer-to-peer electronic cash system. Cryptography Mailing list (2009). https://metzdowd.com

6. Asma K. A blockchain-based smart contract system for healthcare management. Electronics **9**, 94 (2020)

7. Ding, Q., Gao, S., Zhu, J., Yuan, C.: Permissioned blockchain-based double-layer framework for product traceability system. IEEE Access **8**, 6209–6225 (2020). https://doi.org/10.1109/ACCESS.2019.2962274

8. Yan Huang, F.: Drugledger: a practical blockchain system for drug traceability and regulation. In: 2018 IEEE International Conference on Internet of Things (iThings) and IEEE Green Computing and Communications (GreenCom) and IEEE Cyber, Physical and Social Computing (CPSCom) and IEEE Smart Data (SmartData), pp. 1137–1144 (2018)

9. Paik, M., et al.: The case for Smart-Track. In: 2009 International Conference on Information and Communication Technologies and Development (ICTD), pp. 458–467 (2009)

10. Jansen-Vullers, M.-H., van Dorp, C.A., Beulens, A.J.M.: Managing traceability information in manufacture. Int. J. Inf. Manag. **23**(5) (2003). https://doi.org/10.1016/S0268-4012(03)00066-5

11. Kang, Y.-S., Lee, Y.-H.: Development of generic RFID traceability services. Comput. Indust. **64**, 609–623 (2013).https://doi.org/10.1016/j.compind.2013.03.004

12. Bergman, S., Asplund, M., Nadjm-Tehrani, S.: Permissioned blockchains and distributed databases: a performance study. Concurr.-Pract. Exp. **32**(12) (2020)

13. Prince Waqas Khan, F.: A data verification system for CCTV surveillance cameras using blockchain technology in smart cities. Electronics **9**(3) (2020)

14. Quorum. https://github.com/ConsenSys/quorum

15. Ripple. https://ripple.com/

16. Stellar. https://www.stellar.org/

17. Corda. https://www.corda.net/

18. Elli Androulaki, F.: Hyperledger fabric: a distributed operating system for permissioned blockchains. In: Proceedings of the Thirteenth EuroSys Conference, EuroSys 2018, New York, NY, USA (2018)

19. Pelc, A.: Fault-tolerant broadcasting and gossiping in communication networks, Netw. Int. J. **28**(3) (1996)

20. Musamih, A., et al.: A blockchain-based approach for drug traceability in healthcare supply chain. EEEE Access **9**, 9728–9743, 202 (2019)

21. Watanabe, H., et al.: Enhancing blockchain traceability with DAG-based tokens. In: 2019 IEEE International Conference on Blockchain (Blockchain), pp. 220–227 (2019). https://doi.org/10.1109/Blockchain.2019.00036(2019)

22. Omar, I.A., et al.: Automating procurement contracts in the healthcare supply chain using blockchain smart contracts. IEEE Access **9**, 37397–37409 (2021) https://doi.org/10.1109/ACCESS.2021.3062471

23. Sharma, S., Kaushal, S., Gupta, S., Kumar, H.A.: Blockchain-based solution for electronic medical records system in healthcare. In: Luhach, A.K., Jat, D.S., Hawari, K.B.G., Gao, XZ., Lingras, P. (eds.) Advanced Informatics for Computing Research. ICAICR 2021. Communications in Computer and Information Science, vol. 1575. Springer, Cham (2022). https://doi.org/10.1007/978-3-031-09469-9_20

24. Computing (CPSCom) and IEEE Smart Data (SmartData), pp. 1137–1144 (2018). https://doi.org/10.1109/Cybermatics2018.2018.00206
25. Hyperledger Fabric. https://www.hyperledger.org/projects/fabric
26. Ethereum. https://www.ethereum.org

A Deep Learning Model for Breast Cancer Diagnosis Using Mammography Images Classification

Nourane Laaffat[1]([✉]), Ahmad Outfarouin[2], Walid Bouarifi[1], and Abdelilah Jraifi[1]

[1] Mathematics Computer Science and Communication Systems Research Laboratory, National School of Applied Sciences, Cadi Ayyad University, 46000 Safi, Morocco
nourane.laafat@gmail.com
[2] Management and Decision Support Laboratory, National School of Business and Management Dakhla, Ibn Zohr University, 73000 Agadir, Morocco

Abstract. Medical image classification is a technique for categorizing images to the appropriate pathological stage. It is an essential step in computer-aided diagnosis, which was developed to assist the radiologist in the reading and analysis of medical images and to help in the early diagnosis of tumors and other diseases. Deep convolutional neural network (DCNN) models have proven to be widely used in the medical field in recent years and achieve great feats in image classification, especially in terms of high performance and robustness. In this paper, we suggest two approaches using the Transfer Learning (TL) technique to classify Breast Cancer (BC) using mammography images from the Mini-MIAS dataset. The first approach adopts the pre-trained models as Feature Extractors then the classification is performed by two classifiers: SVM (Support Vector Machine) and RF (Random Forest), the second approach performs the classification process using the model fine tuning technique by adjusting the hyperparameters. The DCNNs models chosen for this study are: DenseNet121, MobileNetV2 and InceptionResNetV2. The data augmentation technique performed during preprocessing is of great importance for the improvement of our small dataset. The experimental results shows that MobileNetV2 in the second approach achieves an accuracy of 97% to differentiate abnormal cases from normal ones, and therefore proves to be the most accurate classifier.

Keywords: Breast cancer · Deep learning · Deep convolutional neural networks · Transfer learning · Mammography imaging · Data augmentation · Classification

1 Introduction

Breast cancer (BC) is a cancer that many women get, and according to World Health Organization (WHO) [1], more than 2.2 million cases have been identified in 2020. Approximately 1 in 12 women has a chance of developing breast cancer [1]. It turns out that BC is one of the main types of cancer that can affect a woman and end her life, about 685,000 women died in 2020 [1]. In the face of this malignant pathology, the importance

© The Author(s), under exclusive license to Springer Nature Switzerland AG 2023
F. P. García Márquez et al. (Eds.): ICCIDA 2022, LNNS 643, pp. 411–422, 2023.
https://doi.org/10.1007/978-3-031-27099-4_32

of early detection and diagnosis of the disease becomes apparent. Especially because of its role in the treatment and control of the disease before it develops and affects other organs and becomes difficult to control and therefore contribute in the declining of the mortality rate [2, 3]. Mammography [2, 4, 5] is a radiographic examination that uses X-rays. It is based on the principle of x-raying each breast from the front and from the side in order to visualize the entire mammary gland. Mammography is designed specifically for the women's breast and aims to detect abnormalities if they exist before any symptoms appear [6], diagnose and examine symptoms of breast pathology, such as breast swelling, breast discharge, change in breast size and shape or skin wrinkles, nipple inversion, itching, crusty sores or rash around the breast, etc., and detect BC. Each year, a large volume of mammographic images must be analyzed, which requires intense work, enormous time and several interventions by different radiologists in order to help each other in the decision-making process, especially since this is a sufficiently delicate and essential task for the survival or not of a woman. To this end, several research studies have been directed toward automating mammography reading and decision making. Early work on automatic mammography image processing systems was aimed at providing radiologists with a second interpretation to help them detect/diagnose abnormal lesions at an early stage, regardless of their nature [7]. Specially, Computer-Aided Diagnosis (CADx) systems refer to a complete mammographic image processing system from preprocessing to classification and decision making. It aims to characterize and classify the detected anomaly as benign or malignant or as abnormal or normal.

Artificial intelligence is achieving great success in the medical field. For the classification of breast lesions, several Machine Learning (ML) and Deep Learning (DL) algorithms exist. Deep Convolutional Neural Networks are used in recent years in an intense way due to its success in the efficient classification of anomalies.

This work gives a presentation on the method of application of (DCNN) on the Women's breast dataset using the TL technique in two different ways.

The first one adopt the pre-trained models as Feature Extractors then the classification is performed by two ML classifiers: SVM (Support Vector Machine) and RF (Random Forest), and the second approach performs the classification process using the model fine tuning method by adjusting the hyperparameters. The DCNNs models chosen for this study are: DenseNet121, MobileNetV2 and InceptionResNetV2. Data augmentation technique is a necessary preprocessing step for the improvement of our small dataset.

The remainder of this document goes through the following structure. Section 2 is a presentation of related work. Section 3 is the methodology. Section 4 deals with the experimental results. Section 5 presents a discussion of the experimental results obtained. Finally, Sect. 6 presents a conclusion of the paper.

2   Related Work

Image classification in DL involves extracting image features directly from raw images by adjusting convolution and pooling layer parameters. DL models have made great progress in medical image classification and preventing disease such as BC, Cervical Cancer, Pneumonia, Skin disease…, especially with the emergence of TL techniques and the data augmentation that have evolved the performance rates of classification models.

Transfer learning techniques have been able to remedy several problems, among which we cite the lack of data (medical images) available, since it is difficult to have large datasets, therefore, DCNN find themselves unable to better learn small datasets and end up finding themselves in the phenomenon of overfitting. The transfer learning exploits the pre-trained DCNN in different ways. Either to extract the features, and keep the important knowledge and pass it to a classification model to use it. Either to make specific modifications to it in order to obtain good results.

Alantari and Kim [8] contribute with a breast lesion detection and classification system. The detection phase is performed through the YOLO detector and obtains F1 scores of 99.2% and 98.02%, successively, for the DDSM and INbreast datasets. The classification phase is performed using regular feedforward CNN, ResNet-50, and InceptionResNet-V2 classifiers. For DDSM and INbreast datasets, successively, CNN classification models achieve an accuracy of 94.5% and 88.7%, ResNet-50 achieves 95.8% and 92.5%, and InceptionResNetV2 achieves 97.5% and 95.3%.

N. S. Ismail and C. Sovuthy [9] apply VGG16 and ResNet50 on the IRMA dataset and compare their performance to find the best breast cancer detector and classifier. VGG16 proves to be the top classifier, its accuracy is 94% while that of ResNet50 is 91.7%. In reference [10], the authors contribute with a CNN-based breast cancer mass classifier. They compare the performance criterion of the different models AlexNet and GoogleNet, which vary according to the design and hyperparameters, to find the best classifier using images from four mammogram databases named CBIS-DDSM, INbreast, MIAS and images from Egyptian National Cancer Institute (NCI). The classification models of AlexNet and GoogleNet, achieve accuracies of 100%, and 98.46%, respectively, for the CBISDDSM database, and 100%, and 92.5%, respectively, for the INbreast database, and 97.89% with AUC of 98.32%, and 91.58% with AUC of 96.5%, respectively, for NCI images, and 98.53% with AUC of 98.95%, and 88.24% with AUC of 94.65%, respectively, for MIAS database. Thus, AlexNet model proves to be the most accurate classifier. In reference [11], the authors present a new system idea to segment and classify breast cancer images. They performed the classification into malignant and benign using the InceptionV3, DenseNet121, ResNet50, VGG16 and MobileNetV2 models on MIAS, DDSM and CBIS-DDSM datasets. The proposed model with the classification phase performed by the Inception v3 model, especially with the DDSM dataset, achieves the best result with an accuracy of 98.87%, an AUC of 98.88%, a sensitivity of 98.98%, a precision of 98.79%, 97.99% of F1-score and 1.2134 s for computation time.

In reference [12], the authors contribute with a Covid-19 classifier through pre-trained DCNN models Xception, InceptionV3, MobileNetV2, VGG16, VGG19 and NasNetLarge. The study is done on a dataset consisting of chest x-ray images and formed by three classes: Normal, Covid and Pneumonia. During the training, the hyper-parameters used are: learning rate (1e-3), epochs (35) and batch size (8). The fine-tuned pre-trained VGG-16 model succeed to achieve the highest accuracy of 95.88%.

3 Methodology

Our proposed system consists of 4 steps, we illustrate its architecture in Fig. 1. The first step is to collect the mammographic images of the woman's breast, in our case we used Mini-Mias dataset which has been available for free for scientific research, then we move to the data processing to improve the size of the dataset and the quality of its images. Then, the second step is to split our dataset into three (training, testing and validation). After, the step 3 which concerns the extraction of the features which is carried out by pre-trained DCNNs models through the technique of transfer learning. The models that we have adopted are: DenseNet121, MobileNetV2 and InceptionResNetV2. Finally, the fourth step is carried out in two different ways. The first way is to classify the extracted features through ML classifiers, in this case SVM and RF, which gives rise to approach 1. And the second way, which presents the second approach, consists in fine-tuning the pre-trained models, while implementing the softmax function at the last classification network layer (FC) to specify the predicted output probabilities for the mammography class determination, in this case it is two outputs/classes: Normal or Abnormal.

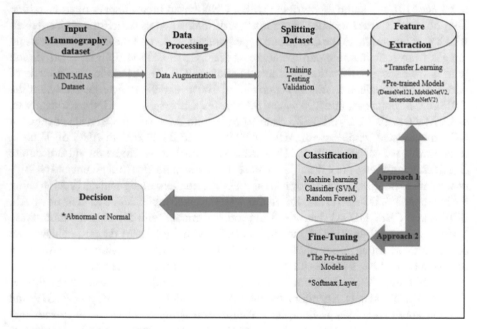

Fig. 1. Overview of our proposed system.

3.1 Data Collection and Pre-processing

We use the Mini-MIAS dataset [13] which has been made available free of charge for scientific research, through the Pilot European Image Processing Archive (PEIPA) of the University of Essex. It was collected and generated by a group of researchers from the UK at the Mammographic Image Analysis Society (MIAS), who are interested in understanding mammographic images and detecting breast lesions. The dataset consists of 322 mammographic images of the mediolateral oblique views of 1024*1024 pixels, including 113 images with abnormalities that may be benign or malignant and 209 without abnormalities (Normal) and divided into 2 classes (35, 1% abnormal and 64, 91% normal). Given the small size of our dataset, we necessarily go through the preprocessing technique to overcome the overfitting and improve the learning capability of the proposed model. For this, we use data augmentation to upsurge the samples number, especially those of the training set. It includes random image rotation, zooming and shearing, horizontal and vertical random flipping.

3.2 Splitting the Breast Imaging Dataset

We split our original dataset into three sets: Training, Validation and Testing sets.

4 Experimental Results

It is always a question of finding powerful and robust models that will allow an accurate diagnosis as much as possible. We thus evaluate the performance of the models adopted, through the parameters measuring the effectiveness and efficiency. For BC classification, performance evaluation is performed by some commonly used measures such as accuracy, precision, recall and sensitivity [14]. We mention that for the second model based on fine-tuning, we fix, during the training process, the different hyper parameters used. We present them in Table 1.

Table 1. Hyperparameters values.

Hyperparameters	Values
Optimizer	Adam
Learning rate	0.00005
Batch size	32
Epoch	10

4.1 Effectiveness

This section evaluates the efficiency of all adopted classifiers through the model speed prediction, correctly classified instances, misclassified instances and accuracy. Our Testing set consists of 32 images. We present the results of the first approach in Table 2 and Table 3, and the results of the second approach in Table 4.

Simulation errors are also taken into account to better evaluate the performance of the classifiers. In this study, the classifiers effectiveness is also evaluated through:

- Kappa statistic (KS),
- Mean Absolute Error (MAE),
- Root Mean Squared Error (RMSE),
- Relative Absolute Error (RAE),
- Root Relative Squared Error (RRSE).

KS, MAE and RMSE are in numeric values. RAE and RRSE are in percentage. The results for the first approach are shown in Table 5 and Table 6 and those for the second approach are presented in Table 7.

Table 2. Performance of the classifiers with SVM in the first approach.

Evaluation Criteria	Classifiers with SVM		
	DenseNet121	MobileNetV2	InceptionResNetV2
Speed of Prediction (s)	0.00107	0.0013	0.00127
Correctly Classified Instances	30	28	30
Incorrectly Classified Instances	2	4	2
Accuracy (%)	94	88	94

Table 3. Performance of the classifiers with RF in the first approach.

Evaluation criteria	Classifiers with RF		
	DenseNet121	MobileNetV2	InceptionResNetV2
Speed of Prediction (s)	0.0096	0.0095	0,0155
Correctly Classified Instances	28	27	28
Incorrectly Classified Instances	4	5	4
Accuracy (%)	88	84	88

Table 4. Performance of the classifiers in the second approach.

Evaluation criteria	Classifiers		
	DenseNet121	MobileNetV2	InceptionResNetV2
Speed of Prediction (s)	1.28	0.68	2.24
Correctly Classified Instances	23	31	29
Incorrectly Classified Instances	9	1	3
Accuracy (%)	72	97	91

Table 5. Simulation error for the first approach using the SVM classifier.

Evaluation criteria	Classifiers		
	DenseNet121	MobileNetV2	InceptionResNetV2
Kappa Statistic (KS)	0,85	0,71	0,85
Mean Absolute Error (MAE)	0,06	0,125	0,06
Root Mean Squared Error (RMSE)	0,25	0,353	0,25
Relative Absolute Error (RAE) %	13,85	27,70	13,85
Root Relative Squared Error (RRQE) %	52,63	74,43	52,63

Table 6. Simulation error for the first approach using the RF classifier.

Evaluation Criteria	Classifiers		
	DenseNet121	MobileNetV2	InceptionResNetV2
Kappa Statistic (KS)	0,93	0,61	0,93
Mean Absolute Error (MAE)	0,125	0,156	0,125
Root Mean Squared Error (RMSE)	0,35	0,39	0,35
Relative Absolute Error (RAE) %	27,70	34,63	27,70
Root Relative Squared Error (RRQE) %	74,43	83,22	74,43

Table 7. Simulation error for the second approach.

Evaluation criteria	Classifiers		
	DenseNet121	MobileNetV2	InceptionResNetV2
Kappa Statistic (KS)	0,30	0,93	0,78
Mean Absolute Error (MAE)	0,28	0,03	0,09
Root Mean Squared error (RMSE)	0,53	0,18	0,30
Relative Absolute Error (RAE) %	62,33	6,92	20,78
Root Relative Squared Error (RRQE) %	111,65	37,22	64,46

4.2 Efficiency

This section evaluates the adopted classifiers efficiency through precision, recall, TPR (true positive rate), TNR (true negative rate), FPR (false positive rate), and FNR (false negative rate) for all proposed classifiers, directly after building the predictive model. We summarize the results of this comparison in Tables 8, 9, 10, 11, 12 and 13. We present the confusion matrices that also serve as evaluators of the classifiers. Each row in Tables 14, 15 and 16 represents the rates in an actual class while each column refers to the predicted class.

Table 8. Comparison of TP, TN, FP, RN Rates for the first model using SVM Classifier.

Classifiers	Evaluation Criteria			
	TPR	TNR	FPR	FNR
DenseNet121	1	0,91	0,08	0
MobileNetV2	0,88	0,87	0,13	0,11
InceptionResNetV2	1	0,91	0,08	0

Table 9. Comparison of TP, TN, FP, RN rates for the first proposed model using RF classifier.

Classifiers	Evaluation criteria			
	TPR	TNR	FPR	FNR
DenseNet121	0,88	0,87	0,13	0,11
MobileNetV2	1	0,81	0,19	0
InceptionResNetV2	0,88	0,87	0,13	0,11

Table 10. Comparison of TP, TN, FP, RN rates for the second proposed model

Classifiers	Evaluation Criteria			
	TPR	TNR	FPR	FNR
DenseNet121	0,66	0,73	0,27	0,22
MobileNetV2	1	0,95	0,04	0
InceptionResNetV2	0,9	0,91	0,09	0,1

Table 11. Comparison of accuracy measures for the first model using SVM classifier.

Classifiers	Evaluation criteria				Class
	Precision	Recall	F1-score	AUC	
DenseNet121	1,00	0,82	0,90	0,91	Abnormal
	0,91	1.00	0,95		Normal
MobileNetV2	0,89	0,73	0,80	0,84	Abnormal
	0,87	0,95	0,91		Normal
InceptionResNetV2	1,00	0,82	0,90	0,91	Abnormal
	0,91	1,00	0,95		Normal

Table 12. Comparison of accuracy measures for the first model using RF classifier

Classifiers	Evaluation Criteria				Class
	Precision	Recall	F1-score	AUC	
DenseNet121	1,00	0,64	0,78	0,84	Abnormal
	1,00	1.00	0,91		Normal
MobileNetV2	1,00	0,55	0,71	0,77	Abnormal
	0,81	1,00	0,89		Normal
InceptionResNetV2	1,00	0,64	0,78	0,84	Abnormal
	0,84	1,00	0,91		Normal

Table 13. Comparison of accuracy measures for the second proposed model.

Classifiers	Evaluation Criteria				Class
	Precision	Recall	F1-score	AUC	
DenseNet121	0,67	0,36	0,47	0,63	Abnormal
	0,73	0.90	0,81		Normal
MobileNetV2	1,00	0,91	0,95	0,95	Abnormal
	0,95	1,00	0,98		Normal
InceptionResNetV2	0,90	0,82	0,86	0,88	Abnormal
	0,91	0,95	0,93		Normal

Table 14. Confusion matrices for the first model using SVM

Classifiers	Actual Class		Predicted class
	Abnormal	Normal	
DenseNet121	9	2	Abnormal
	0	21	Normal
MobileNetV2	8	3	Abnormal
	1	20	Normal
InceptionResNetV2	9	2	Abnormal
	0	21	Normal

Table 15. Confusion matrices for the first model using RF

Classifiers	Actual class		Predicted class
	Abnormal	Normal	
DenseNet121	8	3	Abnormal
	1	20	Normal
MobileNetV2	6	5	Abnormal
	0	21	Normal
InceptionResNetV2	8	3	Abnormal
	1	20	Normal

Table 16. Confusion matrices for the second model

Classifiers	Actual Class		Predicted Class
	Abnormal	Normal	
DenseNet121	4	7	Abnormal
	2	19	Normal
MobileNetV2	10	1	Abnormal
	0	21	Normal
InceptionResNetV2	9	2	Abnormal
	1	20	Normal

5   Discussion

From Table 2 to Table 4, we can see that DenseNet121 used in the first proposed model takes about 0.00107 s to make its predictions on the test dataset, unlike InceptionRes-NetV2 in the second proposed model shown in Table 4 which takes 2.24 s. On the other hand, the accuracy obtained by MobileNetV2 in the second model (97%) is better than the accuracy obtained by MobileNetV2 in the first model (88% with SVM Classifier and 84% with RF Classifier). For the first model, InceptionResNetV2 and DenseNet121 with SVM classifier succeed to achieve the highest accuracy (94%). It can also be seen that MobileNetV2 in the second model correctly classifies the largest number of instances with the smallest number of misclassified instances compared to the first model.

From Table 5 to Table 7, we can see that MobileNetV2 used in the second proposed model (Table 7) achieves the best classification (0.93%) with the lowest error rate (0.03). While for DenseNet121 used in the second model, its highest error rate value justifies its misclassified instances (9) compared to MobileNetV2 in the second model (1).

It is possible to determine whether a model works well or not by its ability to detect the anomaly. For this, the values of TPR and TNR must be high, and those of FPR and FNR must be as low as possible. The capacity of correct identification of the cases with the anomaly is produced by TPR which represents the sensitivity, as well as that of the cases without the anomaly is produced by TNR which represents the specificity. Table 10 shows that the second model with MobileNetV2 obtained the highest value of TPR and TNR (100% and 95% respectively) and the lowest value of FPR and FNR (0.04 and 0 respectively).

The area under the curve (AUC) measures how well a classifier is able to discern between existing classes. The chance of having a model that performs well and has a high ability to distinguish between normal and abnormal or benign and malignant or negative and positive anomaly (different classes) depends on a high AUC value. From Table 11 to Table 13, we can see that the second model with MobileNetV2 has the highest AUC value (0.95), unlike the other models.

Table 14, Table 15 and Table 16 show the confusion matrices, and from which we compare the actual class and the predicted results obtained. The second model with MobileNetV2 correctly predict 31 instances out of 32 instances (10 Abnormal instances that are indeed Abnormal and 21 Normal cases that are in fact Normal), and 1 poorly predicted case (1 Normal case predicted as Abnormal and 0 cases of Abnormal class predicted as Normal). This justifies why the accuracy of the second model with MobileNetV2 is better than the other classifiers in the two proposed models, with lower error rate.

In summary, MobileNetV2, in the second proposed model, was able to distinguish itself through its effectiveness and efficiency compared to other models in the two proposed approaches.

6 Conclusion

To analyze biomedical images in order to detect anomalies and classify them, there are several techniques and models in DL and ML. The challenge still remains to achieve a robust model: Fast, stable and accurate. In this work, we proposed an architecture of a CADx for BC diagnosis using the Mini-MIAS mammographic images dataset. It is based on two different proposed approaches and uses the transfer learning technique of the pre-trained models DenseNet121, MobileNetV2 and InceptionResNetV2. The first proposed model consists of using the models as feature extractor, then the classification process is performed through two ML classifiers: SVM and RF. DenseNet121 and

InceptionResNetV2 with SVM classifier achieved an accuracy of 94% and proved to be the best classifiers for the first model. The second proposed model relies on the fine-tuning of the pre-trained models. It achieved an accuracy of 97% with MobileNetV2. Therefore, the fine-tuned MobileNetV2 model surpasses all the classifiers in terms of performance. In general, the Fine-tuning technique is well suited for small datasets. The objective of this work is to provide radiologists with diagnostic assistance that is as accurate as possible and to contribute to the detection of BC. Our next goal is to improve the accuracy obtained at higher value by exploring other deep learning techniques and models, and to work on several datasets of several diseases.

References

1. WHO Homepage https://www.who.int/news-room/fact-sheets/detail/breast-cancer
2. NCI (National Cancer Institute). Annual report to the nation: cancer death rates continue to decline; Increase in Liver Cancer Deaths Cause For Concern, 9 March 2016. https://www.cancer.gov/news-events/pressreleases/2016/annual-report-nation-1975-2012
3. Kavitha, T., Mathai, P.P., Karthikeyan, C., et al.: Deep learning based capsule neural network model for breast cancer diagnosis using mammogram images. Interdiscip. Sci. Comput. Life Sci. **14**, 113–129 (2022). https://doi.org/10.1007/s12539-021-00467-y
4. Rampun, A., et al.: Breast pectoral muscle segmentation in mammograms using a modified holisticallynested edge detection network. Med. Image Anal. **57**, 1–17 (2019). https://doi.org/10.1016/j.media.2019.06.007
5. Punitha, S., Amuthan, A., Joseph, K.S.: Benign and malignant breast cancer segmentation using optimized region growing technique. Future Comput. Inf. J. **3**(2), 348–358 (2018). https://doi.org/10.1016/j.fcij.2018.10.005
6. Cheikhrouhou, I.: Description et classification des masses mammaires pour le diagnostic du cancer du sein. (Description and classification of breast masses for the diagnosis of breast cancer). Doctoral Thesis. University of Évry Val d'Essonne, France (2012). https://dblp.org/rec/phd/hal/Cheikhrouhou12
7. Astley, S., Gilbert, F.: Computer-aided detection in mammography. Clin. Radiol. **59**(5), 390–399 (2014)
8. Al-antari, M.A., Kim, T.-S.: Evaluation of deep learning detection and classification towards computer-aided diagnosis of breast lesions in digital X-ray mammograms. Comput. Methods Program. Biomed. **196**, 105584 (2020)
9. Ismail, N.S., Sovuthy, C.: Breast cancer detection based on deep learning technique. In: 2019 International UNIMAS STEM 12th Engineering Conference (EnCon), pp. 89–92 (2019). https://doi.org/10.1109/EnCon.2019.8861256
10. Hassan, S.A., Sayed, M.S., Abdalla, M.I., Rashwan, M.A.: Breast cancer masses classification using deep convolutional neural networks and transfer learning. Multimedia Tools Appl. **79**(41–42), 30735–30768 (2020). https://doi.org/10.1007/s11042-020-09518-w
11. Salama, W.M., Aly, M.H.: Deep learning in mammography images segmentation and classification: Automated CNN approach. Alexandria Eng. J. **60**(5), 4701–4709 (2021). ISSN 1110–0168. https://doi.org/10.1016/j.aej.2021.03.048
12. Makris, A., Kontopoulos, I., Tserpes, K.: COVID-19 detection from chest X-ray images using deep learning and convolutional neural networks. In: Proceedings of the 11th Hellenic Conference on Artificial Intelligence (2020). https://doi.org/10.1101/2020.05.22.20110817
13. http://peipa.essex.ac.uk/info/mias.html
14. Murtaza, G., et al.: Deep learning-based breast cancer classification through medical imaging modalities: state of the art and research challenges. Artif. Intell. Rev. **53**(3), 1655–1720 (2019). https://doi.org/10.1007/s10462-019-09716-5

Highly Secure and Robust Forensic System: Fordex Forensic Chain

Faruk Takaoğlu[1] , Mustafa Takaoğlu[1](✉) , Taner Dursun[1] , Erkan Demirci[1] ,
Adem Özyavaş[2] , Firas Ajlouni[3] , and Naim Ajlouni[2]

[1] The Scientific and Technological Research Council of Türkiye, Kocaeli, Türkiye
`{faruk.takaoglu,mustafa.takaoglu,taner.dursun,`
`erkan.demirci}@tubitak.gov.tr`
[2] İstanbul Atlas University, İstanbul, Türkiye
`{adem.ozyavas,naim.ajlouni}@atlas.edu.tr`
[3] Lancashire College of Further Education, Blackburn, UK
`firas@lcfe.org`

Abstract. Highly secured forensic document examiners are devices of great demand with the advancement of artificial intelligence and processing power. In most cases, it can be seen that there exists a rush from law enforcement agencies and criminals to utilize new methods for the discovery of fraudulent acts or for the achievement of a perfect fraudulent act. This work aims to extend the abilities of the forensic document examiner device Fordex by proposing the use of Blockchain technology to eliminate trust issues in the field of forensics. Fordex is a device that is currently used in forensic document analysis. It is developed by TÜBİTAK, BİLGEM, UEKAE, and Bioelectronics Systems Laboratory. It is intended to use Hyperledger Fabric, a permission Blockchain platform, in the Blockchain environment. In the proposed system, the Fordex software and the Fordex-Forensic-Chain (FFC) Blockchain system will interact within the Hyperledger Fabric platform in a reliable and scalable manner. The proposed architecture allows the system administrator to access and examines records of case studies tested by the Fordex device. The designed control mechanism protects the forensic images using the SHA256 hash algorithm while keeping them in the traditional database and alerts the system administrator in case of any unauthorized change in the recorded data. To the best of our knowledge, the FFC will be the first Blockchain application in which forensic devices are used.

Keywords: Forensic Blockchain · Hyperledger fabric · Forensic document analysis · Forensic examination device · Fraud & Falsification detection

1 Introduction

Bitcoin, which was introduced in 2009 by Nakamoto [1], was the base on which Blockchain technology was formed in 2014. Its cryptologic foundations go back to Merkle trees and timestamps [2]. It is a very sophisticated amalgamation of ongoing cryptologic and computer science studies. The idea of Bitcoin, which was put forward

F. P. García Márquez et al. (Eds.): ICCIDA 2022, LNNS 643, pp. 423–437, 2023.
https://doi.org/10.1007/978-3-031-27099-4_33

with the dream of decentralized finance, attracted the global attention of society because of its financial success and enabled the development of the technology behind it. Today, it is realized that Blockchain technology's importance is far beyond its widespread use in cryptocurrencies. For example, various studies were carried out on applying Blockchain to distributed ledger technology. During the pandemic, Blockchain technology has gone through different tests in every field, such as crypto investment and the metaverse. As it is explained in the following sections, because of its structural features, systems using Blockchain technology as a plugin have become more reliable and powerful [3].

Forensics is another field of study in which Blockchain technology can make great contributions [4]. Forensic evidence collection, examination, and storage is an attractive research topic. The digitization of evidence and its storage in the computer environment has specific issues as well as requirements. Even though the probability of tampering with forensic data is low, manipulations of forensic data are still possible in conventional databases. In this context, blockchain technology is fundamental when working with forensic data. In the literature, there are studies on protecting forensic data with Blockchain [5–10]. However, no study focuses on forensic devices and blockchain integration.

Special devices are produced for fraud and falsification cases like the forgery of signatures and similar cases that require forensic examination. There are very few countries in the world that carry out studies in this field and produce products. Forensic document examination device is produced by four companies in 4 different countries today. These countries are England, Switzerland, Belarus, and Turkey. The first device made in this field in Turkey was named "Forensic", which TÜBİTAK, BİLGEM in 2004 produced. BİLGEM has exported this device to 14 different countries. In 2018, the R&D processes of the existing device were restarted in the TÜBİTAK, BİLGEM, UEKAE, Bioelectronics Systems Laboratory, and a new generation document verification device was introduced in 2022 under the name Fordex [11].

The results of the procedures performed in the forensic examination laboratory have a critical impact on the lives of the people involved. Forensic issues studied by experts can be stressful and can affect experts' psychology. For this reason, human errors are expected to have a negative effect on forensic investigations in the long run. Despite the level of maturity of the technology in forensic studies, it still requires ongoing studies under the control of experts. For this reason, forensic examination laboratories, especially document review units, need sustainable and controllable systems, considering that human-based errors will also occur. In addition, retrospective immutable systems are required to re-examine the previous forensic reports challenged by the judiciary to check whether there are any errors in previous reports.

The Fordex device results from years of ongoing R&D activities; many innovative detection methods, especially mobile modules that enable optical forensic examination for long periods, can be used. Considering forensic documents in particular, although the advancement of technology allows the production of better forensic devices, innovative techniques are being developed in this regard in forgery methods. For this reason, fraud detection devices should be regularly subjected to R&D activities.

Fordex document examination device, unlike other devices, has a measurement infrastructure that includes movable sub-modules and the exact coordinate information of these modules. The height of the movable lighting sources in relation to the examination floor is one of this infrastructure's unique variables/features; it is recorded for each examination. In addition, the panels containing the lighting sources can make tilt-oscillation movements between 90° and 150°. It can give the instant position of the document placed on the movable floor table (open frame stage) in the device to the camera and the floor table. A number of lighting sources can be used in combination with the device software. Various device lighting sources can be chosen from a combination of different heights and angles, which is beneficial in revealing the material's absorbance, fluorescence, and reflectance properties. This device will also be able to test different forensic needs that will be developed in the future.

In forensic informatics laboratories, many similar analysis processes are stored as physical or digital reports. The reports are recorded and stored in a closed intranet network and can be shared among employees. The laboratory director or supervisor observes all sharing instances. Employees' performances can be evaluated using digital laboratory document management software produced for these tasks. It should be highlighted that important variables such as moving lighting, angles, heights, and ground position values are not being recorded in detail by similar devices in the market.

Fordex device is one of the best forensic examiner devices, and it's constantly being improved according to new user requirements. In this context, the results of the document examination should be protected from unauthorized manipulations. Blockchain technology is ideal for adding this structural feature to the desired systems. The Fordex device is an upgradable device where these add-ons can be made.

In this study, a system with high security and controllability has been proposed so that the forensic data produced by the Fordex device are stored in the Blockchain system, which is tamper-proof.

The rest of the paper is arranged as follows. Section two introduces information about the systems and platforms used in Blockchain technology. Section three provides detailed information about forensic document examination devices and document analysis. Section four presents the proposed system. Finally, section five provides information about the positive contributions of the study.

2   Blockchain Technology

According to the idea of Bitcoin, Blockchain technology was a decentralized, anonymous, secure, and distributed financial system that removed third parties from the system [1]. However, with the testing of Blockchain technology outside of finance and the introduction of smart contracts, it is seen as a safe option that can address the needs of people and organizations [12].

Blockchain systems, in short, can be seen as a digital ledger of transactions duplicated and distributed throughout the network. Data transfers realized and recorded in Blockchain systems are called transactions. Each recorded transaction is in a structure that cannot be deleted once it has been recorded in the system. In other words, Blockchain systems' data are immutable. Transactions are recorded in blocks when they occur. When each block is full or when the block creation time comes, it is approved according to the consensus algorithm used by the validators in the system, and block generation takes place. These recorded blocks are connected like a ledger using hash algorithms, and this system is called Blockchain, a subtype of distributed ledger technology [13].

Hash algorithms and digital signatures are cryptologic mechanisms that provide security in Blockchain systems. Hash algorithms are structures that work one way and produce unique values. Generally, the SHA256 hash algorithm is widely used in today's Blockchain systems. The data summarized using the SHA256 algorithm is transformed into 32-bytes structures. The result obtained will change in the slightest change that may occur in the hashed data. This is due to the hash mechanism, if any transaction or block approved in the Blockchain is changed or tampered with, the hash value will change, and the change will be noticed because it will not match the value in the system. One of the properties of hash algorithms is that the chance of collision of hashes is almost zero. Also, hashing functions produce very different hash values for similar or almost identical data; it becomes next to impossible to guess the actual data given its hash. Because of the hashing algorithms' properties, Blockchain systems gain a tamper-proof structure [14].

One of the essential components of Blockchain systems is the consensus algorithm. Consensus algorithms, which are used to write the transactions created in the Blockchain to the blocks, are an approval mechanism in the system and are chosen in accordance with the developed Blockchain system. Today, many consensus algorithms are used, especially Proof-of-Work (PoW), Proof-of-Stake (PoS), Practical Byzantine Fault Tolerance (PBFT), Proof-of-Capacity (PoC), and Proof-of-Elapsed Time (PoET) [15, 19].

Blockchain systems are divided into three categories in terms of accessibility; public, private, and consortium (permissioned). In public Blockchain, anyone can access the system, make transactions, and become a miner. Public Blockchain is distributed transparent system. Since they are anonymous systems, the transaction history of the people can be followed but cannot be associated with its owner. Private Blockchain is a closed structure to which access is granted upon invitation. Private systems are more secure, scalable, and have high throughput. However, private Blockchain is centralized, and on-chain data and transactions can be altered by the network operator, so their immutability aspect is weak. Permissioned Blockchain is preferred when an additional layer of security, identity maintenance, and permission management features are needed. One of the best features of the permissioned Blockchain is the access control layer, which enables the network operator to limit participants' access and assign different roles. For this reason, the Hyperledger Fabric (HLF) v2.x platform, a permissioned Blockchain system, was chosen for the proposed FFC study [16].

Another critical component of Blockchain is smart contracts. Smart contracts are codes developed for specific cases that make Blockchain systems more dynamic. With the applicability of smart contracts in Blockchain systems, DLT has started to be applicable in many areas other than finance solutions [17].

The Hyperledger Fabric Blockchain platform is an open-source, permissioned, and modular system. It allows the use of different programming languages in smart contract development. It also allows the governance and versioning of smart contracts. The HLF platform is a Blockchain system that can be considered faster than competitor platforms and provides high interoperability [18].

3 Fordex – New Generation Forensic Document Examination System

Analysis methods in forensic document examination processes are divided into two groups. The first group is the examinations made by interfering with the integrity of the object, which has the quality of evidence to carry out the examination process. In general, analyses that harm the integrity of forensic documents are not preferred, but there are examination procedures that have to be used due to the lack of technological development and the type of examination process in use, such as ink age and paper age analysis.

Another forensic document examination method is an optical-based method that does not harm the integrity of the examined sample. The main purpose of these methods is to observe the damage and falsifications of the document by using sources that illuminate at different wavelengths in combination with different types of optical lenses. This observation process basically works according to the logic of observing the energy that light emanates from a lighting source of the material loads into the examination area. Materials can exhibit different properties according to the wavelength of the loaded energy, and it is necessary to be able to detect these properties by remote inspection methods. At this stage, optical lenses and camera sources are used with sensitivity at different wavelengths. These elements are used to measure how much a substance absorbs, reflects, and the light energy it is exposed to after being excited and its reaction at different wavelengths. The obtained data are converted from analogue to digital form and are then passed through different analysis algorithms, and the differences in the documents are examined by experts. This examination method is reproducible compared to other methods and has the advantage of not causing irreversible damage to the integrity of the samples to be examined.

Due to the increasing number of violent cases, the need for forensic documents and evidence analysis laboratories is increasing. The technological development competency of forensic document examination laboratories is directly proportional to the development of the technical capabilities of the devices and the software used. Users are trained on regular bases to cope with technological advances in this field. In forensic laboratories, the time the employees spend at a device and the results obtained should be reported and recorded. Then these cases should be re-examined by the official authorities. Mostly this type of examination is performed by laboratory-type document examination devices.

Fordex Forensic Document Examiner is a supporting device used by forensic experts, developed and commercialized by TÜBİTAK, BİLGEM, UEKAE, and Bioelectronic Systems Laboratory. It is one of today's most advanced forensic devices.

Fordex Next Generation Document Examiner is a system consisting of hardware and software modules to increase the capabilities of document review experts for forensic purposes. It is designed to detect forensic frauds visually by using a video camera, lighting sources, optical tools, and software. The Fordex device transforms 2D images to 3D using surface conversion software, enabling forensic evidence to be examined without destruction. Fordex device takes measurements using optical spectroscopy parameters that can be measured physically. In addition, because of its hyperspectral analysis technique, it can display the details of the reflection, transmittance, or fluorescence spectrum at any point on the image by obtaining image and spectrum information at the same time. The Fordex device uses powerful optoelectronic elements that capture slight differences in micro meter structures [11]. Figure 1 shows the Fordex device, and Fig. 2 shows the general scheme of the device.

Fig. 1. Fordex new generation forensic document examination device

The basic operation of the Fordex device is to obtain material properties or differences in the forensic case examined by optical means. For this purpose, different wavelengths of illumination sources and optical filters are used. In addition to the digital images obtained, they can be passed through image processing software or if the obtained findings are sufficient, every piece of evidence is recorded in the digital environment and reported. In the reporting process, the device's current state, which is used during the process of obtaining the findings, is automatically recorded by the device. These features include the location of the object under investigation, its X and Y coordinates on the testing table, the height of the lighting source relative to the ground, the lighting source used and its wavelength, and the angular slope values according to the examined sample.

The Fordex device creates all these metrics and angular values, and with the developed Blockchain-supported software system, the test results reported by forensic experts are preserved [11].

3.1 Illustrative Example of Typical Fordex

An example of the capabilities of the Fordex device is shown in Fig. 3 and Fig. 4; the document contains vital information that somebody is trying to hide by a layer of ink. Figure 3 is the document being examined by the device.

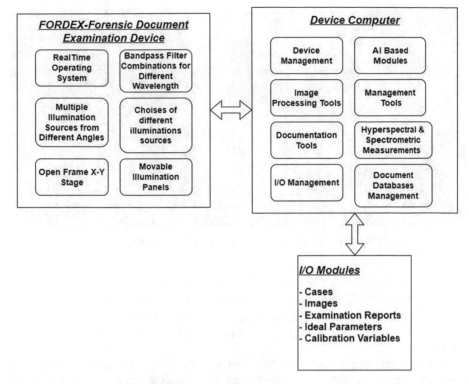

Fig. 2. General scheme of Fordex device

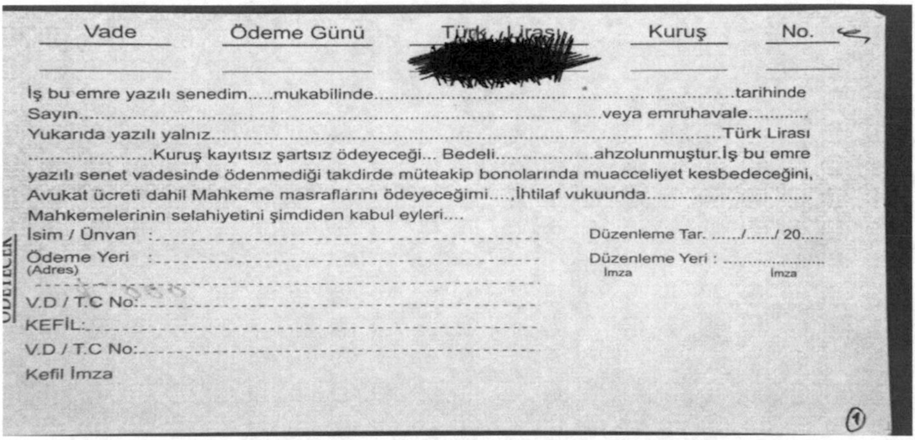

Fig. 3. Fordex forensic document examination case 1

Figure 4 shows the test result; the results show that the device can separate the layer of ink covering the data from the data itself.

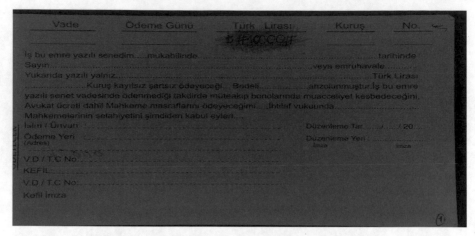

Fig. 4. Fordex forensic document examination case 1 result

The Fordex parameters produced and used for this test are recorded. The data are obtained when the lighting panels are located 30 mm above the floor and look at a 60-° angle towards the floor. The linear permeable filter was fixed at 695 nm and the illumination sources at 515 nm and 640 nm on the panel were believed to be more suitable.

In the second test, it is believed that the document shown in Fig. 5 is manipulated. Therefore, a forensic scientist uses the Fordex device tests to check if the document has been manipulated in any way.

Vade	Ödeme Günü	Türk Lirası	Kuruş	No.

₺ 50000₦

İş bu emre yazılı senedim.....mukabilinde...tarihinde
Sayın...veya emruhavale.............
Yukarıda yazılı yalnız...Türk Lirası
....................................Kuruş kayıtsız şartsız ödeyeceği... Bedeli...................ahzolunmuştur.İş bu emre
yazılı senet vadesinde ödenmediği takdirde müteakip bonolarında muacceliyet kesbedeceğini,
Avukat ücreti dahil Mahkeme masraflarını ödeyeceğimi.....,İhtilaf vukuunda...............................
Mahkemelerinin selahiyetini şimdiden kabul eyleri....
İsim / Ünvan :
Ödeme Yeri
(Adres)
V.D / T.C No:
KEFİL:
V.D / T.C No:
Kefil İmza

Düzenleme Tar. 06 / 02 / 2019
Düzenleme Yeri :
İmza   İmza

Fig. 5. Fordex forensic document examination case 2

The results of the second test are shown in Fig. 6; as seen, the test indicates that somebody has manipulated the document by changing its original variable. The Fordex parameters produced and used for this test are recorded. The data are obtained when the lighting panels are located 50 mm above the floor and centered at an angle of 45° against the floor IR long pass filter was used during the examination, and broadband halogen lamps were used for illumination.

Fig. 6. Fordex forensic document examination case 2 result

A third test was conducted using the Fordex device; in this test, another criminal piece of evidence is tested to see if it has been manipulated in any way. Figure 7 shows the document to be tested. In this case, the number $1,230 is changed to $42,800 and the number 9 is changed to 9632.

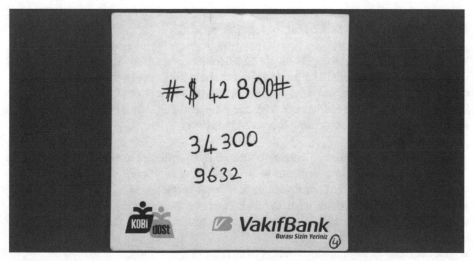

Fig. 7. Fordex forensic document examination case 3

The Fordex test output image shows the manipulation act very clearly again. Highlight the actual changes that have taken place. The Fordex test parameters related to this test are recorded.

Fig. 8. Fordex forensic document examination case 3 result

The Fordex device recorded parameters for this test include illumination sources between 545–675 nm using a 725 nm linear permeable filter position at a similar illumination height and angle. The examinations kept the zoom, focus, iris, integration time, and RGB channel values constant. All the test record values are test related and would be changed for different cases, and their instantaneous values can be recorded.

4   Proposed Fordex Forensic Chain System

Forensic studies carried out in Forensic laboratories are regularly reported and archived using traditional methods. The proposed blockchain-supported Fordex-Forensic-Chain (FFC) system has been prepared for use in forensic laboratories for Fordex devices. Forensic investigations carried out with the proposed FFC system are recorded in the Blockchain environment. Among the saved parameters,

- digital copy of the visual's version obtained from different analyses of the forensic evidence examined, instantaneous location of the object under investigation,
- the floor table X and Y coordinates of the object being tested,
- the type of lighting source used and its related wavelength,
- the height of the lighting source relative to the ground if the lighting source is on the device lighting panels,
- and the relative angular slope values of the examined sample.

Fordex devices can make more effective measurements due to their mobile hardware facilities that are not available in similar devices. However, during measurement, the information on the positions of the device must be recorded to be used in re-enactment scenarios.

The possibility of human-based errors in the forensic investigation processes, which is also mentioned in the introduction, has not been ignored in this study. For this reason, in the proposed FFC system, there is the possibility of re-enactment of previous forensic investigations. It is possible to re-enact the forensic examination made by calling the parameters stored in the external database and Blockchain platform of the examinations made by the forensic technician using the Fordex device. Therefore, when it is thought that there is an error in the forensic report, it allows the old report to be examined and checked by a second expert. It is important that the examinations made in the laboratories can be re-examined. In an environment with advanced and multiple devices, experts or supervisors must have the opportunity to supervise and control the analyses automatically. The proposed FFC system allows for a complete re-enactment of old forensic studies recorded using the Fordex device.

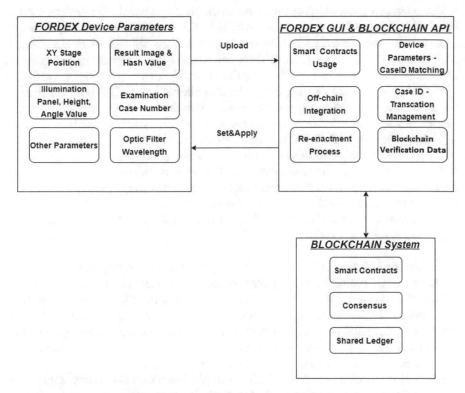

Fig. 9. Fordex blockchain general scheme

Figure 9 shows that the addition of Blockchain networks will enable users to carry out the re-enactment of any transaction registered within the system with very high accuracy. The re-enactment process will look at the transaction requested for re-enactment, it will extract the data related to this test from the Blockchain verification data (Transaction ID and Transaction Hash) will compare them with the test information if they match then the truncation data is used to re-enact the document test and compare the original results with the re-enactment test.

In the proposed study, Hyperledger Fabric (HLF), a permissioned blockchain environment, was utilized as the blockchain platform. The images produced from the evidence examined in the Fordex device are kept in an external database (MySQL) in order not to avoid scalability problems in the Blockchain system. The hash value of the forensic image and its address in the database are stored in the Blockchain. In case of a forensic report for which re-enactment is requested, the image in the database is examined. Suppose the hash value of the image in the database does not match its corresponding value in the Blockchain. In that case, the information of the responsible person is reported to the relevant authorities, and the re-enactment process is not authorized. With this mechanism, tampering attempts that may occur in the database are deterred, and the security of this data source outside the blockchain system is ensured.

There are many nodes in the proposed blockchain system. Fordex devices used in forensic laboratories in every province within Turkey are included in the system as a node. (Istanbul node, Izmir node, Ankara node, etc.). Users with appropriate authorizations (Technicians and Supervisors of Technicians) using the Fordex device can generate transactions. Ministry of Justice, Supreme Courts, and Ministry of Interior are planning to join as a partner organization to this blockchain network. TÜBİTAK Blockchain Research Laboratory (TBRC) is the root organization that sets up the Blockchain, organizes the joining of other nodes to the system, and operates the Orderer and Certification Authority services of HLF. TBRC also provides the necessary technical support to maintain the blockchain system. TBRC is not authorized to access Fordex device and external database. TBRC can monitor transactions that occur in the blockchain environment of the FFC, but cannot perform the re-enactment process. However, TBRC guarantees that the blockchain system is reliable and distributed.

End user types who can use the Fordex device in the forensic lab. Blockchain nodes are planned in three categories:

1. Laboratory Manager (LM): The user designated as Laboratory Manager has the reader role in the blockchain system. It sees the transactions taking place in the blockchain system and can pull the case study from the Blockchain and external database and re-enactment it, in cases where it deems necessary or when the judicial review is requested to be rechecked with a court decision. LM has no right to enter, change or delete data on the system. But LM can examine the forensic case reports produced in any nodes when the necessary situation arises.

2. Forensic Technician Supervisor (FTS): Forensic Technician Supervisor is the supervisor of technicians using Fordex devices in the laboratory. The proposed blockchain system, FTS has writer and reader roles. FTS can see case studies completed by forensic technicians and re-enactment them. At the same time, since FTS is a technician itself, it can perform the forensic tasks assigned using a Fordex device. FTS has no right to change or delete data on the system.

3. Forensic Technician (FT): Forensic Technicians are actors in the role of the writer in the Blockchain system who have the right to use the Fordex device. They perform the forensic work assigned to them with the help of the Fordex device and can enter the case study results into the system. They do not have the authority to process, change or delete data on the system.

In addition to the above role types, the Auditor type of users can access the blockchain network from other peer organizations.

The FFC system consists of a Fordex device, re-enactment software (including a control mechanism), an external database, and a permissioned blockchain. The general architecture of the system using a single Fordex device is shown in Fig. 8.

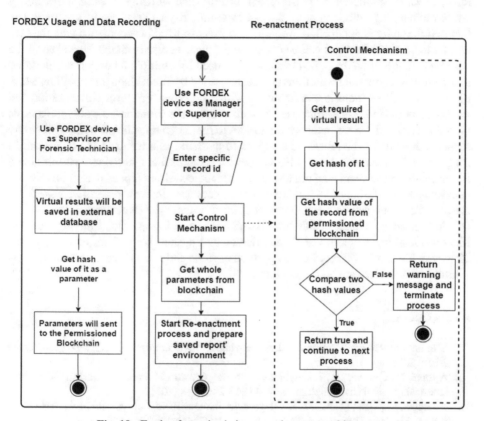

Fig. 10. Fordex forensic chain general system architecture

The re-enactment process, which can be seen in Fig. 10, is assigned a record ID. The relevant request form will be approved by the Lab Manager and Forensic Technician Supervisor, and the result related to the test will be recorded in the blockchain system. Incoming parameters are handled primarily in the Control Mechanism process. The hash values of the images kept in the external database are compared with the hash values stored in the Blockchain. In case of consistency, the data, the re-enactment process continues, and using the parameters pulled from the Blockchain; the Fordex device creates the same analysis environment as before. At this stage, where the Manager or Supervisor makes the control, the evidence is placed on the Fordex device, and controls are made on the old analysis environment.

5   Conclusions

The actual use of Blockchain technology for the protection of forensic document review devices has not been reported by any entity to date. This means that the proposed system is the first of its kind. The system basically provides a technological infrastructure opportunity in which studies can be performed more often and with higher speed and accuracy while maintaining both laboratory environment integrity and test results accuracy. The fact that the study is not open to everyone and decentralized in accordance with the idea of Blockchain technologies can be considered a general disadvantage. However, in the very near future, systems to be developed and used by multiple laboratories belonging to different government institutions can be supported by Blockchain and auxiliary security that will allow them to communicate with each other for closer collaboration and exchange of evidence. The proposed system reduces the time for the result and speeds up the delivery of the test analysis document to the judiciary; plus, it will significantly reduce the time for the repetitive analyzes to be made as a result of the objection. In the proposed blockchain-based FFC system, forensic data are stored in a permissioned Blockchain environment, and old test results are reviewed and repeated accurately when necessary. Since the conditions under which the contested forensic reports were prepared are recorded, they can be recalled to repeat the test under the same conditions.

Further advancement in this field is intended to utilize Artificial Intelligence and Machine learning to grant the Fordex device and system some form of intelligence. However, this work will require the approval of some end users since the actual datasets available in this field are highly confident.

References

1. Bitcoin: a peer to peer electronic cash system. https://bitcoin.org/bitcoin.pdf. (2008). Accessed 20 Jul 2022
2. Ajlouni, N., Özyavaş, A., Takaoğlu, M.: A survey of artificial intelligence driven blockchain technology: blockchain intelligence. MJAIAS **2**(2), 1–8 (2021)
3. Takaoğlu, M., Özer, Ç., Parlak, E.: Blokzinciri Teknolojisi ve Türkiye'deki Muhtemel Uygulanma Alanları. Uluslararası Doğu Anadolu Fen Mühendislik ve Tasarım Dergisi **1**(2), 260–295 (2019)
4. Lone, A.H., Mir, R.N.: Forensic-chain: blockchain based digital forensics chain of custody with PoC in hyperledger composer. Digit. Investig. **28**, 44–55 (2019)
5. Shah, M.S.M.B., Saleem, S., Zulqarnain, R.: Protecting digital evidence integrity and preserving chain of custody. J. Digital Forensics, Secur. Law **12**(2), 121–130 (2017)
6. Lone, A.H., Mir, R.N.: Forensic-chain: ethereum blockchain based digital forensics chain of custody. Sci. Pract. Cyber Secur. J. **1**(2), 21–27 (2017)
7. Liza, A., Salam, K, Farkhund, I., Faouzi, K.: Blockchain-based chain of custody: towards real-time tamper-proof evidence management. In: Proceedings of the 15th International Conference on Availability, Reliability and Security, ARES 2020, pp.1–8. Association for Computing Machinery, New York (2020)
8. Bonomi, S., Casmi, M., Ciccotelli, C.: B-CoC: a blockchain-based chain of custody for evidences management in digital forensics. arXiv preprint, arXiv:1807.10359 (2018)
9. Tian, Z., Li, M., Qiu, M., Sun, Y., Su, S.: Block-DEF: a secure digital evidence framework using blockchain. Inf. Sci. **491**, 151–165 (2019)

10. Billard, D.: Weighted Forensics Evidence Using Blockchain. Proceedings of the International Conference on Computing and Data Engineering, ICCDE 2018, pp.57–61. ACM, Shanghai (2018)
11. Fordex. https://biyoelektronik.bilgem.tubitak.gov.tr/fordex/. Accessed 26 Jun 2022
12. Pop, C., Cioara, T., Antal, M., Anghel, I., Salomie, I., Bertoncini, M.: Blockchain based decentralized management of demand response programs in smart energy grids. Sensors **18**(1), 162 (2018)
13. Takaoğlu, M., Özyavaş, A., Ajlouni, N., Alshahrani, A., Alkasasbeh, B.: A novel and robust hybrid blockchain and steganography scheme. Appl. Sci. **11**(22), 10698 (2021)
14. Lam, D.K., Le, V.T.D., Tran, T.H.: Efficient architectures for full hardware scrypt-based block hashing system. Electronics **11**(7), 1068 (2022)
15. Ullah, M.A., Setiawan, J.W., ur Rehman, J., Shin, H.: On the robustness of quantum algorithms for blockchain consensus. Sensors **22**(7), 2716 (2022)
16. Pieroni, A., Scarpato, N., Felli, L.: Blockchain and IoT convergence-a systematic survey on technologies. Protoc. Secur. Appl. Sci. **10**(19), 6749 (2020)
17. Zhang, L., et al.: SPCBIG-EC: a robust serial hybrid model for smart contract vulnerability detection. Sensors **22**(12), 4621 (2022)
18. Westphall, J., Martina, J.E.: Blockchain privacy and scalability in a decentralized validated energy trading context with hyperledger fabric. Sensors **22**(12), 4585 (2022)
19. Nijsse, J., Litchfield, A.: A taxonomy of blockchain consensus methods. Cryptography **4**(4), 32 (2020)

Enhancing Vehicle Networks Performance by Using Deep Learning Techniques for Artificial Intelligence

Abdullah Saad Zeki$^{(\boxtimes)}$ (iD) and Muhammad Ilyas

Altinbas University, Istanbul, Turkey
203720140@ogr.altinbas.edu.tr, muhammad.ilyas@altinbas.edu.tr

Abstract. The Quality of Services (QoS) in-vehicle networks are facing a significant challenge because of the proliferation of heterogeneous wireless devices and the rising demand for cutting-edge applications like autonomous driving and vehicle internet. High spectrum efficiency, high reliability, and low latency are related to these difficulties. Artificial intelligence is considered one of the most important technical factors that can overcome these challenges. In the 5G network, the cache-aided Non-Orthogonal Multiple Access (CA-NOMA) system is divided into two phases, the cache phase and deliver phase. In this scientific paper, we will focus on the delivery phase. We trained two models (MobileNetV2, NasNETmobile) of transfer learning models of convolutional neural networks (CNN) to classify the images of traffic signs that the vehicle needs in self-driving at peak times from the base station (BS) and compared the two models using the traffic sign image dataset of 5000 images divided into 15 classes from the Kaggle website and the work application on MATLAB program. NasNETmobile accuracy of 0.98 percent outperformed MobileNetV2.

Keywords: 5G · Vehicle network · Convolutional neural network · MobileNetV2 · NasNETmobile

1 Introduction

With the rapid development and qualitative leaps in the field of communications and the huge increase in the number of subscribers that reached 2 billion in connection with the use of the internet [1], and the almost complete dependence on it for commercial transactions and e-government in addition to the huge increase in new applications, it has become necessary to work on increasing the volume of data in networks connection [2]. The emergence of new technologies such as smart cities and the development in the field of medicine to include virtual reality, augmented reality, the internet of things, and the development of vehicle networks such as simulation between vehicles and self-driving cars has led to the urgent need to find new technologies to deal with these challenges [3]. We need high data transmission speed and low latency to get a suitable quality to deal with these new applications, some of which may have a serious error area, such as medicine and self-driving. The 5G network introduced new technologies

© The Author(s), under exclusive license to Springer Nature Switzerland AG 2023
F. P. García Márquez et al. (Eds.): ICCIDA 2022, LNNS 643, pp. 438–447, 2023.
https://doi.org/10.1007/978-3-031-27099-4_34

that can overcome these obstacles, providing speeds of up to 10 gigabytes per second and a response time of 0.1 ms. Among the new technologies in the 5G network is the CA-NOMA technology, Where NOMA works to increase the number of subscribers significantly by exploiting the communication channels based on the principle of non-orthogonality, as well as contributes to increasing the rate of data transfer and taking advantage of the geographical distribution of users in one cell and works to serve users at the edge of the cell better and provides the advantage of fairness between users where he is based by sending a high power to the distant user and low power to the near user. The presence of cache in the vehicles also reduces the load on the base station, as the vehicles benefit from the information in other vehicles directly without the need to return to the base station, which helps improve response time [4]. As mentioned earlier, the massive increase in the number of subscribers and other applications has led to a massive increase in the volume of data at the base station in-vehicle networks. One of the promising solutions to access the required information at the base station at a high speed is the introduction of deep learning in this field that helps to exploit the time and reduce the response time significantly, especially at the peak time when the load on the traffic of base station is large, and neural networks in this field are very useful.

In order to identify the type of signal installed on the road, convolutional neural networks are used in image recognition [5]. The outcomes and classification of the training of a CNN improve with the number of images or data employed. In various visual processing applications, such as autonomous driving, image classification, and the medical industry, such as COVID-19 object detection and other classifications, convolutional neural networks have demonstrated highly good performance [6, 7]. We installed the data from a dataset of road traffic signs from previously trained Kaggle models in MATLAB for this study [8]. The accuracy, loss, and time of the two trained models (MobileNetV2 and NasNETmobile) were compared [9, 10]. The proposed models were tested for a learning transfer mechanism [11].

2   Related Works

2.1   Neural Network

The field of artificial intelligence is one of the modern fields, it showed its signs in 1983 [12]. Artificial intelligence aims to learn a computer to simulate the intelligence processes that take place inside the human mind so that the computer can solve problems, make decisions in a logical and orderly manner, and perform some work and functions that simulate the work of the human mind [13]. Through the work of biological neural networks, the idea of artificial neural networks has been revived by imitating the biological neural network in the computer in what is known as the artificial neural network, where designed models that simulate the way, the human brain works using the computer to solve some problems that use traditional methods to solve them[14]. Artificial neural networks learn in a way like human learning through examples and training, and artificial neural networks are configured and organized for specific applications such as model discrimination, perception, and data classification through the learning process [15]. Artificial neural networks are characterized that they depend on a strong mathematical basis, as they accept any type of quantitative or qualitative data and could store the knowledge acquired through the cases that are run on the network, which can treat the behavior of nonlinearity so that it can find nonlinear relationships between variables and take them into consideration in giving results [16].

2.2 Convolutional Neural Network

A convolutional neural network (CNN) is a kind of multi-layer neural network, which is good at dealing with machine learning problems related to images, especially large images. Through a series of methods, the convolutional network successfully reduces the dimensions of the image recognition problem by the huge data volume, and finally enables it to train. CNN simulates feature differentiation through convolution, reduces the order of magnitude of network parameters through weight sharing and convolution aggregation, and finally completes tasks such as classification through traditional neural networks [17, 18]. In practice, CNNs learn the values of the filters by themselves during the training process (considering that we must pre-determine some parameters before training the model such as the number of filters, filter size, network architecture, etc.) [19]. The more filters there are, the more features can be extracted from the input image, so we have a better network at identifying patterns in new images that the network hasn't seen before [20].

3   Proposed System

The main goal of using a transfer learning mechanism is to make use of well-known models that have already been trained on a large amount of data. This will enable the researcher to use and retrain these models by altering the final layers of designing a previously trained model and utilizing our layers. After processing activities, we train 15 classes of MobileNetV2 and NasNETmobile models until we reach the necessary working accuracy. Figure 1 shows the working stages of all models. In this study, 5000 photos were used to recognize traffic signs put on the road using the MATLAB simulation tool. Data from the Kaggle website was utilized to train the two networks. This data contains different images of the road traffic signs, which vary in clarity to train the network on all weather conditions that may affect the clarity of the image on the road, as these images are very important for the self-driving of vehicle because the vehicle needs information on the road continuously, which must be updated on an ongoing basis, which the base station provides this information.

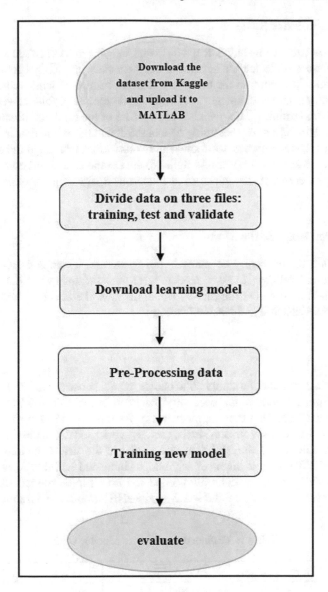

Fig. 1. A proposed system

3.1 Import Data Set

The information was imported from Kaggle and entered into the *MATLAB*© program. After downloading and extracting the data. At this stage, we imported the data from the Kaggle website. The Kaggle website and the working application on the MATLAB program were used to divide the 5000-image dataset into 15 classes. The data was then divided into three sections: one for training, one for validation, and one for testing.

3.2   The Preliminary Stage

In the course of this, we installed deep Network Designer in MATLAB, which includes a number of up-to-date learning modules for networks including MobileNetV2 and NasNETmobile. We increase the data we utilize in this study, which comprises 5,000 photos distributed across 15 categories. After installation, three folders were created for this data; one had training data, which made up 70% of the total, additionally, there is a folder named data validation that holds 20% of the data. 10% of the data is located in the last folder, Test Data, which is not equivalent in numbers. Each file in the collection has an image of a different kind of traffic light. To ensure the successful completion of the self-driving procedure, these signs feature numerous images with various clarity levels for training.

3.3   Valuation Stage of the Model

After achieving the necessary accuracy in train progress, which consists of two types of data: train and validate, the two models were assessed using test data. We use the Valuation function to assess the model, which displays the accuracy and loss for two networks, (MobileNetV2 and NasNETmobile).

4   Result

After processing the data of roughly 5000 photos, which consisted of 15 classes of traffic signs, and training to obtain the necessary precision in the task, the two models were trained using the suggested training mechanism. We trained 913 layers in the NasNET-mobile, one of the newest convolutional neural network models, to obtain the necessary accuracy. We found that the MobileNetV2 model is the fastest in terms of training time, but NasNETmobile is the most accurate, as indicated in Table 3, as compared to the MobileNetV2 models, which only needed 154 layers to be trained. Tables 1 and 2 display the development of MobileNetV2 and NasNETmobile, two training models.

Table 1. Explain the training of MobileNetV2.

	Train-Accuracy	Val-Accuracy	Train-loss	Val-loss
Epoch1	5.00%	15.70%	3.0857	2.9291
Epoch1	80.00%	52.97%	0.8561	1.4409
Epoch1	75.00%	74.22%	0.8450	0.8031
Epoch1	90.00%	77.58%	0.2008	0.3589
Epoch1	90.00%	88.44%	0.4795	0.3567
Epoch2	95.00%	91.54%	0.1297	0.1202
Epoch2	100.00%	91.86%	0.0839	0.0811

(continued)

Table 1. (*continued*)

	Train-Accuracy	Val-Accuracy	Train-loss	Val-loss
Epoch2	95.00%	96.58%	0.0654	0.0646
Epoch2	95.00%	94.64%	0.2744	0.1101
Epoch2	90.00%	87.21%	0.1610	0.0642
Epoch3	100.00%	94.38%	0.0122	0.0400
Epoch3	90.00%	94.90%	0.0668	0.0418
Epoch3	95.00%	96.64%	0.0566	0.0703
Epoch3	100.00%	97.55%	0.0058	0.0689
Epoch3	100.00%	97.22%	0.0253	0.0693
Epoch4	95.00%	97.80%	0.0014	0.0603
Epoch4	100.00%	97.87%	0.0002	0.0644
Epoch4	100.00%	97.87%	0.0007	0.0817
Epoch4	100.00%	97.87%	0.0006	0.0720
Epoch4	100.00%	98.26%	0.0005	0.0724
Epoch5	100.00%	98.39%	0.0002	0.0445
Epoch5	100.00%	98.45%	0.0001	0.0520
Epoch5	100.00%	98.90%	5.2767e-05	0.0610
Epoch5	100.00%	98.64%	2.9324e-05	0.0658
Epoch5	100.00%	98.64%	2.9324e-05	0.0523

Table 2. Explain the training of NasNETmobile.

	Train-Accuracy	Val-Accuracy	Train-loss	Val-loss
Epoch1	5.00%	15.70%	3.0857	2.8292
Epoch1	75.00%	61.97%	0.8571	1.3408
Epoch1	80.00%	75.22%	0.8461	0.7032
Epoch1	90.00%	78.58%	0.2007	0.2588
Epoch1	95.00%	89.55%	0.3790	0.2565
Epoch2	95.00%	92.30%	0.2296	0.1201
Epoch2	100.00%	92.56%	0.0725	0.0721
Epoch2	95.00%	95.57%	0.0525	0.0564
Epoch2	95.00%	94.63%	0.2433	0.1102
Epoch2	90.00%	88.22%	0.1612	0.0541
Epoch3	100.00%	94.38%	0.0121	0.0391
Epoch3	95.00%	95.86%	0.0659	0.0392

(*continued*)

Table 2. (*continued*)

	Train-Accuracy	Val-Accuracy	Train-loss	Val-loss
Epoch3	95.00%	96.64%	0.0557	0.0681
Epoch3	100.00%	97.55%	0.0049	0.0687
Epoch3	100.00%	97.22%	0.0242	0.0643
Epoch4	95.00%	97.80%	0.0013	0.0603
Epoch4	100.00%	97.87%	0.0002	0.0642
Epoch4	100.00%	97.87%	0.0006	0.0816
Epoch4	100.00%	97.87%	0.0005	0.0711
Epoch4	100.00%	98.26%	0.0005	0.0710
Epoch5	100.00%	98.39%	0.0003	0.0434
Epoch5	100.00%	99.45%	0.0001	0.0512
Epoch5	100.00%	99.50%	4.2767e–05	0.0592
Epoch5	100.00%	99.60%	1.9324e–05	0.0523
Epoch5	100.00%	99.60%	1.9324e–05	0.0523

Table 3. Compare two networks during the model valuation.

	MobileNetV2	NasNETmobile
Validation-Accuracy	98.71%	99.42%
Validation-loss	0.1317	0.1214

The relationship between accuracy and validity in MobileNetV2 and NasNETmobile is shown in the figures below. Additionally, describe how loss in train and validation relate to each of the two training models (Figs. 2, 3, 4 and 5).

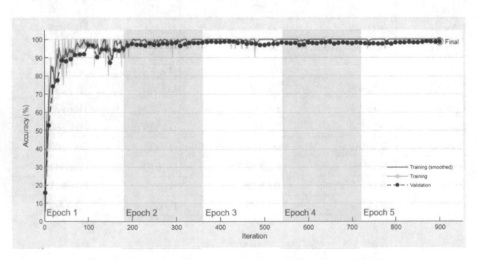

Fig. 2. Compares validate accuracy and train of MobileNetV2.

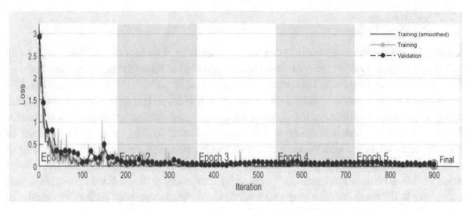

Fig. 3. Compare between in MobileNetV2.

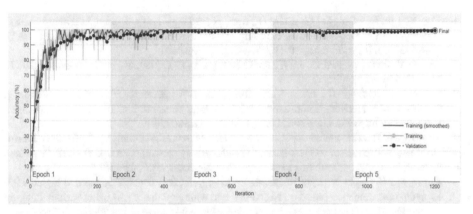

Fig. 4. Compares validate accuracy and train of NasNETmobile.

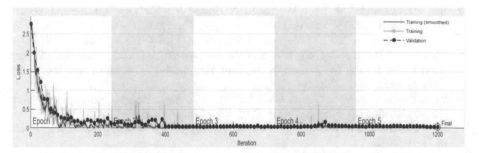

Fig. 5. Compare between loss in NasNETmobile.

5 Conclusion

A MATLAB program was used in this study to train the two models (MobileNetV2 and NasNETmobile), and the models were then compared based on the accuracy and training that each network obtained. This paper showed us that MobileNetV2 and NasNETmobile are sometimes the same in terms of accuracy, but NasNETmobile is more intelligent, and it is possible to increase the number of images and categories to get better results. Even though the aforementioned results are particular to this dataset, they might be generalizable to other sizable datasets, and further research might confirm the impact of layer tweaking to obtain peak performance. Instead of layer-wise fine-tuning, further work in this area may alternatively be done by undertaking block-wise fine-tuning. It is necessary to do research into particular versions of these structures so that the results of this study may be more easily interpreted.

References

1. Yin, Y., Liu, M., Gui, G., Gacanin, H., Sari, H., Adachi, F.: QoS-oriented dynamic power allocation in NOMA-based wireless caching networks. IEEE Wirel. Commun. Lett. 10(1), 82–86 (2021). https://doi.org/10.1109/LWC.2020.3021204
2. Liu, J.: Vehicular Networks. In:Wiley 5G Ref, pp. 1–17 (2019). https://doi.org/10.1002/9781119471509.w5gref091
3. Rezvani, S., Mokari, N., Javan, M.R., Jorswieck, E.A.: Resource allocation in virtualized CoMP-NOMA HetNets: multi-connectivity for joint transmission. IEEE Trans. Commun. 1–54 (2021). https://doi.org/10.1109/TCOMM.2021.3067700
4. Kaneko, M., Randrianantenaina, I., Dahrouj, H., Elsawy, H., Alouini, M.-S.: On the opportunities and challenges of NOMA-based fog radio access networks: an overview. IEEE Access 8, 205467–205476 (2020). https://doi.org/10.1109/access.2020.3037183
5. Ahsan, M.M., Gupta, K.D., Islam, M.M., Sajib Sen, M., Rahman, L., Hossain, M.S.: COVID-19 symptoms detection based on NasNetMobile with explainable AI using various imaging modalities. Mach. Learn. Knowl. Extract. 2(4), 490–504 (2020). https://doi.org/10.3390/make2040027
6. Akay, M., et al.: Deep learning classification of systemic sclerosis skin using the MobileNetV2 Model. IEEE Open J. Eng. Med. Biol. 2, 104–110 (2021). https://doi.org/10.1109/OJEMB.2021.3066097
7. Bega, D., Gramaglia, M., Fiore, M., Banchs, A., Costa-Perez, X.: DeepCog: cognitive network management in sliced 5G networks with deep learning. In: Proceedings of - IEEE INFOCOM, vol. 2019-April, pp. 280–288 (2019). https://doi.org/10.1109/INFOCOM.2019.8737488
8. Bimorogo, S.D.: DeepCog: cognitive network management in sliced 5G networks with deep learning. Int. J. Adv. Trends Comput. Sci. Eng. 9(3), 2824–2833 (2020). https://doi.org/10.30534/ijatcse/2020/53932020
9. Buiu, C., Dănăilă, V.R., Răduţă, C.N.: MobileNetV2 ensemble for cervical precancerous lesions classification. Processes 8(5) (2020). https://doi.org/10.3390/PR8050595
10. Dileep, P., Bolla, B.K., Ethiraj, S.: Revisiting facial key point detection : an efficient approach, pp. 1–16 (2022)
11. Enkvetchakul, P., Surinta, O.: Effective data augmentation and training techniques for improving deep learning in plant leaf disease recognition. Appl. Sci. Eng. Prog. (2021). https://doi.org/10.14416/j.asep.2021.01.003

12. Ethiraj, S., Bolla, B.K.: Classification of astronomical bodies by efficient layer fine-tuning of deep neural networks. In: 2021 5th Conference on Information and Communication Technology, CICT 2021 (2021). https://doi.org/10.1109/CICT53865.2020.9672430

13. Fernández Maimó, L., Huertas Celdrán, A., Gil Pérez, M., García Clemente, F.J., Martínez Pérez, G.: Dynamic management of a deep learning-based anomaly detection system for 5G networks. J. Ambient. Intell. Humaniz. Comput. **10**(8), 3083–3097 (2018). https://doi.org/10.1007/s12652-018-0813-4

14. Indraswari, R., Rokhana, R., Herulambang, W.: Melanoma image classification based on MobileNetV2 network. Procedia Comput. Sci. **197**, 198–207 (2021). https://doi.org/10.1016/j.procs.2021.12.132

15. McClellan, M., Cervelló-Pastor, C., Sallent, S.: Deep learning at the mobile edge: opportunities for 5G networks. Appl. Sci. **10**(14), 4735 (2020). https://doi.org/10.3390/app10144735

16. Nweke, H.F., Teh, Y.W., Al-garadi, M.A., Alo, U.R.: Deep learning algorithms for human activity recognition using mobile and wearable sensor networks: state of the art and research challenges. Expert Syst. Appl. **105**, 233–261 (2018). https://doi.org/10.1016/j.eswa.2018.03.056

17. Reddy, N., Rattani, A., Derakhshani, R.: Comparison of deep learning models for biometric-based mobile user authentication. In: 2018 IEEE 9th International Conference on Biometrics Theory, Applications and Systems, BTAS 2018, November 2018. https://doi.org/10.1109/BTAS.2018.8698586

18. Sandler, M., Howard, A., Zhu, M., Zhmoginov, A.: Sandler_MobileNetV2_Inverted_Residuals_CVPR_2018_paper.pdf," pp. 4510–4520 (2018)

19. Saxen, F., Werner, P., Handrich, S., Othman, E., Dinges, L., Al-Hamadi, A.: Face attribute detection with mobilenetv2 and nasnet-mobile. In: International Symposium on Image and Signal Processing and Analysis, ISPA, vol. 2019-Septeember, no. October, pp. 176–180, (2019). https://doi.org/10.1109/ISPA.2019.8868585

20. Winoto, A.S., Kristianus, M., Premachandra, C.: Small and slim deep convolutional neural network for mobile device. IEEE Access **8**, 125210–125222 (2020). https://doi.org/10.1109/ACCESS.2020.3005161

What Drives Success in Data Science Projects: A Taxonomy of Antecedents

Gonca Tokdemir Gökay[1]($\boxtimes$), Kerem Nazlıel[1], Umut Şener[1], Ebru Gökalp[2],
Mert Onuralp Gökalp[3], Nergiz Gençal[3], Gizemnur Dağdaş[3], and P. Erhan Eren[1]

[1] Graduate School of Informatics, Middle East Technical University, Ankara, Turkey
`{gonca.gokay,knazliel,sumut,ereren}@metu.edu.tr`
[2] Department of Computer Engineering, Hacettepe University, Ankara, Turkey
`ebrugokalp@hacettepe.edu.tr`
[3] Data Analytics Center, Tupras, İstanbul, Turkey
`{MertOnuralp.Gokalp,Nergiz.Gencal,Gizemnur.Surgun}@tupras.com.tr`

Abstract. Organizations have been trying to reshape their business processes and transform them into a smart environment to attain sustainable competitive advantage in their markets. Data science enables organizations to define interconnected and self-controlled business processes by analyzing the massive amount of unstandardized and unstructured high-speed data produced by heterogeneous Internet of Things devices. However, according to the latest research, the success rate of data science projects is lower than other software projects, and the literature review conducted reveals a fundamental need for determining success drivers for data science projects. To address these research gaps, this study investigates the determinants of success and the taxonomy of antecedents of success in data science projects. We reviewed the literature systematically and conducted an expert panel by following a Delphi method to explore the main success drivers of data science projects. The main contributions of the study are twofold: (1) establishing a common base for determinants of success in data science projects (2) guiding organizations to increase the success of their data science projects.

Keywords: Data science · Project management · Project success · Critical success factors

1 Introduction

Digital Transformation (DX) has been increasing its popularity over the last two decades and reshaping economies, industries, and organizations by introducing tremendous opportunities yet encountering non-negligible threats. Businesses that managed to leverage the momentum by developing new business models, providing better offerings, transforming processes, and improving capabilities experienced growth and efficiency; and secured their position against the competition. Though, those who failed to act successfully lagged behind and were left unguarded in their dynamic environment.

Every interaction in the digital era generates data, and this makes data a crucial part of the DX [1]. For organizations completing their analog-to-digital transition, the next step is concentrating on the data and analytics aspects of their DX [2, 3]. However, data science projects launched by organizations to achieve higher agility, visibility, efficiency in their operations, and their interactions with the environment do not have high rates of success [4–6]. According to VentureBeat [7], "*87% of these projects never make it to production*". Therefore, it is crucial for businesses to understand the sources of success for data science projects and to put into practice the well-fitted project environment for achieving the targeted outcomes.

Due to the ever-changing nature of data, processes in developing data science projects necessitate unique approaches that cannot be met only by traditional project management practices. Hence, what drives success in these projects should be explored in a systematic way. Accordingly, this study investigates the taxonomy of antecedents of success of data science projects. Towards this aim, a Systematic Literature Review (SLR) is conducted, and then a taxonomy is developed by incorporating the results of the SLR with expert opinions via a Delphi study.

The remainder of the study is structured as follows: background and related studies are explained in Sect. 2. In Sect. 3 and Sect. 4, the methodology and findings of the SLR are presented, respectively. Then, the proposed taxonomy for data science project success is described in Sect. 5, followed by the final section, which is dedicated to the discussion and conclusion of the study.

2   Literature Review

The background of the study covering the explanation of data science and project success is given in this section. Additionally, related studies in the literature are discussed.

2.1   Background of the Study

Data science is the application of scientific methodologies and procedures to extract knowledge and value from massive amounts of structured, semi-structured, or unstructured data [8, 9]. With the solutions enabled by data science, businesses can quickly accumulate enormous amounts of data from several sources and extract valuable insights to improve their data-driven decision-making capabilities. Despite various competitive advantages, high failure rates are reported for data science projects [7, 10, 11]. The challenges faced within organizations in putting data science projects into practice go beyond being analytical [12]. Team management, project management, and data & information management are topics under which Martinez et al. [8] organize the real-world difficulties in data science projects. Hence, it is worth drilling down data science projects in terms of determinants of success to take full advantage of the benefits.

Project Management Institute defines a project as "*a temporary endeavor undertaken to create a unique product, service, or result*" [13]. Ensuring project success started to get more attention from researchers and practitioners by the 1980s. Efficiency, effectiveness, achievement of intended objective, delivering within time, cost and quality, perceived quality, meeting stakeholder requirements, and customer satisfaction [14, 15] have been the major terms used for defining project success so far. However, research in the area has not reached a consensus on a definition [16, 17]. Still, two dimensions are brought to the fore to elaborate on the ambiguous and multidimensional [18] nature of the topic: project success factors and project success criteria.

Success criteria are the *"variables that measure the success"* of projects; the most popular ones being the triplet of time, cost, and quality, also known as the iron triangle. On the other hand, success factors are the facts, conditions, or occasions that must occur to achieve the goals and that increase the chances of success. Factors can be considered as the causes of success, and the criteria as the results. Therefore, in search of understanding the drivers of project success, a closer look at these factors is needed, especially at the critical success factors (CSFs) that, if missing, would end up causing the project to fail [19].

Scholars investigating CSFs have approached the concept from two perspectives: universal and project-specific. The universal perspective covers that researchers explored success factors that are applicable to any project. One of the most referenced studies is [20], which includes the development of the "Project Implementation Profile" model, made up of 10 CSFs, which also served for empirically testing earlier remarkable frameworks. Similarly, [21] introduced a new scheme expressing interrelated groups of factors, their effects on the system, and results as success or failure. Even the research in this direction continues mainly in the project management domain, there are other types of studies that are based on the view that dimensions of success might vary significantly from one project to the next [17, 22], and the subject should be approached on a more project-specific basis [18]. In this respect, scholars studied specific domains such as construction, Information Technology (IT), and education. They examined drivers of success in projects of certain types (e.g., enterprise resource planning (ERP)). Software development projects, have been studied by many researchers in terms of their success (e.g., [23]), as their practices show similarities with data science projects [24].

2.2 Related Studies

As a result of investigating studies related to the determinants of project success in data science projects, it was observed that only a few studies exist on the success of projects in sub-fields of the data science domain. Big data analytics [25–28] is the most investigated field, parallel to its popularity in recent years. Agile analytics [29], artificial intelligence [15, 30], and business intelligence [31] projects are also examined, and classifications for CSFs are proposed as models. Yet, to the best of our knowledge, CSFs applicable to any data science project have not been comprehensively explored, and there is no mature taxonomy accepted by the data science community for identifying the drivers of success.

3 Methodology

An SLR was conducted by following the methodology proposed by Kitchenham [32, 33] and the steps specified for scientific studies in [34] to systematically review the existing academic literature and determine the success factors of data science projects. Details referring to each step are presented below.

Research Questions. The main research question is identified as: What are the CSFs identified in the literature that are applicable to data science projects?

Search Terms. Considering the immaturity of existing literature in terms of the objective of this research, besides data science, we have also included information systems (IS) and software development domains in our search. The following keywords were identified under five groups: Domain (data science", "information systems", "software development", "software"), Scope ("project"), Context ("success", "performance", "quality"), Task ("evaluating", "assessing", "measuring"), Unit ("model", "framework", "factor", "criteria", "metric").

Search String and Scope. For the preliminary search, the search string was composed as: ("data science" OR "information system*" OR "software development" OR software) AND (project) AND (success OR performance OR quality) AND (model OR framework OR criteria OR factor OR metric OR assess* OR measur* OR evaluat*). Scopus and Web of Science were selected as the academic databases for this study. The SLR was carried out in July 2022 and included studies from the last 20 years.

Initial Search and Selection Process. Running the search string in titles, abstracts, or keywords of publications in the targeted databases resulted in a total of 55,459 studies. After applying the language, publication, and date criteria (Table 1), the total number was reduced to 19,469. Then the relevance criterion was introduced. We filtered the studies that mentioned the type of project in its title, abstract, or key (e.g., "data science project") and whose title contained the keyword "project" together with one keyword from context (e.g., success), and one from task or unit (e.g., "factor" or "assessment"). Removing the duplicates revealed a meaningful set of 86 articles.

Table 1. Initial search

Selection Process	Databases
Language: English Publication: Article Date ≥ 2003 Date ≤ 2022	Scopus Web of Science
Relevance (Title, Abstract or Keyword)	

Primary Pool and Snowballing. 86 articles in the initial pool were further examined by reading and evaluating the content of their abstracts. As the selection criteria, the articles containing information relevant to the success of projects were qualified for the next step. Accepted articles of 65 were taken to full-text review in the second pass. Only the accessible articles identifying, synthesizing, or modeling CSFs and applying a sound research methodology were selected. Adding four new articles to the accepted 15 articles in compliance with the selection criteria with backward and forward snowballing resulted in a primary CSF pool of 19 articles (refer to Fig. 1 below).

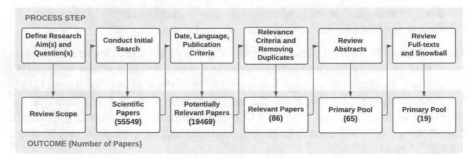

Fig. 1. The SLR process

4   Findings

To provide a broad analysis and review, 19 articles in the primary CSF pool were investigated in several aspects (e.g., project type). An exhaustive list (Table 2) was compiled from the CSFs mentioned in the articles in the primary CSF pool.

Type of projects. More than half of the projects mentioned in the articles are software (development) projects [35–38]. Among these, four articles specifically focus on agile software projects. Other projects mentioned are (in descending order of frequency) IS, IT, ERP, and large-scale agile transformation projects. However, none of the papers covered data science projects, proving this domain's immaturity in terms of research on project success.

Country. Five articles [36, 37, 39–41] concentrated on projects in a specific country (India, Jordan, Norway, Turkey) or group of countries (Former Yugoslavia), which might imply some limitations in terms of generalizability.

Research Method. Findings in each of the articles were based on previous literature. Additionally, some of the researchers included surveys in their studies to identify CSFs [35, 37, 40–42] or for validation [35, 38, 39, 41, 43]. Other sources of information are interviews [36], and analysis of real-world IS projects [44].

CSF Categorization. In most of the studies, success factors are categorized under high-level constructs (i.e., people, process, technical), for proposing models or frameworks. Some of the studies utilized already proposed models by other researchers (e.g., [38] combined people factors from two studies [23, 35] and conducted a survey to validate).

Table 2. CSFs (with more than 50% frequency), based on the SLR results

CSF	Frequency	[45]	[43]	[39]	[44]	[46]	[47]	[40]	[35]	[48]	[49]	[50]	[51]	[41]	[52]	[36]	[37]	[53]	[38]	[42]
Top-level management support	79%	x	x			x	x	x		x		x	x	x	x	x	x	x	x	x
Organizational culture and management style	68%	x	x			x		x	x	x		x		x	x	x	x		x	x
Leadership	68%	x	x	x	x	x		x		x		x	x	x	x	x			x	
Monitoring and controlling	68%	x	x	x	x		x	x		x		x	x	x		x	x		x	
Communication	68%	x	x	x	x	x	x	x		x			x	x	x	x	x			
Clear objectives and goals	63%	x	x			x	x	x		x		x	x		x	x	x	x		
Planning	63%	x	x	x			x	x	x	x			x	x	x		x	x		
Team commitment	58%	x	x			x	x	x		x		x	x		x	x	x			
Change management skills	53%	x	x			x	x		x	x		x	x		x					x
Team empowerment and composition	53%	x	x	x	x	x		x		x		x		x	x					
Team capability	53%	x	x				x	x	x	x		x		x	x				x	
User (customer) involvement	53%	x	x	x			x	x	x			x		x		x		x		
User (customer) training and support	53%	x	x	x			x	x				x	x	x		x		x		

5   The Proposed Taxonomy

In the light of the literature review findings, the drivers of the success in data science projects were identified by applying the Delphi method [54]. The drivers were categorized under four main dimensions, which are "Data", "Organization", "Technology", and "Strategy" [55] as seen in Fig. 2.

The Delphi method was built on several group meetings in which domain experts expressed their unspecified opinions and freely discussed the topics identified. After constructing the proposed taxonomy's first version for the success drivers of the data science projects, group meetings were set to employ the Delphi approach. First, a meeting moderator was assigned to manage the sessions and distribute the discussion forms indicating the topics to be discussed. Secondly, domain experts shared their knowledge, experiments, and opinions and discussed the conflicts identified throughout the sessions. After several revised versions were created, the final version was approved by all experts, and the proposed taxonomy was formed.

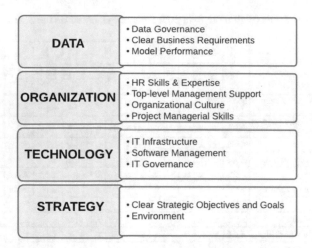

Fig. 2. The taxonomy of antecedents of success in data science projects

5.1 Data Dimension

Data dimension is investigated under topics of data governance, clear business requirements, and model performance.

Data Governance. Data governance in data science projects affects the overall project performance [49]. Data quality requirements [56] and data security & privacy issues should be identified by all stakeholders of the projects. As indicated in [48], the project's success or failure is highly correlated with data quality and security issues. An appropriate data governance policy for data profiling should be formed to support project goals. Moreover, data should be prevented from any third-party unauthorized access and recovered [57]. Data preparation steps should be standardized and documented by indicating how the project team approaches data cleansing, data wrangling, and data fusion operations. Efficient data lifecycle management [58] is found as an essential building block for more mature organizations in terms of data science capabilities.

Clear Business Requirements. Customer involvement is found as a significant CSF for project development to define clear business requirements and objectives [35, 37, 40, 43, 45, 48, 57]. Misra et al. [35] point out a correlation between customer collaboration and project success and a negative correlation with project uncertainty [43]. Project complexity and time restrictions can cause uncertainty and negatively influence project success [45]. Clear and well-documented requirements are among the most critical factors for projects' success [41, 52]. Therefore, customers should be involved in the development phase and should provide continuous feedback [35, 57].

Model Performance. In data science projects, a model is built after the data extraction, analysis, and preparation stages, respectively. The teams may employ different approaches, from descriptive to prescriptive analytics, to study the data science-related problems in the project scope. For example, a data scientist can analyze the project data to retrieve valuable insights about the business domains and make estimations for a short or long-time horizon. Therefore, using different data science approaches, such as supervised learning, clustering, artificial neural networks, etc., and finally building a model are critical stages of these projects. Moreover, the model should be validated to be suitable for the data science project. The quality of the model can be tested against different success metrics and business performance indicators. As for the training and validation stages of the model, these operations need to be applied manually. They can also be automated by applying MLOps values to artificial intelligence structures to support continuous delivery without manual intervention [55] though the automation cannot ensure returns due to characteristics of AI configurations [59].

5.2 Organization Dimension

The organization dimension consists of four sub-dimensions: HR skills & expertise, top-level management support, organizational culture, and project managerial skills.

HR Skills and Expertise. Effective communication, empowerment, expertise, experience, commitment, and composition are vital requirements for a team in software projects. These characteristics determine how quickly the team understands risks and affect the chances of project success [45]. When we compare agile and traditional project management approaches, small, self-organizing, autonomous teams composed of skilled people with commitment, are much more successful with the agile framework [43]. Teams with relatively little or no diversity are more successful with projects that can be done with the existing knowledge [60]. On the other hand, for complex data science projects that require problem-solving skills and creativity, groups with different expertise, experience, and educational background are needed [44].

Top-Level Management Support. Top-level management support is mainly affected by the leadership characteristics, vision and mission determined by the parental organization. [45, 61] state that an agile mindset supported by top-level management with a clear vision raises the success rate of data science projects. A team that receives the support of the top management increases its commitment and involvement in the project. The main risk factors are lack of corporate leadership, deviation from vital pillars of project management (e.g., time and budget), and inadequate skills [62].

Organizational Culture. The utilized project management methodology by an organization may provide important insights into its culture. Since agile methodologies mainly deal with dynamic changes, projects should be capable of adapting to new environments easily. According to [63], change management also plays a crucial role in the success of data science projects. Organizations need to define and implement an effective change management process to adopt emerging technologies and changes effectively [43]. It is also essential to involve all related external parties and stakeholders to gain acceptance of the change. It is recommended for the team to interact with stakeholders and collect regular feedback about the projects' success. Perceptions of the stakeholders should be considered to measure the success of the project.

Project Managerial Skills. Project management practices can be a product of the combination of various methodologies, each having certain fundamentals [45]. A study [49] revealed that management activities such as planning, budgeting, solving conflicts, and controlling requirements have been pointed out as the second most frequent factors, after factors related to people, in determining success [39].

5.3 Technology Dimension

Technology is another essential pillar of the proposed taxonomy, as it plays a critical role in the success of a data science project. The following sub-sections evaluate technology in terms of IT Infrastructure, Software Management, and IT Governance.

IT Infrastructure. IT Infrastructure consists of hardware and software components to run and manage IT operations [64]. The hardware infrastructure comprises of servers, storage disks, and networking equipment like switches and routers. In the case of data science and analytics, these are the resources where data science workloads are developed and deployed. There are different ways of conducting digital deployments: On-Premise, Edge, Cloud, High-Performance Computing, and Hybrid resources [65, 66]. Sufficient and appropriate infrastructure resources are integral to implementing data science practices in daily business routines. In addition, regularly planned hardware maintenance, removing legacy systems, and upgrades according to recent IT trends are critical due to hardware aging and obsolescence in short periods [48, 50, 51]. Also, data science workloads use state-of-the-art machine learning and deep learning models to make inferences and predictions, requiring computational and memory-intensive operations. Using obsolete, improper, and incompetent hardware resources can decrease productivity and create technical debts for the developers [67].

Software Management. According to [40, 43, 48, 51], the selection of the appropriate software development approach is critical to the software projects' success. Pursuing a simple design [37] followed by diligent testing [42], verification and validation [48] are considered a blueprint to success in software development. It is pivotal for the development team to be experienced with these development methods to streamline their development processes. Furthermore, [37] and [42] emphasize using agile-oriented approaches for their software project success. Identifying and selecting suitable technologies for data science from a large pool of alternatives is a complex problem that can significantly affect the course of data science projects and create technical debts [9, 67].

IT Governance. IT Governance, a part of corporate governance, helps organizations to align IT Strategy and business strategy and measure the performance of IT in achieving goals. It helps to ensure that IT investments generate business value. IT governance is critical for the success of software and data science projects. Studies [36, 43, 45] demonstrate that technological uncertainty, improper technology selection, and inadequate IT strategy can lead to failure in software projects. To establish effective IT Governance, implementing one or multiple IT Governance Frameworks, such as COBIT, ITIL, and CMMI [68] is a common practice followed in the industry.

5.4 Strategy Dimension

Strategy is the last critical pillar of the proposed taxonomy. It is evaluated considering two dimensions: strategic goals and environmental factors affecting strategy.

Clear Strategic Objectives and Goals. Project definition includes clear and well-defined objectives, among other criteria [48]. Clear strategic objectives and goals are significant for data science projects as well. Garousi et al. [40] explain the importance of a strategic alignment of project members with the developer. An example case would be when a developer might share the goal of the functional team, and reaching this goal might be sufficient for the developer to consider the project as successful. Whereas this cannot solely define the success of the project [40]. Similar to software projects, the success of a data science project depends on the strategic alignment of the project team. While developing a roadmap for a data science project, the value to be added to the business should be taken into consideration [55].

Environment. Considering the environmental factors of strategy and the data science projects' success, various items such as vendors, customers, outsourced consultants, and external events can be discussed. In a study by Sudhakar [52] in which CSFs are investigated for software projects, occurrences of environmental factors such as customer involvement, vendor partnership, and external environment events in the literature are found to be significant.

6 Discussion and Conclusion

This study aims to investigate the determinants of success and the taxonomy of antecedents of success in data science projects. Since the transformation into data science applications is still in a very early phase, there is an essential need to manage these projects successfully. Our preliminary research on this need revealed that the data science domain is relatively immature in terms of studies focusing on project success. In order to address this gap and explore the determinants of success in data science projects, we initially conducted an SLR of academic studies published within the last 20 years. CSFs identified as an outcome of the SLR set the base for the next step: Applying a Delphi method for constructing the taxonomy with inputs from domain experts.

The proposed taxonomy developed in several iterations in this study classifies drivers of success of data science projects under four main dimensions: Data, Organization, Technology, and Strategy. The main contribution of the study is that the proposed taxonomy generates valuable insights about which factors of projects should be enhanced and which powerful assets the company should build on. Furthermore, this taxonomy encourages managers or decision-makers to form a step-wise action plan and exploit the opportunities for intelligent transformation by evaluating their data science projects.

The significant highlights of this research are explained as follows:

- Although many studies indicate that the "data" factor is a significant CSF for project success, they only consider the data quality aspect. This study reveals that data life-cycle management and data governance steps such as data security, data preparation, data cleansing, data wrangling, and data fusion operations should also be considered for data science project success. That means the identification of the data quality requirements will not be sufficient for the projects' success if the stakeholders of the projects ignore the factors mentioned above.

- Customer involvement plays a significant role in having clear business requirements, which consist of well-documented and specific business requirements.
- Continuous integration and delivery offer many advantages, (i.e., automation of training and model validation, reproducibility of approaches, and scalability).
- Making the necessary maintenance, upgrades, and migrations according to developments and trends in software and hardware technologies is critical for the success.
- Aligning the available and prospective hardware and software technologies with software development approaches and strategies has utmost importance for success.

This study aimed to explore one pillar of project success of data science projects, with the notion that the primary factors of success are needed to be understood first and foremost. The other pillar -success criteria used for assessing the projects' success -deserves equal attention. Several authors (e.g., [17]) expressed the need to link both pillars to manage the success. Further research can investigate this connection and provide practitioners the ability to assess whether their efforts to fulfill the conditions and facts outlined in this study; contribute to the success of their data science projects.

Future studies can focus on overcoming the limitations of the research methods applied in the study (i.e., SLR and Delphi). For example, examining the studies in the gray literature on data science projects can help to reveal different success factors. On the other hand, a survey can be conducted to collect larger audiences' views. Incorporating experts' opinions from various experience levels, companies and countries to the taxonomy can help to improve the generalizability of findings.

References

1. Gökalp, E., Martinez, V.: Digital transformation maturity assessment: development of the digital transformation capability maturity model. Int. J. Prod. Res. 1–21 (2021). https://doi.org/10.1080/00207543.2021.1991020
2. Priyadarshy, S., Krigsman, M.: How to use Data Science for Digital Transformation I CXOTalk, https://www.cxotalk.com/episode/how-use-data-science-digital-transformation. Accessed 05 Jul 2022
3. Gökalp, E., Martinez, V.: Digital transformation capability maturity model enabling the assessment of industrial manufacturers. Comput. Ind. 132, (2021). https://doi.org/10.1016/j.compind.2021.103522
4. Kayabay, K., Gokalp, M.O., Gokalp, E., Eren, P.E., Kocyigit, A.: Data science roadmapping: towards an architectural framework. In: 2020 IEEE International Conference on Technology Management, Operations and Decisions ICTMOD 2020 (2020). https://doi.org/10.1109/ICTMOD49425.2020.9380617
5. Gökalp, M.O., Gökalp, E., Kayabay, K., Gökalp, S., Koçyiğit, A., Eren, P.E.: A process assessment model for big data analytics. Comput. Stand. Interfaces. 80, 103585 (2022). https://doi.org/10.1016/j.csi.2021.103585
6. Gökalp, M.O., Gökalp, E., Gökalp, S., Koçyiğit, A.: The development of data analytics maturity assessment framework: DAMAF. J. Softw. Evol. Process. e2415 (2021). https://doi.org/10.1002/smr.2415

7. VentureBeat: Why do 87% of data science projects never make it into production?, https://venturebeat.com/2019/07/19/why-do-87-of-data-science-projects-never-make-it-into-production/, last accessed 2022/07/14
8. Martinez, I., Viles, E., Olaizola, I.G.: A survey study of success factors in data science projects. In: Proceedings - 2021 IEEE International Conference on Big Data, Big Data 2021. pp. 2313–2318 (2021)
9. Nazliel, K., Kayabay, K., Gokalp, M.O., Gokalp, E., Eren, E.: Data Science Technology Selection: Development of a Decision-Making Approach. 172–178 (2022). https://doi.org/10.1109/TEMSCONEUROPE54743.2022.9802054
10. Gökalp, M.O., Kayabay, K., Gökalp, E., Koçyiğit, A., Eren, P.E.: Assessment of process capabilities in transition to a data-driven organisation: a multidisciplinary approach. IET Softw. 15, 376–390 (2021). https://doi.org/10.1049/sfw2.12033
11. Gökalp, M.O., Kayabay, K., Gökalp, E., Koçyiğit, A., Eren, P.E.: Towards a model based process assessment for data analytics: an exploratory case study. In: Yilmaz, M., Niemann, J., Clarke, P., Messnarz, R. (eds.) EuroSPI 2020. CCIS, vol. 1251, pp. 617–628. Springer, Cham (2020). https://doi.org/10.1007/978-3-030-56441-4_46
12. Gökalp, M.O., Gökalp, E., Kayabay, K., Koçyiğit, A., Eren, P.E.: Data-driven manufacturing: an assessment model for data science maturity. J. Manuf. Syst. 60, 527–546 (2021). https://doi.org/10.1016/J.JMSY.2021.07.011
13. PMI: PMBOK Guide | Project Management Institute (2021)
14. Khang, D.B., Moe, T.L.: Success criteria and factors for international development projects: a life-cycle-based framework. Proj. Manag. J. 39, 72–84 (2008)
15. Miller, G.J.: Artificial intelligence project success factors—beyond the ethical principles. In: Ziemba, E., Chmielarz, W. (eds.) FedCSIS-AIST/ISM -2021. LNBIP, vol. 442, pp. 65–96. Springer, Cham (2022). https://doi.org/10.1007/978-3-030-98997-2_4
16. Hussein, B.A., Ahmad, S.B.S., Zidane, Y.J.T.: Problems Associated with defining project success. Procedia Comput. Sci. 64, 940–947 (2015)
17. Ika, L.A.: Project success as a topic in project management journals. Proj. Manag. J. 40, 6–19 (2009). https://doi.org/10.1002/pmj.20137
18. Shenhar, A.J., Dvir, D., Levy, O., Maltz, A.C.: Project success: a multidimensional strategic concept. Long Range Plann. 34, 699–725 (2001)
19. Çaldağ, M.T., Gökalp, E.: Exploring critical success factors for blockchain-based intelligent transportation systems. Emerg. Sci. J. 4, 27–44 (2020). https://doi.org/10.28991/esj-2020-SP1-03
20. Pinto, J.K., Slevin, D.P.: Critical Factors in Successful Project Implementation. IEEE Trans. Eng. Manage. EM-34, 22–27 (1987). https://doi.org/10.1109/TEM.1987.6498856
21. Belassi, W., Tukel, O.I.: A new framework for determining critical success/failure factors in projects. Int. J. Proj. Manage. 14, 141–151 (1996)
22. Wateridge, J.: How can IS/IT projects be measured for success? Int. J. Proj. Manage. 16, 59–63 (1998). https://doi.org/10.1016/S0263-7863(97)00022-7
23. Chow, T., Cao, D.B.: A survey study of critical success factors in agile software projects. J. Syst. Softw. 81, 961–971 (2008). https://doi.org/10.1016/J.JSS.2007.08.020
24. Aho, T., Sievi-Korte, O., Kilamo, T., Yaman, S., Mikkonen, T.: Demystifying data science projects: a look on the people and process of data science today. In: Morisio, M., Torchiano, M., Jedlitschka, A. (eds.) PROFES 2020. LNCS, vol. 12562, pp. 153–167. Springer, Cham (2020). https://doi.org/10.1007/978-3-030-64148-1_10
25. Eybers, S., Hattingh, M.J.: Critical success factor categories for big data: a preliminary analysis of the current academic landscape. In: 2017 IST-Africa Week Conference, IST-Africa 2017 (2017)

26. Gao, J., Koronios, A., Selle, S.: Towards a process view on critical success factors in big data analytics projects. In: 2015 Americas Conference on Information Systems, AMCIS 2015 (2015)
27. Gómez, L.F., Heeks, R.: Measuring the barriers to big data for development: design-reality gap analysis in Colombia's public sector. Development Informatics Working Paper no. 62, (2016). Available at http://dx.doi.org/10.2139/ssrn.3431745
28. Saltz, J.S., Shamshurin, I.: Big data team process methodologies: a literature review and the identification of key factors for a project's success. In: Proceedings - 2016 IEEE International Conference on Big Data, Big Data 2016, pp. 2872–2879. Institute of Electrical and Electronics Engineers Inc. (2016)
29. Tsoy, M., Staples, D.S.: What are the critical success factors for agile analytics projects? Inf. Syst. Manag. **38**, 324–341 (2021). https://doi.org/10.1080/10580530.2020.1818899
30. Alhashmi, S.F.S., Salloum, S.A., Abdallah, S.: Critical success factors for implementing artificial intelligence (AI) projects in Dubai government United Arab emirates (UAE) health sector: applying the extended technology acceptance model (TAM). In: Hassanien, A.E., Shaalan, K., Tolba, M.F. (eds.) AISI 2019. AISC, vol. 1058, pp. 393–405. Springer, Cham (2020). https://doi.org/10.1007/978-3-030-31129-2_36
31. Ranjbarfard, M., Hatami, Z.: Critical success factors of BI project implementation: An implementation methodology perspective. Interdiscip. J. Inf. Knowl. Manage. **15**, 175–202 (2020). https://doi.org/10.28945/4607
32. Kitchenham, B., Pearl Brereton, O., Budgen, D., Turner, M., Bailey, J., Linkman, S.: Systematic literature reviews in software engineering - a systematic literature review. Inf. Softw. Technol. **51**, 7–15 (2009). https://doi.org/10.1016/j.infsof.2008.09.009
33. Kitchenham, B.: Procedures for performing systematic literature reviews (2004)
34. Çaldağ, M.T., Gökalp, E.: The maturity of open government data maturity: a multivocal literature review. Aslib. J. Inf. Manage. **74**(6), 1007–1030 (2022). https://doi.org/10.1108/AJIM-11-2021-0354
35. Misra, S.C., Kumar, V., Kumar, U.: Identifying some important success factors in adopting agile software development practices. J. Syst. Softw. **82**, 1869–1890 (2009)
36. Siddique, L., Hussein, B.A.: A qualitative study of success criteria in Norwegian agile software projects from suppliers' perspective. Int. J. Inf. Syst. Proj. Manage. **4**, 63–79 (2016)
37. Stankovic, D., Nikolic, V., Djordjevic, M., Cao, D.B.: A survey study of critical success factors in agile software projects in former Yugoslavia IT companies. J. Syst. Softw. **86**, 1663–1678 (2013). https://doi.org/10.1016/j.jss.2013.02.027
38. Tam, C., Moura, E.J. da C., Oliveira, T., Varajão, J.: The factors influencing the success of on-going agile software development projects. Int. J. Proj. Manage. 38, 165–176 (2020)
39. Bhoola, V.: Impact of project success factors in managing software projects in India: an empirical analysis. Bus. Perspect. Res. **3**, 109–125 (2015)
40. Garousi, V., Tarhan, A., Pfahl, D., Coşkunçay, A., Demirörs, O.: Correlation of critical success factors with success of software projects: an empirical investigation. Softw. Qual. J. **27**(1), 429–493 (2018). https://doi.org/10.1007/s11219-018-9419-5
41. Otoom, A.F., Kateb, G. AL, Hammad, M., Sweis, R.J., Hijazi, H.: Success factors importance based on software project organization structure. Inf. **10**, 391 (2019)
42. Zaleski, S., Michalski, R.: Success factors in sustainable management of IT service projects: exploratory factor analysis. Sustain. **13**, (2021). https://doi.org/10.3390/su13084457
43. Ahimbisibwe, A., Daellenbach, U., Cavana, R.Y.: Empirical comparison of traditional plan-based and agile methodologies: critical success factors for outsourced software development projects from vendors' perspective. J. Enterp. Inf. Manage. **30**, 400–453 (2017)
44. Sanchez, O.P., Terlizzi, M.A., de Moraes, H.R. de O.C.: Cost and time project management success factors for information systems development projects. Int. J. Proj. Manage. **35**, 1608–1626 (2017). https://doi.org/10.1016/j.ijproman.2017.09.007

45. Ahimbisibwe, A., Cavana, R.Y., Daellenbach, U.: A contingency fit model of critical success factors for software development projects: a comparison of agile and traditional plan-based methodologies. J. Enterp. Inf. Manage. **28**, 7–33 (2015)
46. Dikert, K., Paasivaara, M., Lassenius, C.: Challenges and success factors for large-scale agile transformations: a systematic literature review. J. Syst. Softw. **119**, 87–108 (2016)
47. Irvine, R., Hall, H.: Factors, frameworks and theory: a review of the information systems literature on success factors in project management. Inf. Res. **20**(3) (2015). https://informati onr.net/ir/20-3/paper676.html#author
48. Durmic, N.: Information systems project success factors: literature review. J. Nat. Sci. Eng. **2** (2020). https://doi.org/10.14706/JONSAE2020218
49. Iriarte, C., Bayona, S.: IT projects success factors: a literature review. Int. J. Inf. Syst. Proj. Manag. **8**, 49–78 (2020). https://doi.org/10.12821/ijispm080203
50. Kirmizi, M., Kocaoglu, B.: The key for success in enterprise information systems projects: development of a novel ERP readiness assessment method and a case study. Enterp. Inf. Syst. **14**, 1–37 (2020). https://doi.org/10.1080/17517575.2019.1686656
51. Nasir, M.H.N., Sahibuddin, S.: Critical success factors for software projects: a comparative study. Sci. Res. Essays. **6**, 2174–2186 (2011). https://doi.org/10.5897/sre10.1171
52. Sudhakar, G.P.: A model of critical success factors for software projects. J. Enterp. Inf. Manag. **25**, 537–558 (2012). https://doi.org/10.1108/17410391211272829
53. Subiyakto, A., Ahlan, A.: implementation of input-process-output model for measuring information system project success. TELKOMNIKA Indones. J. Electr. Eng. **12**, 5603–5612 (2014). https://doi.org/10.11591/telkomnika.v12i7.5699
54. Jones, C.: Estimating Software Costs. McGraw-Hill Education, New York (2007)
55. Kayabay, K., Gökalp, M.O., Gökalp, E., Erhan Eren, P., Koçyiğit, A.: Data science roadmapping: an architectural framework for facilitating transformation towards a data-driven organization. Technol. Forecast. Soc. Change. **174**, 121264 (2022). https://doi.org/10.1016/J.TEC HFORE.2021.121264
56. McLeod, L., MacDonell, S.G.: Factors that affect software systems development project outcomes: a survey of research. ACM Comput. Surv. **43**, 1–56 (2011)
57. Joosten, D., Basten, D., Mellis, W.: Measurement of information system project success in german organizations. Int. J. Inf. Technol. Proj. Manage. **5**, 1–20 (2014)
58. Gökalp, M.O., Gökalp, E., Kayabay, K., Koçyiğit, A., Eren, P.E.: The development of the data science capability maturity model: a survey-based research. Online Inf. Rev. **46**, 547–567 (2022). https://doi.org/10.1108/OIR-10-2020-0469
59. Google: MLOps: Continuous delivery and automation pipelines in machine learning, Cloud Architecture Center, Google Cloud, https://cloud.google.com/architecture/mlops-con tinuous-delivery-and-automation-pipelines-in-machine-learning. Accessed 29 Jul 2022
60. Mannix, E., Neale, M.A.: What differences make a difference? The promise and reality of diverse teams in organizations. Psychol. Sci. Public Interest. **6**, 31–55 (2005)
61. Wan, J., Wang, R.: Empirical research on critical success factors of agile software process improvement. J. Softw. Eng. Appl. **03**, 1131–1140 (2010)
62. Oz, E., Sosik, J.J.: Why information systems projects are abandoned: a leadership and communication theory and exploratory study. J. Comput. Inf. Syst. **41**, 66–78 (2000)
63. Fui, F., Nah, H., Delgado, S., Fui-Hoon Nah, F.: Critical success factors for enterprise resource planning implementation and upgrade. J. Comput. Inf. Syst. **46**, 99–113 (2016)
64. Chanopas, A., Krairit, D., Khang, D.B.: Managing information technology infrastructure: a new flexibility framework. Manage. Res. News. **29**, 632–651 (2006)
65. Netto, M.A.S., Calheiros, R.N., Rodrigues, E.R., Cunha, R.L.F., Buyya, R.: HPC cloud for scientific and business applications: Taxonomy, vision, and research challenges. ACM Comput. Surv. **51** (2018). https://doi.org/10.1145/3150224

66. Ometov, A., Molua, O.L., Komarov, M., Nurmi, J.: A survey of security in cloud, edge, and fog computing. Sensors **22**(3), 927 (2022)
67. Sculley, D., et al.: Hidden technical debt in machine learning systems. In: Advances in Neural Information Processing Systems, pp. 2503–2511 (2015)
68. Betz, C.T.: ITIL, COBIT, and CMMI: Ongoing Confusion of Process and Function. BPTrends. pp. 1–13 (2011)

Performance Evaluation of Clustering Techniques on a Hybrid RFM-Based Scoring Framework

Mona Mosa[1]([✉]) [ID], Nedaa Agami[2] [ID], Ghada Elkhayat[3] [ID], and Mohamed Kholief[1] [ID]

[1] Arab Academy for Science, Technology and Maritime Transport, Alexandria, Egypt
monamosa@hotmail.com, kholief@aast.edu
[2] Cairo University, Giza, Egypt
[3] Alexandria University, Alexandria, Egypt
ghada.elkhayat@alexu.edu.eg

Abstract. In today's dynamic business world, customers are considered the most valuable asset. Thus, enhancing decision-making strategies that focus on availing better services to those customers has a significant impact on gaining a competitive edge across organizations. Achieving this enhancement requires in-depth understanding of customers' behavior by using innovative methodologies like machine learning and advanced analytics techniques to discover valuable insights. Clustering is a machine learning technique that divides customers into meaningful clusters based on their characteristics to comprehend customers' behavior. Similarly, Recency, Frequency, Monetary and Adoption 'RFMA', is a scoring model that is used to simplify customers' behavior representation. Clustering techniques can be integrated with scoring models for obtaining better results. However, there is limited research that evaluates the impact of applying clustering techniques on scoring models to ensure its effectiveness. In this paper, a new method that integrates the Expectation Maximization 'EM' clustering technique with RFMA model is introduced in order to assess the robustness and sensitivity of the RFMA model. This method was implemented on genuine bank data where stakeholders target prospective customers who have the ability to use digital channels. Additionally, the study conducted in this paper evaluates the impact of applying different clustering techniques such as K-means and EM on the RFMA model. The results show that the model is robust across those techniques.

Keywords: Scoring model · RFMA · Clustering · K-means and Expectation maximization

1 Introduction

Nowadays, the retention of existing customers and acquisition of new ones, eventually, provides a competitive edge among organizations (Apte and Phil 2020). Recently, organizations have been following a customer-centric perspective that prioritizes customers to achieve customer's satisfaction, loyalty and long-term relationships with customers. Consequently, organizations face mounting pressure of customers' needs, requirements

© The Author(s), under exclusive license to Springer Nature Switzerland AG 2023
F. P. García Márquez et al. (Eds.): ICCIDA 2022, LNNS 643, pp. 463–478, 2023.
https://doi.org/10.1007/978-3-031-27099-4_36

and high expectations, thus, they seek to enhance the effectiveness of its decision-making strategies concerning those customers (Unal and Schivinski 2018), (Vivek and Selvan 2020). Recently, machine learning and advanced analytics techniques have been developed to support decision-makers to make the right decisions for their customers. These techniques have the ability to find hidden patterns, extract useful relationships, and uncover new insights from customers' data (Guo and Qin 2018). It is important to obtain new correlations, patterns, and trends, which are understandable and valuable to the decision-makers. There have been many researches and applications focusing on different techniques and methodologies. Clustering is an unsupervised learning technique that enables organizations to deeply comprehend customers' behavior. It divides customers into distinct, meaningful, and homogeneous segments or clusters based on various attributes and characteristics (Pahwa et al. 2018). It also identifies groups in which the customers are as much alike as possible and greatly diverse from customers in other segments, (Turkan et al. 2020). It begins with establishing the business objectives and concludes with the conveyance of various strategies per cluster. In the same context, the Recency, Frequency, Monetary and Adoption 'RFMA' is a dynamic scoring model used to simplify customers' behavior representation with a broad view (Wansbeek et al. 1999), (Mosa et al. 2022). Usually, scores are allocated to customers according to the estimates of the model parameters (Aycin et al. 2018). 'R', indicates the recency of customers' transactions, 'F', indicates the number of times in which transactions are made by customers. 'M' indicates the expenditures of customers. Finally, 'A' recognizes customers' interactions across multiple channels.

Clustering techniques and scoring models are constructive on each of their own, thus, there is an opportunity to combine them for obtaining preferable results. Also, it is important to measure the results of this combination.

This paper presents a new method which combines the Expectation Maximization (EM) model-based clustering technique with RFMA scoring model in order to assess the RFMA model robustness and sensitivity. EM clustering technique was chosen due to its power and effectiveness (Appiah et al. 2022). The new method was applied on genuine data from a bank where stakeholders are looking forward to identify potential customers to be better acquainted with digital channels usage. The research contribution is extended to compare the impact of different clustering techniques such as K-means and EM on the RFMA in terms of robustness, sensitivity and performance. The output reveals that the model is fairly robust across those techniques. It also, successfully ascertaines effective behavioral clustering through the usage of the new method. This paper is assembled as follows: Sect. 2 articulates the literature of clustering techniques and RFMA scoring model. Section 3 demonstrates the scoring - clustering hybrid framework. In Sect. 4, a detailed experiment is illustrated including the experiment input, processing and output. In Sect. 5, the experiment result's evaluation is presented. Finally, in Sect. 6, in-depth analysis is conducted, conclusion is summarized and future work is discussed.

2 Related Work

Clustering is a machine learning technique regularly used to group customers into clusters to support the decision-making process, so that data within any segment are similar while data across other segments are different (Kotler and Keller 2006). Kmeans is one of the most popular clustering techniques. It is intensely used in both scientific and industrial applications across various domains (Huang and Chang 2010), (Ashabi et al. 2020). Its input includes some parameters and also the number of clusters which are divided accordingly. K-means calculations are determined based on centroids (Micheline and Jiawei 2006). K-means has several advantages such as simplicity, high speed to access datasets on a very large scale and it is interpretable to stakeholders. However, it has some limitations, for instance, it can only handle numerical data, it forms spherical clusters and its centroids can be dragged by outliers (Chandra and Guptar 2020). Also, the algorithm is inflexible to define a number of appropriate clusters at the beginning of the algorithm for obtaining acceptable results. Accordingly, Elbow method is commonly utilized for recognizing the ultimate number of clusters for maximum effectiveness (Syakur et al. 2018).

Moreover, Expectation Maximization 'EM' is one of the most powerful clustering techniques which generates superior output (Nadif and Govaert 2008). EM depends on the likelihood of data distribution. EM has some advantages; it can detect the concentration of clusters, deals with large datasets and also can be used for the latent hidden variables (Mohamed Ifham 2021). In contrast, it has some disadvantages such as its complexity and requirement of huge computational time more so than other clustering techniques.

Some researchers have attempted to evaluate the performance of diverse clustering techniques to achieve different objectives. For example, for the sake of assessing the quality of red wine, Jung, Kang, & Heo used logistic regression tool to compare the generated clusters from K-means versus EM to test the output accuracy (Jung et al. 2014). Like Messele conducted a comparative study that applied k-means versus EM in order to identify natural grouping of Amharic noun terms based on a set of individual and a combination of high performing features (Messele 2020). Additionally, Kisho & Venkateswarlu proposed hybridizing K-means and EM techniques to accelerate data clustering. The output shows that hybridizing both techniques proved its value. It was observed that the proposed algorithm for hybridization of K-means and the EM techniques was consistently taking less execution time than using each technique separately (Kisho and Venkateswarlu 2016).

In the context of using advanced analytics to support decision-making with regard to customers, RFM has been also used to support customer value analysis and target potential customers based on identified objectives, it represents customers' behavior in a simple way. It is a scoring model which assigns scores to customers based on their prior purchasing history according to three parameters, Recency 'R', Frequency 'F', and Monetary 'M' (Mccarty and Hastak 2007). However, there are some limitations that eliminate the current RFM from making a better contribution to the eventual advancement of the decision-making with regards to the customers. Some of which are the neglect of vital characteristics of customers' behavior (Behrooz 2015), (Firdaus and Utama 2021), ineffective methods to transform RFM parameter values into scores (Miglautsch 2000), and lack of scientific approaches to combine RFM parameter scores to form a final score (Turkan et al. 2020). Thus, there were some researchers who challenged improving RFM model by including additional parameters to emphasize the effectiveness of the model (Amir and Esmaeil 2014), (Guptar and Chandra 2020) or by integrating the model with further clustering techniques. For instance, Huang, He and Zhang suggested adding the community relationship 'C' parameter to the RFM parameters to make it more effective, and then K-means was applied for clustering customers (Huang et al. 2020).

Additionally, Kholief, Elkhayat, Agami and Mosa proposed a hybrid RFM-based framework, in a recent research, to accentuate the effectiveness of the existing RFM scoring model, thus, an innovative parameter, Adoption 'A', was introduced. It mainly compares customers' interactions across multiple channels and then assigns each customer a score. It also provides more comprehensive view to customers' behavior. This proposed framework endorsed that the RFMA enhanced model can be integrated with clustering techniques to obtain meaningful and interpretable results. Accordingly, the RFMA model was applied earlier in integration with K-means clustering technique on authentic data from a bank to improve the use of digital channels (Mosa et al. 2022). The results show that the approach is effective and appropriates clustering to customers' base, which, in turns, assists the bank stakeholders; enhancing their decision-making strategies for each generated cluster.

The mentioned literature indicates some approaches that evaluate the performance of clustering techniques in general. There are also researches that refer to possibility of integrating scoring models such as RFMA with these techniques. Nevertheless, there is still limited research on how to assess the robustness and sensitivity of the scoring models and also how to evaluate the impact of applying different clustering techniques on these scoring models, in particular. Thus, research is still needed in this area.

3   The Scoring – Clustering Hybrid Framework

This section gives an overview on the structure of the earlier proposed hybrid framework (Mosa et al. 2022). It is a generic framework that can be applied by different methods to achieve various objectives. The framework procedure consists of 6 main steps, which can be illustrated in Fig. 1.

The procedure starts with calculating the values of RFM parameters for every single customer based on business rules for the applied study. The next step is to assign scores to 'A' innovative parameter for each customer. Associating 'A' with RFM administers a more thorough insight on customer behavior.

The proposed hybrid framework states that there is a possibility to integrate scoring models with clustering techniques. This integration aids in further comprehending customers' behavior, obtaining more accurate behavioral representation and finding meaningful results. The selection of the applied clustering techniques depends on each research objective and its practical scenarios. The next step is to apply a proper clustering technique to transform the values of the main parameters, independently, into scores. Then, the 4 parameters, 'R', 'F', 'M' and 'A' should be combined to obtain one final score for each customer. The parameter's combination can be done by summing, multiplying or concatenating the scores. After combining all scores to obtain a final one,

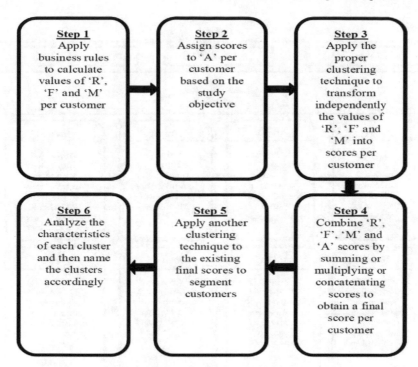

Fig. 1. Hybrid framework procedure

clustering techniques can be applied again to cluster customers' final score. Finally, the characteristics of the generated clusters should be analyzed to provide new insights to support decision-making concerning customers. The rest of this section discusses the full process of integrating the RFMA model with clustering techniques across two different methods, including input, processing and output.

3.1 RFMA with K-means

K-means and RFMA integration was suggested earlier for gaining optimum clusters. The way of integration stated in this method including input, processing and output is summarized in Fig. 2 to explain how it works:

As shown in Fig. 2, in the input phase, customer, product and transaction data enter the model to be processed. Then, in the processing phase, the steps embedded in the hybrid framework procedure, explained previously in Fig. 1, should be followed. K-means was selected, as a powerful clustering technique, to transform parameter values into scores, the technique can also applied to cluster customers' final score. Finally, the output of this procedure is customers' clusters. This method was applied and tested on real data. The findings will be presented in the next sections.

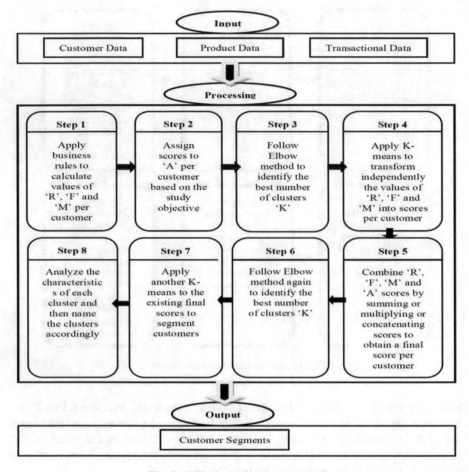

Fig. 2. RFMA and K-means method

3.2 RFMA with Expectation Maximization

A modified structure encompassing the new method that integrates RFMA with the EM is depicted in Fig. 3 below:

The procedure used in that method shown in Fig. 3 is the same as the procedure stated in Fig. 2 except for applying a different clustering technique, EM, stated in the processing phase. This new method is also applied and tested on real data. The findings will be discussed in the next sections.

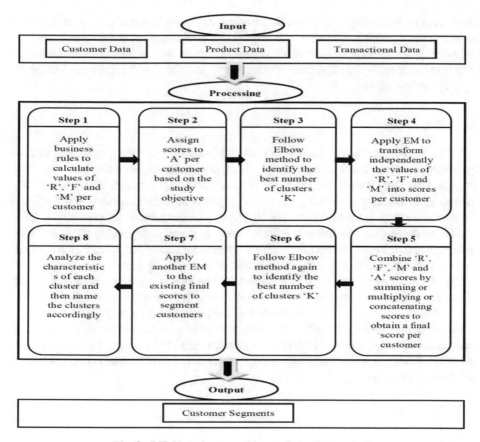

Fig. 3. RFMA and expectation maximization method

4   Experimental Set-Up

In order to assess the RFMA model effectiveness, robustness and sensitivity to different clustering techniques practically, a bank was selected as a case study in which the stakeholders intend to advance the customer engagement for using digital channels.

4.1   Input

A random sample of 138,750 customers was acquired for two years to be analyzed.

The case study includes 12 snapshots in different iterations to monitor customers' behavior over time. Data were uniquely and accurately identified, data meaning was enforced through the definition of data elements. Both profiling data for customers and trended data for products and transactions were described with field ID, field description, applied rules and remarks. Data location with proper access level was identified. Data were checked in cooperation with bank data governance experts to ensure data integrity and accuracy according to the bank data quality matrix standards. Errors were detected, missing values were imported, bad designed fields and duplicated data were removed. Abnormal and ambiguous values were checked. Data format were validated, irrelevant fields were deleted, and duplicated records were removed. Some values were renamed, aggregated and so on.

4.2 Processing

Statistical analysis was conducted for each field including tendency measures such as penetration, mean, median, maximum values, minimum values, null values and zero values. Data were summarized to identify correlations and patterns. Trends were visualized by histograms. Following the previous procedures demonstrated in Figs. 2 and 3, customers data were entered to the RFMA model as an input for processing. RFM values were computed for every single customer per iteration depending on the business rules confirmed by the stakeholders. The 'A' parameter was added to conduct a comparison between digital and non-digital transactions; designating a score to each customer. Different paths were suggested to establish 'A' scores and the best path was selected. Customers who didn't perform any transactions whether digital or non-digital acquired a score of 0. Customers with low number of digital transactions acquired a score of 1. Customers with medium number of digital transactions acquired a score of 3. Finally, customers with high number of digital transactions acquired a score of 5. The optimum number of clusters was achieved by using the Elbow method. 4 was the best number of clusters as shown in Figs. 4, 5 and 6:

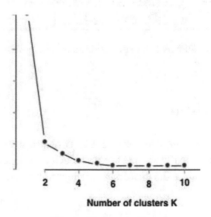

Number of clusters K

Fig. 4. Recency elbow

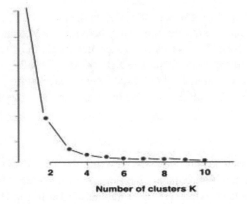

Fig. 5. Frequency elbow

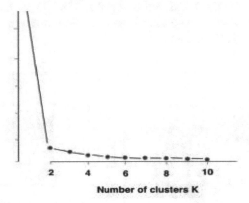

Fig. 6. Monetary elbow

K-means algorithm was applied twice, first, to alter each RFM value into an individual score per customer for every iteration. Elbow method was reapplied to allocate the best number of clusters for obtaining a final RFMA score. Results show that the optimum number of clusters was also 4 as indicated in Fig. 7.

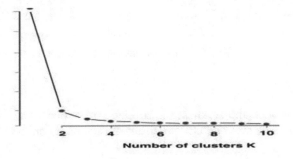

Fig. 7. RFMA elbow

K-means was utilized for the second time, on the final score, to segment customers into clusters. Additionally, another clustering technique, EM, was applied to assess the RFMA model robustness. EM was applied twice, first, to change the values of parameters into scores, second, to segment customers into clusters. Finally, customers are defined by certain clusters. The output for both experiments, integrating RFMA with K-means versus integrating RFMA with EM will be discussed in the next subsection.

4.3 Output

The output of both methods resulted in the same number of clusters which is 4. However, the clusters allocation, which presents the number of customers in each cluster across all iterations, differs from one method to another according to the technique of each procedure. Customers who have similar attributes belong to the same cluster. Customers total number; existing in the 4 clusters are equal to 138,750 per iteration to express the whole representative sample. The allocation of each cluster per iteration for the two methods is illustrated in Figs. 8 and 9.

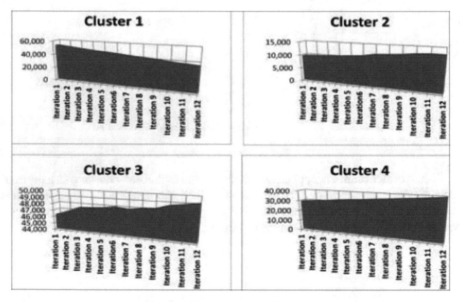

Fig. 8. K-means clusters allocation

On the same note, the following is a brief explanation of the resulted clusters according to the distinctive and characteristics of each based on the calculated RFMA parameters:

1. Non-Digital: is a cluster that encompasses substandard customers in which Recency, Frequency, Monetary and Adoption towards digital usage are absolutely defective.

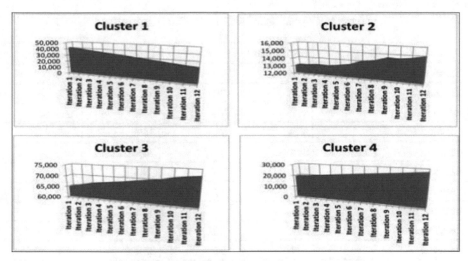

Fig. 9. Expectation maximization clusters allocation

2. Classical: this cluster includes customers whose Recency, Frequency, Monetary and Adoption are mediocre in digital usage.
3. Promising: this cluster encompasses customers whose Recency, Frequency, Monetary and Adoption are above average in digital usage.
4. Champion: it is group of customers with ideal digital Recency, Frequency, Monetary and Adoption.

5 Evaluation

Evaluation is an essential phase in that research. In that phase, result's evaluation is presented for the two methods; assessing its effect on the RFMA model. The evaluation phase encompasses some measures for the descriptive model such as cluster allocation and movement analysis, in addition to stakeholder's feedback.

5.1 Clusters Allocation Evaluation

It shows the stability of clusters allocation in all iterations. It calculates the customers' percentage per cluster for every iteration and monitors how it varies from one iteration to another. Table 1 clarifies the clusters allocation for the first method that uses K-means:

Table 1. Clusters allocation based on k-means method

Customer cluster/Iteration	Non digital	Classical	Promising	Champion
Iteration 1	39%	7%	33%	21%
Iteration 2	37%	7%	34%	22%
Iteration 3	37%	7%	34%	22%
Iteration 4	36%	7%	34%	23%
Iteration 5	34%	8%	34%	24%
Iteration 6	33%	8%	35%	24%
Iteration 7	33%	8%	34%	25%
Iteration 8	30%	9%	35%	26%
Iteration 9	29%	9%	35%	27%
Iteration 10	29%	9%	35%	27%
Iteration 11	28%	9%	35%	28%
Iteration 12	26%	9%	36%	29%

Table 2 illustrates the clusters allocation for the second method that uses EM:

Table 2. Clusters allocation based on expectation maximization

Customer cluster/Iteration	Non digital	Classical	Promising	Champion
Iteration 1	30%	9%	47%	14%
Iteration 2	30%	9%	47%	14%
Iteration 3	27%	10%	48%	15%
Iteration 4	25%	10%	49%	16%
Iteration 5	25%	10%	49%	16%
Iteration 6	24%	10%	49%	17%
Iteration 7	23%	10%	50%	17%
Iteration 8	22%	10%	50%	18%
Iteration 9	20%	11%	51%	18%
Iteration 10	18%	11%	52%	19%
Iteration 11	17%	11%	52%	20%
Iteration 12	17%	11%	52%	20%

By comparing the figures in both tables, it is shown that there is no significant variance of the clustering allocation across all iterations for each of the mentioned methods.

5.2 Movement Analysis

It tracks customer movement stability across the clusters, for each two consecutive iterations. The following equations calculate the stability of each cluster per iteration, the stability of all clusters per iteration, and finally, the overall stability across all iterations, where (i) is the iteration, $(i - 1)$ is the previous iteration, (N_p) is the total number of points, (N_c) is the total number of clusters, (N_i) is the total number of iterations, (C_{pi}) is the cluster membership of point (p) in iteration (i), (S_{ci}) is the stability of cluster (c) in iteration (i), (S_i) is the stability of all clusters in iteration (i) and (S) is the overall stability across all iterations:

$$S_{Ci} = \frac{\sum \begin{cases} if\, C_{pi} = C_{pi-1}.1 \\ Otherwise.0 \end{cases}}{\text{Number of Points of } C_{i-1}} \tag{1}$$

$$S_i = \frac{\sum \begin{cases} if\, C_{pi} = C_{pi-1}.1 \\ Otherwise.0 \end{cases}}{N_p} \tag{2}$$

$$S = \frac{\sum_1^{N_i} S_i}{N_i} \tag{3}$$

Hence, for both methods, the overall stability is summarized in Table 3:

Table 3. Movement analysis overall stability summary

K-means overall stability	EM overall stability
93%	95%

As shown in Table 3, the movement analysis outcome for both methods shows stable results regardless of the clustering technique. Stability reached 93% when K-means was applied whereas reached 95% in case of EM. Moreover, the output pertaining to cluster representations was shared with the stakeholders.

6 Discussion, Conclusion and Future Work

In the previous sections of this paper, a detailed comparison was explained between the two presented methods, RFMA associated with K-means against RFMA associated with EM. The output of the experiment was shared with the stakeholders and gained wide acceptance due to the proven effectiveness of the two methods; apparent in the logical results. Accordingly, they started planning for enhancing digital engagement based on the shared output. The rate of enhancement for the digital usage in each cluster is considered the difference between the customers' percentage in the primary iteration versus the last one. This rate is illustrated in both methods as follows (Table 4).

Table 4. Rates of digital enhancement by clustering techniques

Customer Cluster/ Iteration	Non-Digital	Classical	Promising	Champion
K-means first iteration	39%	7%	33%	21%
K-means last iteration	26%	9%	36%	29%
K-means enhancement rate	**−13%**	**2%**	**3%**	**8%**
EM first iteration	30%	9%	47%	14%
EM last iteration	17%	11%	52%	20%
EM enhancement rate	**−13%**	**2%**	**5%**	**6%**

The previous table shows that there are no gaps between the results. The enhancement rates in the Non-Digital cluster and Classical cluster are identical for the two methods. For the Promising cluster, the enhancement rate in K-means is 3% while in the EM is 5%. Unlike, the Champion cluster, where the enhancement rate in K-means is 8% while in the EM is 6%. On a further note, clusters' similarity measure is considered. It represents the percentage of unchanged customers, who remained existent in the same cluster when applying the two clustering techniques. Table 5 summarizes the clusters' similarity in one iteration:

Table 5. Clusters' similarity summary

Customer cluster	Similarity percentage
Non-Digital	78%
Classical	12%
Promising	100%
Champion	66%

The prior table shows that there is a difference in the similarity percentages in both methods. The logic behind this discrepancy that K-means is based on the mean of data points towards centroids, while in EM data points are Gaussian distributed, it is based on the probability of data distribution and concentration.

In conclusion, RFM is a dominant scoring model that helps in improving decision-making concerning customers throughout representing customers' behavior. 'A' Parameter was added to the model main parameters to create a comprehensive overview of customers' behavior. Following the hybrid framework that integrates RFMA model with clustering techniques, K-means clustering technique was integrated previously on a recent research. In this research, a new method that integrates the RFMA model with EM clustering technique was introduced to assess the robustness and sensitivity of the RFMA model to different clustering techniques. These methods mentioned in the bank case study are especially designed to facilitate decision-making with regards to digital customer engagement. Accordingly, customers were assigned to different clusters to better understand their behavior. The output in both two cases, were deliberated and assessed. It shows that RFMA model is effective and robust towards using different clustering techniques, it was also a platform on which the stakeholders were raised towards their targets. In future research, it is promising to associate RFMA scoring model with different clustering techniques to ensure the effectiveness of the output. Also, the proposed hybrid approach can be applied in different industries such as healthcare, education and insurance.

References

Appiah, S.K., Wirekoh, K., Aidoo, E.N., Oduro, S.D., Arthur, Y.D., Ling, N.: A model-based clustering of expectation–maximization and k-means algorithms in crime hotspot analysis. Res. Math. **9**(1) (2022). https://doi.org/10.1080/27684830.2022.2073662

Apte and Phil.: E-banking challenges and opportunities. In: 6thOne Day International Conference on Commerce, Management and Social Sciences, pp. 0474–9030. Satara (2020)

Ashabi, A., Sahibuddin, S.B., Haghighi, M.S.: The systematic review of k-means clustering algorithm. In: The 9th International Conference on Networks, Communication and Computing, pp. 13–18 (2020). ICNCC. https://doi.org/10.1145/3447654.3447657

Aycin, E., Dogan, O., Bulut, Z.: Customer segmentation by using RFM model and clustering methods: a case study in retail industry. Int. J. Contemp. Econ. Admin. Sci. **8**(1), 1–19 (2018)

Behrooz, N.: An analysis of mobile banking customers for a bank strategy and policy planning. In: Proceedings of International Academic Conferences (2015).\

Firdaus, U., Utama, D.N.: Development of bank's customer segmentation model based on RFM+B approach. ICIC Int. **12**(1) (2021). https://doi.org/10.24507/icicelb.12.01.17

Guptar, M.K., Chandra, P.: A comprehensive survey of data mining. Int. J. Inf. Technol. **12**, 1243–1257 (2020)

Huang, S., Chang, E.: Using k-means method and spectral clustering technique in an outfitter's value analysis. Qual. Quant **44**, 807–815 (2010). https://doi.org/10.1007/s11135-009-9240-0

Huang, Y., Zhang, M., He, Y.: Research on improved RFM customer segmentation model-based on k-means. In: 5th International Conference on Computational Intelligence and Applications (ICCIA), IEEE (2020). https://doi.org/10.1109/ICCIA49625.2020.00012

Kisho, R., Venkateswarlu, N.: Hybridization of expectation-maximization and k-means algorithms for better clustering performance. Cybern. Inf. Technol. **16**(2) (2016). https://doi.org/10.1515/cait-2016-0017

Kotler, P., Keller, K.L.: Marketing Managementn 12th edn. Prentice Hall, Hoboken (2006). https://www.biblio.com/book/marketing-management12th-edition-philip-kotler/d/581511378

Mccarty, J., Hastak, M.: Segmentation approaches in data mining: a comparison of RFM, CHAID, and logistic regression. J. Bus. Res. **60**(6), 656–662 (2007). https://doi.org/10.1016/j.jbusres.2006.06.015

Messele, A.M.: Comparing k-means and expectation maximization algorithm to cluster amharic noun terms. J. Multidimension. Res. Rev. **1**(1), 31 (2020)

Micheline, K., Jiawei, H.: Data Mining Concepts and Techniques, 2 edn. Elseveir, San Francisco (2006). https://www.elsevier.com/books/data-mining-southeast-asia-edition/han/978-0-12373584-3

Miglautsch.: Thoughts on RFM scoring. J. Database Mark. **8**(1), 67–72 (2000). https://doi.org/10.1057/palgrave.jdm.3240019

Mohamed Ifham, K.B.: Unsupervised learning approach for clustering source code based on functionalities. In: 2021 International Conference on Decision Aid Sciences and Application, IEEE (2021)

Mosa, M., Agami, N., Elkhayat, G., Kholief, M.: A novel hybrid segmentation approach for decision support: a case study in banking. Comput. J. (2022). https://doi.org/10.1093/comjnl/bxac009

Nadif, M., Govaert, G.: Algorithms for model-based block gaussian clustering. In: The 2008 International Conference on Data Mining, Las Vegas, vol. 2, pp. 14–17 (2008)

Pahwa, B., Taruna, S., Kasliwal, N.: Role of data mining in analyzing consumer's online buying behavior. Int. J. Bus. Manage. Inven. **6**(11), 45–51 (2018)

Qin, H., Guo, F.: Data mining techniques for customer relationship management. J. Phys. Conf. Ser. **910**(1), (2018). https://doi.org/10.1088/17426596/910/1/012021

Schivinski, B., Unal, G.: Literature review on conceptualisation of online consumer. Handel Wewnętrzny **371**(6), 353–362 (2018)

Syakur, M., Khotimah, B., Rohman, E., Satoto, B.: Integration k-means clustering method and elbow method for identification of the best customer profile cluster. In: IOP Conference Series Materials Science and Engineering, vol. 336 (2018). https://doi.org/10.1088/1757-899X/336/1/012017

Turkan, E., Sabuncu, I., Polat, H.: Customer segmentation and profiling with RFM analysis. Turkish J. Market. **5**(1), 22–36 (2020). https://doi.org/10.30685/tujom.v5i1.84

Vivek, N., Selvan, S.: Customer perception towards the digital transformation in banking industry with special reference to Madurai City. In: International Conference on Digitalisation of Banking Operations (2020)

Wansbeek, T., Leeflang, P., Spring, P.: The combination strategy to optimal target selection and offer segmentation in direct mail. J. Mark. Focused Manage. **4**(3), 187–203 (1999). https://doi.org/10.1023/A:1009899802421

A Systematic Literature Review on Data Provenance Visualization

Ilkay Melek Yazici$^{(\boxtimes)}$ and Mehmet S. Aktas

Computer Engineering Department, Yildiz Technical University, Istanbul, Turkey
{ilkay.yazici,aktas}@std.yildiz.edu.tr

Abstract. Data provenance is being one of the emerging needs for the domains and technologies to grant end-user to analyse and evaluate data life-cycle. Particularly in Big Data world with the help of Internet of Things environments, data amount increases rapidly each day. With the growth of data, the metadata also overgrows on origin, process and life-cycle of data. Innovations and approaches using data provenance are required to provide better interpretation and understanding of data. Efficient data visualization addresses this need as an important instrument for making complex data open, accessible and available. Data visualization proposes significant approaches to establish an instinctive interpretation by maximizing the user's perception. This research introduces a Systematic Literature Review (SLR) on Data Provenance Visualization for describing the studies that explicitly researched Data Provenance Visualization identifying which visualization approaches were considered mostly in the literature. A comprehensive and rigorous process is followed for giving confidence to the review study. Relevant primary studies are selected, and analysed in a systematic way. By proposing a significant systematic review to the literature, this study is being regarded as a reference aiming to examine particular researches on data provenance visualization approaches and technologies.

Keywords: Data provenance · Data Visualization Systematic Literature Review (SLR)

1 Introduction

Provenance is a data record describing the agents, activities and entities which are involved in data generation or releasing a part of it [1]. Understanding of the data provenance, data lineage, is an essential activity to assess data quality and trust, to enhance reproducibility, to debug complex data systems efficiently. Data provenance generating, recording and analysis activities are essential steps for past, present and future investigations [2]. Metadata is commonly linked with data provenance for enhancing the comprehending of the data and processing details. Especially, provenance of research data has a critical importance on reproducement and trust in scientific outputs [3] enabling scientists to view data in a derived view and to make observations about reliability and data quality, hence provenance monitoring is highly regarded as a significant functionality for visualization systems.

Data visualization suggests numerous approaches and methodologies to serve data preparation instinctively and it should be perfectly integrated to the analysis tools to have a complete data view, to find the insights and to make complex data more open, accessible and available. Especially interactive data visualization brings data exploration and analytical results by enabling expert and non-expert users in a significant way. In provenance discovery process, visualization facilitates the provenance data understanding through explanation and exploration visualization methods allowing users to interact with it. Visualization designers propose a wide range of functionality such as clustering, reproducibility and collaboration for provenance metadata to ease data mining and to increase user's perception in a large-scale data.

The amount of fine-grained provenance data can be overwhelming for end-user and data analysts. Along with supportive queries over provenance, this problem can be solved by generating proper, interactive visualizations to empower the end-user to explore and analyse provenance information [4].

In last decade, several significant studies address data provenance visualization to explore large-scale, multi dimensional data set enabling users to choose the visualization approaches and data set for their analysis process, to trace data path to investigate the data arrival at a certain time and define starting points and to reproduce existing paths or create alternative paths from paths' previous points. Most of the studies allow users to conduct multi step analysis based on high amount files generated by numerous workflows and implemented new visualization approaches in their researches. In the significant study of Kunde et al. [2], end-users' requirements were evaluated on an abstract level as a basis of data provenance visualization researches to develop new visualization technologies or to improve existing ones.

The common goal of this research is presenting a Systematic Literature Review (SLR) to describe studies that explicitly addresses data provenance visualization in general or in different domains defining visualization approaches and techniques. By following a rigorous process, a comprehensive screening was facilitated with primary studies in the literature.Section 2 describes the background summary while Scct. 3 gives details of SLR research methodology. Section 4 addresses the SLR findings and Sect. 5 answers the research questions defined at the beginning of this SLR study. Section 6 explains the overall conclusion.

2   Background

2.1   Systematic Literature Review

Systematic Literature Review (SLR) is an important tool to synthesize reference information in a research area as a rigorous approach. The aim is to investigate all relevant evidence set on defined research questions. First SLR is presented for medicine domain [5] and Kitchenham proposed a guideline to perform SLR for Software Engineering by referencing from medical research guidelines. In this significant study, SLR is regarded as an essential activity for identifying, assessing and interpreting possible literature to a set of particular questions. SLR study is called secondary study while the individual studies are being issued of this review are called primary studies.

Due to its rigorous manner, SLR is more scientific than a simple literature review. In literature, SLR studies have several reasons to be performed such as summarizing the existing evidences concerning a technology, identifying gaps in current research area to be able to suggest further investigations providing a common background to support new research activities appropriately.

2.2 Systematic Literature Review on Data Provenance Visualization

In the literature, several studies were addresses literature review or SLR study on Data Provenance Visualization with a subset of questions to be asked about visualization features and approaches.

In their research [6], Xu et al. introduced a survey on provenance data analysis methodologies including user-centric approaches and system-centric approaches such as adaptive systems. A three stage process was defined to conduct literature review by asking why, what, how, where questions to the selected studies by addressing the papers between 2009–2019. Main goal was the analysis of the user interactions as a multi disciplinary area of the artificial intelligence, machine learning, human computer interface fields to improve provenance analysis innovations. The literature study provides a multi disciplinary awareness of provenance analysis for the development of new technologies and some visualization opportunities accordingly. The goal holds the potential to support visualization systems to expand its role in data science and to contribute to the fields like explainable AI, while our current study address directly a well-formed Systematic Literature review (SLR) directly on the data provenance visualization systems by issuing more recent papers, between 2012–2022. Especially in data science, visualization techniques are often integrated with analysis methods for supporting human and machine capabilities to explore data by improving sense-makings of visually interactive systems [7]. Hence, identification of possible visualization techniques mostly relies on provenance analysis studies.

In the survey study of Suh. et al., simulation provenance systems' visualization features were investigated besides modeling, capturing, querying aspects for scientific simulations and workflows. Visualization approaches of selected studies were examined by considering the stored provenance data querying and visualization of an issued query result. Querying aspects were classified as low, medium or high for the graph visualization tools and system in the proposed literature review document [8].

In the significant study, [25], Oliveira et al. realized a partial SLR study by including 10 selected studies investigating some basic attributes such as Filtering, Viewing and Searching including the study itself to give a background and to propose the motivation of the study. In [17], Menin et al. conducted a surveyed comparison between Data Visualization Tools by looking to the attributes like Temporal, Distributed, Relational features of five selected visualization tools.

With all of the studies mentioned above, literature reviews on data provenance visualization were conducted partially as a part of another research. In this study, our aim is to introduce a comprehensive SLR study to extract the research data to be referenced by the future studies in this area following a well-defined SLR methodology.

3   Research Methodology

In this SLR study, the guideline steps below are derived from Kitchenham [5] studies:

1. Review Planning
   Description of the requirement for a literature review
2. Review Process
   - Identifying the review research
   - Selecting reference studies
   - Studying quality evaluation
   - Data extraction and screening
   - Data interpretation
3. Review Reporting
   According to the Kitchenham, the background itself explains the need for the review by giving the rationale for the survey. The Background section is given in Sect. 2. The following Research methodology sections includes the Review Process details and a report is generated after the SLR study presented in this paper.

3.1   Research Questions

During the SLR study, the main question is proposed as following "Which studies in the literature focuses on data provenance visualization to meet the analysis needs in domains?". To answer this main question a set of research questions are defined as illustrated in Table 1.

Each defined question addresses a research question context specifically as shown in the third column. Primary studies are the issue of this SLR study for knowledge synthesis.

3.2   Data Sources and Search Strings

A literature study was conducted to build the common search strings and to select the databases for the literature papers. Specifically we focus on data provenance visualization by performing the advance specific searches on the well-known digital libraries to obtain the primary studies.

Table 1. Research questions.

ID	Research question	Main focus
R01	How is the distribution of the publications in years?	Distribution
R02	What was the venue of publications?	Venue
R03	Which domains data provenance is proposed/used?	Domain
R04	What provenance ontology/standards are used in data visualization?	Interoperability
R05	Which visualization approaches are needed to meet visualization needs?	Purpose
R06	How is end-user validation handled in studies?	Usability

To build the search string, the common keywords were identified at first place. All of the issued digital library sources support searches on digital papers, by having the operator 'OR' for synonyms of search terms, and 'AND' operator for linking synonyms set together. As a result, the search string used in this SLR study was: (DATA PROVENANCE VISUALIZATION OR PROVENANCE VISUALIZATION)

Referenced digital data sources are illustrated in Table 2. Data sources have their own search pecularity then the search sentences needed to adapt for each different search activity keeping their essences.

Table 2. Digital data sources

Source	URL
IEEE Explore	https://ieeexplore.ieee.org/Xplore
ACM Digital	https://dl.acm.org
Web of Science	https://www.isiknowledge.com/
Springer	https://link.springer.com/
Science Direct	https://www.sciencedirect.com/
Scopus	https://www.scopus.com/

3.3 Study Selection and Data Extraction

Due to the fact that the first official draft for the reference provenance data definition called PROV Document Family [1] was announced in December 2012 by World Wide Web Consortium (W3C), papers before the date were discarded. In addition some inclusion and exclusion criteria were introduced below to select relevant works to be reviewed in this SLR study.

Inclusion Criteria

IC1- Automatically identified research articles and papers focusing on data provenance visualization in electronic data sources.

Exclusion Criteria

ExC1- Duplication papers
ExC2- Secondary papers
ExC3- Papers are not written in English
ExC4- Review Articles,Surveys,Book Chapters, Tutorials,Reports

The research articles were exported in BIBLEX, RIS or CSV format from the electronic databases with respect to the search engine's type support and formatted using free Jabref and Zotero tools. Results were imported into the tool called stART that supports Systematic literature Revirew (SLR) Process properly. stART is a Free Of Charge tool specialized for performing SLR processes according to the steps defined by Kitchenham [5].

The literature references were organized and SLR steps were managed such as Data selection and Data Extraction. As a conclusion the SLR results were exported to a spreadsheet.

4   Data Extraction

It is not a convenient way inspecting and reading all the selected data to facilitate the literature review. First data was classified by the following fields, title, writers, publication database, type and year of the publication.

Our SLR process is defined in Fig. 1 in detail. The literature search was conducted in 07.2022 returning 2923 papers in total. At first step, the papers were rejected regarding to Exclusion Criteria defined previously. Then in Phase 1 shown in the Figure, each paper were inspected individually by reading tile, the keywords and abstract fields and 305 papers were selected as candidates for data extraction phase. In Phase 2, a comprehensive reading took place for each selected papers, and 18 papers were extracted in this process to answer our SLR research questions shown in Table 1.

The final 18 studies are the subjects of the current literature review listed in Table 3 including type (CN: Conference, JR:Journal) and reference information. The synthesis of the papers are done in the following headings accordingly for answering the research questions as motivations of this SLR study.

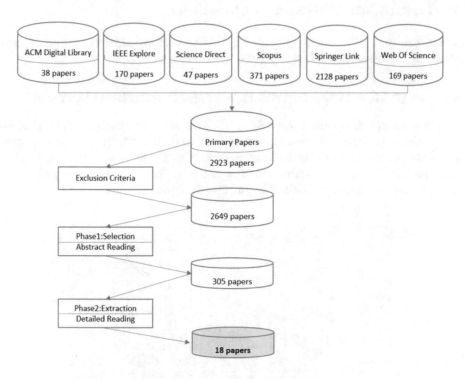

Fig. 1. SLR process steps

Table 3. Categorical data and reviewed studies reference mapping

Study	Reference	Type	Year
SP-01	[9]	CN	2016
SP-02	[10]	JR	2021
SP-03	[11]	CN	2020
SP-04	[12]	JR	2013
SP-05	[13]	CN	2017
SP-06	[14]	JR	2016
SP-07	[15]	JR	2019
SP-08	[16]	CN	2012
SP-09	[17]	CN	2021
SP-10	[18]	JR	2013
SP-11	[19]	CN	2016
SP-12	[20]	CN	2021
SP-13	[21]	JR	2019
SP-14	[22]	CN	2017
SP-15	[23]	CN	2015
SP-16	[24]	CN	2021
SP-17	[25]	JR	2022
SP-18	[26]	JR	2018

5   Discussion of Research Questions

Literature review in detail is given in this section by addressing the specific research objectives we defined in Table 1.

5.1   RQ1: How is the Distribution of the Publications by Year?

The studies are distributed from 2012 till 2022 and the distribution seems balanced between years. We found that the 2016 and 2021 have the majority, three and four papers respectively, as illustrated in Fig. 2. The years 2012, 2015, 2017, 2020 and 2021 has less papers which is one. On the other hand in year 2014, there is not any paper published in our search topic.

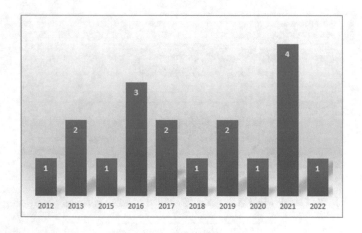

Fig. 2. Publications in years

5.2   RQ2: What was the Venue of Publications?

The goal of the research question is to clarify what are the data sources of the issued papers. Almost the same amount of publications were released in conferences, 10 papers and in journals, 8 papers. These conference papers indicate that they are not still mature and have enough details for publishing in journals.

The list of journals and conferences that the SLR papers were published within are illustrated in Table 4 below.

Table 4. Journals and conferences that reviewed papers published under

Journal/Conference Name	Digital Database
IEEE Int. Conference on Trust, Security and Privacy in Computing	IEEE Explore
Concurrency and Computation: Practice and Experience	Wiley Online
IEEE Visualization Conference (VIS)	IEEE Explore
IEEE Transactions on Geoscience and Remote Sensing	IEEE Explore
Biennial Conference on Innovative Data Systems Research	Conference Resource
Computer Graphics Forum	Wiley Online
Workshop on Massive Data Analytics on Scalable Sys. (DataMASS)	IEEE Explore
International Conference Information Visualisation (IV))	IEEE Explore
IEEE Transactions on Visualization and Computer Graphics	IEEE Explore
International Provenance Annotation Workshop	Springer Link
Working Conference on Software Visualization (VISSOFT)	IEEE Explore
IEEE Computer Graphics and Applications	IEEE Explore
USENIX Workshop on the Theory and Practice of Provenance	ACM Digital
International Provenance and Annotation Workshop	Springer Link
ACM IKDD CODS and COMAD	ACM Digital
Knowledge and Information Systems	Springer Link

Three papers were published in IEEE Transactions on Visualization and Computer Graphics journals and as seen in the table, most of the papers are related with IEEE journals. This is a significant finding in our research since it would direct the researchers to the environments which publish articles about data provenance visualization mostly. Springer Link, ACM Digital and Wiley Online Library are the other publish platforms.

5.3 RQ3: Which Domains Data Provenance is Proposed/used?

This research question is asked to evaluate the domains and use cases that provenance visualization needs are met within the studies.

Most of the papers state that the provenance visualization approaches are studied for general provenance visualization purposes but specific use-cases are demonstrated to highlight proposed visualization features. Studied domains and use-case areas of the selected studies are listed in Table 5.

Table 5. Study and domain/use case relation

Study ID	Domain or use case area
SP-01	Cloud Computing, Security
SP-02	E-Science
SP-03	Web application
SP-04	GeoScience, E-Science, Cyber Security
SP-05, SP-13	Data management, data wrangling
SP-06	Biomedical, E-Science
SP-07	data retrieval, biomedical
SP-08, SP-09, SP-10, SP-15, SP-17	Network, File Systems
SP-11	Georeferencing
SP-12	Software Development
SP-14	Health Data
SP-16	Social Data
SP-18	Time Series

Network management and E-Science domains are the ones which data visualization is studied as an emerging need.

5.4   RQ4: What Provenance Ontologies/standards are Used in Data Visualization?

The goal of this research question is to assess the data model specifications or languages facilitated in the studies to represent provenance data. The studies were analysed for the sake of interoperability. It is observed that the most of the studies focus on W3C provenance models and in other studies, customized provenance languages are defined or JSON is used as data type. The details can be seen in the Table 6 below.

Table 6. Provenance data model specification proposed in the studies

Study ID	Provenance data model specification
SP-02, SP-11, SP-12, SP-14, SP-15, SP-17	W3C PROV-O Family
SP-04	RDF, OWL
SP-03, SP-05, SP-06, SP-13, SP-16, SP-18	Customized data model
SP-01, SP-07	JSON,CSV
SP-08	W3C OPM
SP-09	RDF
SP-10	Provenance-Aware Storage System, PASS

As seen in the Table 6, the W3C provenance ontology models are the first ones chosen by the studies since the standardization meet the provenance visualization requirements defined by Kunde et al. [2]. The first three requirements, related to process results and relationship are satisfied by an accepted provenance representation model.

5.5 RQ5: Which Visualization Approaches are Studied to Meet Visualization Needs?

The research question is asked to investigate visualization attributes and studied visualization approaches in the reviewed papers. Visualization features are compared in logical groups in order to support intelligibility, so called visualization features and interactive visualization attributes. Each is handled in a subsection with a detailed SLR table.

In each table, the studies that has X-mark in respective fields are the studies that indicate this feature is proposed in these studies, blank fields are either not proposed or not specified.

Visualization Features: Visualization of provenance is still an open area and being studied by researches recently and the visual attributes are changing from conventional methodologies to new technologies and approaches.

The visualization features specified as functional needs in the reviewed publications have been examined and presented in detail in the table shown in Table 7.

Since data visualization is directly related to human perception, human-computer interaction principles become more important in these studies. As an example, Garae et al. [9] aimed to increase usability efficiency by incorporating Gestalt principles into the system at the system design stage.

Interactive Visualization Components: Interactive visualization approaches and their details presented in the reviewed publications are examined under this section.

All of the reviewed studied support easy navigation for metadata tracking such as actions and entities. The interaction components and features presented to the end user in the studies were reviewed systematically and proposed in the Table 8 in detail.

5.6 RQ6: How is End-User Validation Handled in Studies?

The goal of to ask this question is to investigate the user validation methodology of the studies. Since the end-user is at the heart of such systems, it is important that the system is evaluated by the user/user group. Some of the studies conducted a user study for the validation and the rest planned to do a user evaluation process in the future.

In the following studies [9,10,18,21,22,25] planned user studies carried out by user groups and the evaluations were presented accordingly. Especially in the studies [10] addresses a SUS usability metric applied for e-Science domain using e-Science documents [27] and in [10,18,25], a comprehensive usability assessment steps were defined ans users were asked to answer several questions related to systems' usability, using Likert Scale defined in the studies. According to the responses, the systems' validation were assessed using average and mean values.

Table 7. Visualization feature comparison

		Basic Features			Operational		Quality)		Implementation		
ID	Name	Visual* Representation.	Temporal	Workflow Support	Abstract (AV) Multiple (MV) Views	Collaboration (CO) Web Based (WB) Configurable Parameters (CF) Annotation (AN) Action Recovery(RC)	Gather Insight	Reproduction	User Centric Design	Open Src	Tools GUI Libs
SP1	UVisP	TM,HM,BD,FD	X		MV				X		D3
SP2	N/A	GV	X	X	AV	AN		X	X	X	Cytoscape
SP3	Ttrack	GV		X	AV	CO,WB,CF,AN,RC	X	X	X	X	Javascript
SP4	GePWProv	WM		X	AV,MV	WB		X			GeoPWT, GeoPOM
SP5	DVMS	PD,BC	X	X	AV,MV	CF,AN,RC	X				
SP6	Avocado	GV		X	AV,MV	CO		X		X	Dagre,D3
SP7	KnowledgePearls	GV,PD,BC	X		AV,MV	CO,AN	X			X	Gapminder
SP8	N/A	GV,FD		X	AV,MV	AN			X		Cytoscape
SP9	MGExplorer	GV,MT,CV,BC	X		AV,MV	CO,AN,RC					
SP10	InProv	RTL	X		VM				X		Processing
SP11	ProvViewer	GV	X	X	AV,MV	CF,RC				X	Jung
SP12	N/A	GV,BC,PD,CM			AV,MV	WB	X			X	GIT-LAB2PROV
SP13	DQProv Explorer	GV,BC	X	X	AV,BC	CO,AN,RC	X				OpenRefine
SP14	ProvComics	GV	X	X	AV	WB,AN			X	X	ProvStore JQuery
SP15	Prov-O-Viz	GV	X	X	AV	WB					
SP16	N/A	GV			AV					X	
SP17	Visionary	GV	X	X	AV,MV	WB	X			X	
SP18	SomFlow	GR,CM,DM FD,MT,HM.	X	X	AV,MV	CF,AN	X		X		Java 2D

* GV: Graph View, CV: Circular View, FD: Force Direct graph.
RTL: Radial based Tree Layout
BD: Bubble Diagram, PD: Plot Diagram
BC: Bar Chart, GR: Grid, MT: Matrix
WM: Web Map, TM:Tree Map, HM:Heat Map, CM: Color Map

Table 8. Interactive visualization components comparison

ID	Clustering/ Grouping	Comparison	Graph Merge	Details on Demand	Stream Visualization	Filter (F) Highlight (H) Sort (SO) Search (SE) Zoom (Z)	Query by Definition	Query by Example	History Retrieving
SP1				X				X	X
SP2	X	X		X	X	S,F,SE,Z		X	X
SP3	X							X	X
SP4		X		X		SE,F,H			X
SP5	X				X	F,H	X		
SP6	X	X				F,H,Z,SE			X
SP7	X	X		X		SE,F,H		X	X
SP8	X	X				Z			X
SP9	X	X		X		F	X		X
SP10	X			X		F,SE,Z			X
SP11	X		X	X		F,Z,H			
SP12						F			X
SP13		X		X		F,H			X
SP14	X			X					
SP15						F,H			
SP16	X								X
SP17		X				F,SE,H			
SP18	X	X		X		F,Z,H			

6   Conclusion

All papers issued in this review include data provenance visualization solutions with several visualization approaches in general or in specific domains. 18 studies are investigated that were published between 2012 and 2022.

Although some literature reviews conducted in some of the main provenance visualization researches as a minor SLR studies, this paper can be regarded as a first comprehensive SLR study in this area. The studied have shown that provenance visualization is one of the key research area to meet metadata analysis needs especially in e-Science domain. It seems that the area is still under investigation since each study proposes future plans showing the numerous open points to be researched in this area.

References

1. Missier, P., Belhajjame, K.: A PROV encoding for provenance analysis using deductive rules. In: Groth, P., Frew, J. (eds.) IPAW 2012. LNCS, vol. 7525, pp. 67–81. Springer, Heidelberg (2012). https://doi.org/10.1007/978-3-642-34222-6_6
2. Kunde, M., Bergmeyer, H., Schreiber, A.: Requirements for a provenance visualization component. In: Freire, J., Koop, D., Moreau, L. (eds.) IPAW 2008. LNCS, vol. 5272, pp. 241–252. Springer, Heidelberg (2008). https://doi.org/10.1007/978-3-540-89965-5_25
3. Magagna, B. Goldfarb, D. Martin, P. Atkinson, Malcolm Koulouzis, S. Zhao, Z.: Data provenance. In: Liu, L., Ozsu, M.T. (eds.) Encyclopedia of Database Systems. Springer, Boston (2009). https://doi.org/10.1007/978-0-387-39940-9_1305
4. Glavic, B.: Data provenance. Found. Trends Datab **9**, 209–441 (2021). https://doi.org/10.1561/1900000068
5. Kitchenham, B., Brereton, O.P., Budgen, D., Turner, Mark, Bailey, J., Linkman, S.: Systematic literature review in software engineering. Inf. Sw. Tech. **51**(1) (2009)

6. Xu, K., Alvitta O.y, Conny Walchshofer, M., Remco C., J. E. W. Survey on the analysis of user interactions and visualization provenance. Comp. Graphics Forum **39**(3), 757–783 (2020)

7. Madanagopal, K., Ragan, E., Benjamin, P.: Analytic provenance in practice: the role of provenance in real-world visualization and data analysis environments. IEEE Comp. Graphics App. 1–1. https://doi.org/10.1109/MCG.2019.2933419

8. Young-Kyoon, S., Ki Yong, L.: A survey of simulation provenance systems: modeling, capturing, querying, visualization, and advanced utilization. Hum. Centric Comput. Info. Sci. **8**(1), Article 150 (2018)

9. Garae, J. Ko, R. Chaisiri, S.: UVisP: user-centric visualization of data provenance with gestalt principles. In: 2016 IEEE Trustcom/BigDataSE/I SPA (1923–1930). https://doi.org/10.1109/TrustCom.2016.0294

10. Yazici, I., Aktas, M.: A novel visualization approach for data provenance. Concurr. Comput. Pract. Exp. **34**(9) e6523 (2021). https://doi.org/10.1002/6523

11. Cutler, Z. Gadhave, K. Lex, A.,: Trrack: a Library for Provenance-Tracking in Web Based Visualizations. in: 2020 IEEE Visualization Conference (VIS), pp. 116–120 (2020). https://doi.org/10.1109/VIS47514.2020.00030

12. Sun, Z. Yue, P. Hu, L. Gong, J. Zhang, L. Lu, X.: GeoPWProv: interleaving map and faceted metadata for provenance visualization and navigation. Geosci. Remote Sens. IEEE Trans. **51**, 5131–5136. https://doi.org/10.1109/TGRS.2013.2248064

13. Wu, E., Psallidas, F., Miao, Z., Zhang, H., Rettig, L., Wu, Y., Sellam, T.: Combining design. performance in a data . In: CIDR (2017)

14. Stitz, H., Luger, S., Streit, M., Gehlenborg, N.: AVOCADO, visualization of workflow-derived data provenance for reproducible. Biomed. Res. (2016). https://doi.org/10.1101/044164

15. Stitz, H., Gratzl, S., Piringer, H., Zichner, T., Streit, M.: KnowledgePearls-provenance-based visualization retrieval. IEEE Trans. Vis. Comp. Graps. PP. 1-1 (2010). https://doi.org/10.1109/TVCG.2018.2865024

16. Chen, P., Plale, B., Cheah, Y., Ghoshal, D., Jensen, S., Luo, Y.V.: Network data provenance. In: 19th International Conference on HiPC (2012). https://doi.org/10.1109/HiPC.2012.6507517

17. Menin, A., Cava, R., Freitas, C., Corby, O., Winckler, M.: Towards a Visual approach for representing analytical provenance in exploration processes. In: 2021 - 25thInternational Conference Information Visualisation, July 2021, Melbourne/Virtual, Australia, pp. 21–28 (2021). https://doi.org/10.1109/IV53921.2021.00014

18. Borkin, M., et al.: Evaluation of filesystem provenance visualization tools. IEEE Trans. Vis. Comp. Graps. **19**, 2476–85 (2013). https://doi.org/10.1109/TVCG.2013.155

19. Kohwalter, T., Oliveira, T., Freire, J., Clua, E., Murta, L.: Prov viewer: a graph-based visualization tool for interactive exploration of provenance data. In: Mattoso, M., Glavic, B. (eds.) IPAW 2016. LNCS, vol. 9672, pp. 71–82. Springer, Cham (2016). https://doi.org/10.1007/978-3-319-40593-3_6

20. Schreiber, A., Kurnatowski, L., Meinecke, A., de Boer, C.: An interactive dashboard for visualizing the provenance of software development processes. In: 2021 Working Conference on Software Visualization (VISSOFT), pp. 100–104 (2021). https://doi.org/10.1109/VISSOFT52517.2021.00019

21. Bors, C. Gschwandtner, T. Miksch, S., 2019. Capturing and Visualizing Provenance From Data Wrangling. IEEE Comp. Graphs. Apps. PP. 1–1 (2019). https://doi.org/10.1109/MCG.2019.2941856

22. Schreiber, A., Struminski, R.: Visualizing the provenance of personal data using comics. Computers **7** (2017). https://doi.org/10.20944/preprints201712.0153.v1
23. Hoekstra, R., Groth, P.: PROV-O-Viz - understanding the role of activities in provenance. In: Ludäscher, B., Plale, B. (eds.) IPAW 2014. LNCS, vol. 8628, pp. 215–220. Springer, Cham (2015). https://doi.org/10.1007/978-3-319-16462-5_18
24. Rani, A., Goyal, N., Gadia, S.: Provenance framework for Twitter data using zero information loss graph database. In: CODS COMAD 2021: 8th ACM IKDD CODS and 26th COMAD, pp. 74–82. https://doi.org/10.1145/3430984.3431014
25. Oliveira, W., Braga, R., David, J.M., Ströele, V., Campos, F., Castro, G.: Visionary, a framework for analysis and visualization of provenance data. Knowledge and Inf. Syst. **64**. 1–33 (2022). https://doi.org/10.1007/s10115-021-01645-6
26. Sacha, D., et al.: SOMFlow, guided exploratory cluster analysis with self-organizing maps and analytic provenance. IEEE Trans. Visual. Comput. Graphics (99), 1-1 (2017)
27. Yazici, I. Aktas, M. (2021). Usability Study on Data Provenance Visualization Approaches. 1–6. https://doi.org/10.1109/UYMS54260.2021.9659779

Effects of Online Educational System on Personal Health of Students and Teachers in COVID-19 Crises

Fariha Shaikh[1]([✉]) [iD], Shafiq-ur-Rehman Massan[2] [iD], Sania Bhatti[1] [iD],
Shafqat Shahzoor Chandio[3] [iD], and Muhammad Mujtaba Shaikh[3] [iD]

[1] Department of Software Engineering, Mehran University of Engineering and Technology,
Jamshoro, Pakistan
fairy.shaikh2@gmail.com
[2] Department of Computer Science, Newports Institute of Communications and Economics,
Karachi, Pakistan
[3] Department of Basic Sciences and Related Studies, Mehran University of Engineering and
Technology, Jamshoro, Pakistan

Abstract. Healthful living is the central characteristic to live a wonderful life which is badly affected when people are physically inactive. Physically unfit people are restricted in several walks of life. Therefore, physical fitness and activeness is key to live a healthy life. COVID-19 restricted people due to lock-downs. Several physical activities were seized. Along with human lives, all businesses, import-export, economies of countries and many other areas were affected. Similarly, educational system was badly affected. Due to lockdowns, it was not possible to carry traditional educational system (TES) in COVID crises, therefore, everyone had to move towards online educational system (OES). By adopting OES, going to academies, attending physical classes, performing practical sessions, sports, face-to-face interactions with friends, being ready for outdoor and several other physical activities were eliminated from student's and teacher's routine. Whereas, excessive screen time, continuous sitting, increased laziness, weight gain, spine issues and other health problems were reported. This work focuses higher educational institutes – universities – of Pakistan, and attempts to investigate seriousness of health risks, diseases and chances of being unhealthy during such COVID times. The aim is to investigate the unsettled health issues by gathering the perceptions of students and teachers through surveys. With the help of different statistical techniques and comparisons, responses are analyzed to identify the health problems. To avoid and prevent students and teachers from recognized health issues in future, some suitable and relevant suggestions have been provided to audience with the help of existing studies and doctors' opinion, which will help academia to adopt OES smoothly and more effectively.

Keywords: COVID-19 · Online education · OES · TES · Healthcare

1 Introduction

Online educational system (OES) is not a recent concept. It had been adopted in several countries even many years before the COVID-19. But, the difference matters now as it became mandatory to be adopted to carry education everywhere nowadays and during COVID crises. It is no more our choice [1, 5]. As OES is carried out by using several software tools and classes are conducted purely online with the help of computers, laptops and other digital devices, therefore, physical activities are eliminated from lives of students and faculty members. Physical inactivity from daily routine and excess use of digital screens cause several and dangerous health issues including computer vision syndrome (CVS), high sugar levels, high blood pressure, obesity, different types of cancers and even early deaths [7, 8]. In traditional educational system (TES), students and teachers are active and punctual due to their fixed schedules of theoretical classes, practical sessions and other physical activities. But all these physical activities were discontinued suddenly due to COVID-19 from their daily routine. All people were locked in their houses as lock-downs were imposed by governments in almost every country by observing exponentially growing cases of COVID positives [4]. Due to increase in screen time, continuous sitting and physical inactivity, students' and teachers' health was badly affected.

Audience using computer screens on daily basis were affected from the symptoms of CVS [9–11, 18]. By adoption of OES, students and teachers were bound to face computer screens for almost 6 h daily except weekend for participating in the online classes. Apart from that, mobile phones, tablets, iPads, and other devices were also used for other purposes such as for bill payments, gaming, entertainment purposes, streaming, communication, shopping, etc. By using digital screens for long while being in wrong postures and physical inactivity surely resulted in spinal cord abnormalities which would create severe spine pain and other major diseases in future [9–11]. Similarly, several other issues were reported in literature, which are discussed in detail in Sect. 2.

This research intends to conduct a survey on several health issues of students and faculty members of different higher educational institutes of Pakistan which were caused and became severe by adoption of OES. The three main categories are selected for respective causes or effects of each health issue and nine different survey questions (SQs) are defined as shown in Fig. 1. Of SQ1-SQ9, each SQ addresses some specific health problem associated with respective category. By carrying this research, unsettled health issues can be identified and with the help of literature survey and experts' opinions, suitable suggestions will help students and teachers to prevent themselves from further health problems.

This paper is organized as follows. Section 2 presents several similar existing studies from literature review which helped to design survey questions for this research. In Sect. 3, nature of data, sample size, survey questions, and techniques of data analysis are discussed. Section 4 illustrates results in form of table and charts, and discusses percentage variations, descriptive statistical parameters, and overall difference in the responses of students and faculty members that how the excessive use of digital screens in COVID-19 times with OES has affected human health. By observing the drastic increment in health cases, suitable suggestions to academia and users of online educational tools are provided in Sect. 5. Finally, Sect. 6 includes main findings of the study.

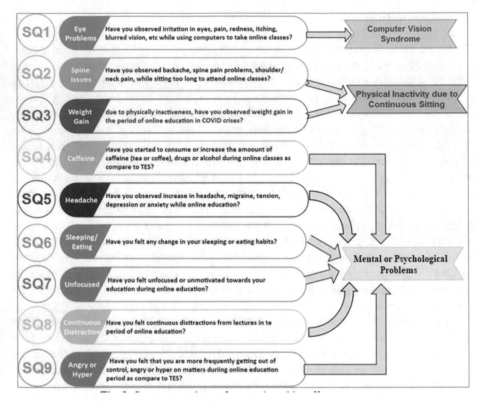

Fig. 1. Survey questions of research and its effects or causes

2   Literature Review

Several studies have been carried out by different authors to evaluate the online tools used to conduct online classes [23], to compare OES and TES [25], to evaluate teaching experiences [24], challenges, problems, limitations and improvements [1, 26, 28, 30, 31] faced by academic members, comparison of perceptions of students [27, 29, 32], teachers or both [22] and to monitor research activities [34] during and after COVID-19 period. While the main focus is to highlight the studies which were purely related to the effects of OES on health, the following papers help to identify the unsettled health related issues even in 2022 in Pakistan by using online tools.

Several studies were carried out by different authors at different places to recognize, report, avoid and reduce the causes of CVS [9–11, 18]. It has been one of the major reasons leading to different eye problems i.e. SQ1 even before COVID-19. But its effects are increased drastically because OES became so common due to this pandemic time.

In 2007 (before COVID-19), a similar and extended study for the effects of OES was carried out in Pakistan by evaluation of the Virtual university [2] where audience reported SQ1, SQ2 and SQ5 health-related problems. Along with this, several other categories and their parameters were included in survey for which responses from students were collected towards virtual education.

In 2020, another detailed study, carried out in India [3], was based on responses from 358 students of higher educational institutes. To analyze the collected data, correlation and co-variance between several selected factors for online education were used along with descriptive statistical techniques. Six different categories were selected for survey for addressing general concerns, content delivery, interaction, assessment, health and social issues. Thus, SQ1, SQ5 and SQ6 were directly assessed in health category in [3]. According to their analysis, more than 74% students felt that the extreme use of screen time had become part of their routine due to online learning. Also, this way of education affected their sleeping routine due to stress.

The report [6] highlighted the impact of OES on mental health of students. It carried most of the health related issues of Fig. 1: SQ1, SQ2, SQ5, SQ6, SQ7 and SQ8, in their report and assured that students of Brunei Darussalam University were facing these health problems due to OES. In such situation, government of Brunei provided tele-counselling facilities to citizens through which students and other citizens who were mentally affected due to COVID-19 could talk to professionals and follow their advices to prevent themselves from other psychological problems.

Further in 2020, another study was conducted for checking the anxiety level of Chinese university students in China [19] after adoption of OES due to COVID-19 which is actually SQ5 in this research. The study involved 3611 male and female students. After collecting data, analyzing that and comparing different aspects between genders, grades and batches of students, it was concluded that anxiety level was relatively higher in Chinese university students as compared to normal residents in COVID period. Whereas, anxiety of female students was much higher than male students. Anxiety was assessed by authors by using self-rating anxiety scale.

In September 2020, [20] investigated the effects by adopting the OES on the experience to use online tools, enjoyment, computer anxiety (somehow related to SQ1 because of CVS and SQ5 due to anxiety) and self-efficacy of students of a single university in Poland [20]. In total enough sample size i.e. 1692 students have been considered. Study concludes by analyzing data using Smart-PLS software tool as students have accepted and adopted the OES and their views are reported medium for their productivity, effectiveness and self-efficacy but they want to go back towards TES.

So far, no any detailed and similar research has been conducted in Pakistan after COVID crises. Where student's and teacher's perspectives have been considered. It is important to identify the unsettled health issues by observing the current circumstances of Pakistan to direct users for optimal use of technology by lowering the health risks.

3 Materials and Methods

To observe effects on health of OES, this qualitative research is focusing to gather student's and teacher's perceptions by conducting a survey in different higher educational institutes of Pakistan. Stages shown in Fig. 2 have been followed by authors to find the unsettled health related issues and also to observe the difference in between the collected responses.

Survey parameters or questions for this study are already mentioned in Fig. 1 which are collected with the help of literature review and expert opinions. Based on those

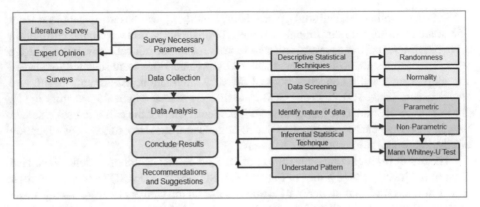

Fig. 2. Flow diagram of methodology

parameters, the questionnaire is designed. Other stages mentioned in Fig. 2 are discussed below.

3.1 Sample and Survey (Data Collection)

To avoid the biased thoughts and fuzzy texts, authors have decided to take the survey in multiple choice form to get specific data about selected health issues. Selected Likert scale for this research is divided into 5 options; strongly disagree (SD), disagree (D), neutral (N), agree (A) and strongly agree (SA). The options of scale are divided into three categories further to evaluate the responses of respondents as; negative side (containing SD and D options), positive side (having A and SA options) and neutral (which is middle phase of the scale). Further, coding has been done by authors to apply statistical techniques for converting available options into numbers to import dataset in SPSS for analysis. Weightage for SD, D, N, A, and SA is done as 1, 2, 3, 4 and 5 respectively through coding shown in Fig. 3. Prepared questionnaire is consists of total 9 SQs from selected health problems.

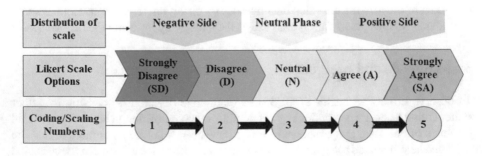

Fig. 3. Likert scale options, distribution and coding

To conduct this survey, 14 universities have been selected across Pakistan. This survey was conducted in the months of March-April 2022. Authors believed that 2022 is an ideal time to conduct this research because previous studies were carried out in the time when audience i.e. students and teachers have recently adopted OES and they were totally new to SW tools. While in 2022 same audience is experienced and used to towards online tools and have adopted new system. At this level the unsettled issues can easily be pointed by gathering their views. Surveys were distributed by sharing Google Forms link (to geographically distant universities) and by sharing the hard copies (in nearby universities). After conducting the survey in different universities of Pakistan, total 775 responses have been gathered. Where 122 teachers and 653 students have been participated. Main keywords of each SQ is shown in Table 1 along with nature.

Table 1. Survey questions of research, and its nature

Serial	Variables	Selected questions related to health issues	Nature
01	SQ1	Problem in eyes?	Qualitative
02	SQ2	Backache or spinal-cord issues?	Qualitative
03	SQ3	Gained weight?	Qualitative
04	SQ4	Use of caffeine (tea or coffee), drugs or alcohol?	Qualitative
05	SQ5	Increase in headache, tension or anxiety?	Qualitative
06	SQ6	Observed any change in sleeping and eating?	Qualitative
07	SQ7	Felt unmotivated or unfocused towards education?	Qualitative
08	SQ8	Felt continuous distraction from lectures?	Qualitative
09	SQ9	More frequently getting out of control, angry?	Qualitative

3.2 Data Analysis

To analyze the data, understand the insights of collected responses and find relationship in between perceptions of students and teachers some descriptive and inferential statistical techniques are used with the help of IBM SPSS (statistical package and social sciences) software tool. Before moving towards any statistical approach, dataset was formatted properly and coding has been done by authors for all available options of SQs shown in Fig. 3. The framework to apply statistical techniques for this research is followed from [21].

Descriptive Statistical Techniques. Collected data is analyzed by different descriptive statistical techniques like: frequencies, percentage, median, mode, variance, minimum, maximum, range and standard-deviation to see how data is scattered or deviated towards its central point, how responses of students and teachers are varying from each other, what percentage of reported health issues have been found from collected responses etc. Further, results are discussed in Sect. 4.

Data Screening. At this stage, randomness and normality of data have been checked by using RUNs and 1-Sample Kolmogorov Smirnov (K-S) tests respectively in SPSS environment where level of significance (α) was set to 0.05 with possible null and alternate hypothesis as follows.

H_0: *Data is random and normal*

H_1: *Data is not random and/or normal*

It was observed from the obtained results that data is random (as different universities have been selected and random audience have taken part in it also audience was having random choice for each SQs from 5 Likert scale options). But data was not normally distributed towards its mean.

Inferential Statistical Tests. As all nine selected SQs are qualitative in nature and data is not normal as well. Therefore, non-parametric test found to be suitable for this type of data. Responses are collected from students and teachers of various universities. So, Mann Whitney test for 2 independent samples is used to get further results and to observe the difference in between the perception of students and teachers. Initially suitable null and alternate hypothesis are decided as follows:

H_0: *There is no difference in the responses of students and teachers.*

H_1: *There is difference in the responses of students and teachers.*

And obtained p-values are compared with significance level (α) at levels 0.05 and 0.1.

4 Results and Discussion

As total 775 responses have been collected from which is 84.3% students i.e. 653 students and 15.7% teachers i.e. 122 faculty members and have participated. This percentage shows that majority students have participated in responses. First of all, percentage variation of students and teachers have been observed.

Figure 4 illustrates the percentage of responses of students only. Fair agreement can be seen in the given Fig of students for several health issues. Because, mostly respondents have selected A and SA options for all survey questions except SQ4. Which shows most of the students have faced health issues during the period of online education in COVID crises. Furthermore, more than 53% students have selected A and SA for survey questions SQ1, SQ2, SQ5, SQ6, SQ7 and SQ8. Which indicates most of the students have faced eyesight issues, spinal-cord problems, headache or migraine, change in their sleeping habits, unfocused towards their studies and continues distractions from lectures. Whereas, more than 50% students have selected D and SD for SQ4. Which means, they have not increased the amount of caffeine or drugs due to OES. Results of SQ3 shows that around 45% of students have gained weight because of online classes which may lead to other serious health problems due to overweight and obesity in future [7].

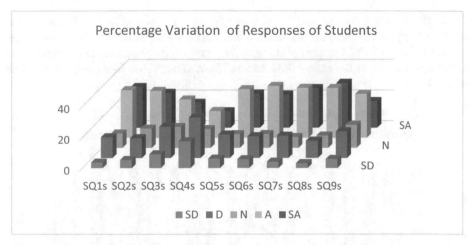

Fig. 4. Percentage variation of responses of students.

Figure 5 shows the percentages variations of responses of teachers only. It can be seen in the graph that higher number of teachers i.e. around 45% of teachers have agreed that health issues described in SQ1, SQ2, SQ3, SQ5, SQ6, SQ7, and SQ8 were faced by them in the period of OES. Specifically for SQ4, more than 40% of teachers have increased amount of caffeine whereas, around 27% were neutral and 33% were not satisfied with this statement. Apart from this, balanced percentage of faculty members for SQ9 have been selected for positive and negative sides of Likert scale. Which indicates almost 40% of teachers said they are getting out of control and 40% of teachers have disagreed from this point. So, no clear difference in the responses of teachers have been observed for SQ9.

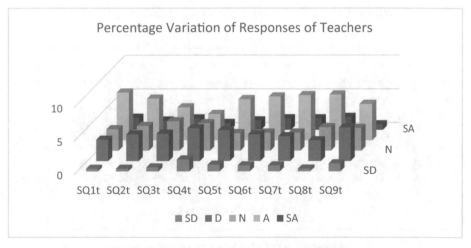

Fig. 5. Percentage variation of responses of teachers

In Fig. 6, responses are distributed based on the types of audience. It illustrates the results of responses separately for students and teachers with their individual percentage. It can be observed that the ratio of students and teachers for positive side of Likert-scale is dominant for every SQ except SQ4. Most of the audience have faced health issues due to OES.

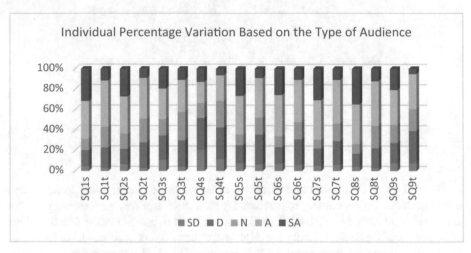

Fig. 6. Individual percentage variation based on the type of audience

Whereas, the graph shown in Fig. 7 demonstrate the shared percentage of students and teachers i.e. 84.3% and 15.7% respectively based on individual SQs.

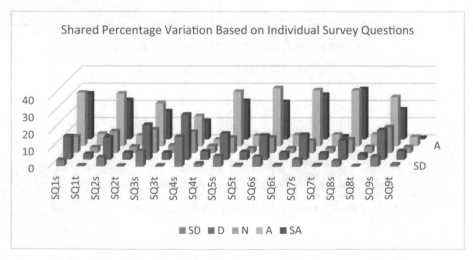

Fig. 7. Shared percentage variation based on individual survey questions

To further know about the data, some descriptive statistical techniques have been applied by authors. Median, mode, standard deviation, variance, range, minimum and maximum parameters are measured for each SQ to have an idea about how data is scattered. It can be seen in Fig. 8 that median of all SQs is same i.e. 4 (Agree) except SQ3 and SQ4. Whereas, the frequent selected option i.e. mode is again 4 (Agree) except SQ4. As all SQs are having same sequence of options which are in ordinal form and scaled from 1 to 5 as SD = 1, D = 2, N = 3, A = 4 and SA = 5. So minimum and maximum observations can simply be observed as 1 and 5 respectively. Whereas their difference indicates range for each SQs i.e. Max-Min = Range. So, 5–1 = 4. Which indicates range of all SQs is 4. So, for all SQs the range, minimum and maximum values are constant as 4, 1 and 5 respectively.

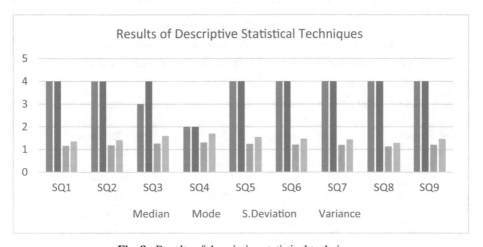

Fig. 8. Results of descriptive statistical techniques

As all selected SQs are qualitative in nature. So, main focus of authors were towards their median instead of mean. Figure 9 illustrates the median values for each SQ separately. Three bars for each SQ indicates the median of students' responses, median of teachers' responses and median of merged responses. It could be observed that there is no difference in the medians of students and teachers for SQ1, SQ3, SQ6, SQ7, and SQ8. Whereas in SQ2, SQ4 SQ5 and SQ9 teachers were neutral.

Table 2 shows the results of data screening phase. At level of significance (α) 0.05, authors got p-values. Based on which test type for each SQ is decided. It can be seen from the given table that data is almost random but not normal. So non-parametric test is applied for further analyses.

From Fig. 10, it can be seen that p-values for every SQ except SQ3 and SQ4 are zero. Therefore, alternate hypothesis is true which indicates except SQ3 and SQ4 there is no difference in responses of students and teachers. Even from above results dominant and major number of respondents reported health problems described in SQ1, SQ2, SQ5, SQ6, SQ7, SQ8 and SQ9. Therefore, it is evidenced by observing clear visuals from Fig. 1, Fig. 2 and Fig. 4 that respondents have assured that by adopting OES their health is affected. Whereas, for SQ3 and SQ4 null hypothesis is true. As shown in Fig. 7 that

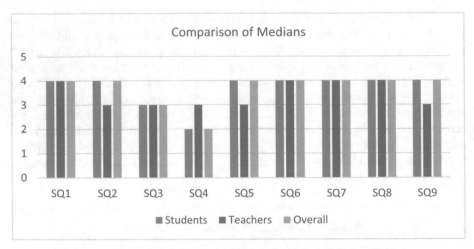

Fig. 9. Comparison of medians of each SQ in terms of students and teachers

Table 2. Randomness and normality of each SQ along with suitable test type.

Variable name	Randomness	Normality	Test type
	$\alpha = 0.05$	$\alpha = 0.05$	
SQ1	No	No	N-Parametric
SQ2	Yes	No	N-Parametric
SQ3	Yes	No	N-Parametric
SQ4	Yes	No	N-Parametric
SQ5	Yes	No	N-Parametric
SQ6	Yes	No	N-Parametric
SQ7	Yes	No	N-Parametric
SQ8	Yes	No	N-Parametric
SQ9	Yes	No	N-Parametric

observed p-values are lying in confidence region at $\alpha = 0.05$ and $\alpha = 0.1$. Which directs that there is difference in the responses of students and teachers unlikely other SQs than SQ3 and SQ4. As Fig. 4 indicates major number of students have gained weight whereas there is minor difference in positive and negative sides of scale for teachers in SQ3. However, there is no any major difference for SQ4 in positive and negative sides of scale for teachers.

Authors found very limited similar studies to identify effects on OES on health during COVID-19. Each study has been carried out for different sample sizes and different parameters. But only common observation for all studies has been found that more than 55%–80% of audience is affected from several health problems by adopting OES. Similarly, in Pakistan, more than half of the sample size is affected from same problems.

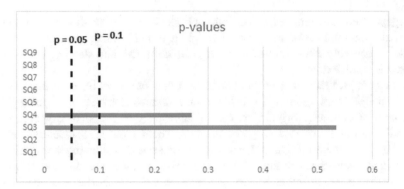

Fig. 10. Mann Whitney-U test results with p-values for each survey question

It is the time to adopt cares and precautions while using digital screens to avoid health problems.

5 Suggestions and Recommendations

By observing major positive responses for some health issues from most of the collected responses, it is suggested by authors that academies should arrange some awareness sessions in which all individuals should be guided properly. Following cares and precautions must be made compulsory for every individual during online classes as CVS, spine issues, headaches and other problems can be avoided to some extent. The awareness session must contain importance of being physically active, nutritional knowledge like; balanced diet (to avoid obesity), necessity of best sitting postures while using digital screens to avoid spine problems, neck pain, eye strain and headache [9–11], cares to use computers for eyes etc.

It is suggested by authors and proved by several experts that best sitting posture always help computer user to avoid neck pain, eye strain, spine issues and headache problems. A good posture is necessary to avoid tiredness and computer anxiety [33] for academic members when they are using online tools to conduct classes. Optimal sitting positions while using computers, laptops and digital screens are already discussed in several studies. Where thighs of user must be parallel to floor, there should be comfortable chairs for laptop users with back support, keyboard must be placed at elbow height, digital screens need to be placed at eye level as shoulder and neck pain can be avoided, there should be padding to give support and relaxation to wrists of users and footrest is also necessary to avoid heavy pain in feet during long sitting position to attend online classes for hours [32].

To maintain BMI, avoid weight gain and prevention of obesity, two important aspects needs to be involved in awareness sessions as all individuals can understand its seriousness, those are **Diet** and **Physical Activity** [14, 16]. For that, eating healthy and balanced diet; consumption of enough amount of carbohydrates, fats, protein and fiber on daily basis is very important [13, 15–17]. It is also necessary for students and teachers to engage themselves in highly physically active workouts or jobs and adopt regular physical exercise [13, 15–17] after attending online classes such as; home chores, walking, jogging, yoga etc. For females, being highly physically active for almost 60 min in a day is needed to maintain BMI and prevent weight gain [12]. Being physically active will definitely positively effect on health through which mood, and sleeping disorders will be cured to some extent.

To prevent problems in eyes and headache, frequent breaks from the digital screens [9–11], enough room lightning [9, 10], special computer glasses (blue fighter) [10, 11], adjustment of computer brightness, use of lubricant eye drops for dry eyes (by the suggestion of optometrist), anti-glared filters for the screens of computers, laptops or PCs to be safe from harmful rays [10] are very necessary. By following these instructions students and teachers can be saved from several abnormalities and diseases of eyes. Eye strain usually creates different types of headaches which will also be reduced when these cares are adopted.

6 Conclusion

This study focused on exploring the health related issues with causes and effects while going through OES as compared to the conventional TES, especially during and after the COVID'19 crises. The analysis and investigation was based on survey of the literature to identify health related issues in such situations, and then the perceptions of students and teachers on similar issues. The perceptions have been analyzed statistically using tools of descriptive and inferential statistics. Thus, relevant indicators have been attained to provide an overall overview and severity based on comparative analysis. By going through the detailed analyses and observations of all results after collecting responses from different higher educational institutes of Pakistan, it has been concluded that by adoption of OES, the drastic increment in different health problems such as: CVS, spine issues, headache, weight gain, unfocused towards studies, getting angry, change in sleeping order and continuous distraction were excessively reported by students and teachers. These issues may lead to other dangerous diseases if not controlled, prevented or treated on time. From this study, it was observed that recognized problems in students and faculty members were not new, and match with several existing studies, as highlighted in Sect. 2, which were carried out at different places to report such issues. While everyone is lacking to adopt the optimal and efficient solution to reduce or avoid such issues, it is suggested that if some formal awareness programs or sessions are arranged in which seriousness of these issues and importance of healthful living is highlighted in universities and some safeguards and careful measures mentioned in Sect. 5 are made mandatory for all individuals to be adopted during online classes, then these problems can be avoided to some extent. As technology is embedded in such a way to complete several tasks of daily routine in our lives that it cannot be removed or separated [10]. Therefore, users should avoid the incorrect and excess usage of such devices and guidelines mentioned in Sect. 5 should be practiced, adopted and shared with others as well to prevent serious problems as weak eye sight (due to CVS), restricted towards several physical activities (due to spine problems), to become irritate and unfocused (due to headache and change in sleeping order) and diabetes, blood pressure, inactiveness and laziness (due to obesity caused by continuous increase in weight). Furthermore, these problems may lead to its worst towards different types of cancers and early deaths. Hence, it is mandatory to control such problems at initial phases to avoid further difficulties.

References

1. Dhawan, S.: Online learning: a panacea in the time of COVID-19 crisis. J. Educ. Technol. Syst. **49**(1), 5–22 (2020)
2. Hussain, I.: A study of student's attitude towards virtual education in Pakistan. Turk. Online J. Distance Educ. **8**(2), 69–79 (2007)
3. Chakraborty, P., et al.: Opinion of students on online education during the COVID-19 pandemic. Hum. Behav. Emerg. Technol. **3**(3), 357–365 (2021)
4. Chandra, Y.: Online education during COVID-19: perception of academic stress and emotional intelligence coping strategies among college students. Asian Educ. Dev. Stud. (2020)
5. Bisht, R.K., Jasola, S., Bisht, I.P.: Acceptability and challenges of online higher education in the era of COVID-19: a study of students' perspective." Asian Educ. Dev. Stud. (2020)
6. Alam, M.: Mental health impact of online learning: a look into university students in Brunei Darussalam. Asian J. Psychiatr. **67**, 102933 (2022)
7. Lee, I.-M., et al.: Effect of physical inactivity on major non-communicable diseases worldwide: an analysis of burden of disease and life expectancy. Lancet **380**(9838), 219–229 (2012)
8. Das, P., Horton, R.: Rethinking our approach to physical activity. Lancet (London, Engl.) **380**(9838), 189–190 (2012)
9. Kumar, N., Sharma, N.: To determine the prevalence of computer vision syndrome among computer users: a descriptive study. Eur. J. Mol. Clin. Med. (EJMCM) **7**(10), 2020 (2020)
10. Blehm, C., et al.: Computer vision syndrome: a review. Surv. Ophthalmol. **50**(3), 253–262 (2005)
11. Adane, F., Alamneh, Y.M., Desta, M.: Computer vision syndrome and predictors among computer users in Ethiopia: a systematic review and meta-analysis. Trop. Med. Health **50**(1), 1–12 (2022)
12. Lee, I.-M., et al.: Physical activity and weight gain prevention. JAMA **303**(12), 1173–1179 (2010)
13. Strong, K.A., et al.: Weight gain prevention: identifying theory-based targets for health behavior change in young adults. J. Am. Diet. Assoc. **108**(10), 1708–1715 (2008)
14. Laska, M.N., et al.: Interventions for weight gain prevention during the transition to young adulthood: a review of the literature. J. Adolesc. Health **50**(4), 324–333 (2012)
15. Levine, M.D., et al.: Weight gain prevention among women. Obesity **15**(5), 1267–1277 (2007)
16. Fulton, J.E., et al.: Interventions for weight loss and weight gain prevention among youth. Sports Med. **31**(3), 153–165 (2001)
17. Baranowski, T., et al.: Are current health behavioral change models helpful in guiding prevention of weight gain efforts? Obes. Res. **11**(S10), 23S-43S (2003)
18. Rosenfield, M.: Computer vision syndrome: a review of ocular causes and potential treatments. Ophthalmic Physiol. Opt. **31**(5), 502–515 (2011)
19. Wang, C., Zhao, H.: The impact of COVID-19 on anxiety in Chinese university students. Front. Psychol. **11**, 1168 (2020)
20. Rizun, M., Strzelecki, A.: Students' acceptance of the COVID-19 impact on shifting higher education to distance learning in Poland. Int. J. Environ. Res. Publ. Health **17**(18), 6468 (2020)
21. Sultan, M., Shaikh, M.M., Chowdhry, N.P.: Comparative analysis of knee joint replacement and stem cells therapy treatment for knee osteoarthritis using statistical techniques. Res. Med. Eng. Sci. **10**(4), 887–897 (2020)
22. Lei, S.I., So, A.S.I.: Online teaching and learning experiences during the COVID-19 pandemic–a comparison of teacher and student perceptions. J. Hosp. Tour. Educ. **33**(3), 148–162 (2021)

23. Foz, J.B.C., et al.: An application of analytical hierarchy process in the comparison of zoom, Google meet, and MS teams (2021)
24. Ma, L., et al.: Practice and thinking of online teaching during epidemic period. In: 2020 15th International Conference on Computer Science and Education (ICCSE). IEEE (2020)
25. Behzadi, Z., Ghaffari, A.: Characteristics of online education and traditional education. Life Sci. J. **8**(3), 54–58 (2011)
26. Mukhtar, K., et al.: Advantages, limitations and recommendations for online learning during COVID-19 pandemic era. Pak. J. Med. Sci. **36**(COVID19-S4), S27 (2020)
27. Muthuprasad, T., et al.: Students' perception and preference for online education in India during COVID-19 pandemic. Soc. Sci. Human. Open **3**(1), 100101 (2021)
28. Adedoyin, O.B., Soykan, E.: Covid-19 pandemic and online learning: the challenges and opportunities. Interact. Learn. Environ. 1–13 (2020)
29. Hussein, E., et al.: Exploring undergraduate students' attitudes towards emergency online learning during COVID-19: a case from the UAE. Child Youth Serv. Rev. **119**, 105699 (2020)
30. Adnan, M., Anwar, K.: Online learning amid the COVID-19 pandemic: students' perspectives. Online Submission **2**(1), 45–51 (2020)
31. Schneider, S.L., Council, M.L.: Distance learning in the era of COVID-19. Arch. Dermatol. Res. **313**(5), 389–390 (2020). https://doi.org/10.1007/s00403-020-02088-9
32. Peper, E., Lin, I.-M., Harvey, R.: Posture and mood: implications and applications to therapy. Biofeedback **45**(2), 42–48 (2017)
33. Epstein, R., et al.: The effects of feedback on computer workstation posture habits. Work **41**(1), 73–79 (2012)
34. Shaikh, M.M., Dahri, A.S.: Effect of COVID-19 epidemic on research activity of researcher in Pakistan Engineering University and its solution via technology. 3C Technol. Innov. Gloss. Appl. SMEs 249–263 (2020)

Testing the Performance of Feature Selection Methods for Customer Churn Analysis: Case Study in B2B Business

Semanur Sancar[1(✉)] and Meryem Uzun-Per[1,2]

[1] BiletBank R&D Center, Akdeniz PE-TUR A.S., Istanbul, Turkey
{semanur.sancar,meryem.uzunper}@petour.com
[2] Department of Computer Engineering, Istanbul Health and Technology University, Istanbul, Turkey
meryem.uzunper@istun.edu.tr

Abstract. Churn analysis has recently become one of the favorite topics of marketing teams with the development of machine learning models. This study aims to discover the most suitable feature selection (FS) model for churn analysis by using the databases of BiletBank, a business-to-business (B2B) company. It was found that some categorical data such as agency type and currency used by customers, along with periodic flight sales data, are also meaningful features for churn analysis in the BiletBank customer portfolio. This feature selection study in the database will be a source for future churn analysis studies.

Keywords: Customer churn analysis · Feature selection · B2B · Sequential forward selection · Sequential backward selection · Classification · Logistic regression · Support vector machines · Random forest classifier · Extra tress classifier

1 Introduction

After the effects of digitization in the industry, one of the most critical agenda items has become data science. Cleaning, analyzing, and making sense of the data kept in the databases of businesses can provide significant benefits. One of the most important of these analyzes is the customer churn analysis. Churn analysis makes it possible to look at customers' various characteristics and interactions with the system and report churn probability. Businesses that can detect the churn status of customers can develop their marketing strategies accordingly because retaining existing customers is less costly than acquiring new customers.

Various methods are used for churn analysis in the literature. These methods evolve from classical statistical approaches to machine learning (ML) models. ML models need big data to train. Therefore, the literature is more advanced for business-to-customer (B2C) [1,2] firms with a rich customer pool than for business-to-business (B2B) firms [3]. B2B companies have fewer customers and their customer behavior can be different than B2C companies.

F. P. García Márquez et al. (Eds.): ICCIDA 2022, LNNS 643, pp. 509–519, 2023.
https://doi.org/10.1007/978-3-031-27099-4_39

The relationship of the feature set with the prediction target is the most crucial factor affecting the performance of ML algorithms. Therefore, both the collection and cleaning of the data and the correct selection of features directly affect the accuracy of the ML algorithms. Sequential Forward Selection (SFS) and Sequential Backward Selection (SBS) methods are popular feature selection methods, also known as wrapper methods [4]. In the SFS model, the features are added to the model one by one, while in the SBS model, the features are removed from the model one by one [5]. Thus, the best feature set that maximizes model performance can be obtained. Various models can be used to measure the performance of feature sets when applying the SFS and SBS methods. Models such as Random Forest (RF) [6], SVM [7], and ETC [8] took place in the literature.

The ML methods used for the churn analysis of B2B companies are diverse. Tamaddoni Jahromi et al. [3] used a cost-sensitive decision tree, simple decision tree (DTs), and boosting methods for churn analysis in a B2B firm. However, they used the logistic regression (LR) method as a benchmark. In [9], LR, DT analysis, artificial neural network (ANN), and support vector machine (SVM) approaches were used for churn analysis in a B2B company. Similarly, in study [10], LR, SVM, and boosting models were used to make a churn analysis of a B2B firm in the hotel industry.

In this study, it is aimed to discover the optimal feature selection method for the churn analysis of BiletBank B2B business within the scope of R&D center research. SFS and SBS methods were used to find the proper feature set. Tests of SFS and SBS models were performed with SVM, LR, RF, and ETC models. The LR model, generally used as a benchmark in the literature, was used for the churn analysis performance analysis of feature sets suggested by feature selection approaches. The contributions of the study to the literature can be listed as follows:

- A rich data set was created by collaborating with various business units of BiletBank.
- Time-dependent interactions of customers were evaluated separately for different periods.
- SFS and SBS methods were compared with four different models and eighth different feature sets were evaluated.
- The model's feature sets were compared with the LR model.
- The most important churn analysis features in B2B companies have been identified.

Section 2 describes the methodology of this study. Along with a detailed explanation of SFS and SBS methods, ML models that provide binary classification and metrics used to measure accuracy performance are explained. Section 3 includes results and discussion. The dataset is clearly defined in this section. The features suggested by the feature selection models are discussed and their accuracy performances are compared. Finally, we conclude the study in Sect. 4.

2 Methodology

BiletBank is a company that provides services in the field of tourism at an international level. It has large databases that contain both the information of tourism agencies and the records of consumers such as flight ticket purchases. Big data, machine learning and deep learning studies are carried out with this database such as [11,12], and [13]. Churn analysis is performed based on various activities of customers. A study was also conducted [14] using a portion of the customers' flight dataset. Indeed, one of the most critical parts of churn analysis is data collection and feature selection. Increasing the feature set may affect the model positively in all machine learning models. However, overfitting problems may occur after a point. As part of the churn analysis of BiletBank, a B2B company, feature sets were evaluated with various feature selection methods.

These feature selection methods are SFS and SBS, known as wrapper methods. As a working principle, SFS measures the accuracy performance of the model by adding the features one by one. It continues by adding the feature that will provide the highest performance increase in the model [15]. On the other hand, SBS first starts to measure model performance with all the features in the data set. Then, it makes feature selection that will increase the accuracy performance of the model by extracting features one by one.

The ML models in the "sklearn" library can be used as a function to measure the performance of adding-removing features in SFS and SBS models. In this study, SVM, LR, RF, and ETC models were used for performance tests.

With the feature selection models, it was envisaged to determine the optimal number of features and to select the efficient features to be used from the data set. The logistic regression model is often used as a benchmark in binary classification problems such as churn analysis [16,17]. This study aims to test the performance of the feature sets used in the churn analysis model with logistic regression.

Various metrics were used when comparing churn analysis results according to data sets. The metrics used are "accuracy", "F1-score", "precision", and "recall". "Churn" was represented as a positive label, and "non-churn" was represented as a negative label when calculating "F1-score", "precision", and "recall". The general structure of this study, which includes the evaluation of feature sets for customer churn analysis, is summarized in Fig. 1.

3 Results and Discussion

The data sets were collected from different business units of BiletBank to make sense of the customers with the goal of Churn analysis. In this context, "Search-Count" represents the flight ticket search of the customers, "SaleCount" represents the flight ticket sales, "sale/search" obtained from the ratio of these two data sets, the chat times of the customers in the "onlinechat" system, the customer call method "Helpdesk- Ticket", "SessionCounts" representing the number of customers' log in, and "HotelSearch", which represents the number of customers searching for hotel reservations, were collected from BiletBank database

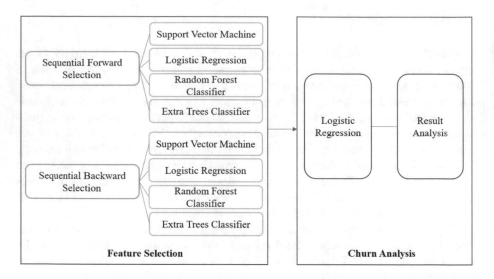

Fig. 1. General structure of the study.

in terms of the daily transaction number. In addition, some categorical data representing the countries-cities where the customers are located, the currencies they use, the year-month they are included in the system, and their activities were also added to the data set collected for churn analysis as the features of the customers.

The "Churn" label was created by evaluating the customers' churn status with the marketing team. The average number of flight ticket sales in the first six months of 2014–2021 and the number of flight ticket sales in the first six months of 2022 were calculated for each customer. First, the customer was designated "Churn" if the number of flight ticket sales in the first six months of 2022 fell below a specific rate. In the next step, the marketing team re-evaluated the "Churn" labeling in the client list. The marketing team changed the label of "Churn" for customers whose sales potential has decreased for different reasons but who are not churning. All these data sets were combined to create a rich feature pool for churn analysis.

It was aimed to determine the optimal feature set for churn analysis by establishing eight models based on SFS and SBS methods. First, the SFS and SBS methods were integrated with the SVM, LR, RF, ETC models. Due to these eight models, the features that affect the churn analysis from the whole data set were tried to be evaluated. Exemplarily, Fig. 2 and Fig. 3 show periodic performance changes of the RF model over SBS and SFS, and Fig. 4 and Fig. 5 represent the performance profiles of ETC models. Performance increases periodically as the feature is removed in the SBS model. This can be interpreted as the low performance of the full feature set due to overfitting, and the performance increase as the feature is reduced. The estimation performance decreases when the feature set falls below a certain number. The SBS method's SVM,

LR, RF, and ETC models reached the peak performance value with 10, 8, 19, and 10 features, respectively. In the SFS method, on the other hand, the model accuracy performance moves in the opposite direction of the SBS method as the number of features increases. While an increase in performance is observed from adding the first feature to a certain number of features, accuracy performance starts to decrease due to overfitting after a certain number of features. With the SFS method, the SVM, LR, RF, and ETC models suggested 12, 46, 9, and 10 features for the highest performance. The SFS and SBS methods help find the optimal number of features and the most accurate combination of features. In this study, the most compelling feature combinations were determined with eight different models.

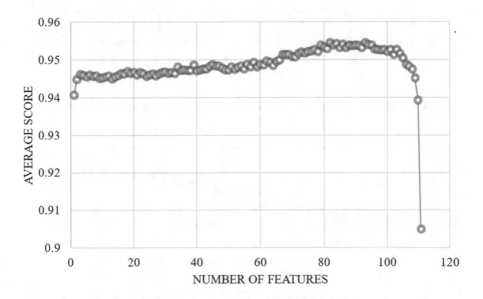

Fig. 2. Performace of the SBS-RF model based on the number of the features.

The selected features and how frequent these features are selected by the feature selection algorithms are listed in Table 1. The maximum number for a feature to be selected was four. The most selected features were "2018 SaleCount", "2022 SaleCount", "LastThreeMonth SaleCount", "LastMonth sales/search", and "DefaultCurrency". "SaleCount" is periodic data representing the number of customers' flight ticket sales. Since BiletBank is a company that focuses on flight ticket sales, it is expected that flight sales will affect loyalty of the customers. The only categorical data selected by the four models was the "DefaultCurrency" feature. "DefaultCurrency" consists of the Turkish lira, euro, and dollar currencies and can provide indirect information about the customer's region.

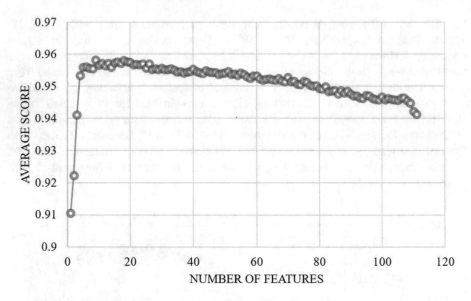

Fig. 3. Performace of the SFS-RF model based on the number of the features.

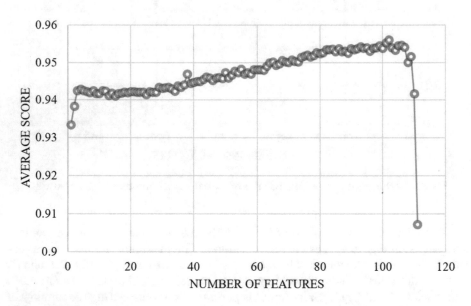

Fig. 4. Performace of the SBS-ETC model based on the number of the features.

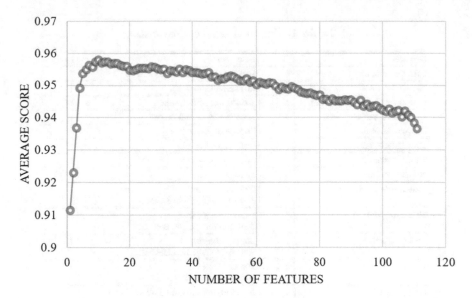

Fig. 5. Performace of the SFS-RF model based on the number of the features.

The feature set selected by the three models generally consists of the number of flight sales of the customers. In addition, "Status IfActive", "AgencyType", and "GroupCode" categorical features were also chosen by three models. "Status IfActive" gives if the customer is still working with our company or not in the current date. While the "churn" tag isn't exactly determined by this feature, it's reasonable to see an impact on the CA. "AgencyType" defines the characteristics of the customer, such as being a B2C or a member of The International Air Transport Association. "GroupCode" was determined by the BiletBank marketing team considering the domestic and international sales intensity of the customers. Determining the effect of the aforementioned categorical data on churn analysis will greatly contribute to future churn analysis studies.

"LastMonth_HotelSearch" and "CreationYear" were the features removed from the model in the top 20 of three SBS models. As mentioned before, since BiletBank is a flight ticket sales-oriented company, it can be expected that the "LastMonth_HotelSearch" feature will have little impact on CA. However, the "CreationYear" feature, which represents the year customers join BiletBank systems, shows that the customer churn status is independent of the customer registration date. "CreationYear" was also one of the last features in SFS models. Another feature not preferred to be selected in SFS models was the "Session-Counts" features. "SessionCounts" gives the number of times the customer has logged into the BiletBank system. However, in the past, if the customer was

Table 1. Common selected features.

Features	Count	Features	Count
2018_SaleCount	4	LastSixMonth_SearchCount	1
2022_SaleCount	4	2014_SaleCount	1
LastThreeMonth_SaleCount	4	LastMonth_SaleCount	1
LastMonth_sales/search	4	LastTwoYear_SaleCount	1
DefaultCurrency	4	2014_sales/search	1
2019_SaleCount	3	2015_sales/search	1
2020_SaleCount	3	2017_sales/search	1
2021_SaleCount	3	2020_sales/search	1
LastOneYear_SaleCount	3	2019_onlinechat-Acente_Sohbet_icin_Bekledi	1
2022_sales/search	3	2020_onlinechat-Acente_Sohbet_icin_Bekledi	1
Status_IfActive	3	2021_onlinechat-Acente_Sohbet_icin_Bekledi	1
AgencyType	3	2022_onlinechat-Acente_Sohbet_icin_Bekledi	1
GroupCode	3	LastThreeMonth_onlinechat-Acente_Sohbet_icin_Bekledi	1
2016_SearchCount	2	LastTwoYear_onlinechat-Acente_Sohbet_icin_Bekledi	1
2022_SearchCount	2	Total_onlinechat-Acente_Sohbet_icin_Bekledi	1
LastMonth_SearchCount	2	2022_onlinechat-Toplam_Sohbet_Suresi	1
LastThreeMonth_SearchCount	2	LastMonth_onlinechat-Toplam_Sohbet_Suresi	1
LastOneYear_SearchCount	2	LastThreeMonth_onlinechat-Toplam_Sohbet_Suresi	1
LastTwoYear_SearchCount	2	LastSixMonth_onlinechat-Toplam_Sohbet_Suresi	1
LastSixMonth_SaleCount	2	LastTwoYear_onlinechat-Toplam_Sohbet_Suresi	1
Total_SaleCount	2	2019_Helpdesk-Ticket-Count	1
2019_sales/search	2	2020_Helpdesk-Ticket-Count	1
LastThreeMonth_sales/search	2	2022_Helpdesk-Ticket-Count	1
LastSixMonth_sales/search	2	LastMonth_Helpdesk-Ticket-Count	1
LastOneYear_sales/search	2	LastThreeMonth_Helpdesk-Ticket-Count	1
2018_Helpdesk-Ticket-Count	2	LastSixMonth_Helpdesk-Ticket-Count	1
2021_Helpdesk-Ticket-Count	2	LastOneYear_Helpdesk-Ticket-Count	1
2015_SessionCounts	2	2020_SessionCounts	1
2022_HotelSearch	2	LastMonth_SessionCounts	1
LastOneYear_HotelSearch	2	LastThreeMonth_SessionCounts	1
Level	2	LastSixMonth_SessionCounts	1
Status_IfForbidden	2	LastMonth_HotelSearch	1
2014_SearchCount	1	LastThreeMonth_HotelSearch	1
2015_SearchCount	1	LastSixMonth_HotelSearch	1
2017_SearchCount	1	LastTwoYear_HotelSearch	1
2020_SearchCount	1	Location_City	1
2021_SearchCount	1	Location_Country	1

inactive for a while in the BiletBank system, the customer was logged out. According to the new structure, even if the customer is inactive for a long time, there is no need to log in again unless the customer logs out. Finally, it was observed that the previous "SessionCounts" data was more effective for churn analysis than the recent "SessionCounts" data.

The LR model was applied to the dataset containing all the features, and the model's "accuracy" was 0.8184, "F1-score" was 0.8942, "precision" was 0.9868, and "recall" was 0.8175. After that, the LR model applied to different feature sets selected by SFS and SBS methods are given in Table 2 and Table 3, respectively. Performance metrics used for comparison were determined as "accuracy", "F1-score", "precision", and "recall". Based on the performance results, it can be said that SFS models show higher performance than SBS models. "Accuracy" calculates the accuracy of predictions. However, "F1-score" comes to the fore for classification algorithms. The "F1-score" is the harmonic mean of the "precision" and "recall". "Precision" indicates how many of the customers estimated as "churn" are actually "churn". "Recall" represents how many of the customers who are actually "churn" correctly predicted as "churn". Finally, the "F1-score" indicates how balanced these two metrics are and is used to measure the performance of classification algorithms. When all metrics are evaluated, it is seen that the ETC model has the highest performance among both SFS and SBS methods. When churn analysis was performed with the help of the LR model based on the entire data set, the "F1-score" was calculated as 89.42%. This rate increased to 93.98% after feature selection with the SBS-ETC model. Thanks to the feature selection made with the SFS-ETC model, the "F1-score" performance increased by approximately 7% and became 96.24%.

In both feature selections using the ETC model, ten features were selected. Five of these features are the same in both models. These features, which cover some periods of 2018, 2020, and 2022 are composed of "SaleCount" data. As mentioned before, the "churn" labeling was developed with the cooperation of the marketing team by evaluating the flight ticket sales. Within the scope of this identification approach, the effect of the number of flight sales in the previous three months, the previous six months, the last one year, and the recent years on the churn estimation is a admissible result.

Table 2. LR churn analysis performance results on features selected by SFS method.

Feature selection model	SFS-ETC	SFS-LR	SFS-RF	SFS-SVM
Accuracy	0.9414	0.9253	0.8785	0.8829
F1-Score	0.9624	0.9525	0.9261	0.9283
Precision	0.9652	0.9633	0.9793	0.9755
Recall	0.9597	0.942	0.8784	0.8855

Table 3. LR churn analysis performance results on features selected by SBS method.

Feature selection model	SBS-ETC	SBS-LR	SBS-RF	SBS-SVM
Accuracy	0.9026	0.8302	0.8426	0.8536
F1-Score	0.9398	0.8996	0.8926	0.9014
Precision	0.9783	0.9783	0.8409	0.8606
Recall	0.9043	0.8325	0.951	0.9462

4   Conclusion

In this study, data sets that may be meaningful for the churn analysis of a B2B company working in the tourism sector were investigated. Within the scope of the research, categorical features indicating the identity characteristics of customers and datasets representing their periodic interactions were evaluated. SVM, LR, RF, and ETC models were used to compare SFS and SBS feature selection methods. The accuracy performances for churn analysis of eighth data sets obtained from the feature selection processes were tested with the LR model. While the "F1-score" from the whole data set was 0.8942, the "F1-score" increased to 0.9624 with the data set suggested by the SFS-ETC model. Thus, feature selection methods were compared, and optimal feature sets were obtained for future churn analysis studies. It is planned to discover the most suitable estimation model with the selected feature sets for churn analysis by comparing LR, SVM, RF, neural networks with different layers, decision tree classifier, and boost models.

References

1. Xiahou, X., Harada, Y.: B2C E-commerce customer churn prediction based on k-means and SVM. J. Theor. Appl. Electr. Commerce Res. **17**(2), 458–475 (2022). https://www.mdpi.com/0718-1876/17/2/24
2. Huang, E.Y., Tsui, C.j.: Assessing customer retention in B2C electronic commerce: an empirical study. J. Mark. Anal. **4**(4), 172–185 (2016). https://doi.org/10.1057/s41270-016-0007-x
3. Tamaddoni Jahromi, A., Stakhovych, S., Ewing, M.: Managing B2B customer churn, retention and profitability. Ind. Mark. Manage. **43**(7), 1258–1268 (2014)
4. Karunakaran, V., Rajasekar, V., Joseph, S.I.T.: Exploring a filter and wrapper feature selection techniques in machine learning. In: Smys, S., Tavares, J.M.R.S., Bestak, R., Shi, F. (eds.) Computational Vision and Bio-Inspired Computing. AISC, vol. 1318, pp. 497–506. Springer, Singapore (2021). https://doi.org/10.1007/978-981-33-6862-0_40
5. Ibrahim, N., Hamid, H., Rahman, S., Fong, S.: Feature selection methods: Case of filter and wrapper approaches for maximising classification accuracy. Pertanika J. Sci. Technol. **26**, 329–340 (2018)
6. Genuer, R., Poggi, J.M., Tuleau-Malot, C.: Variable selection using random forests. Pattern Recogn. Lett. **31**(14), 2225–2236 (2010)

7. Patel, A.K., Chatterjee, S., Gorai, A.K.: Development of a machine vision system using the support vector machine regression (SVR) algorithm for the online prediction of iron ore grades. Earth Sci. Inf. **12**(2), 197–210 (2018). https://doi.org/10.1007/s12145-018-0370-6

8. Gaur, V., Kumar, R.: Analysis of machine learning classifiers for early detection of DDoS attacks on IoT devices. Arab. J. Sci. Eng. **2**, 1–22 (2021). https://doi.org/10.1007/s13369-021-05947-3

9. Chen, K., Hu, Y.H., Hsieh, Y.C.: Predicting customer churn from valuable B2B customers in the logistics industry: a case study. Inf Syst E-Bus Manage **13**(3), 475–494 (2015)

10. Wit, D.: An analysis of non-contractual churn in the B2B hotel industry. Master's thesis, Tilburg University (2017)

11. Uzun-Per, M., Gürel, A.V., Can, A.B., Aktas, M.S.: An approach to recommendation systems using scalable association mining algorithms on big data processing platforms: a case study in airline industry. In: 2021 International Conference on Innovations in Intelligent SysTems and Applications (INISTA), pp. 1–6 (August 2021)

12. Uzun-Per, M., Gurel, A.V., Can, A.B., Aktas, M.S.: Scalable recommendation systems based on finding similar items and sequences. Concurr. Comput. Pract. Exp. **34**(20), e6841 (2022), https://onlinelibrary.wiley.com/doi/abs/10.1002/cpe.6841, _eprint: https://onlinelibrary.wiley.com/doi/pdf/10.1002/cpe.6841

13. Can, A.B., Uzun-Per, M., Aktas, M.S.: A Novel Sequential Pattern Mining Algorithm for Large Scale Data Sequences. In: Gervasi, O., Murgante, B., Misra, S., Rocha, A.M.A.C., Garau, C. (eds.) Computational Science and Its Applications - ICCSA 2022 Workshops, pp. 698–708. Springer International Publishing, Cham, Lecture Notes in Computer Science (2022). https://doi.org/10.1007/978-3-031-10536-4_46

14. Uzun-Per, M., Can, A.B., Volkan Gürel, A., Aktaş, M.S.: Big data testing framework for recommendation systems in e-science and E-commerce domains. In: 2021 IEEE International Conference on Big Data (Big Data), pp. 2353–2361 (December 2021)

15. Rodriguez-Galiano, V.F., Luque-Espinar, J.A., Chica-Olmo, M., Mendes, M.P.: Feature selection approaches for predictive modelling of groundwater nitrate pollution: an evaluation of filters, embedded and wrapper methods. Sci. Total Environ. **624**, 661–672 (2018), https://www.sciencedirect.com/science/article/pii/S0048969717335751

16. De Bock, K.W., De Caigny, A.: Spline-rule ensemble classifiers with structured sparsity regularization for interpretable customer churn modeling. Decis. Support Syst. **150**, 113523 (November 2021). https://www.sciencedirect.com/science/article/pii/S0167923621000336

17. Gattermann-Itschert, T., Thonemann, U.W.: How training on multiple time slices improves performance in churn prediction. Euro. J. Oper. Res. **295**(2), 664–674 (2021). https://www.sciencedirect.com/science/article/pii/S037722172100463X

Multi-agents Path Planning for a Mobile Robot in a Dynamic Warehouse Environment

Mustafa Mohammed Alhassow[(✉)], Oguz Ata, and Dogu Cagdas Atilla

Electrical and Computer Engineering, Altinbas University, Istanbul, Turkey
mustafaalshalhe@gmail.com, {oguz.ata,
Cadas.atilla}@altinbas.edu.tr

Abstract. Route planning in robotic systems is a critical and complex task in any environment. Robotic systems allow multiple robots to accomplish multiple goals simultaneously. Many mobile service robots are now used in warehouses to reduce operating and overhead costs. Large warehouses may have multiple robots to handle a large number of tasks. Route planning means finding the best route, i.e. the route without collisions. Optimizing both parameters can be a daunting task. By properly addressing the problem of route planning between robots, we can improve the efficiency of the operation of the entire warehouse. At the beginning, every robot will navigate to its desired goal by funding the optimal route without collisions with other robots. In this work, a relative study with the notable route plan was presented. The proposed intelligent approach was presented for a multi-robot system that finds the best collision-free path in the warehouse and processes the storage box. This paper proposes a sensible variety metric for multi-robotic structures to intelligently become aware of goals and take the best minimum paths to attain them without encountering collisions. Using an intelligent variety metric to discover the route that we want to reach our goal. The proposed planning path are similar to different works including A *, RNN, PRM and heuristics. Three exclusive times of the warehouse have been taken into consideration to carry out experiments with parameters including route length, common route, and elapsed time. Experiments with 800 pods and sixteen robots have said overall performance enhancements of as much as 2.3%, common route length, and elapsed time of 11%
.

Keywords: Warehouse · Mobile robot · Dynamic environment · Conflict algorithm

1 Introduction

Mobile robot systems need to be properly tuned to perform difficult and complex tasks such as managing goods and services in warehouses [1]. Tring to solve multiple goals at the same time in a common environment can be a daunting task. In such cases, two important parameters are needed to maximize the optimization of some robots. The first is to derive the greatest route for every robotic, and the second one is to hold among the robots. There is an honest quantity of labor performed withinside the region of robot route

© The Author(s), under exclusive license to Springer Nature Switzerland AG 2023
F. P. García Márquez et al. (Eds.): ICCIDA 2022, LNNS 643, pp. 520–534, 2023.
https://doi.org/10.1007/978-3-031-27099-4_40

making plans, however maximum awareness on an unmarried robotic, and a few make bigger those principles to multi-robotic systems. However, now while trajectory-making plans is carried out to a multi-robotic system, there's some other hassle that desires to be addressed thoroughly, known as coordination. Also supplied are a few works proposed by researchers concerning adjustment schemes [2–4]. Merging both problems in parallel may need an intelligent approach to achieving concurrence between these two parameters. Given the complexity of MAPF, the maximum of the studies associated with multi-agent structures have centered on fixing an easy sub-trouble, multi-agent pathfinding (MAPF). The pathfinding trouble operates in a discrete area, instead of the non-stop area taken into consideration in movement-making plans troubles. The assumption of a not unusual place illustration for all dealers and the homogeneity of nation transition periods frequently enforced through MAPF strategies save them from being immediately carried out to non-stop-area MAMP domains. Therefore, many real-international troubles including high-degree-of-freedom manipulators and heterogeneous multi-agent groups cannot be solved through MAPF approach individually, due to the fact dealers rely upon particular international representations. Several MAMP strategies, including Neighborhood Grid Discretization [5], map their motions to a not unusual place community illustration to address disjoint agent nation-area representations. Regrettably, the high-satisfactory of answers for this form of approach is surprisingly dependent on the surroundings illustration. MAMP/PF has general approaches: coupling and isolation. A blended approach can offer a first-class solution however, it searches for a shared nation area. Decoupled strategies discover every robot's person nation area independently, however, do now no longer assure completeness or optimality due to the fact they discover the person agent's nation areas in isolation earlier than combining them later. In this paper we provide an effective multi-agent path MAPP to address the issue that generalizes current path planning issues in a warehouse system. As a re-planning algorithm, Conflict-Based Search (CBS) generates the path and develops the state spaces. We demonstrated how our approach outperforms MAP-CBS solvers in terms of both planning time and solution quality. We compare our work to other pertinent works to validate it. Also, we've tried to provide a detailed definition of all the important aspects related to planning and tuning multi-robot paths. In addition, this section contains some notable contributions and related literature reviews. Our paper will be organized as Sect. 2 illustrate the overview and related works. Section 3 will illustrate the problem definition, Sect. 4 will illustrate our proposed solution, Sect. 5 will illustrate the result analysis and the comparison with the other related research.

2 Related Work

In this part, we tried to collect the nearest research that has a relation to our proposed work. Initially, robotic path planning was carried out only for some industrial production purposes. Since then, robotics has become widely accepted in different area of research, with mutable classification of robots being used to perform tasks and growing works that related to path planning algorithms. Planning algorithms have an extended history. From heuristic tactics to fixing route-making plans troubles to evolutionary and hybrid algorithms to discover the excellent route from supply to vacation spot while robots generally tend to transport in selected configuration space. Multiple-robot systems (MR)are used to perform complex tasks. Most of the simplified tasks can be handled by a single robot. If we need to complete multiple goals at the same time, we will need a multi-robot, which is complicated. Applications such as dynamic mission planning [6], collective construction, multitasking assignments, and environment variable mapping have historically used multiple robots. To work, we need a completely different architecture. Each robotic is assigned a completely unique intention to complete. Managing more than one robot concurrently in unusual place environment is greater dynamic and complex. It is dynamic withinside the experience that every different robot are barriers to all different robots, besides static barriers that exist already withinside the environment. Therefore, due to the fact, that the surroundings are continuously changing, conventional path-making plan algorithms want to act otherwise to discover the high-quality path for every robotic. Xinye et al. In [7], introduced the hassle, of course, planning for a collection of cell robots with a couple of locations exists because of a couple of traveling salesman hassle (MTSP) with one or greater depots. To clear up the hassle, a two-cause algorithmic technique to (ACO) primarily based totally on me-metric algorithms is presented. Therefore, a critical issue is that every robotic has to go to at least one vacation spot, in order that every vacation spot to be visited as soon as via way of means of one robotic. In addition, simultaneous optimization of the full direction period and most direction period of the robotic have to be achieved. This work used an ACO based on an algorithm that uses local search and integrates [8]. The experimental consequences of this technique, examined in a static environment, are compared to different classical algorithms that display that the proposed technique yields higher consequences. Lambert and others' paper [9] describes a technique primarily based totally on stochastic neuro-fuzzy logic. This technique consists of stages of fuzzy structure and is primarily based totally on a chief-follower scenario. The first is stochastic fuzzy control. It handles uncertainty and error, avoids traversal glitches, and comparatively approximates the robotic's position. The neuro-fuzzy inference device layer allows the construction of chief-follower actions among robots. This trouble is supplied by the use of a fuzzy structure diagram. The neuro-fuzzy inference device acts as a navigation controller for every robot. Its enter coordinates, robotic orientation, and output are the robotic's linear and angular velocities. Experimental consequences for this technique may be received via way of means of simulating one-of-a-kind environments. In the experiment, the actions of the chief and followers are observed. The future scope consists of trying out modern procedures in reality complex environments and growth strategies to enhance overall performance and decrease complexity. The study [10] focuses commonly on enhancing the orbital making plans of multi-robotic structures with the usage of the Genetic Algorithm (GA). This is executed through the usage of the memetic method. The proposed progressed memetic set of rules-primarily based totally method (IMA) includes the implementation of GA with variable-period chromosomes the usage of of -factor crossover and bacterial mutation manipulation to keep away from optimum nearby problems. Use a seek approach that mixes nearby neighbour searches and jamming techniques to enhance average convergence. In addition, a method for managing multi-robotic course-making plans in dynamic surroundings has been proposed. Expert-intellectual outcomes acquired via real-time checking out and simulation are presented, demonstrating giant enhancements over not unusual place genetic and memetic approaches. Enteral, etc. The paper [11] objectives to broaden a whole and optimum solution to the multi-robotic course-making plans problem. The present Push and Spin Algorithm (PASp) affords a whole method to the multi-robotic course making plans problems. An improved push-and-spin set of rules (PASp+) has been evolved to get the fine course. It adapts the same old PASp set of rules in ways the usage of a frictionless operation that removes redundant moves withinside the course and trying to find different techniques the usage of heuristic seeks. The pre-ceding set of rules makes use of a push-and-spin operation while robots are concurrently captured at a not unusual place vertex role alongside the trajectory. Instead of a spin operation, an evaluation of heuristic values with different to-be-had paths is completed and a clean operation is implemented at the quilt of the listing of retrieved solutions. The proposed approach has been tested and compared to standard algorithms, demonstrating that the proposed approach is more powerful. To reduce the work involved, this doc-ument focuses on warehouse applications that implement multi-robot systems. In this regard, Table 1 gives a brief overview. Warehouse applications are one of the most used areas for multi-robot systems to meet and meet their needs. Again, we find that most of the proposed solutions are centralized or decentralized. Jianya Yuan et.al [12] also proposed a novel work based on Artificial neural network (ANN) to work in dynamic environment. On the other hand, Hannah Lee et.al [13] proposed a hierarchal multi agent path finding for a mobile robot, but they haven't mentioned the dynamic issue that may occurred at any moments.

3 Problem Definition

A multi-agent pathfinding (MAPF) problem has been specified by undirected unweighted graph $G = (V, E)$ and set $\{a1$ of k agents.... $ak\}$, where ai has a start node si $\in$ V destination node gi $\in$ V. Time is discretized into time steps. Each agent can switch between successive time steps wait at the adjacent vertex or the current vertex. With movement wait actions have a unit price unless the agent waits last destination node with zero cost. AI road the result move and wait for the action that leads ai from si To Gi, Aj, v, ti are node conflicts when ai and aj are equal. Vertex v of time step t and tap, Aj, u, v, ti is edges conflict only if ai and aj pass the same edge (u, v) in opposite Direction between time steps t and t + 1, nodes while minimizing the total cost of these paths. The trouble is an installation in a dynamic warehouse surrounding more than one robot with a preliminary configuration described and a hard and fast of goal places specified. Therefore, the trouble right here may be visible because the most desirable goal mission to the robotic and next direction making plans so as to increase a collision-loose direction for every robotic to attain the goal. In a multi-robotic device, path-making plans call for understanding to reap a collision-loose path that may be carried out with minimum robotic movement. The static garage surroundings right here include static barriers that the robotic can't skip through, and the ultimate loose area is passable (see Fig. 1). Target mission performs a vital position right here due to the fact the records wanted for direction making plans is the goal region to approach. In general, a multi-robotic trajectory-making plans device has a described mapping of dreams for every robotic to attain its dreams. Therefore, the vacation spot mission is a further challenge that impacts the very last direction retrieved. The main goal of goal assignment is to assign goals to minimize the path length and calculation time of the future paths taken. Using a random order column for target assignment can significantly increase path length and calculation time. Therefore, an appropriate approach that considers the path length in advance is desirable.

Table 1. Literature survey on the path planning in warehouse environments.

Author	Method	Environment type
Kmur et al.	Heuristic	Normal map based WH
Han et al.	Heuristic	Warehouse Env.
Kaushlendra Sharm	BFS	Warehouse
Jiaoyang Li1	Cbs	Normal map
Yang	Ant-colony	Warehouse
Bolu et al.	A*	Warehouse
Jianya Yuan et al.	RNN	Dynamic environment
Hannah Lee et al.	Hierarchical CBS	Static environment

Conflict-Tree (CT) with regard to node solution cost, the CT is searched using the best-first method. This leads to the best possible solutions. in the beginning, there are no limitations at the CT root node. The collection of paths node allows for free computation. The root node then serves. As the foundation for expanding the CT. The primary steps in this process are expansion, validation, and addition in CT. Number of conflict detection and Resolution are two crucial activities in this process of resolving disputes. Conflict detection planner assesses CT node solutions for path conflicts. The conflict detection process loops through all of the different paths and compares the times and locations traveled by each agent. The CT node is designated as a goal node if no conflict is found. Resolution of Conflicts the high-level planner creates a path constraint for each agent involved whenever a conflict (c) is found. The vertex v and collision time t together make up an agent ai constraint. Each restriction generates a child CT node. Additionally, every child node copies the parent node's constraints in full. Discovered goal nodes incorporate the first-rate solution. Taking into account a set of n robots in a multi-robot system where every robot is represented as Ri where i = 1, 2,..., n shows the robot ID. A matrix display between the robot and target D. According to the Eq. (1).

$$C_i(t) = (x_i, y_i, t) \tag{1}$$

After that, considering if there is a conflict between two robots Ri and Rj will be as Eq. (2).

$$\text{Collision } (i, j) \rightarrow \exists t : C_i(t) = C_j(t) \text{ where } i \neq j \tag{2}$$

To select the desired path, then we will produce the maximum path length (μ pl) as in Eq. (3).

$$\mu p1 = \sum_{i=1, j=1}^{n} pij \tag{3}$$

These situations are to be glad to make certain the answer is optimal, which may be mathematically represented, as cited withinside the following equation. Firstly, collision-loose implies wherein n is the variety of robots withinside the gadget and pij denotes the period of the course traced from Robot R(i) to the aim Gj. The goal is to decide on a collision-loose course with a minimum course period for every robotic withinside the gadget to attain the aim state. Equation (4)

$$\text{Collision } (i, k) = \phi \forall R_i, R_k \text{ where, } i \neq k \tag{4}$$

considering every pair of robots whose paths are concerned in a collision must be empty. Secondly, μ pl must be minimum. Consider the set of ideal paths (S) to all the agents withinside the gadget given with the aid of using the following

$$S = \left[P_{ij,...} \right] \forall i, j \text{ where } D_{i,j} = 1 \tag{5}$$

and, min (μ pl) and collision (i, k) = μ for each pair pair of Ri, Rk robots in the system.

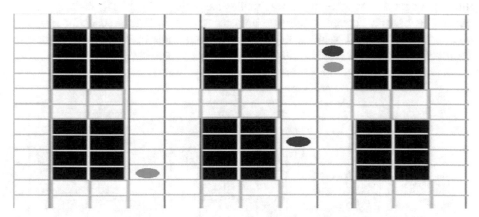

Fig. 1. Our MAP-CBS storage environments that consist of static pods and mobile robots where the red block represent the mobile robot, and the green block represent the dynamic human workers.

4   The Proposed Solution

In this work, we would like to signify a progressed clever choice of pairs to discover the goal closest to every robotic Ri. This progressed clever choice is advanced primitive choice due to the fact primitive choice makes use of conventional distance metrics that don't forget barriers, whilst the previous considers barriers. Figures 2 illustrate the clever choice mechanism of the robotic. The sub-figure in Fig. 2 represents an incremental example of a current problem. Figure 2i indicates the region of all robots and objectives withinside the surroundings. The block subsequent to it indicates that every one of the robots withinside the surroundings is linked to the goal. Then every distance is calculated. The subsequent step is to calculate the gap between all robots from all to-be-had objectives. Then mating occurs. In this manner, every robotic is chosen to skip a particular goal. This is achieved by deciding on the minimal distance from every robotics' goal. Figure 2 additionally indicates the primary barriers of this current approach. This is due to the fact the gap calculation does now no longer don't forget the barriers that exist between the robotic and the goal. Taken into consideration while calculating the gap of every robotic from the goal. The intersection is the region of the second factors at the back of the impediment. This enables us to calculate the real distance, thinking that the impediment is between the robotic and the goal. The distance ratio enables to discovery of the goal for every robotic. This ratio enables us to discover the great distance to journey through overcoming the primary choice of now no longer thinking about the presence of barriers while calculating the gap. The distance ratio μ enables to discovery the great collision-unfastened course. Then practice the proposed idea to resolve issues in warehouse packages wherein more than one robot proportion the identical surroundings and feature more than one objective.

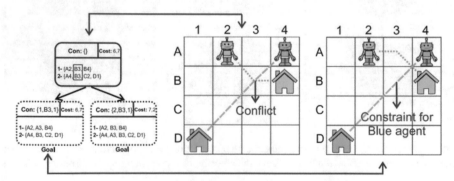

Fig. 2. The various instances of execution represent existing problems with the system. Here is an example of a constraint tree (CT). Two of those nodes are displayed graphically.

4.1   The Method of Path Planning

The environment map must first be converted to binary format, which is a crucial step in the solution. The second stage then generates a source-to-destination pair based on the destination assignment's distance ratio. To guarantee that collision-free paths are obtained, the following step is to apply a distance metric approach for path design. Here, collision avoidance with reserved tables is used to handle robot mutual collisions. A habitat with multiple robots is first constructed. To discover the path that is the dynamic warehouse environment map, a reference map is given as input. White pixels represent acceptable locations, while black pixels indicate impassable storage boxes in the image that was obtained during conversion. Each robot has a preset beginning point, and an intelligent selection technique is used to assign goals or targets that enhance path planning with regard to path length. The goal mapping matrix'D is obtained as a result of the wise choice made and is then employed for additional path planning. For each target Gi, there are numerous distance ratios I according to the smart selection strategy. The robot chosen has the greatest distance to the obstacle and the shortest distance to the goal point, therefore the pair with the lowest value is chosen. This enhances the estimation of the path's free space traveled. In a static warehouse setting, assigning destinations is done by calculating the move's cost. This cost is calculated by adding the overall path length and all collisions on the intended path for each target among the available robots. At the lowest possible cost, targets are given to the robot; this process is repeated for each target in the system. Target assignment scenarios are shown in Figs. 3 and 4.

4.2 Producing the Destination and Initial Points

In a dynamic warehouse environment, target assignment is carried out by calculating the path cost of each target using the available robots. By adding the overall path length and the collisions along the planned course, this cost may be calculated. All targets in the system are assigned to the robot using a minimally expensive technique. Target assignment is demonstrated in the case in Fig. 3. This depicts a warehouse setting with numerous robots. It includes the beginning position, target position, storage, and depot of the robot.

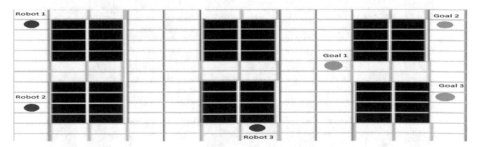

Fig. 3. Shows a scenario that demonstrates target assignment.

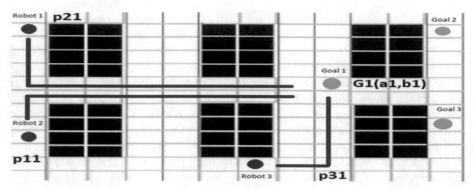

Fig. 4. Shows that p11 is a collision-free path from robot R1 to G1 and is the shortest path. Similarly, p21 is the path from the non-collision robot R2 to G1 and p31 is the path from the non-collision robot R3 to G1.

5   The Analysis of the Result and Discussion

One of the industries with the greatest growth is e-commerce, so warehousing is essential to accommodate this demand. They respond to inquiries accurately and quickly using robots. A fantastic illustration of an intelligent warehouse is the Alibaba system. The path planning idea that was putted in this article was implemented in a warehouse application. After receiving the right ordering job, each robot in a smart warehouse application often needs to plan a route from its current location to the target pod and then from the pod's location to the workstation. There must be no conflicts between these paths and other robots' paths, and there must be no disconnections when the robots are moving (Fig. 5)
.

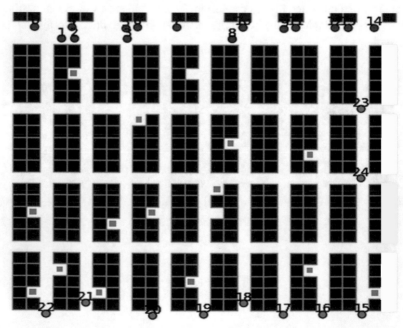

Fig. 5. The proposed system applies to amazon storage with dynamic obstacles inside the environment map with a group of storage bins in the system.

The suggested method is applicable to a static storage environment map with a team of 16 robots, 400 storage bins, and dynamic barriers. The implementation makes the assumption that the robot is a universal item, and the experimental findings are displayed in Table 2, and. By changing the number of robots and pods and using the algorithm on another multi-robot system, the results were achieved. The average path length and total elapsed computation time are used as parameters to evaluate the findings (Fig. 6).

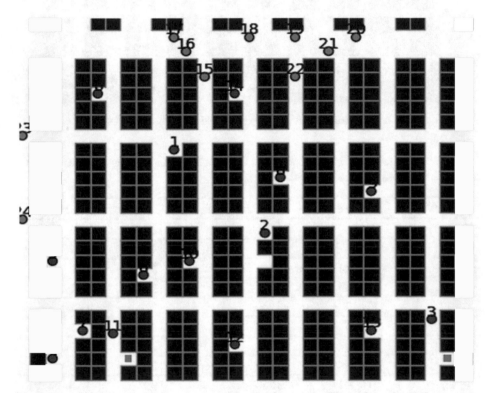

Fig. 6. Simulation preview of a warehouse application where 800 pods and 16 robots move through the warehouse environment and are in close proximity to each other to reach their goals.

Value of success rate during the same experiment the algorithm does not change significantly after the environment changes. Therefore, we felt it was appropriate to set only the success rate. In principle, both algorithms are capable of this Find a quick solution by experimenting with 48 * 48 cards. The success rate values are high in these experiments. On the other hand, in order to cope well with other environments, we managed the environment with a single agent for multi-dynamic failures. As shown in Fig. 7. The intelligent approach concepts are used in this work to implement tailored path planning in a warehouse application. Each robot in an intelligent warehouse application often has to plan a route from its current location to the target pod and then, after receiving the matching order job, from the pod location to the workstation. There is no stopping point between moving robots, and these paths must not cross those of other robots (Figs. 8, 9, 10 and Table 3) .

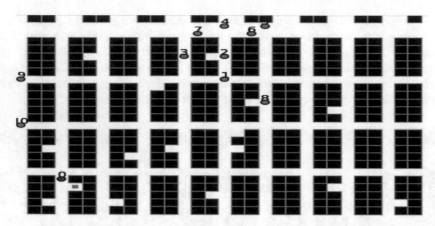

Fig. 7. Environment with a single agent and multi-dynamic obstacles.

Table 2. The achieved experimental results for a multi-robot system in a warehouse environment with 400pods using the suggested methodology.

Source(S)	Destination	Heuristic CBS	RNN	PRM
(0, 0)	(5, 5)	0.2	1.3	0.5
(0, 3)	(9, 10)	1.1	4.3	1.5
(0, 4)	(15, 16)	1.1	3.5	1.2
(0, 8)	(24, 28)	2.1	2.3	3.2
(0, 7)	(24, 2)	2.2	2.3	4.5
(0, 11)	(17, 2)	2.1	6.5	7.2
(0, 12)	(18, 8)	1.0	1.2	6.5
(0, 15)	(22, 4)	2.5	2.7	4.5
(0, 16)	(11, 17)	2.6	3.2	3.5
(0, 19)	(12, 23)	3.2	4.5	6.2
(0, 20)	(17, 11)	2.3	3.2	3.2
(0, 24)	(24, 7)	1.5	3.5	2.3
(0, 27)	(23, 14)	2.9	4.2	3.2
(0, 28)	(22, 23)	3.2	5.2	3.2
(0, 30)	(5, 14)	3.0	6.2	3.1
(28, 28)	(1, 2)	2.4	6.2	3.2
(28, 25)	(1, 7)	2.3	4.2	2.5
(28, 22)	(1, 10)	3.6	5.2	4.5

(continued)

Table 2. (*continued*)

Source(S)	Destination	Heuristic CBS	RNN	PRM
(28, 18)	(1, 15)	2.6	3.2	3.2
(28, 14)	(1, 18)	3.7	4.2	6.2
(28,10)	(1, 22)	1.4	4.2	3.2
(28, 6)	(1, 25)	3.8	4.5	6.2
(28, 2)	(1, 28)	3.5	4.8	4.9
(8, 30)	(8, 0)	5.2	9.5	6.5
(14, 30)	(14, 0)	2.9	4.6	3.2

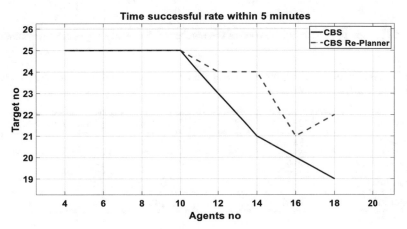

Fig. 8. Plot of mean path length variation for 2 different robot systems with different numbers of pods.

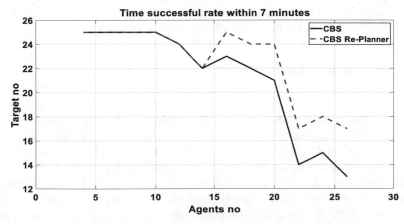

Fig. 9. Plot of mean path length variation for 2 different robot systems with different numbers of pods.

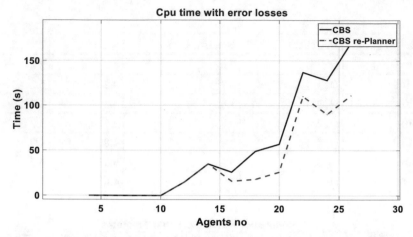

Fig. 10. Cpu time with the increase of the number of the agent's angst the environment.

Table 3. A multi-robot system that uses the approach proposed in a warehouse environment with 16 robots and different numbers of pods.

Warehouse	Method	Time (sec)	Path length average
Kmur et al.	GA and A*	190.3	33.4
Jianya Yuan et al.	RNN	187.5	45.5
Hannah Lee et al.	Hierarchical CBS	170.6	40.2
Our method	Heuristic CBS	165.545	31.2

During the simulation we have the ability to use different shape of the warehouse system and generate a random map. Our work satisfied the problems of path planning in a warehouse environment using our robotino mobile robot as an agent.

6 Conclusion and Future Work

This contribution addressed the problem of multi-robot system path planning. Both topics are crucial in and of themselves. It can be difficult to combine these two problems and solve them simultaneously. Only tweaks are considerably simpler to complete if we put the path length issue to one side. Similar to this, if there are no limits that robots cannot collide with one another or share path segments with other robots, we can optimize the path length of individual robots by avoiding issues. But to effectively automate each application, these two problems must be tackled simultaneously in a multi-robot system. As the other factors are adjusted, each one needs to be strictly managed. In this document, the issue was resolved using a warehousing application. The suggested smart choice method is used to track and verify the outcomes. The main objective is to design and tune distance metric paths utilizing target assignment and intelligent selection to identify the best, collision-free path. For orbital planning with several robots in a static warehouse setting, custom algorithms have been developed, facilitating system integrity and adjustment. Further changes have been made to the algorithm to produce a source-to-destination pair from a certain configuration based on the distance ratio value and to produce a path in accordance with that while retaining adjustments. The suggested method exhibits enhanced outcomes when tested under more than three distinct warehouse application scenarios. Our future work considers as adding some complexity to the environment, like planning as both the surface and the robot that will applied to this environment. Our environment has different complexity considered as the concave shape of the pods and the station.

References

1. D'Andrea, R.: Guest editorial: a revolution in the warehouse: a retrospective on kiva systems and the grand challenges ahead. IEEE Trans. Autom. Sci. Eng. **9**(4), 638–639 (2012)
2. Yang, Y., Zhang, J., Liu, Y., Song, Xin: Multi-agv collision avoidance path optimization for unmanned warehouse based on improved ant colony algorithm. In: Pan, L., Liang, J., Qu, B. (eds.) BIC-TA 2019. CCIS, vol. 1159, pp. 527–537. Springer, Singapore (2020). https://doi.org/10.1007/978-981-15-3425-6_41
3. Canny, J.: The Complexity of Robot Motion Planning. MIT Press, Cambridge (1988)
4. Cho, D.-H., Jang, D.-S., Choi, H.-L.: Memetic algorithm-based path generation for multiple dubins vehicles performing remote tasks. Int. J. Syst. Sci. **51**(4), 608–630 (2020)
5. Bakdi, A., Hentout, A., Boutami, H., Maoudj, A., Hachour, O., Bouzouia, B.: Optimal path planning and execution for mobile robots using genetic algorithm and adaptive fuzzy-logic control. Robot. Auton. Syst. **89**, 95–109 (2017)
6. Wu, Q., Lin, H., Jin, Y., Chen, Z., Li, S., Chen, D.: A new fallback beetle antennae search algorithm for path planning of mobile robots with collision-free capability. Soft. Comput. **24**(3), 2369–2380 (2019). https://doi.org/10.1007/s00500-019-04067-3
7. Al-Jarrah, R., Shahzad, A., Roth, H.: Path planning and motion coordination for multi-robots system using probabilistic neurofuzzy. IFAC-Papers OnLine **48**(10), 46–55 (2015)
8. Ni, J., Wang, K., Cao, Q., Khan, Z., Fan, X.: A memetic algorithm with variable length chromosome for robot path planning under dynamic environments. Int. J. Robot. Autom. **32**(4), 414–424 (2017)
9. Han, S.D., Yu, J.: Effective heuristics for multi-robot path planning in warehouse environments. In: 2019 International Symposium on Multi-robot and Multi-agent Systems (MRS), pp. 10–12. IEEE (2019)
10. Chen, X., Zhang, P., Du, G., Li, F.: Ant colony optimization based memetic algorithm to solve bi-objective multiple traveling salesmen problem for multi-robot systems. IEEE Access **6**, 21745–21757 (2018)
11. Almadhoun, R., Taha, T., Seneviratne, L., Zweiri, Y.: A survey on multi-robot coverage path planning for model reconstruction and mapping. SN Appl. Sci. **1**(8), 1–24 (2019). https://doi.org/10.1007/s42452-019-0872-y
12. Roldán, J.J., Garcia-Aunon, P., Garzón, M., DeLeón, J., DelCerro, J., Barrientos, A.: Heterogeneous multi-robot system for mapping environmental variables of green, **16**(7), 1018 (2016). Author, F.: Contribution title. In: 9th International Proceedings on Proceedings, pp. 1–2. Publisher, Location (2010)
13. Cui, R., Guo, J., Gao, B.: Game theory-based negotiation for multiple robot's task allocation. Robotica **31**(6), 923–934 (2013)

14. Wawerla, J., Sukhatme, G.S., Mataric, M.J.: Collective construction with multiple robots. In: IEEE/RSJ International Conference on Intelligent Robots and Systems, vol. 3, pp. 2696–2701. IEEE (2002)
15. Yuan, J., et al.: A novel GRU-RNN network model for dynamic path planning of mobile robot. IEEE Access **7**, 15140–15151 (2019)
16. Solis, I., et al.: Representation-optimal multi-robot motion planning using conflict-based search. IEEE Robot. Autom. Lett. **6**(3), 4608–4615 (2021)

Classification of SCADA Alarms and False Alarm Identification Using Support Vector Machine for Wind Turbine Management

Ana Maria Peco Chacon$^{(\boxtimes)}$ ⓘ and Fausto Pedro Garcia Marquez ⓘ

Ingenium Research Group, Universidad Castilla-La Mancha, 13071 Ciudad Real, Spain
{anamaria.peco, faustopedro.garcia}@uclm.es

Abstract. The efficiency of a wind turbine depends on operations and maintenance activities. Large amounts of complex data are efficiently categorized by machine learning-based classifiers. The implementation of several support vector machine algorithms for the prediction and detection of false alarms in wind turbines is the novelty proposed in this research. A reliable tool for assessing the effectiveness of classification algorithms is K-Fold cross validation. The proposed methodology is verified by using SCADA data from an actual wind turbine. The outcomes show a 98,6% accuracy rate for the quadratic support vector machine classifier. The analysis of the misclassifications obtained from the confusion matrix provides the necessary information, together with the alarm log and maintenance record, to determine whether it is a false alarm. The classifier can reduce the number of false alarms referred to as misclassifications by 25%. These results demonstrate that the suggested methodology is effective at identifying false alarms.

Keywords: Machine learning classification · Support vector machine · False alarm · Wind turbine · Cross validation

1 Introduction

Meeting the aims of climate change agreements depends in large part on renewable energy. The advantages of renewable energy include cost effectiveness and environmental responsibility [1]. The largest source of renewable energy is wind energy [2]. Due to advancements in wind energy technology and governmental economies, wind power generation is increasing gradually and the tendency is anticipated to continue in the future [3]. The installed wind power capacity currently globally stands at 837 GW, including 93.6 GW of new wind energy capacity installed in new wind farms in 2021 [4].

In order to ensure optimal operation of Wind Turbines (WTs), maintenance management seeks to minimize probable component failures. Between 25% and 30% of the overall cost of producing wind energy is spent on maintenance and repairs [5, 6]. The majority of operation and maintenance cost techniques focus on enhancing system dependability by analyzing important components [7, 8]. Consistent information on

F. P. García Márquez et al. (Eds.): ICCIDA 2022, LNNS 643, pp. 535–547, 2023.
https://doi.org/10.1007/978-3-031-27099-4_41

important WT components is provided by condition monitoring systems (CMS), that also enables sufficient analysis and maintenance [9, 10]. The Supervisory Control and Data Acquisition (SCADA) system and the CMS increase the wind efficiency and profitability of wind farms [11, 12]. Due to the size of the signal and alarm datasets provided by SCADA systems, several signal processing techniques are required [13].

The alarm design can help operators determine the actual condition of WTs [14]. A false alarm occurs when a fault is indicated but there is not actually a problem with the WT; this results in large financial losses because unneeded maintenance procedures are carried out [15]. The false alarm detection requires powerful algorithms to extract reliable data from the SCADA dataset [16].

Machine learning algorithms employ computer strategies to recognize patterns in a dataset by learning from the data [17]. There are two varieties of machine learning: supervised and unsupervised. The primary objectives of supervised machine learning are the creation of classifications and the enhancement of forecasting models [18].

Support Vector Machine (SVM) is a supervised machine learning algorithm [19]. SVM models have been used to resolve problems in classification and regression. Figure 1 illustrates the increasing evolution of publications on the SVM algorithm.

Fig. 1. Evolution of SVM publications in latest years based on [20]

Arcos et al. [21] used SVM and other classifiers to identify blade delamination in WTs. In reference [22],the SVM is employed to identify and diagnose intelligent faults in WTs. Hübner et al.[23] had employed SVM to identify mass imbalance in the rotor of WTs with more than 84% of accuracy. In [24], SVM with a 10-fold cross validation, and adaptive threshold were used to find anomalies in the gearbox of WTs.

The approach had an accuracy of 91.11% and the false positives were 3.5%.

The main contribution of this work is the application of a data-based analysis to predict and detect false alarms with SVM classifiers. This method can be used to increase the reliability of WTs with historical SCADA data. The innovative approach put out in this work is based on the identification of false alarms, where literature analysis of failures. The analysis of misclassifications has been performed by providing values for the different types of alarms. These points are also examined together with the alarm log and the maintenance log, this provides significant information to determine the causes of false alarms.

The remainder of the paper is organized as follows: Sect. 2 describes the approach used in this study, the different SVMs applied and the validation method used are explained, as well as the different metrics classifiers. Section 3 reports the results obtained with different SVM algorithms using SCADA data from an actual turbine. Lastly, Sect. 4 shows the conclusions of this study.

2   Approach

The summary of methodology is shown in Fig. 2.

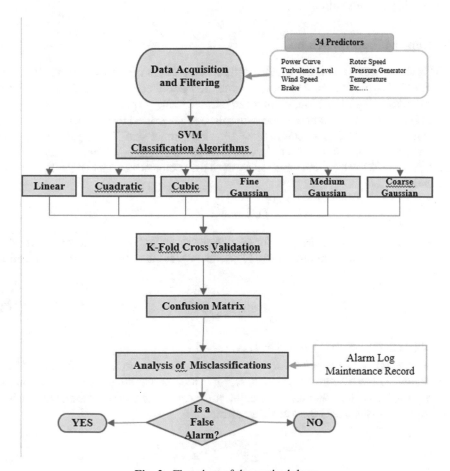

Fig. 2. Flowchart of the methodology.

The proposed methodology uses SCADA variables to detect and classify the different alarms in the alarm log. The alarm log collects data of the different types of alarms and the times during that they are activated, as well as the alarm code and description. In this research, there are more than 22,000 observations available for 34 SCADA variables that were measured at a frequency of every 10 min. The SCADA variables and the data from the alarm log are synced on a same time frame. This is the data acquisition and filtering phase, because if any "nan value" in SCADA dataset needs to be removed.

The next step is sorting the dataset with different SVM classifiers. In this case, the validation has been carried out with 5-Fold Cross Validation. Subsequently, the SVM algorithm is chosen with the best accuracy and less processing power.

The third phase is the analysis of misclassifications from the confusion matrix. The alarm log and the maintenance log are used to analyze misclassifications to identify whether they are false alarms.

2.1 Support Vector Machine

SVM is an algorithm that is frequently used for complex data analysis and fault detection [25]. SVM classifier seeks to identify boundaries between several data classes with the greatest possible margin of separation [26]. Considering a training set of $\{(x_i, y_i)\}_{i=1}^{N}$, Eq. (1) is used for the linear case; the hyperplane is discovered by increasing the margin between classes. The input vectors have a V-dimensional size. The input vectors have a V-dimensional size and $x \in R^V$. In this instance, S refers to a single observation of SCADA data [27]. The vector that determines the vector orthogonal to the hyperplane is called d.

$$f(x) = S'x + d \tag{1}$$

The distance between the parallel hyperplanes equals to $2/\|S\|$. To maximize this margin, $1/2\|S\|$ must be reduced [28]. Consequently, solution becomes an optimization issue.

$$min_{S,d} y \frac{1}{2} \|S\|^2 \tag{2}$$

Subject to $y_i \left(S'x_i + d \right) \geq 1, i = 1, \ldots, N$

The generalized expression for the linear kernel is given in Eq. 3. There must be a slack factor, β, and a control parameter, R, for instances that are not linearly separable. The control parameter specifies the weight of the second term for minimization, while the slack factor represents the allowable error.

$$min_{S,d} y \frac{1}{2} \|S\|^2 + R \sum_{i=1}^{N} \beta_i \tag{3}$$

Subject to $y_i \left(S'x_i + d \right) \geq 1 - \beta_i, \beta_i > 0, i = 1, \ldots, N$

The general scenario of SVM is shown in Eq. 4, where it is impossible to directly define the hyperplanes in the original feature space. For the non-linear transformation of the data, the "kernel trick" is employed [29]. The hyperplanes with the kernel transformation are obtained using the inner product and a function.

$$min_{S,d,\beta} y \frac{1}{2} \langle S, S \rangle^2 + R \sum_{i=1}^{N} \beta_i \tag{4}$$

Subject to $y_i \left(S', \mu_x > +d \right) \geq 1 - \beta_i, \beta_i > 0, i = 1, \ldots, N$

To prevent the unlimited dimensions that inner products may have, SVM is employed in dual form. The kernel function, or K, is specified for vectors with different parameters (x_i, x_j) is $K(x_j, x_j) = \langle \mu_{x_j}, \mu_{x_j} \rangle$ [30].

$$min_{\alpha} \sum_{i=1}^{N} \sum_{j=1}^{N} \alpha_i \alpha_j y_i y_j K(x_i, x_j) - \sum_{i=1}^{N} \alpha_i \tag{5}$$

Subject to $. \sum_{i=1}^{N} y_i \alpha_i = 0, 0 \leq \alpha_i \leq R$

Equation 6 is found by using the Aronszajn Theorem [31].

$$K(x_i, x_j) = \langle \mu(x_i), \mu(x_j) \rangle \forall x_i, x_j \in X \tag{6}$$

Only the matching kernel functional form must be known to solve the dual problem; the set of transformation basis functions need not be understood. The nature of the data set determines the type of kernel function to be implemented. Table 1 shows the different kernel functions formulas to be applied in this research. The standard deviation parameter is denoted by δ, while the parameter P specifies the degree of the polynomial.

Table 1. The fundamental Kernel functions.

Types of functions	Kernel functions formulas
Lineal	$K(x_i, x_j) = \langle x_i, x_j \rangle$
Polynomial	$K(x_i, x_j) = \langle 1 + x_i, x_j \rangle^P$
Gaussian	$K(x_i, x_j) = \exp\left(\frac{\|x_i - x_j\|^2}{2\tau^2} \right)$

It is classified into the following categories for Gaussian Kernel cases:

- Fine Gaussian: Class distinctions are made to a fine degree. N is the numbers of predictors and the kernel scale is equal to $\sqrt{N}/4$.
- Medium Gaussian: As the name implies, it is less distinct than the previous case, hence its kernel scale is $\sqrt{N}$.
- Coarse Gaussian: The class distinctions are less accurate in this scenario and the kernel scale is $4 \cdot \sqrt{N}$.

2.2 Cross Validation

The purpose of validation is to evaluate how well the used model is working. The K-fold cross-validation (CV) method first separates the data set into k comparable subsets, or folds. Each subset is acquired using hierarchical sampling to retain the consistency of the data distribution as much as feasible [32]. The partitioned folds are trained and tested in K iterations, leaving one-fold per iteration for testing and training the model on the remaining K-1 folds, as shown in Fig. 3. The accuracy achieved throughout each iteration is averaged to determine the model accuracy [33]. The frequent use of randomized subsamples for training and validation, as well as the validation of results after each iteration, are its key benefits [34].

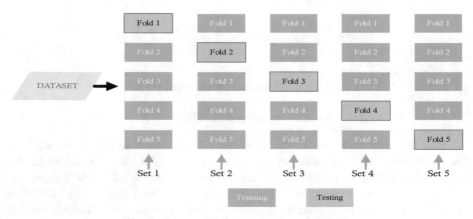

Fig. 3. Cross validation method based on [35].

2.3 Confusion Matrix and Indicators for False Alarms

In statistical classification issues, the confusion matrix is a commonly used technique [36]. The effectiveness of categorization models is displayed by the confusion matrix. The confusion matrix is a set of rows and columns in tabular form to display the classification outcomes of the classifier [37]. The columns correspond to the classes the model predicts, and the rows represent the actual classes. Correctly classified cases are displayed on the diagonal, whereas the rest of the cells are misclassifications.

There are many indicators to measure the effectiveness of classification methods. Accuracy, sensitivity and false negative rate, are some of the performance metrics used in this study.

The accuracy is the ratio between the number of samples that were successfully classified and the total number of instances.

$$Accuracy = \frac{TP + TN}{TP + TN + FP + FN} \tag{7}$$

True Positive Rate (TPR), also known as sensitivity, is the proportion of successfully classified observations in each true class [38].

$$\text{True Positive Rate(TPR)} = \frac{TP}{TP + FN} \tag{8}$$

False Negative Rate (FNR) is the percentage of incorrectly classified negative observations in each negative class, Eq. (9).

$$\text{False Negative Rate(FNR)} = \frac{FN}{TP + FN} = 1 - TPR \tag{9}$$

3   Case Study and Results

This research uses SCADA data from a real WT. Every 10 min, SCADA signals were monitored, yielding more than 22,000 observations. Classification techniques look for a correlation between the response variable and independent variables (also known as predictors). The 34 SCADA variables, in this case, are the predictor variables, while the alarm log is the response variable. The alarm log has 17 values according to the type of alarm, as shown in Table 2.

Table 2. Different types of alarms

Assigned number	Types of alarms	Occurrences	Assigned number	Types of alarms	Occurrences
0	Deactivated alarm	21.795	9	Power converter alarms	27
1	States specification	401	10	Hydraulic unit alarms	3
2	Control box system alarms	2	11	Grid connection alarms	0
3	Vibration monitors alarms	0	12	Various alarms	8
4	Ambient conditions alarms	185	13	Fire detection alarms	0
5	Speed sensors alarms	6	14	GMS system alarms	0
6	Yaw system alarms	71	15	Active power and frequency alarms	0
7	Gearbox system alarms	3	16	Combination of alarms	106
8	Generator system alarms	0			

Table 3 shows the 5-fold CV accuracy for the different SVM classification models. In this research other values of K-fold CV have been applied, and the accuracy values obtained are similar. Another important factor is the validation calculation time, the more K value the more computing time. Consequently, a satisfactory relationship between accuracy and computation time is attained using the 5-Fold CV. The quadratic SVM classifier shows the best accuracy with the lowest training time and number of misclassifications.

Table 3. Comparison for different classifiers.

Model	Accuracy	Misclasifications	Training time (seconds)
Linear	98,1%	420	593
Cuadratic	**98,6%**	**319**	154
Cubic	98,6%	322	287
Fine Gaussian	97,8%	498	747
Medium Gaussian	98,2%	401	691
Coarse Gaussian	97,5%	575	767

The Receiver Operating Characteristic (ROC) curve displays the ratio of true positive rate and false positive rate. The accuracy of the classifier is displayed by the area under the ROC curve [39]. Performance of the classifier improves as area under the curve (AUC) increases [40], an effective model has an AUC value that is near to 1. In this instance, the period when no alarm is activated is regarded to be in the positive class, and the interval where an alarm is triggered is in the negative class. SVM quadratic model can correctly assign 100% of the observations to the positive class (when no alarm is triggered) and is only able to correctly classify 65% of the negative classes (Fig. 4).

The Fig. 5 shows the performance of the classifier for each alarm type and identifies the types of alarms where the classifier has performed poorly. The diagonal elements display the cases where the predicted class and true class accord. The blue cells show the percentage of correctly classified cases (TPR). Otherwise, the orange cells are the proportion of observations incorrectly classified by true class (FNR).

In this case the classifier obtains the best results for the types of alarms that occur more times, and that represent 99% of the sample, and are the types of alarms: alarm deactivated, states specification, ambient conditions alarms and combination of alarms. Despite some alarm types being difficult to classify accurately, this is due to the small number of cases that are pressed in the data sample, it has no bearing on the overall accuracy of classifier, that is 98.6 percent.

The Fig. 6 shows the number of cases of each type of alarm in the study sample.

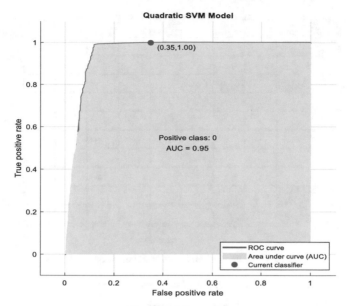

Fig. 4. ROC curve.

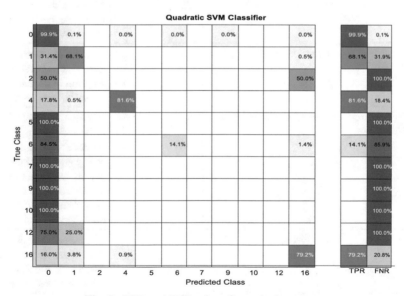

Fig. 5. TPR and FNR values for each alarm type.

Then, misclassifications are examined along with the maintenance log and the alarm log. Table 4 shows the main causes of these misclassifications. The total number of false alarms considered is also counted and 80 cases are detected, this means that 25.07% of false alarms are identified with the proposed methodology.

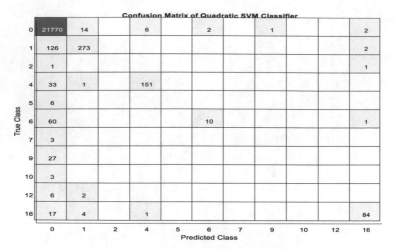

Fig. 6. Confusion matrix with number of observations.

Table 4. Analysis of misclassifications.

Assigned number	Type alarm	Misclassification	Cause
0	No alarm activated	25	They could be viewed as false negatives brought on by the SVM
1	States specification	128	29 cases are considered false alarms because the duration of the alarm activation was seconds
4	Ambient condition alarms	33	10 cases are level of turbulence alarm, the classifier can not to recognise it. 9 cases are considered false alarms as alarms occurred before and/or after the alarm was predicted
6	Yaw system alarms	60	17 cases are considered false alarms because the duration of the alarm activation was seconds
9	Power converter alarm	27	8 cases are considered false alarms because the duration of the alarm activation was seconds

(*continued*)

Table 4. (*continued*)

Assigned number	Type alarm	Misclassification	Cause
16	Combination of alarms	17	Desynchronization between the start or the end of the alarm times in relation to periods predicted

4 Conclusions

The market for wind energy is expanding as a result of enhanced monitoring systems, less downtime, and appropriate preventive maintenance. Condition monitoring system and supervisory control and data acquisition have been used to manage wind turbines, that produce large amounts of data.Wind farms become more dependable and cost-effective thanks to the detection and potential reduction of false alarms. Machine learning offers techniques with a high degree of accuracy. This paper presents a novel approach bases on support vector machine with k-fold cross validation for classifying different types of alarms. In addition, it allows the identification of false alarms of the control and data acquisition system by analyzing misclassifications.

The accuracy of the quadratic support vector machine model with 5-fold CV is 98,6%. The true positive rate obtains the highest values of 99.9% to identify the time intervals that the alarms are deactivated and 81.6% for ambient conditions alarms. The number of false alarms can be decreased by 25,07% percent using the quadratic SVM classifier. It can also be used as an additional tool to the alarm log to identify alarms.

For future research it is advisable to use samples with more alarm events to check that the classification algorithm is efficient in other cases. Utilizing these techniques in real-time to identify false alarms and reduce wind turbine downtimes would be recommended.

Acknowledgement. Authors would like to acknowledge the support of the Department of Science and Technology (DST), Govt. of India, and GIET University, Gunupur, Odisha, India for the financial & technical support for this work. The work reported herein was also supported financially by the Ministerio de Ciencia e Innovación (Spain) and the European Regional Development Fund, under Research Grant WindSound project (Ref.: PID2021-125278OB-I00).

References

1. Marugán, A.P., Márquez, F.P.G., Perez, J.M.P., Ruiz-Hernández, D.: A survey of artificial neural network in wind energy systems. Appl. Energy **228**, 1822–1836 (2018)
2. Ramirez, I.S., Muñoz, C.Q.G., Marquez, F.P.G.: A condition monitoring system for blades of wind turbine maintenance management. In: Xu, J., Hajiyev, A., Nickel, St., Gen, M. (eds.) Proceedings of the Tenth International Conference on Management Science and Engineering Management. AISC, vol. 502, pp. 3–11. Springer, Singapore (2017). https://doi.org/10.1007/978-981-10-1837-4_1

3. Márquez, F.P.G., Karyotakis, A., Papaelias, M.: Renewable Energies: Business Outlook 2050. Springer, Cham (2018). https://doi.org/10.1007/978-3-319-45364-4
4. Global wind energy council. https://gwec.net/global-wind-report-2022/. 14 July 2022
5. Márquez, F.P.G., Tobias, A.M., Pérez, J.M.P., Papaelias, M.: Condition monitoring of wind turbines: techniques and methods. Renew. Energy **46**, 169–178 (2012)
6. Márquez, F.P.G., Muñoz, J.M.C.: A pattern recognition and data analysis method for maintenance management. Int. J. Syst. Sci. **43**, 1014–1028 (2012)
7. García Márquez, F.P., PecoChacón, A.M.A.: review of non-destructive testing on wind turbines blades. Renew. Energy **161**, 998–1010 (2020)
8. García Márquez, F.P., García-Pardo, I.P.: Principal component analysis applied to filtered signals for maintenance management. Qual. Reliab. Eng. Int. **26**, 523–527 (2010)
9. Peco Chacón, A.M., Segovia Ramírez, I., García Márquez, F.P.: State of the art of artificial intelligence applied for false alarms in wind turbines. Arch. Comput. Methods Eng. 1–25 (2021)
10. de la Hermosa González, R.R., Márquez, F.P.G., Dimlaye, V., Ruiz-Hernández, D.: Pattern recognition by wavelet transforms using macro fibre composites transducers. Mech. Syst. Sig. Process. **48**, 339–350 (2014)
11. Gómez Muñoz, C., De la HermosaGonzalezCarrato, R., TraperoArenas, J., Garcia Marquez, F.: A novel approach to fault detection and diagnosis on wind turbines. Glob. NEST J. **16**, 1029–1037 (2014)
12. de la Hermosa González, R.R., Márquez, F.P.G., Dimlaye, V.: Maintenance management of wind turbines structures via MFCS and wavelet transforms. Renew. Sustain. Energy Rev. **48**, 472–482 (2015)
13. GarcíaMárquez, F.P., SegoviaRamírez, I., PliegoMarugán, A.: Decision making using logical decision tree and binary decision diagrams: a real case study of wind turbine manufacturing. Energies **12**, 1753 (2019)
14. Qiu, Y., Feng, Y., Infield, D.: Fault diagnosis of wind turbine with scada alarms based multidimensional information processing method. Renew. Energy **145**, 1923–1931 (2020)
15. PecoChacón, A.M., García Márquez, F.P.: False alarms management by data science. In: García Márquez, F.P., Lev, B. (eds.) Data Science and Digital Business, pp. 301–316. Springer, Cham (2019). https://doi.org/10.1007/978-3-319-95651-0_15
16. Shitharth, S.: An enhanced optimization based algorithm for intrusion detection in scada network. Comput. Secur. **70**, 16–26 (2017)
17. Jiménez, A.A., Zhang, L., Muñoz, C.Q.G., Márquez, F.P.G.: Maintenance management based on machine learning and nonlinear features in wind turbines. Renew. Energy **146**, 316–328 (2020)
18. Stetco, A., et al.: Machine learning methods for wind turbine condition monitoring: a review. Renew. Energy **133**, 620–635 (2019)
19. Mahesh, B.: Machine learning algorithms - a review. Int. J. Sci. Res. (IJSR). [Internet] **9**, 381–386 (2020)
20. D.S.R.S. Dimensions Inc. https://app.dimensions.ai/analytics/publication/overview/tim eline?search_mode=content&search_text=SVM&search_type=kws&search_field=text_s earch. 18 July 2022
21. ArcosJiménez, A., Zhang, L., Gómez Muñoz, C.Q., García Márquez, F.P.: Maintenance management based on machine learning and nonlinear features in wind turbines. Renew. Energy **146**, 316–328 (2020)
22. Wang, J., Liang, Y., Zheng, Y., Gao, R.X., Zhang, F.: An integrated fault diagnosis and prognosis approach for predictive maintenance of wind turbine bearing with limited samples. Renew. Energy **145**, 642–650 (2020)

23. Hübner, G., Pinheiro, H., de Souza, C., Franchi, C., da Rosa, L., Dias, J.: Detection of mass imbalance in the rotor of wind turbines using support vector machine. Renew. Energy **170**, 49–59 (2021)

24. Dhiman, H.S., Deb, D., Muyeen, S., Kamwa, I.: Wind turbine gearbox anomaly detection based on adaptive threshold and twin support vector machines. IEEE Trans. Energy Convers. **36**, 3462–3469 (2021)

25. Islam, M.M., Kim, J.-M.: Reliable multiple combined fault diagnosis of bearings using heterogeneous feature models and multiclass support vector machines. Reliab. Eng. Syst. Saf. **184**, 55–66 (2019)

26. Burman, I., Som, S.: In predicting students academic performance using support vector machine. In: 2019 Amity International Conference on Artificial Intelligence (AICAI), pp. 756–759. IEEE (2019)

27. Chacón, A.M.P., Ramirez, I.S., Márquez, F.P.G.: In Support vector machine for false alarm detection in wind turbine management. In: 7th International Conference on Control, Instrumentation and Automation (ICCIA), pp. 1–5. IEEE (2021)

28. Santos, P., Villa, L.F., Reñones, A., Bustillo, A., Maudes, J.: An svm-based solution for fault detection in wind turbines. Sensors **15**, 5627–5648 (2015)

29. Jakkula, V.: Tutorial on support vector machine (SVM), vol. 37, p. 3. School of EECS, Washington State University (2006)

30. Patle, A., Chouhan, D.S.: In SVM kernel functions for classification. In: 2013 International Conference on Advances in Technology and Engineering (ICATE), pp. 1–9. IEEE (2013)

31. Micchelli, C.A., Pontil, M.: Feature space perspectives for learning the kernel. Mach. Learn. **66**, 297–319 (2007)

32. Wang, G., Jia, R., Liu, J., Zhang, H.: A hybrid wind power forecasting approach based on Bayesian model averaging and ensemble learning. Renew. Energy **145**, 2426–2434 (2020)

33. Refaeilzadeh, P., Tang, L., Liu, H.: Cross-validation. Encyclopedia Database Syst. **5**, 532–538 (2009)

34. Room, C.: Cross-validation. Algorithms **7**, 15 (2022)

35. Ren, Q., Li, M., Han, S.: Tectonic discrimination of olivine in basalt using data mining techniques based on major elements: a comparative study from multiple perspectives. Big Earth Data **3**, 8–25 (2019)

36. Shen, H., Jin, H., Cabrera, Á.A., Perer, A., Zhu, H., Hong, J.I.: Designing alternative representations of confusion matrices to support non-expert public understanding of algorithm performance. Proc. ACM Hum.-Comput. Interact. **4**, 1–22 (2020)

37. PecoChacón, A.M., García Márquez, F.P.: False alarm detection in wind turbine management by tree model. In: Xu, J., García Márquez, F.P., Ali Hassan, M.H., Duca, G., Hajiyev, A., Altiparmak, F. (eds.) ICMSEM 2021. LNDECT, vol. 78, pp. 543–553. Springer, Cham (2021). https://doi.org/10.1007/978-3-030-79203-9_42

38. Hong, C.S., Oh, T.G.: Tpr-tnr plot for confusion matrix. Commun. Stat. Appl. Methods **28**, 161–169 (2021)

39. Rizwan ul, H., Li, C., Liu, Y.: Online dynamic security assessment of wind integrated power system using SDAE with SVM ensemble boosting learner. Int. J. Electr. Power Energy Syst. **125**, 106429 (2021)

40. Kotu, V., Deshpande, B.: Chapter 8 - Model evaluation. In: Kotu, V., Deshpande, B. (eds.) Data Science, 2nd edn, pp. 263–279. Morgan Kaufmann, Burlington (2019)

Intelligence Based Approach for Obtaining Trade-Off Solution Between Minimal Pollution and Cost of a Dynamic System

Srikant Misra[1] , Pratap Kumar Panigrahi[1] , Bishwajit Dey[1] ,
and Fausto Pedro Garcia Marquez[2(⊠)]

[1] GIET University, Gunupur, Odisha, India
[2] University of Castilla-La Mancha: Ciudad Real Campus, Rectorado UCLM, C. Altagracia, 50,
13001 Ciudad Real, Spain
faustopedro.garcia@uclm.es

Abstract. Microgrid can be considered as a miniature power system network wherein the generation, transmission and distribution of power takes place within a limited geographical area. Apart from encouraging the maximum penetration of renewable energy sources (RES), microgrid also have advantages like reduction of transmission losses and costs pertaining to such losses. Combined economic emission dispatch (CEED) study balances the economic and clean performance of a power system. Five objectives namely economic dispatch, emission dispatch, penalty-factor (PPF) based CEED, fractional programming (FP) based CEED and proposed environmental constrained economic dispatch (ECED) are evaluated in this paper, on a dynamic test system. The subject dynamic system considered for the study are both linear and non-linear in nature. A strong and powerful hybrid swarm-intelligence algorithm amalgamating the features of conventional grey-wolf optimizer, sine-cosine and crow-search algorithm was used as the optimization tool for the study. Detailed analysis on the results obtained supports the fact that ECED approach delivers a balanced and compromised solution between diminished generation cost and pollutants emission compered to FP and PPF based approach. The novel hybrid algorithm was found to be robust, efficient and outperformed a long list of algorithms in consistently providing better quality solutions.

Keywords: Combined economic emission dispatch · Price penalty factor ·
Renewable energy · Grey wolf optimizer · Sine cosine algorithm

1 Introduction

Economic Load Dispatch (ELD) utilizes renewable energy sources. According to ELD, if multiple units associated, they don't deliver the equal quantity of load for the equal quantity of capital, i.e., the generation cost per unit of supply will fluctuate from one unit to another. The overall production, on the other side, must match with the total energy requirement [1]. In the context of load requirement, ELD is categorized into two types:

F. P. García Márquez et al. (Eds.): ICCIDA 2022, LNNS 643, pp. 548–563, 2023.
https://doi.org/10.1007/978-3-031-27099-4_42

static ELD (SELD) which pertains to ELD difficulties where the load remains constant during the day; the other one is dynamic ELD (DELD) which refers to ELD where the load demand fluctuates over time. SELD are less complicated and restricted limits such as ramp rates, forbidden operation zones, and so on [2, 3]. In contrast, DELD is a much more complicated problem. DELD, in addition to the limits of SELD, has many more constraints related with the DERs and time periods [4]. Some of the most important and challenging limitations that must be overcome in order to solve DELD are the start-up and shutdown timeframes of DERs and the charging and discharging states of energy storage devices. The most cost-effective distribution of renewable energy generation has been the subject of intense research, which might have a significant impact on the economy of the power producing process. It is being important given the substantial increase in the gasoline price and the environment. Therefore, a thorough research of ELD is important and pertinent. Only when the cost function is continuous, smooth, differentiable, and non-convex can conventional computing techniques, such as reference [5], be employed. However, the equation becomes non-smooth and convex [6] when actual physical restrictions are attempted to be integrated into it, and standard approaches fail. For a very long time, ELD has been solved using conventional optimization techniques. Iterative lambda and gradient approaches, along with active programming, are commonly used to tackle smooth ELD problems [7]. Numerous restrictions must be taken into account in order to effectively represent an ELD situation in practice. But the goal function must be quite complex to get this great level of precision. These conventional techniques failed as a result of the target function's non-smoothness [8]. The use of these algorithms was constrained by the characteristics of the cost curve. Dynamic programming is impacted by dimensionality and takes a long time in big generating systems, while being independent of the cost curve's form [9, 10]. As a result, more meta-heuristic techniques are used, which may reduce the non-linearity of the ELD issue [11]. In these circumstances, evolutionary algorithms—which are based on diverse processes that have contributed to the development of life—proved to be highly helpful [12]. A resilient algorithm, the genetic algorithm (GA) is an evolutionary method based on global search that borrows its concept from genetic and cell reproduction [13]. According to the reduction of costs, the energy management strategy (EMS) of a microgrid is classified as DELD, being more sophisticated than SELD. Conventional generators, renewable energy sources (RES), such as micro-turbines and fuel cells, energy storage systems (ESS), such as batteries and flywheels, and other components make up distributed energy resources (DERs) [14].

2 Formulation of Fitness Function

2.1 Active Power Generation Cost Calculation of the Microgrid System

The fitness function formula for a microgrid system with grid-connectivity is represented by the Eq. (1).

$$ECD = \sum_{t}^{24}\sum_{j=1}^{n}(a_jG_{j,t}^2 + b_jG_{j,t} + c_j) \qquad (1)$$

The valves of steam boilers connected to the fossil fueled generators are operated gradually to maintain the flow of steam thereby controlling the generation of electricity. This phenomenon known as the valve point effect (VPE) is mathematically expressed by incorporating a sinusoidal term in the ECD equation as stated in Eq. 2 where d_j and e_j are VPE coefficients of j_{th} generating unit:

$$ECD = \sum_{t}^{24} \sum_{j=1}^{n} (a_j G_{j,t}^2 + b_j G_{j,t} + c_j) + |d_j * \sin(e_j(G_{j,\min} - G_{j,t}))| \tag{2}$$

2.2 Pollution Detection Formula for the Microgrid System

To evaluate the volume of carbon-di-oxide distributed in the atmosphere at the end of the day from the fossil fueled conventional producing sources we use Eq. (3),

$$EMD = \sum_{t=1}^{24} \sum_{j=1}^{n} (x_j G_{j,t}^2 + y_j G_{j,t} + z_j) \tag{3}$$

The emission dispatch equation of generating units considering VPE contains an exponential term alongside the EMD equation and is mathematically expressed in Eq. 4.

$$EMD = \sum_{t=1}^{24} \sum_{j=1}^{n} (x_j G_{j,t}^2 + y_j G_{j,t} + z_j + u_j * \exp(v_j * G_{j,t})) \tag{4}$$

2.3 Combined Economic Emission Dispatch Using PPF Method

A penalized amount is fixated upon the emission of every unit volume of carbon-di-oxide from the conventional fossil fueled generating sources. This somewhat reduces the harmful pollutants emission but increases the power generation expense of the system. The equation is mentioned in Eq. (5).

$$CEED_{ppf} = Min \left[\sum_{t}^{24} \sum_{j=1}^{n} [(a_j G_{j,t}^2 + b_j G_{j,t} + c_j) + ppf_j * (x_j G_{j,t}^2 + y_j G_{j,t} + z_j)] \right] \tag{5}$$

where 'ppf' is the price penalty factor of individual generators. There are various types of price penalty components which are mentioned in detail in [15–18].

2.4 CEED Using Fractional Programming (FP) Method

This method of evaluating CEED minimizes the overall ratio to get a balanced and negotiated resolution between reduced generation price and contaminants radiated. Mathematically it is represented as shown in Eq. (6).

$$CEED_{ppf} = Min \left[\frac{EMD}{ECD} \right] \tag{6}$$

2.5 Environment Constrained Economic Dispatch (ECED)

Authors [19–21] achieved a good compromise between the two separate fitness functions provided by Eqs. (1) and (3) mentioned above. They provided a simple equation for combining two objective functions with differing goals and achieving a higher-quality compromise solution, given by Eq. (7).

$$ECED = Min\left[\mu * \left[\frac{ECD - ECD_{min}}{ECD_{max} - ECD_{min}}\right] + (1 - \mu) * \left[\frac{EMD - EMD_{min}}{EMD_{max} - EMD_{min}}\right]\right] \quad (7)$$

where μ remains in the range of 0 and 1,

The optimal values found by minimizing Eqs. (1) and (3) are ECDmin and EMDmin, respectively. By replacing the best possible constraints of ECDmin in Eq. (3) and EMDmin in Eq. (1), the maximum quantities of ECD and EMD are found.

2.6 Operating Constraints

The DERs and grid are constrained to some operating constrains as showed in mathematical Eqs. (8) to (10)

$$\sum_{j=1}^{n} G_{j,t} = D_t \quad (8)$$

$$\sum_{j=1}^{n} G_{j,t} + P_{RES,t} = D_t \quad (9)$$

$$G_{j,min} \leq G_j \leq G_{j,max} \quad (10)$$

$$G_{grid,min} \leq G_{grid} \leq G_{grid,max} \quad (11)$$

where D_t is the demand of t_{th} hour, $P_{RES,t}$ is RES output in terms of power.

2.7 Stochastic Modelling

The unpredictable and stochastic character of RES makes evaluating the uncertainty in their creation deriving from their anticipated values a necessary step. Formulae (12) and (13) demonstrate the mathematical equations needed to calculate the uncertainty. (13) [22–24]:

$$PV_{un}^t = dPV_{un} * n_1 + PV_{fc}^t$$
$$dPV_{un} = 0.7 * \sqrt{PV_{fc}^t} \quad (12)$$

$$W_{un}^t = dP_w * n_2 + W_{fc}^t$$
$$dP_w = 0.8 * \sqrt{W_{fc}^t} \quad (13)$$

where dPV_{un} is the PV output standard deviation, is the PV output considering the uncertainty, and is the PV output projected for the day ahead. Similarly, is wind power uncertainty, is wind power standard deviation, and is anticipated wind output for the day ahead. And a conventional distribution function with a average of 1 and a standard deviation of 0 are randomly assessed.

3   Hybrid Grey Wolf Optimizers

This paper employs GWO, modified GWO (MGWO), hybrid MGWO-SCA, hybrid MGWO-CSA and hybrid MGWO-SCA-CSA for optimizing the EMS problem on microgrid systems using renewable energy power stations. MGWO-SCA-CSA have recently been used by authors in solving recent and trending power system optimization problems which includes energy management of microgrid [25], unit commitment of microgrid energy management [24] and electricity market pricing problem [16, 17, 26]. The same have also been realized in solving benchmark problems before being implemented on the aforementioned power system optimization problems. To maintain the brevity of the paper the algorithm has not been discussed din detail in this article.

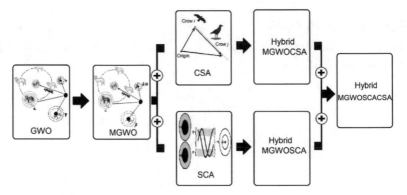

Fig. 1. Formation of proposed MGWOSCACSA from GWO

4   Case Studies

4.1   Overview of the subject test systems

The fitness functions stated in Sect. 2 are evaluated on a 5-unit dynamic system (Table 1). The fitness function of the 5-unit system are linear and quadratic in nature. The system parameters of both the systems are gathered from literatures and displayed in Tables 2. The 24 h active power demand of both the test systems can be seen in Table 3. The wind power generated from wind speed and parameters of the wind energy system have been gathered from [15]. Henceforth the power from the wind turbine have been re-evaluated using Eqs. (18) and (19) to incorporate the effect of uncertainty.

Some additional parameter related to the computer system and the algorithms are:

Personal Computer specifications: Intel Core i5 8th Gen processor 8GB RAM on a MATLAB R2013a software.

Optimization tools parameter settings: Population size- 80; Maximum Iterations- 500. fl: 2.

Table 1. System parameters of the fossil fueled generators of 5-unit system [15]

Generators	Operating limits		Fuel cost coefficients			Emission coefficients		
	min	max	a	b	c	x	y	z
	kW	kW	USD/kW^{2}h	USD/kWh	USD/h	kg/kW^{2}h	kg/kWh	kg/h
G1	100	400	0.005	3.51	44.4	0.01378	1.2489	173.37
G2	50	200	0.006	3.89	40.6	0.00767	0.8051	363.705
G3	50	300	0.004	2.78	66.9	0.0905	0.756	198.5
G4	75	500	0.0026	2.86	87.67	0.0127	1.1677	11.67
G5	150	600	0.003	2.45	105	0.01265	1.3552	22.983

Table 2. Hourly load demand of 5- and 6-units systems. [15]

Hour	1	2	3	4	5	6
5 unit (kW)	800	900	1000	1100	1250	1100
Hour	7	8	9	10	11	12
5 unit (kW)	1150	1200	1350	1450	1500	1450
Hour	13	14	15	16	17	18
5 unit (kW)	1400	1100	1200	1200	1250	1300
Hour	19	20	21	22	23	24
5 unit (kW)	1400	1450	1200	1100	950	750

4.2   Descriptive Analysis on the Results Obtained for 5-Units System

Stepwise detailed description of the study is as follows:

a. To start the study, ECD was evaluated for the linear 5-unit system using proposed MGWOSCACSA as optimization tool for various trials and the best value was recorded in Table 4. It can be noticed that the lowest amount generation price of the system was 103468 USD as achieved using MGWOSCACSA which outperformed a long list of algorithms mentioned in the literature.

b. The minimum amount of pollutants emitted for the 5 units test system was found to be 133930 kg. Since EMD for this test system was not calculated in any of the literatures before, various other hybrid and modified algorithms were implemented to minimize EMD. It can be realized from Table 5 that the least value of EMD was achieved using MGWOSCACSA.

Table 3. Minimum values of ELD and CEED for 5 units system.

Algorithms	GA [15]	PSO [15]	DE [15]	SCA [15]	TLBO [15]	GWO [15]	MGWO [15]	WOA [15]	MWO [15]	CSA [15]	MGWOSCACSA
ELD	103625.2777	103474.3922	103600.8322	103477.1104	103478.823	103486.5133	103482.164	103732.5368	103662.4691	103469.3322	103468.5678
CEED	120305.987	120218.726	120293.963	120224.386	120254.919	120225.316	120221.594	120600.956	120464.711	120219.072	120213.596

Table 4. Microgrid emission dispatch of 5-units system using optimization techniques.

Algorithms	EMD (kg)
GWO	134038.8227
MGWO	134119.5790
MGWOSCA	133964.8185
WOASCA	133937.7333
MGWOSCACSA	**133930.9561**

c. The PPF values for the 5-unit test system gathered from [15] are 0.1001 USD/kg, 0.1547 USD/kg, 0.3007 USD/kg, 0.0251 USD/kg and 0.0840 USD/kg for G1, G2, G3, G4 and G5 individually. Subsequently, PPF centered CEED was performed on the assessment structure and the generation cost was documented for the same. Table 4 displays that generation cost of the 5-units system when PPF based CEED was evaluated using MGWOSCACSA was 120213.5961 USD. This value was better and lesser than those obtained using 10 different algorithms as listed in Table 4. Also the corresponding amount of emitted pollution was found to 157191 kg.

d. When FP based CEED was assessed for the 5-units test system, the generation cost was 117630.6 USD and the corresponding amount of pollutants emitted 137880.3 kg. These results are displayed in Table 5.

Table 5. Fitness function values for 5-units system using MGWOSCACSA

	ECD	EMD	CEED$_{PPF}$	CEED$_{FP}$	ECED
Fitness function value	103468.5678	133930.9561	120213.5	1.039	0.23398
Cost (USD)	**103468.5678**	112847.1929	**120213.5**	**117630.6**	**105559.57**
Emission (kg)	230484.2835	**133930.9561**	**157191.2**	**137880.3**	**157587.41**
CPI	100	0	24.09054	4.0903375	24.500917
EPI	0	100	178.54367	151.00278	22.295374
|CPI~EPI|	100	100	154.4531	146.9124	2.2055

e. Henceforth, ECED was performed for the linear test system with the necessary data obtained in the previous steps (discussed in Sect. 2.5) which are listed below.

Parameters	Values obtained using MGOWCSACSA
ECD_{min} (USD)	103468.5678
ECD_{max} (USD)	112847.1929
EMD_{min} (kg)	133930.9561
EMD_{max} (kg)	230484.2835

The value of ω was altered in the range of 0 to 1 with an increment of 0.1 and the generation cost and pollutants emitted were recorded for every value of ω. It can be seen from Fig. 1 that the best negotiated solution was achieved at $\omega = 0.5$ and is (157587 kg, 105559 USD) (Fig. 2).

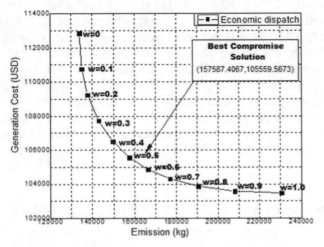

Fig. 2. Cost vs. Emission using MGWOSCACSA for 5-units system

Thereafter the values of ECED at various values of ω ranging from 0.1 to 0.9 were evaluated using DE, GWO and proposed MGWOSCACSA and the generation cost and emission value were studied carefully. It was noticed that the ECED fitness function emphasized more on generation cost when ω ranged from 0.1 to 0.4. Likewise when ω ranged from 0.6 to 0.9 more emphasis was given to minimize the pollutants emission of the system. A balanced and compromised value of both generation cost and pollutants emission was obtained when ω was 0.5. This can be viewed from Fig. 3.

Authors in [14] stated explicitly a better and balanced solution is usually yielded if the absolute difference between CPI and EPI, say 'Difference' is minimum. In order to justify and correlate this statement with the present work, Difference was evaluated for various values of ω while calculating ECED and also for all of FP based CEED, PPF based CEED and ECED. It can be seen from Fig. 3 that least difference between CPI and EPI (2.2055) was obtained when the value of ω was 0.5 (Fig. 4). Also when the Difference was evaluated for different methods of CEED, it can be seen from Fig. 5 that the smallest difference was obtained for proposed ECED compared to FP based CEED

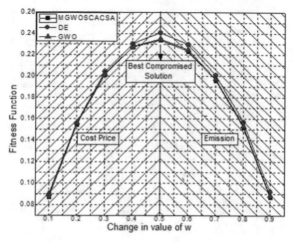

Fig. 3. ECED values for 5-units system evaluated with different values of ω

and PPF based CEED. This study again proves that the proposed ECED approach with $\omega = 0.5$ is significantly better than the other two methods of conducting CEED.

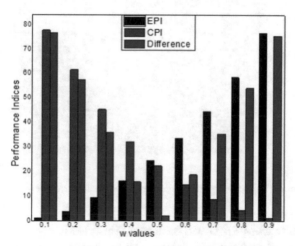

Fig. 4. Indices EPI and CPI for 5-units system evaluated with different values of ω

To display the hourly outputs of the generators in tables corresponding to the best value of fitness function as evaluated by MGWOSCACSA, it would acquire large spaces as each table would be of dimension 24×5 (without headings and titles). To attend this issue, UP as mentioned in Eq. (17) is evaluated of every generator for all the fitness functions and is displayed in Fig. 6.

From the figure it can be depicted that generators G3 and G4 were utilized more while evaluating ECED meaning that these generators have least values of cost coefficients. Generators G1 was utilized 65% to its capacity and G2 was almost completely utilized

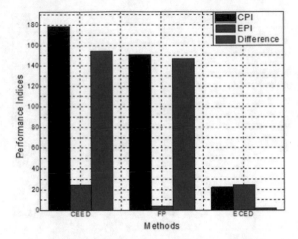

Fig. 5. Difference between CPI and EPI for different methods of evaluating CEED

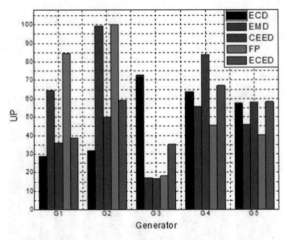

Fig. 6. Utilization percentage of generators for the 5-units system

throughout the day when EMD was evaluated. This clearly signifies that G1 has less and G2 has the least emission coefficients. All the generators can be seen to be used moderately used when ECED was evaluated yielding the best compromised solution between ECD and EMD.

The power generated from the wind energy system corresponding to the hourly wind speed is shown in Fig. 7 below. It is to be noted that the vital contribution of the wind power is in minimizing the load demand share on the fossil fueled generators such that less amount of fuels are combusted. This in turn decreases the amount of harmful gases emitted in the atmosphere.

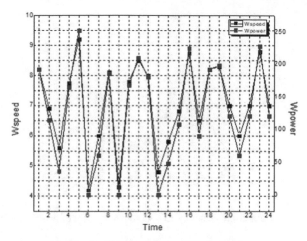

Fig. 7. Wind power vs. Wind speed

4.3 Statistical Analysis on the Results Obtained Using Various Algorithms

The evaluation of the fitness functions were executed with utmost dedication and the results were analyzed from every aspect. DE, GWO and proposed MGWOSCACSA were implemented as the optimization tool to assess ECED on the 5-units system. The value of the fitness function was assessed for 30 specific trials and the elapsed time to attain the maximum number of iterations were recorded for every trial. It can be realized from Fig. 8 that the proposed algorithm consumes 32–37 s to attain the stopping criteria (maximum number of iterations) compared to DE and GWO which needs 38–45 s and 40–45 s respectively. Also it can be pictured from Fig. 9 the value of ECED obtained by the proposed MGWOSCACSA throughout the 30 trials was the least compared to DE and GWO. These analyses aids to the superior and efficient nature of the proposed algorithm.

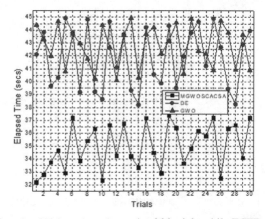

Fig. 8. Elapsed time for 500 iterations for each of 30 trials while ECED for 5 units system

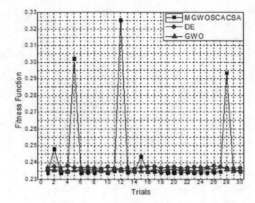

Fig. 9. Fitness function values of ECED for every trial of 5-units test system

Figures 10 shows the convergence curve characteristics of the various algorithms when ECED was evaluated for the test system. A convergence curve characteristics is generally the value of fitness function obtained during each iteration by an algorithm till the maximum number of iterations is reached. It can be seen from both the figures that the least value of fitness function was attained using proposed MGWOSCACSA algorithm.

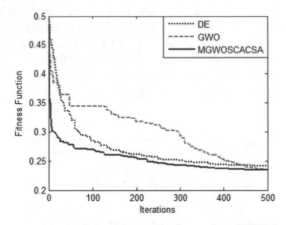

Fig. 10. Convergence curves of algorithms utilized to evaluate ECED for 5 units system

5   Conclusions

The detailed depth-in analysis of the results and discussion studied in the paper points to the following findings:

a. ECED method of obtaining an economic and environmentally clean solution to active power generation in a power system network is by far a better option compared to FP and PPF based CEED given the availability of data.
b. RES support helps in reduction of load share from the fossil fueled generators. This also plays a crucial role in minimizing the emission of harmful pollutants in to the atmosphere.
c. Proposed MGWOSCACSA algorithm outperformed many optimization algorithms in consistently and efficiently delivering providing better quality solutions for both linear and nonlinear objective functions. The hybrid algorithm was also swift enough to handle the large dimensional problems and handling the constraints without even minimum band of error. This elevates the superiority of the proposed MGWOSCACSA algorithm and encourages to be further implemented in solving various optimization problems

Weakness of the Study: An honest and sound approach to conclude the study would also take into account the limitations faced. The fitness function of the proposed ECED method needs to calculate ECD and EMD beforehand to generate the necessary data. This issue is not seen in FP and PPF based approaches.

As far as the proposed hybrid MGWOSCACSA is considered, some initial time consumption is realized while coding the algorithm, but the swiftness and efficiency of the algorithm in delivering a superior quality solution makes up for the same.

Future Scope of Work: To broaden the horizon of the research related to the study of this paper, ECED can be evaluated on a grid-connected microgrid system comprising of different types of distributed energy resources and their complex constraints. This would test the compatibility of the proposed hybrid algorithm in handling those constraints in dedicated amount of execution time.

Declaration of Conflicts of Interest: The authors declare that they have no known competing financial interests or personal relationships that could have appeared to influence the work reported in this paper.

Acknowledgement. Authors would like to acknowledge the support of the Department of Science and Technology (DST), Govt. of India, and GIET University, Gunupur, Odisha, India for the financial & technical support for this work.

References

1. Ghosh, B., Dey, B., Bhattacharya, A.: Solving economic load dispatch problem using hybrid Krill Herd algorithm. In: 2015 International Conference on Energy, Power and Environment: Towards Sustainable Growth (ICEPE), pp. 1–6. IEEE (2015)
2. Yalcinoz, T., Short, M.: Neural networks approach for solving economic dispatch problem with transmission capacity constraints. IEEE Trans. Power Syst. **13**, 307–313 (1998)
3. Dhillon, J., Parti, S., Kothari, D.: Stochastic economic emission load dispatch. Electr. Power Syst. Res. **26**, 179–186 (1993)

4. Dey, B., Roy, S.K., Bhattacharyya, B.: Neighborhood based differential evolution technique to perform dynamic economic load dispatch on microgrid with renewables. In: 2018 4th International Conference on Recent Advances in Information Technology (RAIT), pp. 1–6. IEEE (2018)

5. Kai, S., Qing, L., Jizhen, L., Yuguang, N., Ruifeng, S., Yang, B.: New combination strategy of genetic and tabu algorithm an economic load dispatching case study. In: 2011 Chinese Control and Decision Conference (CCDC), pp. 1991–1995 (2011)

6. Bhattacharya, A., Chattopadhyay, P.K.: Solving complex economic load dispatch problems using biogeography-based optimization. Expert Syst. Appl. **37**, 3605–3615 (2010)

7. Sinha, N., Chakrabarti, R., Chattopadhyay, P.: Evolutionary programming techniques for economic load dispatch. IEEE Trans. Evol. Comput. **7**, 83–94 (2003)

8. Daniel, L., Chaturvedi, K.T., Kolhe, M.L.: Dynamic economic load dispatch using Levenberg Marquardt algorithm. Energy Procedia **144**, 95–103 (2018)

9. Hosseinnezhad, V., Babaei, E.: Economic load dispatch using θ-PSO. Int. J. Electr. Power Energy Syst. **49**, 160–169 (2013)

10. Yang, X., et al.: Multi-objective optimal scheduling for CCHP microgrids considering peak-load reduction by augmented ε-constraint method. Renew. Energy (2021)

11. Nwulu, N.I., Xia, X.: Optimal dispatch for a microgrid incorporating renewables and demand response. Renew. Energy **101**, 16–28 (2017)

12. Coelho, V.N., Coelho, I.M., Coelho, B.N., Cohen, M.W., Reis, A.J., Silva, S.M., et al.: Multi-objective energy storage power dispatching using plug-in vehicles in a smart-microgrid. Renew. Energy **89**, 730–742 (2016)

13. Goldberg, D.E., Holland, J.H.: Genetic algorithms and machine learning (1988)

14. Zheng, Y., Jenkins, B.M., Kornbluth, K., Træholt, C.: Optimization under uncertainty of a biomass-integrated renewable energy microgrid with energy storage. Renew. Energy **123**, 204–217 (2018)

15. Dey, B., Bhattacharyya, B., Srivastava, A., Shivam, K.: Solving energy management of renewable integrated microgrid systems using crow search algorithm. Soft. Comput. **24**(14), 10433–10454 (2019). https://doi.org/10.1007/s00500-019-04553-8

16. Dey, B., Bhattacharyya, B., Devarapalli, R.: A novel hybrid algorithm for solving emerging electricity market pricing problem of microgrid. Int. J. Intell. Syst. **36**, 919–961 (2021)

17. Dey, B., Bhattacharyya, B.: Comparison of various electricity market pricing strategies to reduce generation cost of a microgrid system using hybrid WOA-SCA. Evol. Intel. **15**, 1–18 (2021). https://doi.org/10.1007/s12065-021-00569-y

18. Dey, B., Basak, S., Bhattacharyya, B.: A comparative analysis between price-penalty factor method and fractional programming method for combined economic emission dispatch problem using novel hybrid CSA-JAYA algorithm. IET Smart Grid (2021)

19. Dey, B., Bhattacharyya, B., Márquez, F.P.G.: A hybrid optimization-based approach to solve environment constrained economic dispatch problem on microgrid system. J. Clean. Prod. 127196 (2021)

20. Rajasomashekar, S., Aravindhababu, P.: Biogeography based optimization technique for best compromise solution of economic emission dispatch. Swarm Evol. Comput. **7**, 47–57 (2012)

21. Mandal, S., Mandal, K.K.: Optimal energy management of microgrids under environmental constraints using chaos enhanced differential evolution. Renew. Energy Focus **34**, 129–141 (2020)

22. Dey, B., Bhattacharyya, B., Raj, S., Babu, R.: Economic emission dispatch on unit commitment-based microgrid system considering wind and load uncertainty using hybrid MGWOSCACSA. J. Electr. Syst. Inf. Technol. **7**(1), 1–26 (2020). https://doi.org/10.1186/s43067-020-00023-6

23. Jamshidi, M., Askarzadeh, A.: Techno-economic analysis and size optimization of an off-grid hybrid photovoltaic, fuel cell and diesel generator system. Sustain. Cities Soc. **44**, 310–320 (2019)
24. Li, X., Song, Y.-J., Han, S.-B.: Frequency control in micro-grid power system combined with electrolyzer system and fuzzy PI controller. J. Power Sources **180**, 468–475 (2008)
25. Dey, B., Basak, S., Bhattacharyya, B.: MGWOSCACSA: a novel hybrid algorithm for energy management of microgrid systems. In: Reddy, M.J.B., Mohanta, D.K., Kumar, D., Ghosh, D. (eds.) Advances in Smart Grid Automation and Industry 4.0. LNEE, vol. 693, pp. 669–678. Springer, Singapore (2021). https://doi.org/10.1007/978-981-15-7675-1_67
26. Dey, B., et al.: Optimal scheduling of distributed energy resources in microgrid systems based on electricity market pricing strategies by a novel hybrid optimization technique. Int. J. Electr. Power Energy Syst. **134**, 107419 (2022)

Improving Classification Accuracy of Cyber Attacks in the Banking Sector

Basil Abbas Hasan Hasan$^{(\boxtimes)}$ ⓘ and Muhammad Ilyas ⓘ

Altinbas University, Istanbul, Turkey
203720533@ogr.altinbas.edu.tr, muhammad.ilyas@altinbas.edu.tr

Abstract. Whether using neural networks to classify data or a biometrics application such as handwriting analysis or iris recognition, The Nearest Neighbor Classifier, which achieves classification by locating the closest neighbors to a query sample and utilizing those neighbors to determine the class of the query, is conceivably the most accurate classifier in the stockpile or machine learning techniques. K-NN classification assigns classes to examples depending on how closely those instances resemble those in the training set. This paper presents a variety of output with different distances used in the algorithm, which may help to understand how the classifier responds for the intended application. It also illustrates computational challenges in identifying nearest neighbors and strategies for reducing the dimension of the data. Initial tests using SYN TCP flood data indicate that the kNN classifier can effectively detect intrusive attacks and achieve a high degree of accuracy thar reaches up to 94% in filtered data and 89% in unfiltered data with and error rate of 0.08% .

Keywords: KNN · AI · TCP · Cyber security · Banking sector

1 Introduction

In recent years, there has been a steady growth in the number of Internet users, which has increased in tandem with the development of the network and technology. New types of security vulnerabilities and threats have emerged as a direct result of the rapid expansion. Utilizing a number of inventive methods, hackers may gain access to computer systems and networks. Traditionally, authentication, virtual private networks (VPNs), and firewalls were employed to ensure network security [9]. As a consequence of the evolution of these technologies, intrusion detection systems, or IDSs, have emerged. The basic function of IDSs is to identify network weaknesses and alert the system if an intruder has obtained access. According to the White House, a number of government agencies and commercial enterprises are already using artificial intelligence-powered algorithms. Why? AI is able to quickly scan standardized data and thoroughly examine unstructured data (numbers, speech patterns, and words), so saving time and resources. AI may also examine standardized data (numbers, speech patterns, and sentences). In actuality, artificial intelligence is capable of saving both public dollars and state secrets. In addition to this, there exist informational gaps. In order to get access to the devices, hackers are

© The Author(s), under exclusive license to Springer Nature Switzerland AG 2023
F. P. García Márquez et al. (Eds.): ICCIDA 2022, LNNS 643, pp. 564–576, 2023.
https://doi.org/10.1007/978-3-031-27099-4_43

devising techniques to circumvent security measures of which we were unaware. The time required for a business to notice that a data breach has happened has already gone [11].

The hacker has already vanished, taking the sensitive data he was attempting to get with him. In contrast, artificial intelligence must wait for a hacker to create a mess before it can begin data collection. When a user signs in or inputs their password, artificial intelligence does a scan to identify aberrant hacker behavioral patterns. Because the AI can detect even the most subtle clues, it is feasible that the hacking team may be halted in their tracks. As pointed out by Varughese, any gadget has the potential to be exploited. Because cybersecurity is an endless game of chess, human hackers will constantly seek vulnerabilities in any system, even AI. Even if humans control artificial intelligence, it is still feasible to successfully overcome it. When it comes to connecting and analyzing data, artificial intelligence (AI) is unrivaled [12] (Fig. 1).

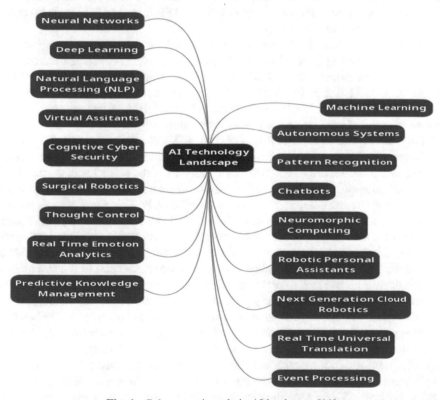

Fig. 1. Cyber security role in AI landscape [11]

As computer hackers grow more adept with AI systems, engineers will be needed to create novel defenses. The cat-and-mouse game will continue, but the arrival of artificial intelligence is a good advance in the effort to keep critical data safe. Google has built a graphical data learning model using the machine learning platform Tensor Flow. The search was conducted on March 9, 2019. The researchers have built an open-source framework known as NSL that use the Neural Graph Learning approach to train neural network data sets and data structures. The implementation of NSL was conducted by the researchers. Tensor Flow, which is used by NSL, is accessible to both experienced and inept machine learning specialists. One approach for creating predictions based on NSL is to use an interactive database. It might be a statistics graph or a medical report. Researchers working on the Tensor Flow project said in a blog post released today that the use of ordered signals during training allows developers to offer increased prediction performance even with a relatively small number of data points. Directly resulting from this, the basics of structured-signal processing are driven to develop more durable models. Google has tried a variety of techniques to increase the performance of their model [14], one of which is learning semantic image embedding. Other methods include In some scenarios, NSL may design representations that take use of visual cues to regularize throughout the development process with less than 10 lines of code; but, in other situations, it may need more than ten lines of code. In addition, the first framework provides developers with data organizing tools and APIs that need little code in order to create vector quantization instances.

In April, Google Cloud introduced new techniques to data management, including connected sheets in Big Query. Google AI, formerly known as Google Research, is the business that open-sourced both the GPT2 and SM3 compilers for large-scale speech recognition models like as Google's BERT [15]. The face recognition capabilities of Siri, Google, and Facebook were made feasible by artificial intelligence (AI). A rising number of payment card issuers are using artificial intelligence (AI) to aid investment banks in avoiding fraud totaling billions of dollars. Alternatively, what applications can they discover for Information Security? Is the use of artificial intelligence advantageous or damaging to the security of digital information? On the other hand, current information management architecture is advantageous since it allows security professionals to review, analyze, and grasp cybercrime more effectively. This is an advantage. With this, companies have a greater chance of defending themselves against cybercrime and protecting both their employees and customers. On the other side, artificial intelligence (AI) is capable of consuming vast quantities of resources [16]. It is possible that no implementation can do this. In reality, it may be a strong weapon in the inventory of cybercriminals who use technology to amplify and improve their attacks against other machines. Due to the fact that every procedure in the economic, industrial, political, and social spheres is directly linked to information resources and the use of information technology, the active informatization of modern society and the increase in confidential information flows have necessitated the implementation of information (data) security (DS) (IT). As information systems and technology continue to advance and become more complicated, it is anticipated that the number of possible threats to these systems will likewise increase. IS's information-protection (IP) systems must be evaluated in order to give DS and precisely foresee how certain circumstances would play out. In order to address this issue, statistical analytic techniques, namely the correlation-regression approach, might be used.

When using these methods, which need complicated calculations and a considerable quantity of experimental data, processing takes a very long time, and it is not feasible to meet the quality standards established by experts. In this specific context, a cognitive approach based on the development of fuzzy cognitive maps (FCM) that was initially presented by Kosco et.al [2] might be considered. When using FCM, one of the key issues is simplifying, building, and elucidating the causal relationships between the many components of a complex system, which may be challenging when employing traditional methods. Adapting to the unpredictability of the original data and conditions is a further important issue [2]. As a last point, it is essential to underline how easy it is to add new components to the cognitive map's graph. This is a crucial aspect. It is crucial to develop methods for describing and classifying the current level of DS service. To successfully implement information protection, it is required to describe and classify the different risks and problems, as well as to design a series of actions that may be performed to counteract these threats and challenges [4]. Using methodologies derived from studies on casual relationships, it is feasible to analyze the number of DS offered. Using these approaches, it is possible to discover the links between a range of risks and dangers, to identify the elements that acted as catalysts for the emergence of particular risk factors, and to build preventive measures. Depending on the needed degree and breadth of protection, a variety of methods may be used to assess the condition of the DS supply. It can now differentiate between various sorts of threats and deliver differing levels of protection based on their severity.

Lee et al. [7] extended the work of Forrest's group and applied RIPPER, a rule learning program, to the audit data of the UNIX send mail program. Both normal and abnormal traces were used. War renders et al. [6] introduced a new data modeling method, based on Hidden Markov Model (HMM), and compared it with RIPPER and simple enumeration method. For HMM, the number of states is roughly the number of unique system calls used by the program. Although HMM gave comparable results, the training of HMM was computationally expensive, especially for long audit traces. Ghosh et.al [8] employed artificial neural network techniques to learn normal sequences of system calls for specific UNIX system programs using the 1998 DARPA BSM data. More than 150 program profiles were established. For each program, a neural network was trained and used to identify anomalous behavior. [5].

In this paper, the author investigated about classification accuracy of cyber attacks in the banking sector. That Threats are constantly evolving, and the cybersecurity landscape is constantly changing. The stakes are very high in the banking and financial industry since there are substantial financial sums at risk and the potential for significant economic upheaval if banks and other banking institutions are attacked. These are the main dangers banks and other financial institutions are expected to face throughout 2022. For many years, the ransom has been a huge nuisance for companies worldwide, and it doesn't appear that this will soon change. This cybercrime locks users out of the system and encrypts user files before asking for money to let users back in.

Organizations hit by ransomware attacks may experience prolonged system crippling, especially if they will not have a backup. It's also not assured that paying these thieves' ransoms will lead to the restoration of access to your systems. Financial institutions now more than ever face potential cybersecurity weaknesses. Extra caution is required because workers are no longer always receiving information on the organization-controlled systems and networks. Banks must ensure that the public cloud is set securely to prevent damaging breaches. Social engineering is among the greatest challenges to banking and finance. People are frequently the weakest link in the security chain since they can be duped into divulging important information and login credentials. Customers and employees of a bank may both be impacted by this. Social engineering can take many different forms, such as sending phony bills that appear to be from a reliable source or engaging in hacking or whale assaults. Keeping your staff updated on social engineering techniques or how these dangers are developing is crucial.

The rest of the structure of the paper is as follows: the proposed methodology is discussed in Sect. 2, Sect. 3 explains the simulation result, and Sect. 4 explains the study's conclusion.

2 Proposed Method

It is likely that the network traffic may change its behavior, so modifying the statistical data, and this will lead to inaccurate inferences being drawn based on past learning and experience. It's also a problem since the data flow is often fast and voluminous, making it practically impossible to label entries due to the limited storage space available in the system. IoT network traffic may be divided into two types: Stream traffic and Data Stream traffic. Because batch learning is not appropriate for these networks, it is necessary to use a different method in this case. Methods of stream classification are highly suggested for networks with this amount of data since they are fast and efficient. Unlike batch learning, these algorithms only get a single sample at a time, which makes them more efficient. [6] In contrast to traditional classification techniques [5], stream classification methods [6] do not need a learning phase, perform well in networks when drift is prevalent, and are especially beneficial when traffic is high. Some of these algorithms, such as [7], are shown in detail.

2.1 Dataset

Only a limited number of nations and a subset of sectors have data available. As a result, when we combine the two databases, some areas' data is missing. Furthermore, the investigation of cross-country variation and the viewed differently across time is somewhat constrained by a lack of consistency across countries and periods in the dataset. Several banking businesses use data mining tools to segment consumers and productivity, authorize credit, forecast financial accounts, advertise, detect bogus activities, etc. The connection between data mining techniques and numerous cybercrimes in banking apps is covered in-depth in this essay. Cyber crime data mining aims to spot trends in criminal behavior to predict, avert, and protect against crime. This work employs special data mining techniques, including K-Means, Influenced Associations Classifier, and J48 Predictions tree, to explore the cyber crime data sets and arrange the engineering problems. The KMeans technique is used for unattended groups inside impacted link classification. Each attack scenario included in the collection is represented by a distinct dataset. In both instances, fabricated IP addresses are generated at random. The target port for TCP flood assaults is TCP port 80. There were eight minutes allocated for each dataset. In each, there is an 80-s rest period followed by a 20-s attack phase. Researchers use variable packet rates to test their detection systems. They provide four attack rates, ranging from 1,000 to 2,500 packets per second. Figure 2 displays the structure of the assault.

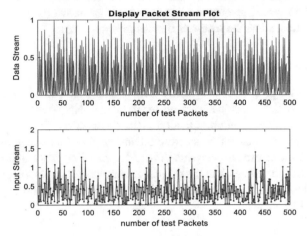

Fig. 2. Ture packet stream (Blue) and Attack Packet stream (Red)

2.2 Attack Model

Using the same host, destination IP address, and port number, a very high number of packets are transmitted. After that moment, the system will fail for one reason or another at some time in the future. This exploit takes use of a TCP protocol weakness and is known as a TCP SYN attack. By sending a high number of SYN requests, the attacker begins contact with the server. The server will send the client a SYN + AFC packet in response to the client's request before waiting for the client to deliver an AFC packet of its own. Though the attacker is unable to transmit an AFC packet, the server will continue to attempt to receive one even if it has not been sent back. Since the server's buffer is already full, any subsequent legitimate requests will be disregarded if the buffer continues to fill. If the information source has n independent symbols each with a probability of choice Pi, the entropy H is defined as follows [8]:

$$H = \sum_{i=1}^{n} P_i \log_2 P_i \tag{1}$$

For the 2 elements $X = \{x1, x2, ..., xn\}$ and $Y = \{y1, y2, ..., yn\}$, $W = \{w1, w2, ..., wn\}$ is the weighted vector and wi is the weight of the component i in the general vector. Then, we compute the similarity between two elements X and Y as follows [8]:

$$Similarity(X, Y) = Cosine(X, Y, W)$$

$$= \frac{\sum_{i=1}^{n} (x_i \times w_i) \times (y_i \times w_i)}{\sqrt{\sum_{i=1}^{n} (x_i \times w_i)^2} \sqrt{\sum_{i=1}^{n} (y_i \times w_i)^2}} \tag{2}$$

An authentication server is susceptible to an attack in which it needs much more resources to verify a false signature provided by an attacker than it does to create a signature. An attacker's enormous number of CGI requests needed a substantial quantity of CPU cycles and system resources. Having laid this groundwork, we will now examine the features of attacks and the methods for recognizing them.

2.3   KNN Workflow

This system's main purpose is to identify network threats. The KNN algorithm is the simplest of all existing machine learning techniques. KNN, sometimes referred to as instance-based learning, is an unmotivated kind of education. [2] The KNN algorithm is a non-parametric method for concealing the underlying data structure. Although the KNN approach may be used for classification and regression, data classification is by far its most prevalent use. Research in the subject of computer security has greatly benefited from efforts made in the field of intrusion detection. Figure 3 depicts the proposed attack detection model of the study.

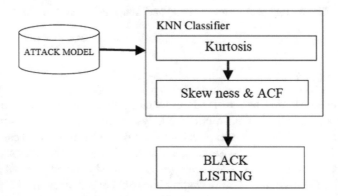

Fig. 3. Diagram of the proposed attack detection model

Currently, anomaly detection and abuse detection are the two most used methods for detecting intrusions. The behaviors of a user are matched to the distinguishing criteria of invasive assaults using a technique known as pattern matching. The acts that fulfill the criteria for classification as an invasion are then labeled as such. Table 1 Shows the types of data input to the system:

In other words, a model-reference approach is important to the process of abuse identification. When it comes to discovering new threats, the findings are not as effective as when using abuse detection to identify known incursion kinds. Nevertheless, abuse detection is fairly successful. When it comes to anomaly detection, on the other hand, you are seeking for patterns that depart from the norm Despite its capacity to identify previously unknown kinds of assault, anomaly detection systems struggle to determine what constitutes "normal." A substantial number of false alarms are often generated by anomaly detection algorithms [3] due to their inability to distinguish between abnormal patterns created by authorized users and those initiated by attackers. Even if a different

Table 1. List of some of the features in CSV dataset

Feature	Data type
FLOW_ID	Categorical
DST_PORT	Integer
TOTLEN_FWD_PKTS	Float
SUBFLOW_FWD_PKTS	Float
PKT_LEN_STD	Float
SYN_FLAG_CNT	Categorical
PSH_FLAG_CNT	Integer

strategy is used, almost all strategies for identifying intrusions depend on some type of user activity trace. A system for detecting intrusions is subject to being misled by hostile actors that vary their activity patterns over time. When a user logs in, he or she has complete control over elements such as the log-in time and command set [4]. In addition to issues around the user's right to privacy, this makes it less enticing to attempt to imitate user behavior [8].

3   Results of Simulation

This program illustrates the Cyber Security Network by using the TCP (Transmission Control Protocol) with Packet Attack Effect & How to Detect this Attack Using KNN Learning algorithm with Statistical Methods such as (Mean, Standard deviation, Kurtosis, & Skew ness & ACF) Also Finding the Blacklist & Preventing Attacks technique.

The mixed stream entries have been plotted as depicted in Fig. 4.

The autocorrelation function, AFC, has been shown in Fig. 5.

The attack stream ACF has been also obtained and plotted in Fig. 6.

Then, the packet stream Skewness measure has been calculated and shown in Fig. 7.

Accordingly, the attack stream Skewness measure has been calculated and shown in Fig. 8.

The packet stream Kurtosis measure has been calculated and shown in Fig. 9.

Similarly, the attack stream Kurtosis measure has been calculated and shown in Fig. 10.

Now, the mixture stream ACF measure has been calculated and shown in Fig. 11.

Also, the mixture stream Skewness measure has been calculated and shown in Fig. 12.

Then, the mixture stream Kurtosis measure has been calculated and shown in Fig. 13.

From the above results, we could conclude that the program has the ability to detect the existing attack stream through the statistical measurements implemented throughout the instruction codes.

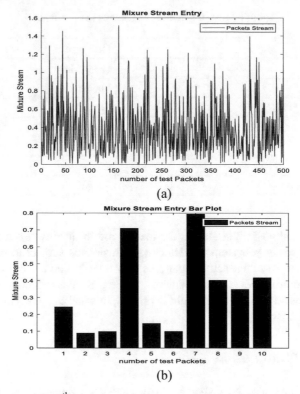

Fig. 4. Results of the 5th plot, the mixed stream, (a) sampled format, (b) bar format.

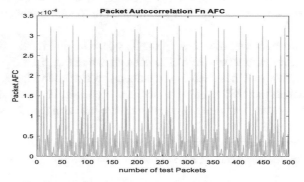

Fig. 5. Results of the mixed stream AFC function test.

4 Conclusion

Based on the k-Nearest Neighbor classifier technique, within the scope of this work. Our first testing, which used data from the SYN Flood audit, indicated that this method is very accurate at detecting potentially invasive software activity. The KNN classifier needs less calculations than other approaches that rely on short system call sequences for

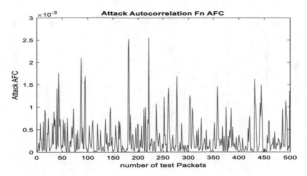

Fig. 6. Results of the attack stream AFC function test.

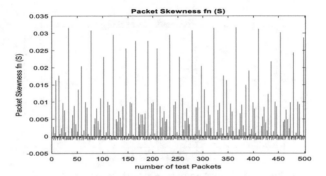

Fig. 7. Results of the packet stream Skewness measure test.

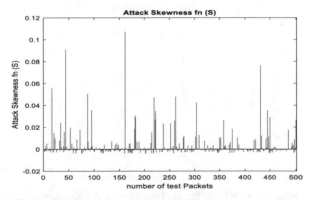

Fig. 8. Results of the attack stream Skewness measure test.

their work since it does not need to create distinct profiles of short system call sequences for each application. In addition, we determined that the probability of obtaining a false-positive result was rather low. Although this finding may not be valid when compared to a data set with a greater degree of complexity, text classification algorithms seem to be a viable option for the intrusion detection industry. Current research focuses on

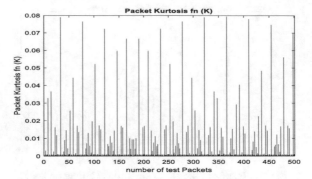

Fig. 9. Results of the packet stream Kurtosis measure test.

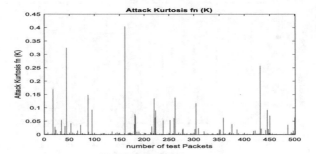

Fig. 10. Results of the attack stream Kurtosis measure test.

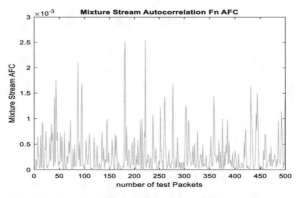

Fig. 11. Results of the mixture stream ACF measure test.

the dependability and scalability characteristics of the KNN classifier method. We are comparing the KNN classifier against a number of machine learning algorithms to see which strategies choose the most relevant system calls for categorization the proposed model has a high accuracy in detecting security threats that reached up to 94% with error rate as low as 0.05%. In the future, our research will include an emphasis on mixed

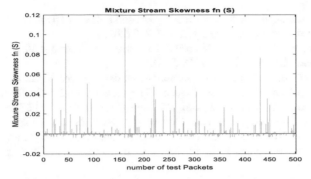

Fig. 12. Results of the mixture stream Skewness measure test.

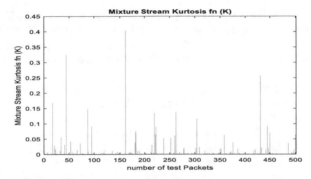

Fig. 13. Results of the mixture stream Kurtosis measure test.

modeling of program behavior. This kind of modeling takes both the local ordering and the frequency of system calls into consideration.

References

1. Shin, S., Gu, G.: Attacking software-defined networks: a first feasibility study. In: Proceedings of 2nd ACM SIGCOMM Workshop Hot Topics Software Defined Netwrk (HotSDN), pp. 165–166 (2013)
2. Fonseca, P., Bennesby, R., Mota, E., Passito, A.: A replication component for resilient OpenFlow-based networking. In: Proceedings of IEEE Network Operations and Management Symposium, pp. 933–939, April 2012
3. Yao, G., Bi, J., Guo, L.: On the cascading failures of multi-controllers in software defined networks. In: Proceedings of 21st IEEE International Conference on Network Protocols (ICNP), pp. 1–2, October 2013
4. Lin, H., Wang, P.: Implementation of an SDN-based security defense mechanism against DDoS attacks. In: Proceedings of Joint International Conference on Economics and Management Engineering (ICEME) and International Conference on Economics and Business Management (EBM), Philadelphia, PA, USA, 2016
5. Yang, J.G., Wang, X.T., Liu, L.Q.: 'Based on traffic and IP entropy characteristics of DDoS attack detection method.' Appl. Res. Comput. **33**(4), 1145–1149 (2016)

6. Saied, R.E.O., Radzik, T.: 'Detection of known and unknown DDoS attacks using artificial neural networks. Neurocomputing **172**, 385–393 (2016)
7. Ye, J., Cheng, X., Zhu, J.: A DDoS attack detection method based on SVM in software defined network. Secur. Commun. Netw. **2018** (2018)
8. Cui, J., Wang, M., Luo, Y.: 'DDoS detection and defense mechanism based on cognitive-inspired computing in SDN.' Future Gener. Comput. Syst. **97**, 275–283 (2019)
9. Yu, Y., Guo, L., Liu, Y., Zheng, J., Zong, Y.: 'An efficient SDN–based DDoS attack detection and rapid response platform in vehicular networks.' IEEE Access **6**, 44570–44579 (2018)
10. De Assis, M.V., Novaes, M.P., Zerbini, C.B., Carvalho, L.F., Abrao, T., Proenca, M.L.: Fast defense system against attacks in software defined networks. IEEE Access **6**, 69620–69639 (2018)
11. Wang, Y., Hu, T., Tang, G., Xie, J., Lu, J.: SGS: safe–guard scheme for protecting control plane against DDoS attacks in software–defined networking. IEEE Access **7**, 34699–34710 (2019)
12. Dong, S., Abbas, K., Jain, R.: A survey on distributed denial of service (DDoS) attacks in SDN and cloud computing environments. IEEE Access **7**, 80813–80828 (2019)
13. Liu, J., Lai, Y., Zhang, S.: FL-GUARD: a detection and defense system for DDoS attack in SDN. In: Proceedings of International Conference on Cryptography, Security and Privacy, pp. 107–111 (2017)
14. Xie, H., Tsou, T., Lopez, D., Yin, H., Gurbani, V.: Use cases for alto with software defined networks, document internet-draft draft-xie-altosdn-extension-use-cases-01. Working Draft, IETF Secretariat (2012)
15. Specht, S.M., Lee, R.B.: Distributed denial of service: taxonomies of attacks, tools, and countermeasures. In: Proceedings of ISCA PDCS, pp. 543–550 (2004)
16. Studer, A., Perrig, A.: The coremelt attack. In: Backes, M., Ning, P. (eds.) ESORICS 2009. LNCS, vol. 5789, pp. 37–52. Springer, Heidelberg (2009). https://doi.org/10.1007/978-3-642-04444-1_3
17. Roodposhti, M.S., Aryal, J., Shahabi, H., Safarrad, T.: Fuzzy Shannon entropy: a hybrid GIS–based landslide susceptibility mapping method. Entropy **18**(10), 343 (2016)
18. Mininet (2014). http://mininet.org [19] (2014). Open Vswitch. http://openvswitch.org
19. Tsai, C.-F., Hsu, Y.-F., Lin, C.-Y., Lin, W.-Y.: Intrusion detection by machine learning: a review. Expert Syst. Appl. **36**(10), 11994–12000 (2009)
20. Li, Y., Xia, J., Zhang, S., Yan, J., Ai, X., Dai, K.: An efficient intrusion detection system based on support vector machines and gradually feature removal method. Expert Syst. Appl. **39**(1), 424–430 (2012)
21. Al-Qaraghuli, M., Ahmed, S., Ilyas, M.: Encrypted vehicular communication using wireless controller area network. Iraqi J. Electr. Electron. Eng. Sceeer. 17–24 (2020). https://doi.org/10.37917/ijeee.sceeer.3rd.3
22. Aldabbagh, M., Ilyas, M.: Smart City GIS

Image Cryptography Using Fibonacci Bit-plane Decomposition and Quantum Chaotic Permutation

Renjith V. Ravi[1]($\boxtimes$) (ID), S. B. Goyal[2] (ID), Chawki Djeddi[3,4] (ID),
and Vladimir Kustov[5] (ID)

[1] Department of Electronics and Communication Engineering,
M.E.A Engineering College, Malappuram, Kerala, India
renjithravi@meaec.edu.in
[2] Faculty of Information Technology, City University, Petaling Jaya 46100, Malaysia
sb.goyal@city.edu.my
[3] Department of Mathematics and Computer Science, Larbi Tebessi University,
Tebessa, Algeria
c.djeddi@univ-tebessa.dz
[4] LITIS Lab, Rouen University, Rouen, France
[5] Saint Petersburg Railway Transport University of Emperor Alexander I,
Saint Petersburg, Russia

Abstract. The use of image encryption to safeguard the privacy and security of images is a good idea. This study presents a unique image encryption technique that encrypts images using Fibonacci Bitplanes as security key images. The Fibonacci numbers are used in this research to present a new encryption technique. Furthermore, a unique bit-plane decomposition for Fibonacci weights is explored, which has cryptographic advantages. The novel lossless image encryption technique provided may encrypt an image for privacy protection utilizing this new decomposition approach. It also offers two layers of encryption, allowing it to handle both pay-preview apps and encrypted communications simultaneously. The algorithm's performance in image encryption is confirmed by modeling and analysis. Finally, the suggested algorithm's encryption performance is shown using results obtained and security assessments.

Keywords: Quantum chaotic permutation · Fibonacci bit-plane decomposition · Image encryption · Image cryptography · Quantum arnold transformation

1 Introduction

The fast advancements in digital multimedia and networks in the contemporary digital age have opened the way for individuals worldwide to acquire, use, and exchange multimedia content. The requirement for intellectual property rights protection has spawned a new study field that includes approaches such as statistics, encryption, information concealing, and computer vision [12] [1] because the

F. P. García Márquez et al. (Eds.): ICCIDA 2022, LNNS 643, pp. 577–588, 2023.
https://doi.org/10.1007/978-3-031-27099-4_44

knowledge that may be utilized to benefit or educate organizations (or people) can also be used to harm them (or individuals). Since then, information security has become a critical and pressing concern for people, corporations, and institutions. Digital image and video scrambling have been a popular issue among digital data security experts.

Information interaction between individuals is becoming more reliant on communication systems. However, although digital information technology and networks provide convenience, they also introduce hidden risks: sensitive information included inside images may be readily stolen, altered, or unlawfully duplicated and spread. Therefore, in the digital era, image security is becoming more important [1]. Therefore, using image encryption to safeguard the privacy and security of images is a good idea.

In order to attain a greater degree of security, we offer a unique approach to encrypting images in this study. This program generates binary key images employing Fibonacci Bit-planes rather than the standard bit-plane technique. The parameterized nature of the Fibonacci Bit-planes and truncated Number Sequences are exploited.

2   Materials and Methods

Image encryption may be accomplished using either the scrambling or permutation encryption technique or the pixel value encryption altering method. The process of applying algorithms to modify the location of pixels is referred to as scrambling encryption. The connection between pixels may be reduced because of the use of encryption, but the values of the pixels themselves cannot be altered. However, this section deals with the algorithms and techniques used for the development of proposed cryptographic algorithm.

There are two inputs for the proposed cryptographic structure: the key image and the plaintext image; the plaintext image is the one to be encrypted. The key image will be initially divided into eight-bit planes, as mentioned in Sect. 2.1. Further, these bit planes will be XORed together to generate the key for diffusion, and the plaintext image will be diffused according to this key. Further, this diffused image will undergo permutation or scrambling according to the sequences generated from the 2D QAT to produce the ciphertext image.

2.1   Truncated P-Fibonacci Bit-Plane (TPFB)

The Truncated P-Fibonacci codes can be used to represent a decimal number N as in Eq. 1. When the p value is 0 when the binary valued sequence is at a particular special instance of the TPF-Sequence.

$$N = \sum_{j=0}^{m-1} t_t * T_r(j) = t_0 * T_r(0) + t_1 * T_r(1) + \ldots + t_{m-1} * T_r(m-1) \qquad (1)$$

Here j the coefficient of weight.

The TPF-Sequence $T_r(j)$ is defined by Eq. 2 [14] [1]

$$T_r(j) = \begin{cases} 0 & j < 0 \\ 1 & j = 0 \\ F_r(j+r) & j > 0 \end{cases} \qquad (2)$$

Here I is the index value of the sequence, and r is a positive integer that serves as a parameter for distance. $F_p(j+r)$ is the P-Fibonacci Sequence defined in Eq. 3 [3] [1].

$$F_r(m) = \begin{cases} 0 & m < 0 \\ 1 & m = 0 \\ F_r(m-1) + F_r(m-b-1) & m > 0 \end{cases} \qquad (3)$$

As a result, the TPF-Sequence code $(t_{m-1}, \ldots, t_2, t_1, t_0)$ may be used to represent a decimal integer. This kind of depiction, however, is not unique. As a result, we'll need a rule to verify that each non-negative integer has its own TPF-Sequence code. In this work, we will use a rule from [1,13].

$$N = T_r(j) + s \qquad (4)$$

where $T_r(j)$ is the jth item of TPF-Sequence with a given r value in the range $0 \leq j < m$; and s is a positive integer that serves as a reminder, $0 \leq s < T_r(j-r)$.

After following the method in Eq. 4, every decimal integer will have just one expression of the TPF code. To provide just one example, the TPF-Sequence code for the decimal value 25 will be $(1,0,0,1,0,0,0,0)$ when r value is 2. The number $25 = (1,0,0,1,0,0,0,0)_2$

An image in grayscale format may be fragmented into many binary bit planes by using TPF codes, which is termed the TPF Bit-plane decomposition, as per the TPFB description in Eq. 1. The Fig. 1 depicts one example of bit plane decomposition of a sample image, which can be used as a key. The number of TPFBs and the decomposition outputs fluctuate when the parameter r values varies [2].

2.2   Quantum Arnold Transformation (QAT)

The Arnold transformation (AT) is a preliminary method for image encryption and watermarking that may jumble transforming a meaningful image into an insignificant image [6]. In. [4] and [5], the authors adapted the classical Arnold transformation to the quantum area of computer vision.

For an image with a size of $2^n \times 2^n$, the traditional Arnold transform is as follows:

$$\begin{pmatrix} a' \\ b' \end{pmatrix} = \begin{pmatrix} 1 & 1 \\ 1 & 2 \end{pmatrix} \begin{pmatrix} a \\ b \end{pmatrix} (\mathrm{mod}\, 2^n) \qquad (5)$$

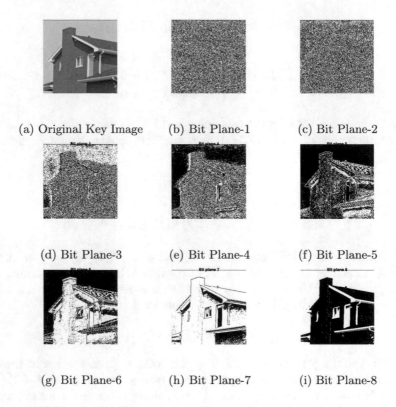

(a) Original Key Image   (b) Bit Plane-1   (c) Bit Plane-2

(d) Bit Plane-3   (e) Bit Plane-4   (f) Bit Plane-5

(g) Bit Plane-6   (h) Bit Plane-7   (i) Bit Plane-8

Fig. 1. Bit Plain decomposition of Sample key image

where a, b represent the pixel locations, and that after transforming the coordinates are altered as:

$$\begin{cases} a' = (a + b) \bmod 2^n \\ b' = (a + 2b) \bmod 2^n \end{cases} \tag{6}$$

The quantum Arnold transform solely modifies the $|ba\rangle$ position information in quantum image $|\bar{I}\rangle$. The QAT's inverse transformation may be calculated as in Eq. 7:

$$\begin{pmatrix} a \\ b \end{pmatrix} = \begin{pmatrix} 1 & 1 \\ 1 & 2 \end{pmatrix}^{-1} \begin{pmatrix} a' \\ b' \end{pmatrix} \bmod 2^n = \begin{pmatrix} 2 & -1 \\ -1 & 1 \end{pmatrix} \begin{pmatrix} a' \\ b' \end{pmatrix} \bmod 2^n \tag{7}$$

The quantum Arnold transform is periodic in nature, and after a particular threshold number of transformations has been reached, the original image may be retrieved precisely as it was before the transform was applied.

2.3 Proposed Cryptography Structure

In this work, we have proposed a cryptographic algorithm with separate architectures for both encryption and decryption. This algorithm has a diffusion-permutation structure and its two main processes in this algorithm are diffusion of plaintext image with the key stream generated from the TPFD and permutation of pixel positions according to the 2D chaotic sequences derived from 2D QAT.

Here there are two images, one is the plaintext image and the other is the key image. We can use any type of grayscale image or any one of the color component of and RGB image as the key image. Initially, the key image will be decomposed into 8 bit plains. Further these bit plains will be XoRed together to get the key for diffusion. For example, if the key input image is H with dimension $M \times N$, then it will be decomposed into 8 bit planes of same dimension.

$$H \rightarrow \{H_1, H_2, H_3, H_4, H_5, H_6, H_7, H_8\} \tag{8}$$

Then, these bit-planes will be XoRed together to get the key matrix H as in Eq. 9

$$H_{MN} = H_1 \oplus H_2 \oplus H_3 \oplus H_4 \oplus H_5 \oplus H_6 \oplus H_7 \oplus H_8 \tag{9}$$

The values in this key matrix H_{MN} will be in te range of $[0, 1]$. Hence, it is necessary to convert these values to the range $[0, 256]$ using [7] [8] Eq. 10.

$$H_{i,j} \leftarrow \left(H_{i,j} \times 10^5\right) \bmod 256 \tag{10}$$

where i and j are the row and column index in the key matrix H.

This key matrix H_{MN} will be XoRed with the plaintext image I_{MN} to get the diffused image C_d as in Eq.

$$C_{i,j} = I_{i,j} \oplus H_{i,j} \tag{11}$$

Further the position all pixels in the diffused image will be altered using the 2D chaotic sequences derived from 2D QAT (Eq. 7) to obtain the ciphertext image. All processes in the encryption procedure is depicted in Fig. 2.

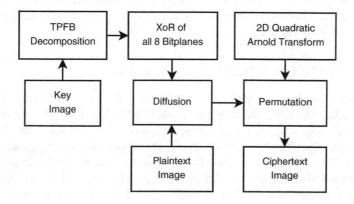

Fig. 2. Proposed encryption method

Decryption is the exact inverse action of encryption. The actions taken during encryption will be reversed during the decryption phase. The R, G, and B colour planes of the ciphertext picture in colour format will be split. The decryption process is depicted in Fig. 3.

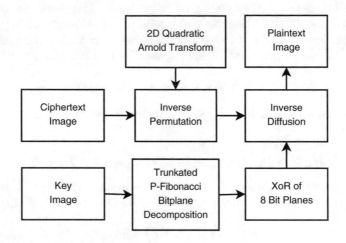

Fig. 3. Proposed decryption method

3   Results and Discussion

The suggested novel technique can encrypt many sorts of images. Numerous investigational results will be presented in this part to demonstrate its encryption effectiveness. In the current investigation, a machine with a 2.60 GHz CPU, 8 GB of RAM, and the OS Windows 10 Pro version has been employed to conduct the suggested encryption and decryption method on the digital images. The standard test images used in the work and its corresponding decryption and encryption results are depicted in Fig 4.

3.1   Anti-statistical Attack Ability Analysis

Images are described using their histograms and the relationships between nearby pixels. The original image's characteristic is clear, and some portions of the image have concentrated distributions of pixel values. The elements in the image are distributed unevenly when this phenomena is represented in the histogram. The connection between neighbouring pixels is quite high when this phenomena is reflected in it. A strong encryption technique may alter the original picture's image distribution properties, uniformize the pixel distribution in the ciphertext image, and weaken the connection between neighbouring pixels. As a result, attackers are unable to decipher cypher images using statistical methods, and also the encryption technique may successfully fend off such

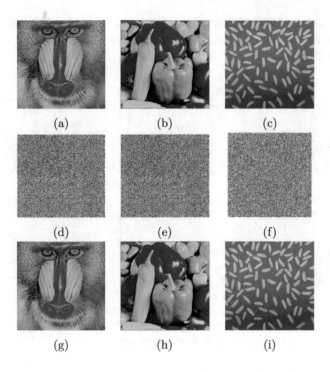

Fig. 4. Results of proposed algorithm in encryption and decryption. (a), (b), (c) Plaintext images, (d), (e), (f) Results of Encryption, (g), (h), (i) Results of Decryption images.

assaults. The correlation across neighbouring pixels and the histograms of the source and cypher images are presented in this study to demonstrate the algorithm's resilience to statistical assaults. To highlight the advantages of the image enciphering method, we also provide some comparative studies with existing image encryption algorithms.

Histogram Analysis: A graph that displays the intensity levels of individual pixels is what's known as an image's histogram. It displays the amount of pixels in the image at various intensity levels. The histogram of an image will illustrate 256 numbers reflecting the frequency of pixels among such grayscale values, as there are 256 distinct potential intensities in an 8-bit image in grayscale format. The distribution of grayscales in the ciphertext image will be quite consistent for a successful encryption.

The histograms of ciphertext images derived from the technique have been developed using grayscale images and colour images of various textures and sizes. Figure 5 illustrates that the histogram of the ciphertext image graphically differs from the histogram of the plaintext image in that it has a consistent distribution of grey values of pixels. Each grayscale value from 0 to 255 is present and equally distributed in the ciphertext, yet none of these values are present in the original image. Therefore, the encrypted image offers no real information about the plaintext image. Due to excellent protection against assaults, this provides encryption.

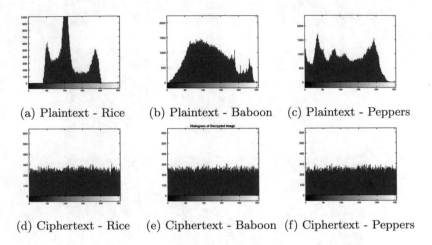

(a) Plaintext - Rice   (b) Plaintext - Baboon   (c) Plaintext - Peppers

(d) Ciphertext - Rice   (e) Ciphertext - Baboon   (f) Ciphertext - Peppers

Fig. 5. Histogram analysis

Auto Correlation Analysis: A correlation is a measurement of the degree to which two pixels within an image are related to each other. The two different pixels have a highly tight correlation if they are adjacent pixels inside an image; otherwise, it is claimed that they have a lower correlation. In an image, this is referred to as neighbouring pixel correlation. The Equations Eq. 12, Eq. 13, Eq. 14, Eq. 15 and Eq. 16 are used to calculate the correlation coefficient (CC), where p and q represents the grey values of two neighbouring pixels in the plaintext and ciphertext image, T represents the total number of neighbouring pixels choose from the image, and E denotes the mean. If the encryption procedure can conceal the specifics of the original image, there is less spatial correlation relationship in the context of an encrypted image. The Table 1 below compares the correlation coefficients between chosen images and the cypher images created during encryption in the three directions, Table 2 compares it with other works in the literature.

$$c = \frac{\text{cov}(p, q)}{\sigma p \times \sigma q} \tag{12}$$

$$\text{where } \sigma p = \sqrt{\text{var}(p)} \tag{13}$$

$$\sigma q = \sqrt{\text{var}(q)} \tag{14}$$

$$\mathrm{var}(p) = \frac{1}{T}\sum_{i=1}^{T}(p_j - E(p))^2 \tag{15}$$

$$\mathrm{cov}(p,q) = \frac{1}{T}\sum_{j=1}^{T}(p_j - E(p))(q_j - E(q)) \tag{16}$$

Table 1. Values obtained for correlation coefficient

Image	Plaintext Image			Ciphertext Image		
	Horizontal (H)	Vertical (V)	Diagonal (D)	Horizontal (H)	Vertical (V)	Diagonal (D)
Baboon	0.9689	0.9316	0.9126	0.0048	0.0011	0.0014
Peppers	0.9842	0.9786	0.9581	−0.016	−0.0051	−0.0007
Rice	0.9766	0.9658	0.9677	−0.0158	−0.0021	0.0062

Table 2. Comparison of correlation coefficient for *Baboon* image

Image	Plaintext image		
	Horizontal	Vertical	Diagonal
Proposed work	0.0048	0.0011	0.0014
Zhang et al. [9]	0.0003	0.0001	0.0015
Zhang et al. [10]	−0.0003	−0.0015	−0.0008

Entropy Analysis: Shannon introduced the idea of information entropy in 1948. The challenge of identifying and quantifying information is resolved by the idea of information entropy, which may also be used to assess the unpredictability of information. We may determine that a piece of information has excellent randomness only when information entropy is near to its optimal value. The degree of unpredictability in the image may be determined using the information entropy of the image. Formula in Eq. 17 illustrates the process for calculating information entropy, where $p(k)$ denotes the likelihood of each possible scenario in a model:

$$H(s) = -\sum_{k=1}^{n}p(k)\log_2 p(k) \tag{17}$$

In an ideal scenario, the entropy of a grey image is 8, because pixel values are dispersed across the range $[0, 255]$, and the likelihood of each instance is. An image has high randomness if its information entropy is near to 8 for a grayscale image. The Table 3 displays the entropy of the cipher images enciphered by this technique and Table 4 displays its comparison with few other algorithms. This algorithm's cipher image has an information entropy that is almost 8, and its output is comparable to that of other algorithms. Accordingly, it can be said that this algorithm's cipher image's unpredictability satisfies the criteria for encryption.

Table 3. Entropy values obtained

Test image	Plaintext	Ciphertext
Baboon	7.6764	7.99837
Pepper	7.5726	7.9031
Rice	7.0130	7.9876

Table 4. Comparison of entropy values for the test image *Baboon*

Test image	Entropy
Proposed Work	7.99837
Zhang et al. [9]	7.99930
Zhang et al. [10]	7.99936

3.2   Anti Differential Attack Ability Analysis

Both the number of pixel change rate (NPCR) and the unified average changing intensity (UACI) are two metrics that are used to quantify the degree of correlation that exists between a cypher image and its corresponding input image, in addition to the anti differential attack capability of the encryption technique. The anti differential attack capability [11] of the encryption method is inversely correlated with the distance between the UACI and NPCR from the ideal values. The Formula in Eq. 18 illustrates the UACI and NPCR calculation procedures:

$$\begin{cases} \text{NPCR} = \frac{\sum_{i,j} |S(P_1(i,j) - P_2(i,j))|}{M \times N} \times 100\%, \\ \text{UACI} = \frac{\sum_{i,j} |P_1(i,j) - P_2(i,j)|}{255 \times M \times N} \times 100\%, \end{cases} \tag{18}$$

$$\text{Sign}(x) = \begin{cases} 1, & x > 0 \\ 0, & x = 0 \\ -1, & x < 0 \end{cases} \tag{19}$$

$S(x)$ is a conceptual function, the formula demonstrates how to calculate it in Eq. 19. The pixel values P_1 and P_2 correspond to the ciphertext image and the encrypted cipher-text image, respectively, following a little adjustment to the plaintext image. The theoretical predictions for UACI and NPCR were 33.4635% and 100%, accordingly. The Table 5 displays the UACI and NPCR values for the two cypher images. From the comparison shown in Table 6, it is clear that this algorithm's cypher image has a good correlation with the input images and is resistant to differential attacks.

Table 5. UACI and NPCR values obtained

Input image	NPCR	UACI
Baboon	99.86%	33.78%
Pepper	99.79%	33.81%
Rice	99.76%	33.62%

Table 6. Comparison of UACI and NPCR Values obtained for *Baboon* Image

Reference	NPCR	UACI
Proposed work	99.86%	33.78%
Zang et al. [9]	99.6159	33.5196
Zang et al. [11]	99.5987	33.5501

4   Conclusion

This work proposes a novel image encryption technique. The security key image is a TPFB of another source image. Any new or current image may be used as a source image to create the TPFBs. Any scrambling method can be used in our suggested encryption process. However, here, we have used a QAT based scrambling for doing the permutation. According to the experimental findings, the suggested technique provides great effectiveness for image encryption. It has adequate keyspace, guaranteeing a high degree of security. It has the potential to be used in the areas of privacy and confidentiality protection.

References

1. Cao, W., Zhou, Y., Chen, C.P.: A new image encryption algorithm using truncated p-fibonacci bit-planes. In: 2012 IEEE International Conference on Systems, Man, and Cybernetics (SMC), pp. 1185–1188. IEEE (2012). https://doi.org/10.1109/icsmc.2012.6377892
2. Cherukuri, R.C., Agaian, S.S.: New normalized expansions for redundant number systems: adaptive data hiding techniques. In: Multimedia on Mobile Devices 2010, vol. 7542, pp. 50–61. SPIE (2010). https://doi.org/10.1117/12.838916
3. Gevorkian, D.Z., Egiazarian, K.O., Agaian, S.S., Astola, J.T., Vainio, O.: Parallel algorithms and vlsi architectures for stack filtering using fibonacci p-codes. IEEE Trans. Signal Process. **43**(1), 286–295 (1995). https://doi.org/10.1109/78.365308
4. Jiang, N., Wang, L.: Analysis and improvement of the quantum Arnold image scrambling. Quantum Inf. Process. **13**(7), 1545–1551 (2014). https://doi.org/10.1007/s11128-014-0749-3
5. Jiang, N., Wu, W.-Y., Wang, L.: The quantum realization of Arnold and Fibonacci image scrambling. Quantum Inf. Process. **13**(5), 1223–1236 (2014). https://doi.org/10.1007/s11128-013-0721-7
6. Liu, X., Xiao, D., Liu, C.: Quantum image encryption algorithm based on bit-plane permutation and sine logistic map. Quantum Inf. Process. **19**(8), 1–23 (2020). https://doi.org/10.1007/s11128-020-02739-w

7. Ravi, R.V., Subramaniam, K.: Image compression and encryption using optimized wavelet filter bank and chaotic algorithm. Int. J. Appl. Eng. Res. **12**(21), 10595–10610 (2017)
8. Renjith, V.R., Subramaniam, K.: Optimized wavelet filters and modified huffman encoding-based compression and chaotic encryption for image data. Int. J. Appl. Eng. Res. **12**(13), 3961–3977 (2017)
9. Zhang, D., Chen, L., Li, T.: Hyper-chaotic color image encryption based on transformed zigzag diffusion and rna operation. Entropy **23**(3), 361 (2021). https://doi.org/10.3390/e23030361
10. Zhang, X., Gong, Z.: Color image encryption algorithm based on 3d zigzag transformation and view planes. Multimed. Tools Appl., 1–33 (2022). https://doi.org/10.1007/s11042-022-13003-x
11. Zhang, X., Wang, L., Cui, G., Niu, Y.: Entropy-based block scrambling image encryption using des structure and chaotic systems. Int. J. Optics **2019** (2019). https://doi.org/10.1155/2019/3594534
12. Zhou, Y., Panetta, K., Agaian, S.: A lossless encryption method for medical images using edge maps. In: 2009 Annual International Conference of the IEEE Engineering in Medicine and Biology Society, pp. 3707–3710. IEEE (2009). https://doi.org/10.1109/iembs.2009.5334799
13. Zhou, Y., Panetta, K., Agaian, S., Chen, C.P.: Image encryption using p-fibonacci transform and decomposition. Optics Commun. **285**(5), 594–608 (2012). https://doi.org/10.1016/j.optcom.2011.11.044
14. Zhou, Y., Panetta, K., Cherukuri, R., Agaian, S.: Selective object encryption for privacy protection. In: Mobile Multimedia/Image Processing, Security, and Applications 2009, vol. 7351, p. 73510F. International Society for Optics and Photonics (2009). https://doi.org/10.1117/12.817699

Author Index

© The Editor(s) (if applicable) and The Author(s), under exclusive license
to Springer Nature Switzerland AG 2023
F. P. García Márquez et al. (Eds.): ICCIDA 2022, LNNS 643, pp. 589–590, 2023.
https://doi.org/10.1007/978-3-031-27099-4

Printed in the United States
by Baker & Taylor Publisher Services